Springer Series in

SOLID-STATE SCIENCES 143

Springer

Berlin
Heidelberg
New York
Hong Kong
London
Milan
Paris
Tokyo

Springer Series in

SOLID-STATE SCIENCES

Series Editors:

M. Cardona P. Fulde K. von Klitzing R. Merlin H.-J. Queisser H. Störmer

The Springer Series in Solid-State Sciences consists of fundamental scientific books prepared by leading researchers in the field. They strive to communicate, in a systematic and comprehensive way, the basic principles as well as new developments in theoretical and experimental solid-state physics.

126 **Physical Properties of Quasicrystals**
Editor: Z.M. Stadnik

127 **Positron Annihilation in Semiconductors**
Defect Studies. By R. Krause-Rehberg and H.S. Leipner

128 **Magneto-Optics**
Editors: S. Sugano and N. Kojima

129 **Computational Materials Science**
From Ab Initio to Monte Carlo Methods. By K. Ohno, K. Esfarjani, and Y. Kawazoe

130 **Contact, Adhesion and Rupture of Elastic Solids**
By D. Maugis

131 **Field Theories for Low-Dimensional Condensed Matter Systems**
Spin Systems and Strongly Correlated Electrons. By G. Morandi, P. Sodano, A. Tagliacozzo, and V. Tognetti

132 **Vortices in Unconventional Superconductors and Superfluids**
Editors: R.P. Huebener, N. Schopohl, and G.E. Volovik

133 **The Quantum Hall Effect**
By D. Yoshioka

134 **Magnetism in the Solid State**
By P. Mohn

135 **Electrodynamics of Magnetoactive Media**
By I. Vagner, B.I. Lembrikov, and P. Wyder

136 **Nanoscale Phase Separation and Colossal Magnetoresistance**
The Physics of Manganites and Related Compounds
By E. Dagotto

137 **Quantum Transport in Submicron Devices**
A Theoretical Introduction
By W. Magnus and W. Schoenmaker

138 **Phase Separation in Soft Matter Physics**
Micellar Solutions, Microemulsions, Critical Phenomena
By P.K. Khabibullaev and A.A. Saidov

139 **Optical Response of Nanostructures**
Microscopic Nonlocal Theory
By K. Cho

140 **Fractal Concepts in Condensed Matter Physics**
By T. Nakayama and K. Yakubo

141 **Excitons in Low-Dimensional Semiconductors**
Theory, Numerical Methods, Applications By S. Glutsch

142 **Two-Dimensional Coulomb Liquids and Solids**
By Y. Monarkha and K. Kono

143 **X-Ray Multiple-Wave Diffraction**
Theory and Application
By S.-L. Chang

144 **Physics of Transition Metal Oxides**
By S. Maekawa, T. Tohyama, S.E. Barnes, S. Ishihara, W. Koshibae, and G. Khaliullin

145 **Point-Contact Spectroscopy**
By Y. Naidyuk and I.K. Yanson

146 **Optics of Semiconductors and Their Nanostructures**
Editors: H. Kalt and M. Hetterich

147 **Solid-State Theory**
An Introduction By U. Rössler

Volumes 1–125 are listed at the end of the book.

Shih-Lin Chang

X-Ray Multiple-Wave Diffraction

Theory and Application

With 183 Figures

Springer

Professor Shih-Lin Chang
Department of Physics
National Tsing Hua University
101, Section 2, Kuang Fu Road
Hsinchu, Taiwan, 300 R.O.C.
E-mail: slchang@phys.nthu.edu.tw

Series Editors:

ISSN 0171-1873

ISBN 3-540-21196-9 Springer-Verlag Berlin Heidelberg New York

Library of Congress Cataloging-in-Publication Data applied for.
Bibliographic information published by Die Deutsche Bibliothek
Die Deutsche Bibliothek lists this publication in the Deutsche Nationalbibliografie;
detailed bibliographic data is available in the Internet at http://dnb.ddb.de.

Springer-Verlag is a part of Springer Science+Business Media

springeronline.com

Typesetting by the author
Final processing: Satztechnik Katharina Steingraeber, Heidelberg
Cover concept: eStudio Calamar Steinen
Cover production: *design & production* GmbH, Heidelberg

Printed on acid-free paper SPIN: 10859948 57/3141/tr - 5 4 3 2 1 0

Dedicated to the late Professor Benjamin Post

Preface

X-ray multiple-wave diffraction, sometimes called multiple diffraction or N-beam diffraction, results from the scattering of X-rays from periodic two- or higher-dimensional structures, like 2-d and 3-d crystals and even quasi-crystals. The interaction of the X-rays with the periodic arrangement of atoms usually provides structural information about the scatterer. Unlike the usual Bragg reflection, the so-called two-wave diffraction, the multiply diffracted intensities are sensitive to the phases of the structure factors involved. This gives X-ray multiple-wave diffraction the chance to solve the X-ray phase problem. On the other hand, the condition for generating an X-ray multiple-wave diffraction is much more strict than in two-wave cases. This makes X-ray multiple-wave diffraction a useful technique for precise measurements of crystal lattice constants and the wavelength of radiation sources. Recent progress in the application of this particular diffraction technique to surfaces, thin films, and less ordered systems has demonstrated the diversity and practicability of the technique for structural research in condensed matter physics, materials sciences, crystallography, and X-ray optics.

The first book on this subject, *Multiple Diffraction of X-Rays in Crystals*, was published in 1984, and intended to give a contemporary review on the fundamental and application aspects of this diffraction. Since then, there has been tremendous progress in both theoretical approaches and experimental developments in this field of research, including the various two-wave approximations for the multiple-wave situation and the methods for measuring and determining the phases of X-ray reflection in macromolecular crystals and less ordered systems. This new volume intends to summarize all these new developments in theory and application and to examine possible future developments of this diffraction technique into advanced methods for structural analysis in the related fields of research.

I am indebted to the late Professor Ben Post of Polytechnic University (Brooklyn) for having taught me X-ray diffraction and crystallography and for having brought me into this field of research. This book is specially dedicated to him. I am also very grateful to Yu. P. Stetsko and Kennith W.K. Shung for reading this manuscript and for useful discussion. The assistance of L.-M. P. Chang, C.-W. Cheng, H.-C. Chang, C.-S. Hwang, C.-Y. Lo,

Y.-R. Lee, H.-H. Wu, W.-S. Sun, W.-C. Sun, Y.-C. Shen, C.-M. Liu, S.-Y. Chen, C.-Y. Chou, and many others is also gratefully acknowledged.

The author is indebted to A.M. Afanas'ev, L.P. Cardoso, G. Faigel, M.V. Kovalchuk, G. Materlik, E. Rossmanith, Q. Shen, Yu.P. Stetsko, J.P. Sutter, G. Thorkildsen, and E. Weckert for kindly giving permission to reproduce their original figures and for providing some of the original photographs and drawings.

Hsinchu, *Shih-Lin Chang*
April 2004

Contents

1. Introduction

X-ray diffraction occurs when a set of atomic planes of a crystal is brought into position to diffract an incident wave. Since the diffraction involves two waves, an incident wave and a diffracted wave, it is usually called two-wave or two-beam diffraction. The interference among the X-ray waves inside the crystal can be constructive or destructive, which results in detectable or null intensity, respectively, depending on the crystal symmetry and the arrangement of the atoms in the crystal unit cell. The crystal can be a periodic arrangement of atoms in one-, two-, three-, or even higher-dimension. For a two-dimensional or higher dimensional crystal, it is possible that two or more sets of atomic planes satisfy Bragg's law simultaneously. Whenever this occurs, more than two diffracted waves are generated. This is called X-ray multiple-wave diffraction. Synonymous terms, such as simultaneous reflection, N-beam diffraction, and multiple diffraction, have been used in the literature.

1.1 Multiple-Wave Diffraction Experiments in the Early Days

In an X-ray multiple-wave diffraction experiment, the interaction among the diffracted waves within a crystal gives rise to variation, either an increase or a decrease, in the intensity of each diffracted wave involved. This variation results mainly from the reallocation of the energy or power of a two-wave diffraction. The energy of the two-wave diffraction may increase through energy transfer from other reflections. Likewise, the energy may also decrease by giving away some energy to other reflections via multiple-wave interaction. The type of diffraction with decreased intensity is called 'Aufhellung', which was first observed by Wagner [1] in 1920 and then by Berg [2] in 1926 on photographic recordings of X-ray reflections from rock salts with white radiation. The brightening of two-wave diffraction images on X-ray films due to this decrease in intensity was recorded. Later in 1928, the 'Aufhellung' effects were detected for the four-wave $(000)(220)(400)(2\bar{2}0)$ diffraction of diamond with Mo$K\alpha$ radiation by Mayer [3]. In his experiments, the crystal was first aligned for the (400) reflection and then rotated around the reciprocal lattice vector of the (400) reflection, while the intensity of the (400) was monitored. The decrease in the diffracted intensity of (400) occurred at the four-wave diffraction position. A scan that employs this type of crystal rotation is called 'Renninger scan' [4] or 'ψ scan'.

The other type of multiple-wave diffraction with increased intensity on a two-wave reflection background was reported by Renninger [4] in 1937. The intensity distribution of the nearly forbidden (222) reflection of a diamond crystal was investigated using the Renninger scan around the reciprocal lattice vector of (222). Tremendous increases in intensity of the (222) reflection were observed for various multiple-wave cases. This intensity gain of the (222) reflection, called the 'Renninger effect', is due to the energy transfer through the detoured excitation from the other reflections involved in the multiple-wave diffraction to the (222) reflection. This detoured excitation was named 'Umweganregung' by Renninger. Similar 'Umweganregung' effects were also observed with X-ray films in a bent crystal of quartz by Cauchois, Hulubei, and Weigle [5], where the darkening of the diffraction images on the films was detected. Extensive investigations along the same direction were carried out in the early 1940s [6,7], including the development of the Kossel diagram to reveal multiple-wave diffraction images [8–11].

Multiple-wave diffraction of electrons was considered in the dynamical theory of electron diffraction by Bethe [12] as early as in 1928. The first few reports on the observation of such diffraction were in 1932 [13] and 1935 [14] for low-energy electrons. For high-energy electrons, multiple-wave diffraction is always observed in electron diffraction experiments, because many reflections are simultaneously detected.

The interference nature of multiple-wave diffraction provides an ideal situation for the determination of the relative phase of an X-ray reflection. The idea of using this diffraction technique for phase determination was first proposed by Lipscomb [15] and conceived independently by Fankuchen and by Ekstein in 1948 [15]. Experiments were carried out by using a Weissenberg camera to record the oscillation pictures for the $[h0l]$ zone of glycine with CuKα radiation, while the crystal was kept at the (040) reflection condition. Very weak intensities of three forbidden reflections, (100), (102) and $(10\bar{2})$, under multiple-wave diffraction conditions were measured. Similar investigation on pentaerythritol had also been pursued by Fankuchen and Ekstein [15]. Unfortunately, no information about phase was obtained. Along the same line of investigation, Kambe and Miyake, however, used the Kosssel-Möllenstedt patterns [16] of electron diffraction from a graphite crystal lamina and succeeded in demonstrating the influence of phase on the intensity distribution of the Kikuchi lines in the vicinity of three-wave diffraction positions [17–19].

Multiple-wave diffraction of neutrons was also observed when the thermal neutron spectrum of a nuclear reactor was analyzed by a crystal monochromator in the late 1950s and early 1960s [20–26]. The spectrum was known to contain a large number of small fluctuations on a background. Some of the small fluctuations on the spectrum curves were identified to be due to the modification of the intensity diffracted from the monochromator crystal resulting from the multiple-wave interaction in that crystal [21]. Some were due to the inelastic scattering of neutrons [23].

Owing to the availability of perfect crystals of semiconductors, such as silicon and germanium, in the 1960s, the anomalous transmission effect at the exact multiple-wave diffraction position for a germanium crystal was discovered by Borrmann and Hartwig [27] in 1965. All the reflections involved in a multiple-wave case were of the transmission (Laue) type, i.e., all the diffracted waves were transmitted through the crystal. In contrast to the conventional two-wave Borrmann effect reported in 1941 [28, 29], the anomalous transmission of X-rays through the crystal is further enhanced. This anomalous phenomenon is called 'multiple-wave Borrmann effect', 'enhanced Borrmann effect' or sometimes 'super Borrmann'.

1.2 Development of Theories for X-Ray Two-Wave Diffraction

Attempts to account for the measured intensity of multiple-wave diffraction had started soon after the observation of 'Umweganregung'. However, before the theoretical development of interpreting diffraction intensities for multiple-wave diffraction is mentioned, it is much more appropriate to briefly review the evolution of the diffraction theories of X-rays in crystals for the two-wave case. In fact, multiple-wave diffraction is a combination of several two-wave diffractions. Thus two-wave diffraction theories serve as the basis for the interpretation of the diffraction intensities of multiple-wave cases. Since there have been many review articles [30–45] and books [46–55] devoting to the dynamical theory, only the important contributions will be mentioned.

About two years before the experimental discovery of X-ray diffraction by Friedrich, Knipping and Laue in 1912 [56], Ewald first theoretically tackled the problem of the optical behavior of the interaction of electromagnetic waves with an anisotropic periodic array of dipoles, aiming to find a quantitative proof for the internal periodicity of crystals. In Ewald's approach, a crystal was considered as a periodic arrangement of dipoles. The excitation of these dipoles by external electromagnetic (EM) waves also generates EM waves. The interaction between the incident waves and the excited waves from the dipoles eventually reaches dynamical equilibrium, thus forming an optical wavefield in the crystal. This self-consistent theoretical approach led to the development of the microscopic theory of dispersion [57], the theory of reflection and refraction [58] of crystal optics, and the theory of X-ray diffraction for two-wave cases [59]. Together with his 1937 article about the search for rigorous solutions of the multiple-wave problem in arbitrary structures [60], Ewald laid the theoretical foundation for the dynamical theory of X-ray diffraction.

The geometrical theory of X-ray diffraction was derived by Laue in his paper reporting on the discovery of X-ray diffraction in crystals [56]. In that paper, the amplitude of the diffracted waves from a crystal was formulated and the phase shift due to the difference in optical path length between the diffracted waves was taken into account. The three Laue equations for diffraction in a vector form were derived in a subsequent report [61]. A rather succinct relation for X-ray diffraction was later given by W.L. Bragg [62–64], where the X-ray reflection from a set of lattice planes [65] was considered. This relation was subsequently called 'Bragg's law'. In addition, with an accurate determination of X-ray wavelength and intensity measurements of reflections from sodium chloride crystals using an X-ray spectrometer, W.H. Bragg and W.L. Bragg reported the first crystal-structure determination by X-ray diffraction in 1913 [65].

In 1914 Darwin considered successive reflections from a set of lattice planes in the reflection (Bragg) geometry and developed the kinematical theory of X-ray diffraction. This theory contained two parts. The first part was merely a geometrical theory similar to Laue's approach [66], where the diffracted amplitude was considered as a small fraction of the incident amplitude. The second part dealt with the interaction between the transmitted and the diffracted wave at each lattice plane [67], which is more dynamical than kinematical. However, this theoretical approach did not include effects due to absorption. Later, Prins [68] and Kohler [69] modified this theory for X-ray diffraction in absorbing crystals.

The discrepancy between the intensity measurements and the theoretical prediction led to the development of extinction theories. In 1922 Darwin [70] proposed a hypothetical model, the so-called 'mosaic model' (the name suggested by Ewald [34]) for a crystal, in which the crystal was composed of many tiny perfect single-crystal blocks slightly misaligned with one another. Intensity attenuation, in addition to the normal absorption, was described as the primary extinction correction due to the decrease of reflecting power by diffraction from large crystal blocks. The attenuation due to diffraction from small crystal blocks was the secondary extinction correction. As a matter of fact, the theory of primary extinction usually approaches the dynamical treatment, while that of secondary extinction is more kinematical.

The dynamical theory was reformulated in terms of Maxwell's equations by Laue [30] in 1931. The crystal was treated as a continuous distribution of electrons around the nuclei and the corresponding dielectric constant for X-rays was assumed to be complex and periodic. The Bloch functions were adopted as the plane-wave solutions to Maxwell's equations, where the involved diffraction vectors satisfied Bragg's law. Laue's dynamical theory is macroscopic, compared with Ewald's microscopic theory, and has been more frequently adopted by investigators in this field of X-ray diffraction. However, Ewald's approach gives a more fundamental view of the excitation of the crystal and the dynamics of X-ray diffraction.

The quantum mechanical theory of X-ray diffraction was developed by Kohler [71] in 1935 to confirm Laue's hypothesis that the diffracted intensity was mainly due to the scattering from the electron density of the crystal unit cell. Other quantum approaches were pursued by Molière [72–74], Ohtsuki [75, 76], Wagenfeld [77], Ashkin and Kuriyama [78], Dederichs [79, 80], Hannon and Trammel [81, 82], Afanas've and Kagan [83] and many others [84–92] for both perfect and imperfect crystals. In these approaches, a generalized current density was derived and the tensor operators were adopted for the dielectric susceptibilities. With these approaches, the effects of resonance at absorption edges, the excitations due to thermal vibration and the Compton scattering could be dealt with. Moreover, the quantum field theory has also been employed to give a more rigorous treatment. With these theories, the description of the optical activity of crystals [93, 94] and X-ray resonance scattering [95–97] among others is possible.

In parallel to the theoretical development, the experimental findings, including the discovery of the Borrmann effect for anomalous absorption and anomalous transmission of X-rays in crystals [28], the observation of the Borrmann fan [98] and double refraction of X-rays in crystals [98–102], and the detection of Pendellösung fringes [103–116], also enhanced our understanding of X-ray dynamical diffraction, especially of the wavefield trajectories.

A spherical-wave theory of X-ray diffraction was derived by Kato [117–122] in 1960 to account for the interference fringes of X-ray section topographs for a wedge-shaped crystal, which had been observed by Kato and Lang [103]. The spherical-wave nature of X-ray diffraction was understood to result from the beam divergence and the coherence length of the X-rays involved in the diffraction experiments.

The dynamical theories for imperfect crystals were developed in the 1960s. These include the ray theories for slightly deformed crystals developed by Penning and Polder [123, 124] and Kato [125–127] and the generalized theories by Takagi [128, 129] and Taupin [130, 131]. The latter consist of the fundamental equations of wavefield in the form of partial differential equations, i.e., the Takagi-Taupin equations. Other developments along this line reported in the literature are given in the references [132–140]. In addition, a generalized extinction theory [141–144] and a statistical dynamical theory [145, 146] were derived by Kato in the late 1970s and the beginning of the 1980s to describe X-ray diffraction from both perfect and imperfect crystals. The ensemble average over various possible routes of X-ray diffraction inside a crystal was considered in the statistical dynamical theory.

Together with the development of theories, several diffraction methods and techniques have been developed for various purposes and applications. For example, X-ray diffraction topography was used for the characterization of defects and lattice deformation (see the references of the pioneering work in [147–158] and the recent reviews [46, 159, 160]). X-ray interferometry was devised by Bonse and Hart [161, 162] using a three-leaf monolithic crystal for

X-ray beam splitting, mirroring, and interfering. The X-ray standing-wave excited fluorescence technique was suggested by Batterman [163,164] for the determination of the positions of impurity atoms in a crystal and the surface registry of atoms and interface structure analysis. Furthermore, X-ray diffraction at grazing incidence was demonstrated to be capable of probing structures of surfaces and interfaces in the late 1970s by Marr, Eisenberger, and Cho [165]. Theoretical interpretations for the grazing incidence diffraction were proposed by Vineyard [166] using a distorted-wave Born approximation and by Afanas'ev and Melkonyan [167] based on the dynamical theory of X-ray diffraction. The so-called 'crystal truncation rods' [168,169] have then become a powerful tool for surface/interface structural investigations. Other diffraction techniques, such as magnetic scattering [170–172] and anisotropic scattering [173,174], have also been developed by using two-wave diffraction.

1.3 Development of Multiple-Wave Diffraction

Ewald's dynamical theory [57,58] was quite general and began with a formulation for a general multiple-wave diffraction. The two-wave diffractions in a (Laue) transmission and a Bragg reflection geometry were considered in detail in Ewald's 1917 article [59]. An attempt to use this theory to account for the observed intensities for multiple-wave diffraction was given by Mayer in 1928 [3]. Later, Renninger developed a semi-quantitative approach for the same purpose. Both attempts turned out to be not very successful in interpreting the measured intensities for multiple-wave cases.

A rigorous solution to the multiple-wave diffraction problem in a general situation was given by Ewald in 1937 [60]. It appeared to Ewald that all the 'Aufhellung' and 'Umweganregung' in multiple-wave diffraction were subject to the detoured excitation in the diffraction process, and therefore all the multiple diffractions are actually of the 'Umweganregung' nature. The dispersion equation and the excitation of wavefields in crystals for three-wave dynamical diffractions were discussed first by Ewald and later by Lamla [175]. A similar approach for electron diffraction based on Bethe's theory was proposed by Kambe and Miyake [17–19] for electron diffraction. However, in the latter approach it's assumed that at least one of the involved reflections in a three-wave case was very weak to justify the approximations used. The multiple-wave dynamical diffraction problem was treated in a matrix form by Kato in 1958 [176]. A three-wave diffraction was described in detail in terms of Laue's dynamical theory in 1960 by Laue himself [50]. The fundamental equations of wavefield were given in that study.

The report on the enhanced Borrmann effect of multiple-wave diffraction in rather perfect crystals of semiconductors interested a lot of researchers in understanding the physics origin of this anomalous transmission of X-rays through solids in the 1960s. Many theoretical and experimental studies had been pursued for multiple-wave diffraction in the (Laue) transmission geometry. For example, Saccocio and Zajac derived the formulation for a symmetric three-wave diffraction with either three equally strong reflections or one of the reflections being forbidden [177,178]. Hildebrandt [179–183] discussed the absorption and wavefields of three-wave Borrmann effects in perfect crystals. Joko and Fukuhara [184] gave definitions of polarization vectors for a symmetric N wave diffraction, where all the reciprocal lattice points lie in a plane. Analytical expressions for the modes of wave propagation, eigenvalues and wavefield amplitudes for symmetric three-, four-, six- and N-wave diffractions were also deduced. In 1968, Ewald and Heno [185,186] presented in detail an analytical form for the dispersion equation and discussed the effect of the reflection phase on the dispersion surface for multiple-wave diffraction. A relevant theory for three-wave diffraction considering the normalized dispersion surface in the reciprocal space was also developed by Penning [187,188] and Penning and Polder [189]. Discussions on the three-wave Borrmann effects and simultaneous diffraction and the analysis of photoelectric absorption were reported by Afanas'ev and Kohn [190–193]. Other theoretical investigations were also reported in the literature [194–197]. The experimental detection of the enhanced Borrmann effect in various multiple-wave diffractions, including the observations of diffraction eclipse and the multiple-wave effects on Pendellösung, has also been investigated by many researchers [194,198–229].

Aside from Mayer's and Renninger's attempts to interpret Bragg-type multiple-wave diffraction, including 'Aufhellung' and 'Umweganregung', kinematical approaches based on the power-transfer equations, were proposed by Moon and Shull [26] in 1964 for neutron diffraction and by Zachariasen [230, 231] in 1965 for X-rays. Both were the extension of Zachriasen's extinction theory [48] to multiple-wave situations. Solutions to the power-transfer equations for more precise description of the reflection powers were given in terms of a higher-order Taylor's series expansion or an exponential function by Caticha-Ellis [232] and many others [233–245].

The fundamental differences between a Laue-type and a Bragg-type multiple-wave diffraction, when their dynamical behaviors are considered, lie in the scattering matrix. The matrix of the former is hermitian, while that of the latter is non-hermitian. In addition, to solve the eigenvalue problem involved in the fundamental equations of wavefield for a Bragg-type diffraction is far more complicated than for a Laue-type, because the procedure of linearizing the eigenvalue equation is not straightforward. Colella developed an algorithm to linearize the eigenvalue equation in 1974, facilitating the finding of correct eigenvalues and eigenvectors using numerical calculations [246]. Other computational approaches have been pursued by Uebach [211], Kohn [247], Chang [248–250], and Hümmer and Billy [251] and many others [252–254]. The details of the differences between the Bragg-type and Laue-type multiple-wave diffractions in relation to the scattering matrix and boundary conditions are given in Pinsker's book [52].

The numerical computational algorithm for multiple-wave dynamical calculations was further developed by Stetsko and Chang [255] in 1997, where a representation of wavefields and eigenvalues in terms of a Cartesian coordinate system was used. The algorithm has wide applications and can be applied to situations involving grazing-angle incidence as well as wide-angle incidence in a general multiple-wave diffraction.

The numerical calculation based on the dynamical theories usually gives exact values of the eigenvalues, wavefield amplitudes, and intensities. However, the dependence of the diffracted intensity on the important physical quantities, such as relative phase and angular position, may sometimes be obscured by those values. Under this circumstance, analytical expressions of diffracted intensity as a function of the physical quantities is desired. However, such expressions are usually unattainable for a general multiple-wave diffraction. An alternative is to find the analytical expressions with approximations. The Born approximation and Bethe approximation then serve this purpose.

The development of the approximation solution in an analytical form, on the other hand, has been closely linked to the attempts to solve the X-ray phase problem using multiple-wave diffraction. Various investigations on direct phase determination were pursued from the 1960s to the 1990s, including the search for solutions to the phase problem for both centrosymmetric [219, 224, 246, 251, 256–276] and noncentrosymmetric crystals [277–326]. Among those, the transmission-type multiple-wave diffraction experiments using three-wave Pendellösung effects were reported by Hart and Lang in 1961 [256] and by Høier and Aanestad in 1981 [224]. The distortion of interference fringes due to relative phase, either 0 or 180 degrees, for centrosymmetric crystals around multiple-wave diffraction positions were observed. The asymmetric intensity distribution of the three-wave Borrmann transmission in centrosymmetric crystals was used for phase determination by Post in 1977 [219, 257]. Similar studies with a different diffraction geometry was carried out by Jagodozinski [258]. All these experimental results could be interpreted by the theoretical considerations of Ewald and Heno [185, 186] and the dynamical calculation for the transmission geometry.

The reflection-type (the Renninger type) multiple-wave diffraction was brought back to attention for phase determination by Colella in 1974 [246]. The intensity asymmetry of the diffracted waves in a Renninger scan due to the effect of phase for a three-wave case of a centrosymmetric crystal was reported subsequently by Chapman, Yoder and Colella [259] in 1981. In the same year, the angle- and the phase-dependence of the intensity asymmetry of Renninger scans were observed and a sign relation of the cosine of the relative phase was deduced from three-wave experiments and the dynamical calculations by Chang [260, 261]. Although this semi-empirical phase relation could be used to describe the intensity asymmetry in terms of the phase and angular position for a Renninger type three-wave diffraction, a rigorous analytical expression of the integrated intensity as a function of the structure-factors, angular position, and polarization was desired. Juretschke adopted the Bethe approximation and treated a three-wave diffraction as a modified two-wave diffraction due to the perturbation of a third wave in 1982 [262, 263]. Similar approaches utilizing the second Bethe approximation were subsequently proposed [53,267,270]. Other perturbative treatments employing the Green's function method, such as the Born approximation [285] and quantum mechanical approach [275], have also been given in the literature.

The phases of reflections from a noncentrosymmetric crystal range from 0 to 360 degrees, in contrast to the phase angle of either 0 or 180 degrees for a centrosymmetric crystal. A theoretical formulation for the quantitative analysis of the cosine of the relative phase from the integrated diffraction intensity of the three-wave diffraction in noncentrosymmetric crystals was derived by Juretschke [263]. Attempts to experimentally determine the phase of reflection, i.e., the values of the sine and cosine of the phase, of non-centrosymmetric crystals were pursued by using either the anomalous dispersion effects in three-wave diffraction [277, 278] or the long-wavelength X-ray diffraction [286, 294]. A theoretical consideration on the experimental determination of phases and enantiomorphs was given by Hümmer and Billy [279] in 1986. A different formalism was derived by Thorkildsen [327] in 1987 from Takagi-Taupin equations for a finite imperfect crystal. Both show the relation between the phase and the integrated diffracted intensity profile for non-centrosymmetric crystals. A detailed theoretical approach based on an iterative Born approximation was elaborated by Chang and Tang [291,292] in 1988 to aim at quantitative phase determination.

The experimental determination of reflection phases on a semi-quantitative and a quantitative basis for noncentrosymmetric crystals was reported by Hümmer, Bondza, and Weckert in 1987 [288, 289, 295] and by Tang and Chang in 1988 [292, 296, 297]. Further investigations [293, 301, 302, 328] indicated that using two symmetry-related three-wave diffractions, proposed by Hümmer et al. [295], was essential for the quantitative phase determination. Other experimental approaches using the Kramers-Kronig relations [329] and an elliptically polarized incident beam [330] were also reported. Applications of multiple-wave diffraction for phasing macromolecular crystals were reported by Hümmer, Schwegle and Weckert [331] and by Chang, King, Huang and Gao [301] in 1991. The improvement on the visibility of intensity asymmetry and the phase sensitivity of the multiple-wave diffraction technique has recently been extensively studied with the use of adjustable linear polarization and elliptical polarization sources of synchrotron radiation [322, 324]. Very recently, the reference-beam technique [309] and the stereoscopic multi-beam imaging [310] have been proposed respectively by Shen and by Chang et al. to experimentally collect a large set of multiple-wave diffraction data from macromolecular crystals. The incorporation of the determined phases with other existing methods, such as direct methods [321] and entropy maximization [311, 317, 325, 332, 333], for structure determination have also been pursued. Along with this experimental development, the distorted-wave Born approximation [166] was applied to the reference-beam technique by Shen [314, 315] to describe the integrated intensity measured. An attempt of phasing partially coherent multiple-wave diffraction with high accuracy has also been pursued [334–336].

An energy-tuning grazing-incidence multiple-wave diffraction technique was developed for phasing the diffraction from surface in-plane reflections of crystal surfaces and interfaces by Chang et al. [276] in 1998. Its applications to thin films and surface overlayers have been demonstrated [337, 338].

Multiple-wave diffraction in the resonance conditions [278, 339] has been investigated very recently. The resonance phases in the near-edge regions were determined by using three-wave diffractions and a resonance perturbation approximation was developed [323].

Multiple-wave diffraction, like the conventional Bragg reflection (the so-called two-wave diffraction), is influenced by the geometric aspects and lattice perfection of the crystal. The geometric aspects are related to the conditions for the occurrence of multiple-wave diffraction, which depend on the lattice constants and the X-ray radiation used. The lattice perfection affects the diffracted intensity. Therefore, the development of this diffraction method follows closely the evolution of the techniques of generating multiple-wave diffraction and of the theoretical interpretation of diffraction intensity. Moreover, the two- and the three-dimensionality of crystals ensure the occurrence of multiple-wave diffraction. The characteristics of this type of diffraction should, in turn, reflect the structure of the crystal in an appropriate dimension. Specific structural problems in crystals, thin films, surfaces and interfaces lead to many new applications of this diffraction technique. The determination of phase is only one of the many applications of multiple-wave diffraction to structure analysis. Other applications include the precise determination of lattice constants of a crystal unit cell [4, 9, 242, 340–348], lattice mismatches between an overlayer or a thin film and a crystal substrate as well as the lattice variation in superlattices [248, 249, 349–362], the characterization and measurement of the wavelengths of radiation sources [363–366], the utilization of the dynamical properties of multiple-wave diffraction for X-ray optics [217, 367–378]; the element-identification and the determination of the positions of atoms and molecules on surfaces and bulk crystals using X-ray standing-wave fluorescence measurement under multiple-wave diffraction conditions [379, 380], and the studies of structural variations of phase transition of materials under the application of external electric field [381–385].

This book is written by following the trend of the development of multiple-wave diffraction. In Chap. 2, a brief introduction to X-ray physics and crystallography is given, which covers a short description of X-ray sources, both the conventional and synchrotron radiation. The interaction between X-rays and matter is briefly described and some of the formulations regarding scattering, emission, and absorption cross sections are outlined for later references. Physical quantities related to diffraction intensity, such as the structure factor, temperature factor, and Lorentz-polarization factors, are given. The concept of reciprocal lattice and crystal symmetry are also briefly reviewed. In Chap. 3, the geometry for two-wave and three-wave diffraction under wide-angle and grazing incidence conditions is illustrated. The corresponding Lorentz-polarization factors are derived for successive reflections involved in multiple-wave diffraction. The geometric aspects of multiple-wave diffraction for multi-crystal systems are also presented.

The experimental methods of generating multiple-wave diffraction by using a well collimated-beam and a divergent-beam are described in Chap. 4. Rotation-crystal methods together with different two-dimensional detectors are mentioned, especially for diffraction from macromolecular crystals. Emphasis is also put on the specially designed diffractometers for acquiring multiple-wave diffraction data with ease.

Chapter 5 covers the fundamentals of the kinematical theories which are valid for diffraction in small crystals and for weak reflections. The power-transfer equations for two-wave and multiple-wave diffractions are described. The solutions to those equations and the approximations are presented. The distributions of the integrated intensity of multiple-wave diffraction in terms of the Lorentz-polarization factors are formulated and discussed.

The fundamentals of the dynamical theory of X-ray diffraction are briefly reviewed in Chap. 6. Laue's dynamical treatment and the cofactor of matrix algebra are adopted to describe the essential parts of the multiple-wave dynamical diffraction. The fundamental equation of wavefield, the equation of dispersion, the energy flow and mode of wave-propagation, the excitation of mode, and the boundary conditions for wavefields are presented and discussed in detail. In addition, the spherical-wave theory and the Takagi-Taupin equations for multiple-wave diffraction in a finite imperfect crystal are also included in this chapter.

Theoretical approaches, including numerical calculation schemes and the approximations for multiple-wave dynamical diffraction, are given in Chap. 7. The linearization of the eigenvalue equation and a Cartesian-coordinate representation of the eigenvalue problem are presented. The quantum theoretical considerations are adopted to explain the multiple-wave interactions. Various modified two-wave approximations, such as the Born, Bethe, distorted-wave Born approximations, and perturbative resonance Bethe approximation, are also described. Details of the applications of the Takagi-Taupin equations for multiple-wave diffraction are given in this chapter.

The properties and behaviors of dynamical diffraction for multiple-wave cases, including two-, three-, four-, six-, eight- and twenty-six-wave diffractions, are discussed in Chap. 8. Various geometric situations, such as the Bragg reflection, Laue transmission, asymmetrical diffraction, grazing incidence diffraction, and back diffraction, are considered. The effects of polarization on the multiple-wave diffraction, such as the indirect excitation of forbidden reflection by polarization and the intensity suppression and enhancement due to polarization, are illustrated.

Applications of the multiple-wave diffraction method to structure analysis, such as the determination of the phases of X-ray reflections for single crystals, crystal surfaces, thin-films, and quasi-crystals are given in Chap. 9. The determination of absolute configuration, the characterization of radiation sources, and the applications to X-ray optics, X-ray fluorescence, and the analysis of phase transformation are also included. The structural analysis of less-ordered systems is also demonstrated via multiple-wave diffraction where the coupling of the diffracted waves from the systems with that from a rather perfect-crystal counterpart exists. The phase determination of fractional reflections due to incommensurate structures, such as those involving charge-density-waves and the like, are also demonstrated in this chapter by using multiple-wave diffraction. A prospect of possible future developments of this diffraction technique for structural studies is tentatively given at the end of this book.

2. Elements of X-Ray Physics and Crystallography

X-ray multiple-wave diffraction, like the usual two-wave Bragg diffraction, is an interference phenomenon between X-rays and a periodic crystal lattice. The detection of this phenomenon relies on the conditions of carrying out the diffraction experiments, which often involve X-ray sources and crystals. In this chapter, the experimental aspects of the currently available X-ray sources are described. The fundamentals of crystallography concerning reciprocal lattice and the factors affecting intensity measurements are briefly reviewed as well. The latter include the structure factor, absorption coefficient, Lorentz-polarization factor due to crystal rotation and the beam polarization. Other inelastic processes associated with X-ray diffraction like absorption and fluorescence are also mentioned.

2.1 X-Ray Sources

Electromagnetic radiation is usually emitted from an accelerated charged particle or from a radioactive source. In the case of X-rays, the radiation is generated when a charged particle is accelerated or decelerated. In conventional X-ray laboratories, X-rays are generated from either sealed X-ray tubes or rotating-anode generators. In that case, electrons emitted from a filament (cathode) are accelerated by an applied voltage and bombard a metal target (anode). At the target, the electrons are decelerated and X-rays are emitted. X-rays can also be generated from accelerated relativistic charged particles like electrons, positrons or the like in the storage ring of a synchrotron facility. These different kinds of X-ray sources are described below.

2.1.1 Conventional Laboratory Sources

The X-ray spectrum of a conventional source consists of characteristic lines and continuum. The lines are due to the atomic transitions of the target element and the continuum arises from the continuous energy loss of the particles during the bombardment.

Figure 2.1 is a typical spectrum of a copper target. The sharp lines are the characteristic CuKα and CuKβ lines and the broad profile underneath is

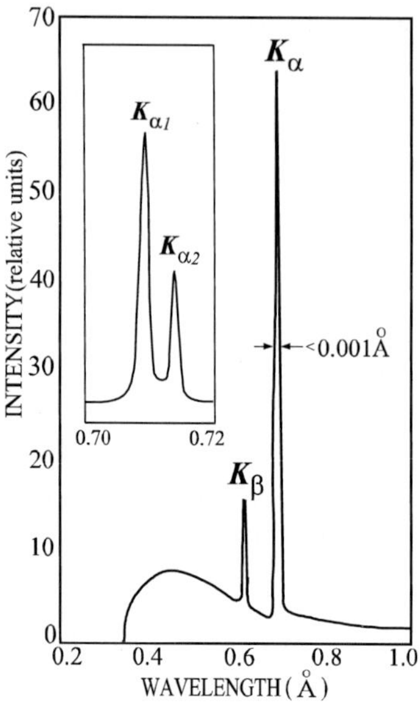

Fig. 2.1. X-ray spectrum from a sealed Cu X-ray tube

the continuum, or the so-called white radiation. The CuKα doublet consists of the CuKα_1 and CuKα_2 lines, which correspond to the transitions of the copper atoms from the L_{III} state to the K state and from the L_{II} state to the K state, respectively. On the other hand, the CuKβ line results from the transition from the M state to the K state. Both the characteristic and the continuous radiation are usually unpolarized. If the applied voltage across the anode and cathode is higher than the corresponding activation energy threshold, the intensities of the characteristic lines will be much stronger than that of the continuum; hence the characteristic lines of X-rays are usually used. However, when the applied electric potential energy is higher than the integer multiple of the activation energy, higher harmonic radiation, like λ/n with integer $n > 1$, may coexist, which contaminates the incident beam. In other words, the incident beam is no longer monochromatic. Strictly speaking, the characteristic lines themselves are not purely monochromatic, because their spectral width is of the order of $10^{-4}\lambda$. Therefore, for some experiments requiring energy resolution better than $10^{-4}\lambda$, an X-ray monochromator made of perfect crystals is employed.

The beam size and beam divergence are determined by the size of the optical projection of the anode target through the beam conditioners onto the plane perpendicular to the beam. The beam conditioners can be a series of slits, pinholes, reflection mirrors, or monochromators arranged along the incident beam path between the anode and the crystal. It is known from electromagnetism that the emitted X-rays from a metal target have a maximum power at about 6 degrees in take-off angle. Sometimes, a take-off angle of 3 degrees may be used for better spatial resolution of the incident beam with some sacrifice in intensity.

The arrangements of X-ray optics for an incident and a diffracted beam follow closely the reciprocity theorem of optics, Liouville's theorem of statistical mechanics [386], and the DuMond diagram [387]. Therefore, the beam size and divergence can be estimated from the geometry aspects of ray tracing for each optical component. According to Liouville's theorem, the product of the beam size and the angular divergence is constant at any position along the beam path. In addition, the energy dispersion versus Bragg angle, or the so-called DuMond diagram, needs to be considered for a double-crystal monochromator (DCM) in order to have an energy dispersive or non-dispersive situation for the incident beam. It is known that the so-called $(+,-)$ arrangements provide energy non-dispersive conditions, while the $(+,+)$ arrangements provide energy dispersive conditions (see Fig. 2.2). The crystal arrangements are sometimes denoted as $(+n,-n)$ and $(+n,+n)$ because the same Bragg reflection n is used for the two crystals. The relationships between the energy and Bragg angle are depicted in Fig. 2.3 for the two arrangements. In most diffraction experiments, a one-to-one correspondence between Bragg angle and photon energy (or wavelength) is desired, hence the $(+,-)$ arrangement for a DCM is frequently employed. If a detailed energy scan is needed for intensity measurements, the $(+,+)$ arrangement may be used for a characteristic radiation. In other words, the small energy range of the characteristic line can be easily divided further into even smaller steps by using this kind of monochromator. If two different reflections, m and n, are used for a DCM monochromator, the arrangement is denoted by $(+m,+n)$ or $(+m,-n)$ for the dispersive or the non-dispersive situation, respectively.

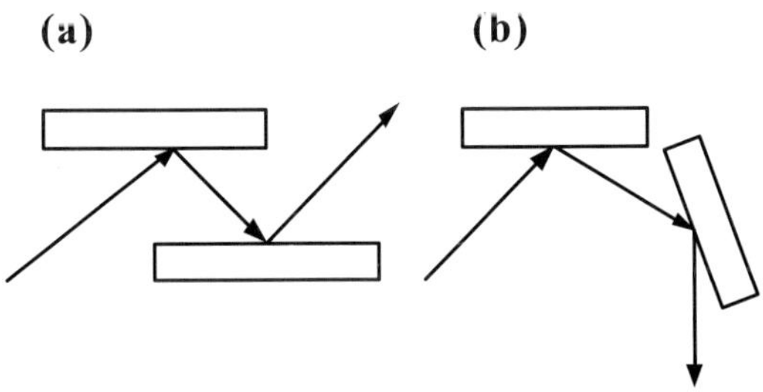

Fig. 2.2. Arrangements of double-crystal monochromator: (**a**) $(+,-)$ and (**b**) $(+,+)$ geometries

2.1.2 Synchrotron Radiation Sources

A synchrotron radiation facility is composed of a particle gun to generate electrons (or positrons), a booster ring or a linear accelerator (LINAC) to

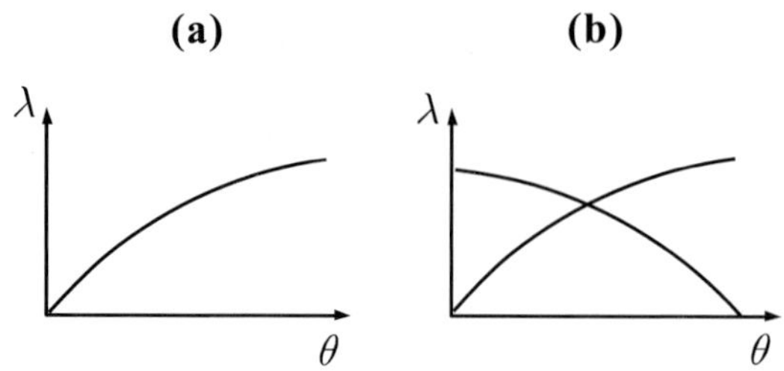

Fig. 2.3. DuMont diagrams for double-crystal monochromators for (**a**) (+, −) and (**b**) (+, +) arrangements

accelerate the electrons to a higher energy, a transport line and an injection system to inject the electrons into a storage ring from the booster. In the storage ring the electrons are confined in a circular path by bending dipole magnets, quadrupole magnets, and sextupole magnets. There are several straight sections along the circular path to accommodate the beam injection, radio frequency (RF) cavities for electric power supply, and insertion devices like wigglers, undulators, and wavelength shifters. Figure 2.4 shows the continuous distributions of synchrotron radiation from the bending magnets and insertion devices of several existing synchrotron radiation facilities. The distributions can be quantitatively described by classical electrodynamics for

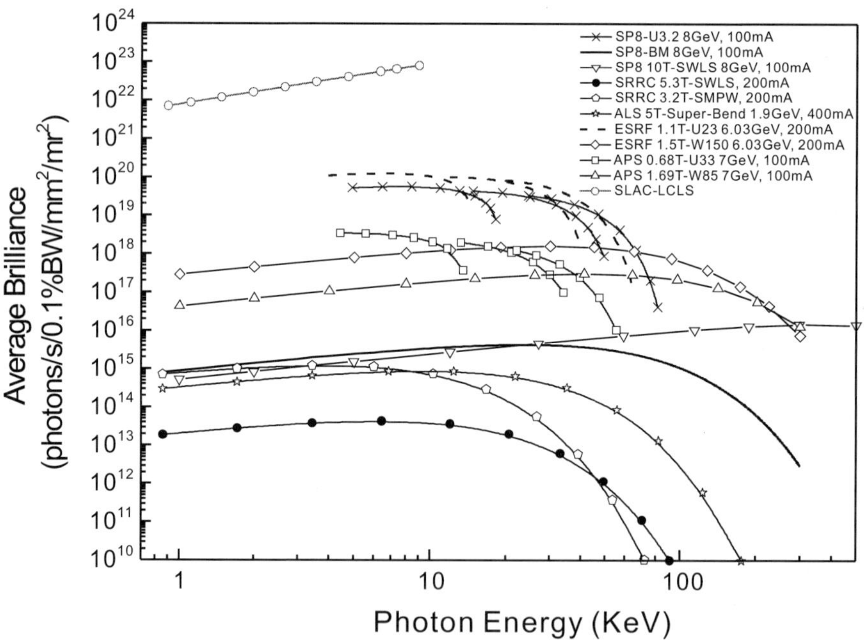

Fig. 2.4. Synchrotron radiation spectrum (courtesy of C.-S. Hwang)

the motion of relativistic electrons (or positrons) in circular orbits. The spectral intensity can be plotted as the number of photons per second per square mrad in the horizontal width per 0.1% bandwidth versus photon energy and wavelength. The so-called brilliance (sometimes called brightness) and flux are also used to quantitatively describe synchrotron radiation. The former, related to intensity per unit area, is defined as the number of photons per second per square mm per square mrad per 0.1% bandwidth. The latter is the integrated intensity over solid angle in the unit of the number of photons per second per 0.1% bandwidth. Figure 2.4 is an example showing radiation brilliance versus photon energy. The critical wavelength λ_c of synchrotron radiation from a bending magnet or insertion device is defined as

$$\lambda_c = \frac{4\pi}{3} R\gamma^{-3} , \tag{2.1}$$

where R is the radius of the electron orbit and γ is the relativistic coefficient equal to $E/(m_o c^2)$, $m_o c^2$ being the electron rest mass energy, c the speed of light, and E the electron energy. The useful spectral range is up to about four times the critical energy E_c, where $E_c = (hc)/\lambda_c$.

The X-ray beam emitted from the storage ring in the orbital plane of the particles is linearly polarized in that plane. However, the beam off the plane is either left- or right-handed elliptically polarized, depending on the direction of particle (electron) circulation being clockwise or counterclockwise around the upward vertical axis. If the beam above the orbital plane is left-handed elliptically polarized, then the beam below the plane is right-handed elliptically polarized. The X-ray emission causes an energy loss of the electrons in the storage ring, which is compensated by an applied RF field. The circulating electrons form bunches in storage rings in synchronization with the RF (radio frequency) field. This leads to a time structure of radiation in pulses of a few pico-seconds (ps) to nano-seconds (ns) due to the bunch lengths. The duration between pulses ranges from ns to µs (micro-seconds), depending on the repetition frequency of the filled bunches. For example, a single-bunch mode gives the lowest repetition frequency and the longest period between pulses. This property makes synchrotron radiation a superb source for time-resolved experiments.

The emission angle in the plane along the tangent to the electron trajectory normal to the orbit plane is equal to $1/\gamma$. The actual beam size and divergence from a storage ring depend on the electron emittances and the photon emittances in both vertical and horizontal directions perpendicular to the photon emission direction. The emittance of a particle is defined as the product of the beam size (in mm) and the beam divergence (in mrad) of the particle. Usually the photon emittances are similar to the electron emittances. The emittance has an elliptical contour in the (y, y') vertical phase space and has elliptical contours extended along a parabola symmetric to x in the (x, x') horizontal phase space. The quantities x and y are the spatial positions, and x' and y' are the angular positions of the photon in the horizontal and vertical directions, respectively [388, 389]. The orbital trajectory is along the z direction. When optical elements, such as slits, mirrors, and monochromators, are introduced into beamlines, the reciprocity theorem and Liouville's theorem need to be followed. Specifically, when double-crystal monochromators are used, the position-angle-wavelength space method in dealing with the photon distributions in x-x'-λ and y-y'-λ phase space should be considered [389]. Here the three-parameter phase space is the combination of the DuMond diagram and the photon emittances. Figure 2.5 shows the horizontal and vertical representations of synchrotron X-rays in this three-parameter space. In Fig. 2.5(a), SX and BX represent the relationships of the photon distributions in the x-x'-λ space at the source position ($z = 0$) and at some distance z from the source along the z-axis, respectively. Similarly, SY and BY in Fig. 2.5(b) represent the situations in the y-y'-λ space at $z = 0$ and $z \neq 0$. The insertion devices, i.e., wigglers and undulators, and low emittances particle-beams and photons are the most important features of a third generation synchrotron facility. These provide excellent conditions for diffraction and scattering experiments. Wigglers and undulators are periodic electromagnetic devices which can force the electrons or positrons to deflect periodically from their usual orbits in a short path. The radiation emitted from these deflected electrons or positrons is incoherent for a wiggler and quasi-coherent, or quasi-monochromatic, for an undulator, because the degree of deflection is larger for the wiggler and smaller for the undulator compared to the natural emission angle, $1/\gamma$, of synchrotron radiation. The intensities from the wiggler and the undulator are therefore proportional to the number of magnet periods N and the square of N, respectively. The spectrum of the wiggler is usually shifted towards higher energies compared to the spectrum from a bending magnet because of the higher field strength of the magnets used for the wiggler. The spectrum of an undulator shows strong peaks at certain energies corresponding to the harmonics, which are governed by the magnet period, the relativistic coefficient γ, the angle of observation, and the so-called deflection parameter K [390]. Planar undulators and helical or

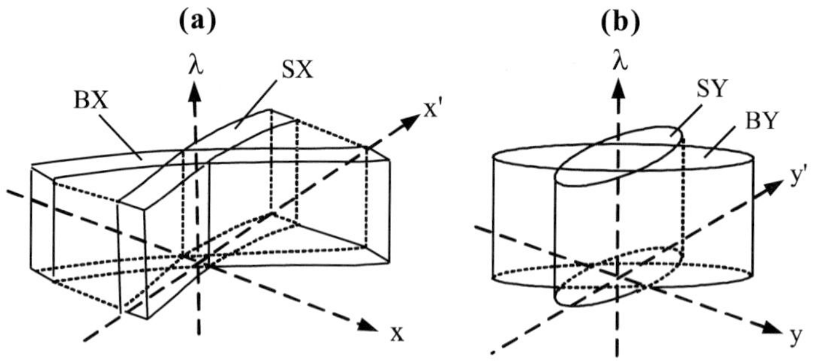

Fig. 2.5. (**a**) Horizontal and (**b**) vertical representations of synchrotron X-rays in the position-angle-wavelength space [389]

elliptical undulators can be designed for specific needs whenever changes in the beam polarization are desired.

The third generation synchrotron radiation sources can provide very high photon flux and spectral brilliance. In terms of brilliance, 10^{14}, 10^{18} and 10^{21} from a bending magnet, wiggler, and undulator, respectively, are obtainable. In addition, low emittances, a few nm-rad in horizontal and a few hundredth nm-rad in vertical, are achievable. Recently the development of the fourth generation synchrotron sources has been launched. Energy recovery LINAC sources and laser-driven X-ray sources, like plasma sources and X-ray free-electron lasers, are expected to produce much higher peak brilliances, from 10^{22} up to 10^{27} or more. Lower emittances and shorter pulses of subpico seconds from these new sources are also anticipated.

2.2 Interaction of X-Rays with Matter

The interaction of X-rays with matter is mainly related to the interaction between the electromagnetic waves and the atomic electrons. Although there are many phenomena associated with the interaction, here we will only consider those closely related to scattering and diffraction, including elastic Thomson scattering, Raleigh scattering, absorption, photoemission, and X-ray fluorescence. In order to describe these phenomena in a more general way, the quantum theory of radiation will be adopted. In the following, we will first consider the X-ray absorption and emission of nonrelativistic electrons in atoms and then extend the formulation to the various scattering mentioned. Since there have been many textbooks and articles giving the quantum mechanical description of the photon-electron interaction [391–393], only the necessary formalism will be outlined in this section.

2.2.1 Emission and Absorption

The electric and magnetic fields, $\boldsymbol{E}$ and $\boldsymbol{H}$, of electromagnetic waves, like X-rays, are governed by Maxwell's equations in Gaussian units, i.e.,

$$\nabla \times \boldsymbol{H} = \frac{1}{c}\left(\frac{\partial \boldsymbol{E}}{\partial t} + 4\pi \boldsymbol{J}_t\right) , \tag{2.2a}$$

$$\nabla \times \boldsymbol{E} = -\frac{1}{c}\frac{\partial \boldsymbol{H}}{\partial t} , \tag{2.2b}$$

$$\nabla \cdot \boldsymbol{E} = 4\pi \varrho_t , \tag{2.2c}$$

$$\nabla \cdot \boldsymbol{H} = 0 , \tag{2.2d}$$

for a given net charge density ϱ_t and current density $\boldsymbol{J}_t$, where c is the speed of light in vacuum. The divergence of (2.2a) and the time derivative of (2.2c) lead to the equation of continuity for ϱ_t and $\boldsymbol{J}_t$:

$$\nabla \cdot \boldsymbol{J} + \frac{\partial \varrho}{\partial t} = 0 \,. \tag{2.3}$$

According to (2.2b) and (2.2d), the electric and magnetic fields can be expressed in terms of the vector and scalar potentials, $\boldsymbol{A}$ and ϕ, as

$$\boldsymbol{E} = -\frac{1}{c}\frac{\partial \boldsymbol{A}}{\partial t} - \nabla\phi \,, \qquad \boldsymbol{H} = \nabla \times \boldsymbol{A} \,. \tag{2.4}$$

Since the radiation field is a transverse field, the vector potential satisfies the transversality condition

$$\nabla \cdot \boldsymbol{A} = 0 \,. \tag{2.5}$$

The interaction between the atomic electrons of mass m and charge e and the radiation field can be described by the time-dependent Schrödinger equation

$$\mathrm{i}\hbar \frac{\partial \Psi}{\partial t} = H\Psi \,, \tag{2.6}$$

where the Hamiltonian H consists of the non-interacting part H_0 and the interaction Hamiltonian H_{int}:

$$H = H_0 + H_{\mathrm{int}} \,, \tag{2.7}$$

where

$$H_0 = \frac{1}{2m}\sum_j \boldsymbol{p}_j^2 + \sum_j V(\boldsymbol{r}_j) + \frac{1}{8\pi}\int \left(E^2 + H^2\right) \mathrm{d}\boldsymbol{r} \,. \tag{2.8}$$

The first two terms are due to the atomic electrons and the third term due to the radiation field. The interaction Hamiltonian H_{int} is composed of the charge and orbital momentum part $H_{\mathrm{int}}^{(\mathrm{e})}$ and the spin part $H_{\mathrm{int}}^{(\mathrm{s})}$ [95, 393]:

$$H_{\mathrm{int}} = H_{\mathrm{int}}^{(\mathrm{e})} + H_{\mathrm{int}}^{(\mathrm{s})} \,, \tag{2.9}$$

where

$$\begin{aligned} H_{\mathrm{int}}^{(\mathrm{e})} &= \sum_j \Big[-\frac{e}{2mc}\left(\boldsymbol{p}_j \cdot \boldsymbol{A}(\boldsymbol{r}_j, t) + \boldsymbol{A}(\boldsymbol{r}_j, t)\cdot \boldsymbol{p}_j\right) \\ &\quad + \frac{e^2}{2mc^2}\boldsymbol{A}(\boldsymbol{r}_j, t)\cdot \boldsymbol{A}(\boldsymbol{r}_j, t)\Big] - \sum_j e\phi(\boldsymbol{r}_j) \,, \end{aligned} \tag{2.10}$$

$$\begin{aligned} H_{\mathrm{int}}^{(\mathrm{s})} &= -\sum_j \frac{e\hbar}{mc}\boldsymbol{\sigma}_j \cdot [\nabla \times \boldsymbol{A}(\boldsymbol{r}, t)]_{\boldsymbol{r}=\boldsymbol{r}_j} - \frac{e^2\hbar}{2\left(mc^2\right)^2} \\ &\quad \times \sum \boldsymbol{\sigma}_j \cdot \left[\frac{\partial \boldsymbol{A}(\boldsymbol{r}, t)}{\partial t} \times \boldsymbol{A}(\boldsymbol{r}, t)\right] . \end{aligned} \tag{2.11}$$

The quantity $\boldsymbol{p}_j$ is the canonical momentum operator, $\boldsymbol{\sigma}_j$ is the spin operator, and V is the potential energy binding the electrons. The third term of $H_{\mathrm{int}}^{(\mathrm{e})}$

is related to Thomson scattering and the first two terms correspond to the charge and orbital momentum contributions to anomalous scattering and magnetic scattering. The $H_{\text{int}}^{(\text{s})}$ terms are due to the interaction between the spin magnetic moment of the electron and the magnetic field.

The vector potential $\boldsymbol{A}(\boldsymbol{r},t)$ is a linear combination of the annihilation and creation operators $b_{k,\epsilon}$ and $b_{k,\epsilon}^{+}$ [391–394]

$$\boldsymbol{A}(\boldsymbol{r},t) = \sum_{k}\sum_{\epsilon} c\sqrt{\frac{2\pi\hbar}{\omega V_o}}\left[\hat{\epsilon} b_{k,\epsilon}\exp(2\pi\mathrm{i}\boldsymbol{k}\cdot\boldsymbol{r}) + \hat{\epsilon} b_{k,\epsilon}^{+}\exp(-2\pi\mathrm{i}\boldsymbol{k}\cdot\boldsymbol{r})\right] , \tag{2.12}$$

where

$$\begin{aligned} b_{k,\epsilon} &= b_{k,\epsilon}(0)\mathrm{e}^{-\mathrm{i}\omega t} , \\ b_{k,\epsilon}^{+} &= b_{k,\epsilon}^{+}(0)\mathrm{e}^{\mathrm{i}\omega t} , \end{aligned} \tag{2.13}$$

$\hat{\varepsilon}$ is the unit-vector of polarization, and $\boldsymbol{k}$ and ω are the wave-vector and angular frequency of the EM wave. c, $\hbar$, and V_o represent the speed of light in vacuum, the Planck constant $h/2\pi$, and the volume irradiated, respectively. $\boldsymbol{A}$ can also be expressed as a function of the position vector $\boldsymbol{r}$ and time t [391–393]:

$$\begin{aligned} \boldsymbol{A}(\boldsymbol{r},t) = &\left(\frac{2\pi c^2 N\hbar}{\omega V_o}\right)^{1/2} \hat{\varepsilon}\exp[\mathrm{i}(2\pi\boldsymbol{k}\cdot\boldsymbol{r} - \omega t)] \\ &+ \left[\frac{2\pi c^2 (N+1)\hbar}{\omega V}\right]^{1/2} \hat{\varepsilon}\exp[-\mathrm{i}(2\pi\boldsymbol{k}\cdot\boldsymbol{r} - \omega t)] , \end{aligned} \tag{2.14}$$

where the first term on the right is related to the absorption of a photon quantum and the second term to the emission of a photon quantum by an electron. N is the photon occupation number in the initial state. The differential cross-section, the absorption probability, and the transition probability can be calculated quantum-mechanically by taking into account the absorption and emission as the scattering of photons by atomic electrons [395]. That is, before scattering, the photon with the wave-vector $\boldsymbol{k}$, angular frequency ω, and polarization vector $\hat{\varepsilon}$ is incident on the atom in its initial state A. After scattering, the atom stays in its final state B and the scattered photon propagates along the $\boldsymbol{k'}$ vector with the polarization vector $\hat{\varepsilon}'$ and angular frequency ω'. In between state A and state B are the excited intermediate states. By neglecting the last term in (2.10) and the spin-magnetic-moment interaction, the Kramers-Heisenberg formula [396], modified to include the effects of radiation damping, gives the differential cross-section, i.e., the derivative of the cross-section σ with respect to the solid angle Ω, as follows [393]:

$$\frac{d\sigma}{d\Omega} = {r_e}^2\left(\frac{\omega'}{\omega}\right)\left|\hat{\varepsilon}\cdot\hat{\varepsilon}'\delta_{AB} - \frac{1}{m}\sum_{I}\left[\frac{(\boldsymbol{p}_e\cdot\hat{\varepsilon}')_{BI}(\boldsymbol{p}_e\cdot\hat{\varepsilon})_{IA}}{E_I - E_A - \hbar\omega - \mathrm{i}\frac{\Gamma_I}{2}}\right.\right.$$

$$+\left.\frac{(\boldsymbol{p}_e\cdot\hat{\varepsilon})_{BI}(\boldsymbol{p}_e\cdot\hat{\varepsilon}')_{IA}}{E_I-E_A+\hbar\omega}\right]\Bigg|^2 , \tag{2.15}$$

where $r_e = e^2/(mc^2)$ is the classic radius of the electron, the subscript I stands for the excited intermediate state, $\boldsymbol{p}_e$ is the electric dipole moment, and δ_{AB} is a Kronecker delta involving the wavefunctions φ_A and φ_B of states A and B. The matrix elements, $(\boldsymbol{p}_e \cdot \hat{\varepsilon})_{BI}$ and $(\boldsymbol{p}_e \cdot \hat{\varepsilon})_{IA}$, involve the wavefunctions φ_A, φ_B and φ_I [393]. The probability of finding the intermediate state I is proportional to $\exp(-\Gamma_I t/\hbar)$, with $\Gamma_I = \hbar/\tau_I$, τ_I being the lifetime of state I. The first term in (2.15) represents the non-resonant amplitude of scattering ($\omega \neq \omega_{IA} = (E_I - E_A)/\hbar$). The second term and the third term have appreciable amplitudes in the resonance condition, where $\omega = \omega_{IA}$. This phenomenon is known as the resonance fluorescence.

The sum of the non-resonant amplitudes, of the order of r_e, is very small compared to the order of c/ω for the resonant amplitude. By ignoring the non-resonant amplitudes, the differential cross-section of a single-level resonance fluorescence process in the vicinity of a non-degenerate resonance state R is given by [393]

$$\frac{\mathrm{d}\sigma}{\mathrm{d}\Omega} = \frac{{r_e}^2}{m^2}\left(\frac{\omega'}{\omega}\right)\left[\frac{|(\boldsymbol{p}_e\cdot\hat{\varepsilon})_{AR}|^2}{(E_R-E_A-\hbar\omega)^2+\frac{{\Gamma_R}^2}{4}}\right]|(\boldsymbol{p}_e\cdot\hat{\varepsilon}')_{RB}|^2 , \tag{2.16}$$

which depends on the product of the probability of finding state R formed by the absorption of a photon and the spontaneous emission probability per solid angle for the transition from state R to state B by emitting a photon. The absorbed photon is characterized by $(\boldsymbol{k}, \hat{\varepsilon})$ and the emitted photon by $(\boldsymbol{k}', \hat{\varepsilon}')$. Namely, the term in the square bracket is proportional to the absorption probability, and the matrix element after the bracket is connected with the emission probability.

Fluorescence. X-ray fluorescence can be considered as either a single-step or a two-step process, depending on the lifetime of the resonance state, namely Γ_R, which is of the order of 10^{-9} eV. According to the uncertainty principles, the lifetime is closely related to the energy resolution ΔE of the X-rays used. For a conventional X-ray and a synchrotron radiation (SR) source, the energy resolution ΔE of an incident X-ray beam is around 10^{-2} to 10 eV [389, 390, 397, 398], except for some special experiments, such as those involving Mössbauer effects, where an energy resolution of 10^{-9} eV is needed [399, 400]. Under normal circumstances, $\Delta E \gg \Gamma_R$ ($\sim 10^{-9}$ eV). This means that the temporal duration of the incident photon (X-ray) beam is shorter than the lifetime of the resonance. In other words, the formation of the metastable resonance state R via absorption and the subsequent X-ray emission can be treated as two independent quantum mechanical processes. In such a case, the emission can be thought of as a spontaneous emission, i.e., the atom undergoes a radioactive transition from state R to state B as if there were no incident electromagnetic waves. The corresponding $\mathrm{d}\sigma/\mathrm{d}\Omega$ then takes the following simple form under the electric dipole approximation [392]:

$$\frac{\mathrm{d}\sigma}{\mathrm{d}\Omega} \cong P_{AB}\frac{\alpha\omega^3}{2\pi c^3}V\left|\boldsymbol{X_{AB}}\right|^2\sin^2\phi\,, \tag{2.17}$$

where P_{AB} is the absorption probability equal to the term in the square bracket given in (2.16), $\left|\boldsymbol{X_{AB}}\right|^2$ is the transmission probability equal to the term after the bracket of (2.16), and α is the fine-structure constant ($\sim 1/137$). ϕ is the angle between the electric dipole moment $\boldsymbol{p}_e$ and the direction of the X-ray fluorescence emitted from the atom. As is well known, this type of fluorescence has the following distinct properties: (1) In general, the fluorescence is directionally isotropic, therefore its distribution is uniform in space. (2) The intensity of the fluorescence has a maximum when the emission direction is perpendicular to $\boldsymbol{p}_e$ and is zero along $\boldsymbol{p}_e$. (3) X-ray fluorescence emitted from different atoms in a sample is incoherent and the emitted photons will not interfere with one another. (4) Whenever the energy of the incident photon is greater than that of an absorption edge, the energy of the emitted fluorescence is determined by the energy difference between the two states involved in the transition, which is independent of the incident energy.

The measurements of the intensity of X-ray fluorescence from a substance as a function of photon energy usually provide the X-ray absorption spectra of the substance studied. This kind of measurement can be easily carried out with synchrotron radiation due to the tunability of photon energy.

Rayleigh Scattering and Thomson Scattering. Elastic scattering occurs when the conditions $A = B$ and $\omega = \omega'$ hold for the Kramers-Heisenberg formula. For $\omega \ll \omega_{IA}$, or $x \gg x_{IA}$, the so-called Rayleigh cross section is obtained from (2.15) as

$$\begin{aligned}\frac{\mathrm{d}\sigma}{\mathrm{d}\Omega} &= \left(\frac{r_e}{m\hbar}\right)^2\omega^4\Bigg|\sum_I\left(\boldsymbol{p}_e\cdot\hat{\varepsilon}'\right)_{AI}\left(\boldsymbol{p}_e\cdot\hat{\varepsilon}\right)_{IA} \\ &\quad +\left(\boldsymbol{p}_e\cdot\hat{\varepsilon}\right)_{AI}\left(\boldsymbol{p}_e\cdot\hat{\varepsilon}'\right)_{IA}\Bigg|^2 \\ &= \left(\frac{r_e m}{\hbar}\right)^2\omega^4\Bigg|\sum_I\left(\frac{1}{\omega_{IA}}\right)^3\left[\left(\boldsymbol{r}\cdot\hat{\varepsilon}'\right)_{AI}\left(\boldsymbol{r}\cdot\hat{\varepsilon}\right)_{IA}\right. \\ &\quad \left.+\left(\boldsymbol{r}\cdot\hat{\varepsilon}\right)_{AI}\left(\boldsymbol{r}\cdot\hat{\varepsilon}'\right)_{IA}\right]\Bigg|^2\,,\end{aligned} \tag{2.18}$$

where the approximation of $1/(\omega_{IA} \mp \omega) \approx [1 \pm (\omega/\omega_{IA})]/\omega_{IA}$ has been employed. Equation (2.18) leads to Rayleigh's law, which states that the scattering cross section for longer wavelengths is inversely proportional to the fourth power of the wavelength.

If the incident photon energy is much greater than the atomic binding energy, i.e., $\omega \gg \omega_{IA}$, and is smaller than the rest energy of the electron, the first term dominates (2.15). This type of scattering is called Thomson scattering. The scattering cross section takes the following simple form:

$$\frac{\mathrm{d}\sigma}{\mathrm{d}\Omega} = {r_e}^2 \left|\hat{\varepsilon} \cdot \hat{\varepsilon}'\right|^2 , \tag{2.19}$$

which is independent of the binding of the atomic electron. The vector product in (2.19) is the polarization factor, which can be calculated by referring to an appropriate coordinate system, for example, the spherical coordinate system. If the scattering angle between the incident and the scattered beam is 2θ, the polarization factor is $[1 + (\cos 2\theta)^2]/2$ for an unpolarized radiation. The scattering cross section for $\theta = 0$ is equal to the square of the atomic scattering factor $f(\omega)$:

$$\left(\frac{\mathrm{d}\sigma}{\mathrm{d}\Omega}\right)_{\theta=0,\varepsilon^{(\alpha)}=\varepsilon^{(\alpha')}} = \left|f(\omega)\right|^2 . \tag{2.20}$$

For an electron in the ground state, f takes the form

$$f(\omega) = -r_e \left[1 - \frac{1}{m}\sum_I \left(\frac{\left|(\boldsymbol{p}_e \cdot \hat{\varepsilon})_{IA}\right|^2}{E_I - E_A - \hbar\omega - \mathrm{i}\Gamma_I/2} + \frac{\left|(\boldsymbol{p}_e \cdot \hat{\varepsilon})_{IA}\right|^2}{E_I - E_A + \hbar\omega}\right)\right] . \tag{2.21}$$

The sign of $f(\omega)$ is positive for attractive interaction and negative for repulsive interaction. Since the scattering of EM waves follows the causality principle, the real and imaginary parts of $f(\omega)$ satisfy the dispersion relations, i.e.,

$$\mathrm{Re}[(f(\omega))] = \frac{2\omega^2}{\pi}\int_0^\infty \frac{\mathrm{Im}[f(\omega')]\mathrm{d}\omega'}{\omega'(\omega'^2 - \omega^2)} , \tag{2.22}$$

and

$$\mathrm{Im}[(f(\omega))] = -\frac{2\omega}{\pi}\int_0^\infty \frac{\mathrm{Re}[f(\omega')]\mathrm{d}\omega'}{\omega'^2 - \omega^2} . \tag{2.23}$$

Equations (2.22) and (2.23) are often called the Kramers-Kronig relations [396, 401–403] and are related to each other by the Hilbert transforms, i.e.,

$$\mathrm{Re}[(f(\omega))] = \frac{1}{\pi}\int_{-\infty}^\infty \frac{\mathrm{Im}[f(\omega')]\mathrm{d}\omega'}{\omega' - \omega} , \tag{2.24}$$

$$\mathrm{Im}[(f(\omega))] = -\frac{1}{\pi}\int_{-\infty}^\infty \frac{\mathrm{Re}[f(\omega')]\mathrm{d}\omega'}{\omega' - \omega} . \tag{2.25}$$

Referring to (2.18), (2.19) and (2.20), the atomic scattering factor f is often expressed as

$$f = f_o + \Delta f' + \mathrm{i}\Delta f'' , \tag{2.26}$$

where f_o is the energy-independent atomic scattering factor, namely $f_o = r_e$. $\Delta f' = \mathrm{Re}[f(\omega)]$ and $\Delta f'' = \mathrm{Im}[f(\omega)]$ are the real and imaginary parts of the correction due to the so-called anomalous dispersion or anomalous scattering.

The atomic scattering factor f, or the scattering amplitude, is constant in all the scattering directions for an ideal point atom where the mass and charge of the atom are concentrated on one point. In reality, the atom, or rather the electron density distribution around the nucleus, occupies a certain volume in space. If the incident X-ray waves are scattered by a small portion of the electron density distribution off the center of the atom, then the scattered waves in the direction deviated by the scattering angle, 2θ, from the incident direction suffer a phase shift. This shift is due to the difference in the optical path between these scattered waves and the waves scattered from the center. The atomic scattering factor of this small portion is therefore equal to the constant scattering amplitude times the phase-shift factor. By integrating this scattering factor over the entire electron density distribution, the resultant atomic scattering factor shows a decaying characteristic as $\sin\theta/\lambda$ increases. Therefore, the atomic scattering factor is also called the form factor.

If the incident photon energy is different from the scattering energy and the final state is not the initial state, the scattering is inelastic. Raman effects and Compton scattering are two well-known examples. The latter shows pronounced characteristic features when the incident photon energy is comparable with the rest energy of the electron. Both scatterings can also be described by the Kramers-Heisenberg formula, (2.15) [396].

2.3 Reciprocal Lattice, Structure Factor, and Electron Density Distribution

2.3.1 Reciprocal Lattice

A crystal is an arrangement of atoms in a two-, three-, or higher-dimensional periodic array. It is represented in real space by a lattice with crystal unit cells as building blocks. A lattice point may be composed of a single atom, many atoms, or molecules. A unit cell of a three-dimensional lattice is defined by three basis vectors, $\boldsymbol{a}$, $\boldsymbol{b}$, and $\boldsymbol{c}$, which involve six cell parameters, i.e., three cell dimensions, a, b, and c, and three angles, α, β, and γ. The angle between $\boldsymbol{b}$ and $\boldsymbol{c}$ is α; the angle between $\boldsymbol{c}$ and $\boldsymbol{a}$ is β; the angle between $\boldsymbol{a}$ and $\boldsymbol{b}$ is γ. Based on the relationships among the cell parameters, there are seven crystal systems: triclinic, monoclinic, orthorhombic, tetragonal, hexagonal, rhombohedral, and isometric (cubic), and fourteen Bravais lattices: triclinic (primitive), monoclinic (primitive and body-centered), orthorhombic (primitive, A (or B, or C)-centered, face-centered, and body-centered), tetragonal (primitive and body-centered), cubic (primitive, face-centered, and body-centered), hexagonal (primitive), and rhombohedral. If one considers the symmetry elements and crystal lattices, there are 32 point groups and 230 space groups for three-dimensional lattices. For two-dimensional lattices, there are 4 plane lattices and 17 plane groups [404]. For quasi-crystals, the lattices are formed by the projection of periodic structures in a hyper-dimension onto three-dimension or lower ones and consist of two different types of unit cells. Therefore, the point group and space group of quasi-crystals are much more complicated than that of traditional crystals.

The Fourier transform of a crystal lattice is the reciprocal lattice, in which a set of atomic planes is represented by a reciprocal lattice point. A reciprocal lattice vector is the vector connecting the origin O of the lattice to the reciprocal lattice point, which is perpendicular to the set of the atomic planes. The basis vectors of a three-dimensional reciprocal lattice are $\boldsymbol{a}^*$, $\boldsymbol{b}^*$, and $\boldsymbol{c}^*$, defined as

$$\begin{aligned}
\boldsymbol{a}^* &= (\boldsymbol{b} \times \boldsymbol{c})/V_c \,, \\
\boldsymbol{b}^* &= (\boldsymbol{c} \times \boldsymbol{a})/V_c \,, \\
\boldsymbol{c}^* &= (\boldsymbol{a} \times \boldsymbol{b})/V_c \,,
\end{aligned}$$

where V_c is the volume of the crystal unit cell. For a given set of atomic planes $G(hk\ell)$, the reciprocal lattice vector, which connects the origin O and the reciprocal lattice point G, is expressed as $\boldsymbol{d}^*_G = h\boldsymbol{a}^* + k\boldsymbol{b}^* + \ell\boldsymbol{c}^*$, i.e. $\boldsymbol{d}^*_G = \boldsymbol{OG}$. With the concept of the reciprocal lattice, the geometry of a Bragg diffraction can be easily described, which will be presented in Chap. 3.

2.3.2 Structure Factor and Electron Density

The scattering amplitude of a given $G(hk\ell)$ from a crystal unit cell is defined as the structure factor F_G:

$$\begin{aligned}
F_G = F(hk\ell) &= \sum_j f_j \exp[2\pi \mathrm{i}(\boldsymbol{d}^*_G \cdot \boldsymbol{r}_j)] \\
&= \sum_j f_j \exp[2\pi \mathrm{i}(hx_j + ky_j + \ell z_j)] \,,
\end{aligned} \tag{2.27}$$

where f_j and $\boldsymbol{r}_j$ are the atomic scattering factor and the position vector of the j-th-atom in the unit cell. $\boldsymbol{r}_j$ is defined as $\boldsymbol{r}_j = x_j\boldsymbol{a} + y_j\boldsymbol{b} + z_j\boldsymbol{c}$, where x_j, y_j, and z_j are the fractional coordinates along $\boldsymbol{a}$, $\boldsymbol{b}$, and $\boldsymbol{c}$, i.e., $-1 \le x_j, y_j, z_j \le 1$. Since F_G is complex, it can be rewritten as

$$F_G = |F_G| \exp(\mathrm{i}\delta_G) \,, \tag{2.28}$$

where $|F_G|$ and δ_G are the magnitude and phase of the structure factor F_G, respectively. The value of F_G depends on the symmetry imposed by the space group to which the crystal belongs [404]. Reflections with a null structure factor are forbidden reflections, which are not measurable in X-ray diffraction experiments. This absence is sometimes called systematic absence or systematic extinction. Experimentally, the determination of a space group relies on the analysis of systematic extinctions and chemical sense.

The Fourier transform of the structure factor is the electron density distribution associated with all the atoms in the unit cell, and vice versa. That is,

$$\varrho(\boldsymbol{r}) = \frac{1}{V_c} \sum_G F_G \exp\left[-2\pi \mathrm{i}\left(\boldsymbol{d}_G^* \cdot \boldsymbol{r}\right)\right] , \tag{2.29a}$$

$$F_G = \frac{1}{V_c} \int \varrho(\boldsymbol{r}) \exp\left[2\pi \mathrm{i}\left(\boldsymbol{d}_G^* \cdot \boldsymbol{r}\right)\right] \mathrm{d}\boldsymbol{r} . \tag{2.29b}$$

The determination of a single-crystal structure means the determination of the electron density distribution, or electron density map, of the crystal. According to (2.29a), both the amplitude and phase are the necessary and sufficient conditions for the determination of a crystal structure.

2.4 Factors Affecting the Diffraction Intensity

The intensity measurement for a Bragg reflection is affected by many factors, including the crystal quality, cell dimension, polarization of the incident beam, crystal rotation, and temperature factor. These factors are described below.

2.4.1 Crystal Perfection

Crystals are usually categorized into ideally perfect and ideally imperfect crystals. Real crystals fall between these two ideal categories. Diffraction from crystals of different categories are different in nature. Diffraction from a perfect crystal is usually considered dynamical, because the interaction between the incident and the diffracted wave plays a major role in the diffraction process. According to Darwin's dynamical approach [66], or from the dynamical theory of X-ray diffraction, the integrated intensity I_G of a Bragg reflection G in a perfect crystal takes the following form [398]:

$$I_G = \left(\frac{I_o}{\omega}\right) \frac{8}{3\pi} r_e \left|F_G\right| \left(\frac{\lambda^2}{V_c}\right) \left[\frac{1 + |\cos 2\theta|}{2 \sin 2\theta}\right] , \tag{2.30}$$

where absorption is neglected. The integrated diffraction intensity is proportional to the wavelength used and the first power of structure factor, and inversely proportional to the unit-cell volume. On the other hand, it is known that the diffraction from ideally perfect crystals suffers intensity decrease, the so-called primary extinction, due to the successive reflections from the perfectly aligned atomic planes in the crystal. The stronger the diffraction is, the greater the extinction. The intensity correction due to primary extinction is proportional to $\tanh A/A$, where A is defined as

$$A = r_e \frac{|F_G|}{V_c} \frac{\lambda T_o}{\gamma_G} , \tag{2.31}$$

and T_o/γ_G is the effective path length of X-rays in the crystal for a symmetric Bragg reflection. T_o is the thickness of the crystal plate, γ_G the direction cosine of the reflected wave with respect to the inward surface normal, r_e the classic radius of electron, F_G the structure factor, and V_c the crystal unit-cell volume.

If the crystal is ideally imperfect, the integrated intensity without considering absorption is governed by [398]

$$I_G = \frac{I_o}{\omega} r_e^2 |F_G|^2 \frac{\lambda^3}{V_c^2} \frac{1 + |\cos 2\theta|^2}{2 \sin 2\theta} \delta V , \tag{2.32}$$

where the interaction between the incident and the diffracted waves in the crystal is not taken into account. Equation (2.32) was derived from Darwin's kinematical theory [67] and is closely related to the Thomson equation for the intensity of coherent scattering by an electron with an unpolarized radiation:

$$I_e = I_o \left(\frac{r_e}{r} \right)^2 \left(\frac{1 + \cos^2 2\theta}{2} \right) . \tag{2.33}$$

Because the integrated intensity in (2.32) was not in a satisfactory agreement with the observed intensities from real crystals, Darwin proposed the theoretical 'mosaic' model in 1922 [70] (named by Ewald [405]), in which the crystal consists of many tiny mosaic blocks with a slight misorientation with one another. Each block is a perfect crystal. The diffraction from this kind of crystals also suffers intensity decrease, the so-called secondary extinction, due to successive reflections in each of the tiny mosaic blocks, although each block diffracts incoherently. Similar to the primary extinction, the secondary extinction is stronger for a stronger reflection. The intensity correction due to secondary extinction is governed by the standard deviation η of the block tilts. The larger the standard deviation, the more severe the correction.

From (2.30), (2.32), and the crystal mosaic model, we see that primary extinction is related to the scattering amplitudes of the individual planes within a block, whereas secondary extinction depends on the scattering power of the individual blocks. Similar situations occur for a multiple-wave diffraction, though the interaction among the diffracted waves of which is more complicated than that of a two-wave diffraction. The details of the kinematical and dynamical theories in multiple-wave cases will be given in Chaps. 5 and 6.

2.4.2 Lorentz-Polarization Factor

The intensity of X-ray diffraction is usually measured by either rocking the crystal about the Bragg angle θ or rotating the detector about the scattering angle 2θ, or even by making a θ-2θ scan. The θ scan, sometimes called ω scan, is performed when the detector is stationary at a given 2θ. A plot of intensity versus θ, the so-called rocking curve, provides the intensity distribution along the transverse direction perpendicular to the reciprocal lattice vector of the Bragg reflection investigated. On the other hand, the θ-2θ scan gives the intensity distribution along the reciprocal lattice vector, i.e., in the longitudinal direction. The 2θ scan at a fixed θ merely provides the intensity distribution of the reflection in 2θ. Both ω scan and θ-2θ scan provide structural information about the crystal. For example, the width of a rocking curve determines the mosaic spread of the crystal, while the width of a θ-2θ profile gives the correlation length of the X-ray scattering in the crystal.

Because either the crystal or the detector, or both, are moving during the measurement, the intensities measured in the various scans are integrated over θ or 2θ. In the case of an ω scan of a given reflection G, the integration depends on the time needed for the reciprocal lattice point G to cross the surface of the Ewald sphere. A factor related to the elapse time is called 'Lorentz factor', which is defined as

$$L_F = \frac{1}{\sin 2\theta} \, . \tag{2.34}$$

The polarization of the incident radiation also affects the diffracted intensity. For a reflection G, the incident and the diffracted beams lie in a plane. This plane is called 'plane of incidence'. The polarization vectors, $\boldsymbol{\sigma}$ and $\boldsymbol{\pi}$, of the incident beam or the diffracted beam are perpendicular to the beam direction. $\boldsymbol{\sigma}$ is normal to and $\boldsymbol{\pi}$ lies in the plane of incidence. For an unpolarized incident X-ray beam, both the σ and π polarizations are effective. The polarization factor p_G, similar to that for the Thomson scattering, is

$$p_G = \frac{1 + \cos^2 2\theta}{2} \, . \tag{2.35}$$

Considering both the ω rotation and an unpolarized incident beam, the combined Lorentz-polarization factor Lp_G is defined as the product of the Lorentz factor and the polarization factor:

$$Lp_G = L_F p_G = \frac{1 + \cos^2 2\theta}{2 \sin 2\theta} \, . \tag{2.36}$$

2.4.3 Temperature Factor

Thermal motion of atoms is another factor that affects X-ray intensities. The atomic scattering factor is the scattering cross section of a stationary atom. At a temperature T, the thermal energy proportional to kT makes the atoms in a crystal vibrate around their equilibrium positions. The magnitude of the vibration causes a phase shift of the scattered waves. The average effect over time and space thus leads to an intensity fall-off as temperature increases:

$$I_G = I_o \exp\left[-2B\left(\frac{\sin\theta}{\lambda}\right)^2\right] , \tag{2.37}$$

where B is the Debye-Waller factor [406–409] which is related to the mean-square amplitude $< u^2 >$ of the atomic vibration by

$$B = 8\pi^2 < u^2 > . \tag{2.38}$$

The atomic scattering factor is thus modified to

$$f = f_o \exp\left[-B\left(\frac{\sin\theta}{\lambda}\right)^2\right] . \tag{2.39}$$

3. Diffraction Geometry

The geometric aspects of X-ray diffraction from crystals can be depicted in real space and reciprocal space. However, the representation of reciprocal lattice, as stated in Chap. 2, facilitates the description of the three-dimensional geometry involved in X-ray diffraction, especially for multiple-wave diffraction. Therefore, in this chapter, the diffraction conditions for the conventional Bragg reflection – the so-called two-wave case – and the cases involving multiple waves for wide-angle and grazing incidence are described in terms of the reciprocal lattice. In addition, the related geometric factors like Lorentz-polarization factors for multiple-wave diffraction are also described.

3.1 Two-Wave, Three-Wave, and Multiple-Wave Diffraction

3.1.1 Geometry Conditions

A Bragg reflection G occurs when a set of atomic planes $(h_G k_G \ell_G)$ is brought into position to diffract an incident wave. The angle θ between the incident wave and the atomic planes is equal to the Bragg angle θ_G so that Bragg's law, holds, i.e.,

$$2d_G \sin\theta_G = \lambda \; , \tag{3.1}$$

where d_G is the interatomic distance and λ is the wavelength of the incident wave. Figure 3.1(a) shows the diffraction in real space. In terms of the reciprocal lattice, two reciprocal lattice points, O and G, are on the surface of the Ewald sphere. The diffraction in the reciprocal space is illustrated in Fig. 3.1(b). The vectors $\boldsymbol{k_O}$ and $\boldsymbol{k_G}$ connecting the center C of the sphere to points O and G are the wavevectors of the incident and diffracted waves, respectively. The angle between the two vectors is $2\theta_G$. The moduli of the wavevectors are equal to $1/\lambda$. The reciprocal lattice point O is also the origin of the reciprocal lattice; hence the vector $\boldsymbol{OG}$ is the reciprocal lattice vector $\boldsymbol{d}^*_{\boldsymbol{G}}$, representing the set of atomic planes $(h_G k_G \ell_G)$. According to Chap. 2, $\boldsymbol{d}^*_{\boldsymbol{G}}$ is perpendicular to $(h_G k_G \ell_G)$ with its modulus equal to $1/d_G$. We can easily find from the triangle COG that $2(1/\lambda)\sin\theta = d^*_G = 1/d_G$, which is nothing but the Bragg diffraction condition. In other words, Bragg's law is always satisfied as long as the reciprocal lattice points O and G lie on the surface of the Ewald sphere. Because the usual Bragg reflection involves two reciprocal lattice points and two wavevectors, it is called two-wave diffraction. If the crystal is rotated around the reciprocal lattice vector $\boldsymbol{d}^*_{\boldsymbol{G}}$, the so-called azimuth ψ rotation, so that an additional reciprocal lattice point, say L, is moved onto the surface of the Ewald sphere, then the three-wave diffraction takes place (Fig. 3.2). The two sets of atomic planes $(h_G k_G \ell_G)$ and $(h_L k_L \ell_L)$ satisfy Bragg's law at the same time (Fig. 3.2(a)). The incident angles with respect to the two planes, θ_G and θ_L, are the Bragg angles. Two diffracted waves are generated along the wavevectors $\boldsymbol{k_G}$ and $\boldsymbol{k_L}$, respectively, in addition to the incident (or direct) diffraction O (Fig. 3.2(b)). Therefore, three reflections are involved in the diffraction process simultaneously. Similarly, N-wave multiple diffraction occurs when N reciprocal lattice points lie on the surface of the Ewald sphere with $N-1$ sets of atomic planes satisfying Bragg's law, i.e.,

$$\boldsymbol{k_O} \cdot \boldsymbol{d}^*_{\boldsymbol{G}} = -\frac{(d^*_G)^2}{2}\,, \tag{3.2a}$$

$$\boldsymbol{k_G} \cdot (\boldsymbol{d}^*_{\boldsymbol{L}} - \boldsymbol{d}^*_{\boldsymbol{G}}) = -\frac{|\boldsymbol{d}^*_{\boldsymbol{L}} - \boldsymbol{d}^*_{\boldsymbol{G}}|^2}{2}\,, \tag{3.2b}$$

where G and L are any two reflections involved in the multiple-wave case. It should be noted that even when no diffracted waves are present, there is

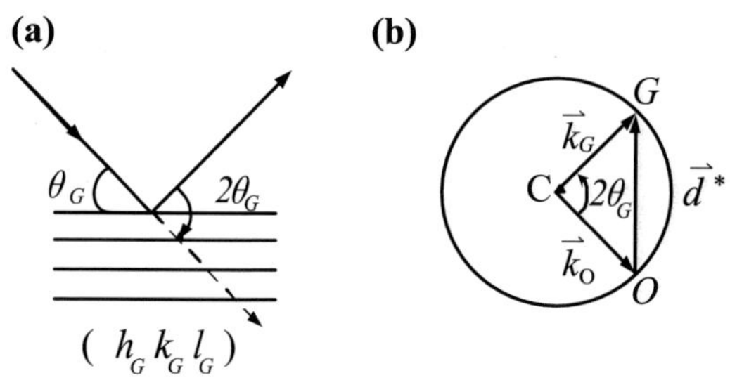

Fig. 3.1. Schematic representation of two-wave diffraction in (**a**) real space and (**b**) reciprocal space

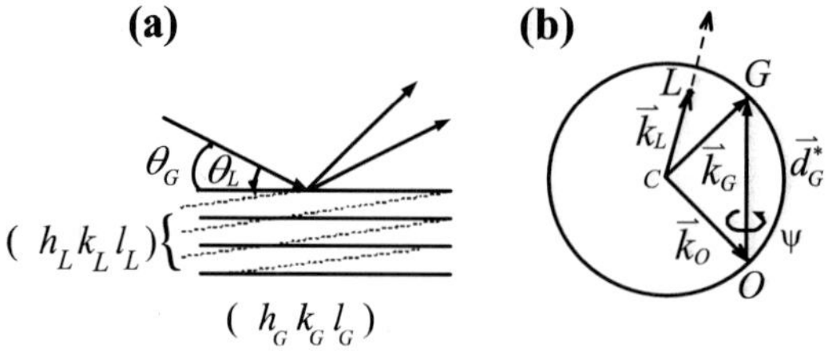

Fig. 3.2. Schematic representation of three-wave diffraction in (**a**) real space and (**b**) reciprocal space

always an incident wave (direct reflection). This diffraction is called one-wave diffraction.

3.1.2 Systematic, Coplanar, and Coincidental Multiple-Wave Diffraction

The reciprocal lattice points O and G and the wavevectors $\boldsymbol{k_O}$ and $\boldsymbol{k_G}$ of the two-wave (O, G) diffraction lie in a plane, as indicated in Fig. 3.1(a). For a three-wave (O, G, L) diffraction, the three reciprocal lattice points are coplanar, while the corresponding wavevectors may or may not be coplanar. For an N-wave diffraction with $N > 3$, both the reciprocal lattice points and wavevectors may or may not be coplanar. The N-wave cases $(N \geq 3)$ with reciprocal lattice points being coplanar but not wavevectors are called 'systematic multiple-wave diffraction'. The condition for this type of diffraction is that the radius r_s of the reflection circle circumscribing all the involved reciprocal lattice points on the surface of the Ewald sphere is smaller than the radius $r_{\mathrm{E}}(= 1/\lambda)$ of the Ewald sphere, namely,

$$r_s < r_{\mathrm{E}} \;, \tag{3.3}$$

where

$$r_s = d_i^* d_j^* \frac{\left|\boldsymbol{d_i^*} - \boldsymbol{d_j^*}\right|}{2\left|\boldsymbol{d_i^*} \times \boldsymbol{d_j^*}\right|} \;, \tag{3.4}$$

and

$$(\boldsymbol{d_i^*} \times \boldsymbol{d_j^*}) \cdot \boldsymbol{d_\ell^*} = 0 \;, \tag{3.5}$$

for all the participating reflections, i, j, and ℓ, in the N-wave diffraction. This type of diffraction depends only on crystal symmetry. It occurs for all wavelengths, provided that (3.3) holds. Besides, all the reflecting planes, i and j, involved have a common zone axis which is perpendicular to all the reciprocal lattice vectors. When

$$r_s = r_{\mathrm{E}} \tag{3.6}$$

and (3.5) holds, all the reciprocal lattice points and wavevectors are coplanar. There is only one specific wavelength at which multiple-wave diffraction can take place. This diffraction is called 'coplanar coincidental diffraction', or simply 'coplanar diffraction' [340]. If the radius r_s equals to r_{E} but (3.5) does not hold, then the coincidental diffraction is noncoplanar. In this case, the involved reciprocal lattice points do not lie in one great circle but are distributed three-dimensionally on the surface of the Ewald sphere.

The occurrence of all the N-wave diffractions with $N \geq 2$, according to the equations given above, depends on the lattice constants, the wavelength of radiation, and the crystal-lattice symmetry. Since all the reciprocal lattice points involved in a systematic diffraction lie on a plane, the lattice symmetry of the plane, namely, the involved plane lattice, plays an important role in the determination of the number of reciprocal lattice points lying in a reflection circle on the surface of the Ewald sphere. According to Burbank [410], general rules can be deduced by considering the possibility of the occurrence of N-wave diffraction from the graphic construction of reflection circles circumscribing a given plane lattice. This finding has been reported [53, 410] for various plane lattices, including oblique, primitive rectangular, centered rectangular, rhombohedral rectangular, hexagonal, and square. The relationship between the plane lattices and the sections of the seven crystal systems has also been described and discussed [53,410]. For non-coplanar coincidental diffraction, however, general rules governing the occurrence of diffraction in relation to crystal lattice symmetry is difficult to derive due to the complexity involved in three-dimensional lattices. Nevertheless, a cubic lattice should, in principle, have the highest possibility of having non-coplanar coincidental diffraction because of its highest lattice symmetry. Moreover, certain reflections, which are systematically forbidden by the space-group symmetry, may be involved in a multiple-wave diffraction. Although their diffraction intensities are null, they still should be considered in the multiple-wave diffraction process. This point will become clear when we discuss the aspects regarding diffraction intensity in Chaps. 5 and 6.

Higher-order coincidental N-wave diffraction with large N can be decomposed into lower-order diffractions of smaller N, where the wavelength is changed so that $r_s \neq r_{\mathrm{E}}$. In most cases, the lower-order diffractions are systematic diffractions and/or simple two-wave diffractions. Similarly, systematic diffractions can also be decomposed into lower-order systematic diffractions due to the changes of lattice parameters caused by external disturbances. In other words, the crystal system undergoes a phase transformation into a different system under temperature, pressure, strain, and compositional gradients, or under applied external fields. To sum up, the decomposition of higher-order diffraction into lower-order ones results from the change in the X-ray wavelength, lattice parameters, and lattice symmetry.

3.2 Grazing Incidence Diffraction Versus Wide-Angle Incidence Diffraction

The diffraction geometry and various kinds of multiple-wave cases were described, in the previous section, in terms of reciprocal lattice without considering crystal boundaries in real space. If, for simplicity, one considers the X-ray diffraction from a semi-infinitely thick crystal bound with a flat surface on the top, then optical reflections from the flat surface result, in addition to the usual Bragg diffraction. These optical reflections are called ‘specular reflections’, which have appreciable intensities only when the incident angle

χ equals to the critical angle for the external total reflection. For hard X-rays the critical angle is of the order of several minutes of arc. If the diffracting atomic planes are perpendicular to the crystal surface, then the incident wave is specularly reflected along $\boldsymbol{k}_O^S$ and simultaneously diffracted along $\boldsymbol{K}_G$ towards the interior of the crystal (Fig. 3.3(a)) [165]. In the meantime, the diffracted wave $\boldsymbol{K}_G$ is also surface specularly reflected along $\boldsymbol{k}_G^S$. Part of the incident wave is also transmitted through the crystal along $\boldsymbol{K}_O$. Both $\boldsymbol{K}_O$ and $\boldsymbol{K}_G$ are the wavevectors inside the crystal. All the transmission, surface specular reflections, and diffraction occur at the same time. This is a two-wave Laue (transmission) diffraction, because the diffracted wave lies on the other side of the crystal surface as compared with the incident wave. The geometric relation between the reciprocal lattice points and the wavevectors inside the crystal is shown in Fig. 3.3(b). For convenience, all the wavevectors inside the crystal will be used to show the diffraction geometry.

If an additional set of planes, say, L, also satisfies the condition for grazing incidence diffraction, then the diffracted wave of the wavevector $\boldsymbol{K}_L$ (not shown in Fig. 3.3(c)) towards the interior of the crystal is specularly reflected by the crystal surface along $\boldsymbol{k}_L^S$. With respect to the crystal surface, the scattered angles of the specularly reflected waves $\boldsymbol{k}_O^S$, $\boldsymbol{k}_G^S$, and $\boldsymbol{k}_L^S$ are α, β, and γ, respectively. In the reciprocal space, the reciprocal lattice points, O, G, and L, lie on the great circle of the Ewald sphere parallel to the surface (Fig. 3.3(d)). Because the radius of the great circle is $1/\lambda$, this three-wave grazing incidence X-ray diffraction (GIXD) is a coincidental multiple-wave diffraction. In other words, this diffraction can only occur at a specific wavelength, which is one of the characteristics of multiple-wave GIXD.

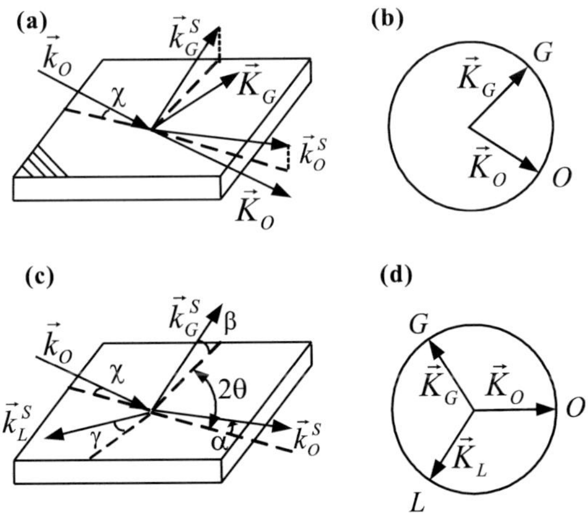

Fig. 3.3. Schematic of grazing incidence X-ray diffraction: Two-wave case in (**a**) real space and (**b**) reciprocal space; three-wave case in (**c**) real space and (**d**) reciprocal space [165]

3.3 Lorentz-Polarization Factor in Multiple-Wave Cases

The Lorentz factor of a two-wave diffraction, as indicated in (2.34), is related to the θ angle rotation of the crystal. For a wide-angle incidence three-wave diffraction, two rotations are usually needed to bring the crystal into the exact position to diffract the incident wave. As stated before, it is convenient to rotate the crystal by varying first the θ angle and then the azimuth angle ψ around $\boldsymbol{d}_G^*(=\boldsymbol{g})$ of the G-reflection. During the ψ rotation, the reciprocal lattice point G is always on the surface of the Ewald sphere. The Lorentz factor L_F is proportional to the time for the reciprocal lattice point L to cross the surface of the Ewald sphere. It is thus related to the reciprocal of the radial component v_r of the velocity $\boldsymbol{v}$, with which point L moves across that surface. Figure 3.4 depicts the ψ rotation in the reciprocal space. The vectors $\boldsymbol{OM}$ and $\boldsymbol{ML}$ are the vector components of the reciprocal lattice vector $\boldsymbol{d}_L^*(=\boldsymbol{\ell})$ of the L reflection parallel and perpendicular to $\boldsymbol{d}_G^*$, respectively, i.e., $\boldsymbol{\ell}_p=\boldsymbol{OM}$ and $\boldsymbol{\ell}_n=\boldsymbol{ML}$. The v_r in the radial direction towards the center of the Ewald sphere is defined as

$$v_r=(\boldsymbol{\omega_1}\times\boldsymbol{\ell})\cdot\frac{\boldsymbol{r}}{r}=\omega_1\ell_n\cos\alpha\;, \tag{3.7}$$

where $\boldsymbol{\omega_1}$ is the angular velocity of the crystal along the reciprocal lattice vector $\boldsymbol{g}$, and α is the angle between $\boldsymbol{v}$ and the radial direction $\boldsymbol{r}(=\boldsymbol{CL})$. The radius r of the Ewald sphere is $1/\lambda$. From the three-dimensional relationship among the reciprocal lattice points G, L, and the center C' of the upper circle shown in Fig. 3.4, it is easy to show that $\cos\alpha=C'N/r$, where $\boldsymbol{C'N}$ is perpendicular to $\boldsymbol{ML}$. By considering $\boldsymbol{C'N}$ and $\boldsymbol{ML}$ in terms of the angles ξ, ψ, and χ, $\cos\alpha$ can be expressed as

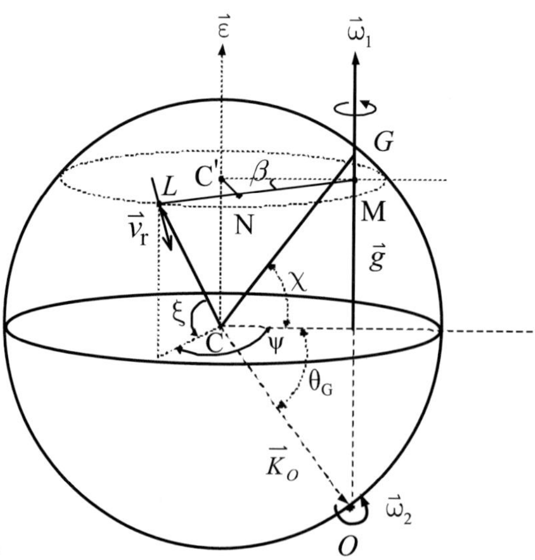

Fig. 3.4. Geometry of azimuthal rotations with the rotation vectors $\boldsymbol{\omega_1}$ and $\boldsymbol{\omega_2}$ parallel and perpendicular to the reciprocal lattice vector $\boldsymbol{g}$ of the primary reflection

$$\cos\alpha = \frac{\sin\psi\cos\chi\cos\xi}{\ell_n\lambda} . \tag{3.8}$$

From (3.7) and (3.8), the Lorentz factor L_F takes the following form:

$$L_F \propto \frac{1}{v_r} = \left(\frac{\lambda}{\omega_1}\right)\frac{1}{\sin\psi\cos\chi\cos\xi} . \tag{3.9}$$

Since λ/ω_1 is a constant for a given wavelength λ and speed ω_1 of rotation, the Lorentz factor is defined as

$$L_F = \frac{1}{\sin\psi\cos\chi\cos\xi} . \tag{3.10}$$

The Lorentz factor is different for the crystal rotation around $\boldsymbol{\omega_2}$ perpendicular to $\boldsymbol{g}$ and the plane of incidence. Referring to Fig. 3.4 and using the Cartesian coordinate system to represent the vectors, we can easily derive the radial component $\boldsymbol{v_r}$ of the velocity of the secondary reciprocal lattice point L for the ω_2 rotation as

$$v_r = (\boldsymbol{\omega_2} \times \boldsymbol{\ell}) \cdot \frac{\boldsymbol{r}}{r} = \left(\frac{\omega_2}{\lambda}\right)[\cos\chi\cos\xi + \sin\chi\cos\xi\cos\psi] . \tag{3.11}$$

The Lorentz factor then becomes

$$L_F = \frac{1}{\cos\chi\cos\xi + \sin\chi\cos\xi\cos\psi} . \tag{3.12}$$

Multiple-wave diffraction can be treated kinematically, as will be described in Chap. 5, as a combination of many successive reflections. The corresponding polarization factors associated with the involved successive reflections are different from and more complicated than that in two-wave cases. To derive an analytical expression for polarization factors in a multiple-wave case, it is necessary to decompose the electric fields of the incident and diffracted waves into σ and π components perpendicular and parallel to the plane of incidence [230, 232, 234–237, 239, 411]. Both the σ and π components of the electric field, denoted by E_σ and E_π, of a given wave $\boldsymbol{K}$ are perpendicular to the wavevector $\boldsymbol{K}$. Let's consider the successive reflections shown in Fig. 3.5. Assume that the incident wave with its σ and π components, $E_{\sigma O}$ and $E_{\pi O}$, propagating along the wavevector $\boldsymbol{K}_O$ is reflected first by the G_1 plane and then by the G_2 plane. The G_1 diffracted wave with $E_{\sigma G_1}$ and $E_{\pi G_1}$

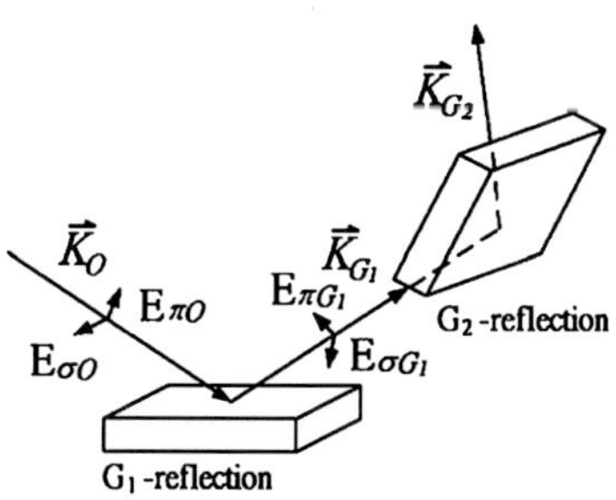

Fig. 3.5. Polarization of successive reflections in a multiple-wave diffraction

is reflected along $\boldsymbol{K}_{\boldsymbol{G_1}}$ and the G_2 diffracted wave with $E_{\sigma G_2}$ and $E_{\pi G_2}$ along $\boldsymbol{K}_{\boldsymbol{G_2}}$, respectively. The σ and π components of the G_1 reflected wave can be expressed in terms of that of the incident wave $\boldsymbol{K}_O$ as

$$\boldsymbol{E}_{G_1} = \begin{pmatrix} E_{\sigma G_1} \\ E_{\pi G_1} \end{pmatrix} = \begin{pmatrix} 1 & 0 \\ 0 & \cos 2\theta_{G_1} \end{pmatrix} \begin{pmatrix} E_{\sigma O} \\ E_{\pi O} \end{pmatrix} = \tilde{P}_{G_1} \boldsymbol{E}_O \,, \tag{3.13}$$

where $\tilde{P}_{G_1}$ is the polarization matrix of the G_1 reflection. The polarization factor P_{G_1} of a kinematical approach is then equal to the sum of the square of the elements of the $\tilde{P}_{G_1}$, i.e.,

$$P_{G_1} = \frac{1}{2}(1 + cos^2 2\theta_{G_1}) \,, \tag{3.14}$$

with the factor $1/2$ being the normalization constant. Similarly, the polarization matrix $\tilde{P}_{G_2}$ and the polarization factor P_{G_2} of the G_2 reflection can be written as

$$\tilde{P}_{G_2} = \begin{pmatrix} 1 & 0 \\ 0 & \cos 2\theta_{G_2} \end{pmatrix} \tag{3.15}$$

and

$$P_{G_2} = \frac{1}{2}(1 + \cos^2 2\theta_{G_2}) \,, \tag{3.16}$$

respectively. Suppose that the angle between the two planes of incidence is ψ. The link between the two coordinate systems defining their corresponding σ and π components of the electric fields is the transformation matrix [411]

$$\tilde{T}_{G_1} = \begin{pmatrix} \cos\psi & \sin\psi \\ -\sin\psi & \cos\psi \end{pmatrix} \,, \tag{3.17}$$

where

$$\psi = \arccos\left[\left(\hat{\boldsymbol{K}}_O \times \hat{\boldsymbol{K}}_{G_1}\right) \cdot \left(\hat{\boldsymbol{K}}_{G_1} \times \hat{\boldsymbol{K}}_{G_2}\right)\right] \,, \tag{3.18}$$

and the $\hat{\boldsymbol{K}}$'s are the unit vectors. The vector product gives

$$\cos\psi = \frac{\cos 2\theta_{G_2-G_1} - \cos 2\theta_{G_1} \cos 2\theta_{G_2}}{\sin 2\theta_{G_1} \sin 2\theta_{G_2}} \,, \tag{3.19}$$

where the relation

$$(\boldsymbol{A} \times \boldsymbol{B}) \cdot (\boldsymbol{C} \times \boldsymbol{D}) = (\boldsymbol{A} \cdot \boldsymbol{C})(\boldsymbol{B} \cdot \boldsymbol{D}) - (\boldsymbol{A} \cdot \boldsymbol{D})(\boldsymbol{B} \cdot \boldsymbol{C})$$

has been employed. If a two-crystal monochromator involving M_1 and M_2 reflections, denoted as (M_1, M_2), is used as the incident beam conditioner, then the polarization matrix for the n-th-order successive reflection is [411]

$$\tilde{P} = \tilde{P}_{M_1}\tilde{T}_{M_1}\tilde{P}_{M_2}\tilde{T}_{M_2}\tilde{P}_{G_1}\tilde{T}_{G_1}\tilde{P}_{G_2}\tilde{T}_{G_2} \cdots \tilde{T}_{G_{n-2}}\tilde{P}_{G_{n-1}} \,, \tag{3.20}$$

where the matrices $\tilde{P}$ and $\tilde{T}$ on the right-hand side of the equation are the polarization matrix and the transformation matrix of each reflection involved.

The corresponding polarization factor is equal to one-half of the sum of the squared matrix elements of $\tilde{P}$. If the two-crystal monochromator is a $(+,-)$ channel-cut type, then $M_1 = M_2$. If a second-order successive reflection, like the one shown in Fig. 3.5, involves a monochromator using the M reflection, then the polarization factor $C_{i,j}(m)$ takes the following form [232]:

$$C_{i,j}(m) = \frac{1}{2}\left[\left(1 - \frac{(\cos 2\theta_m - \cos 2\theta_i \cos 2\theta_j)^2}{\sin^2 2\theta_i}\right)(1 - \sin^2 2\alpha \cos^2 \phi)\right.$$
$$\times \cos^2 2\theta_i + \left(\frac{(\cos 2\theta_m - \cos 2\theta_i \cos 2\theta_j)^2}{\sin^2 2\theta_i} + \cos^2 2\theta_j\right)$$
$$\left.\times (\sin^2 2\alpha \cos^2 \phi + \cos^2 2\alpha)\right] , \qquad (3.21)$$

where $i = G_1$, $j = G_2$, and $m = j - i$. α and ϕ are the Bragg angle of the M reflection of the monochromator and the angle between the normals to the monochromator and the first plane of incidence, respectively. If there is no monochromator ($\alpha = 0$), then the polarization factor becomes [230, 231]

$$C_{i,j}(m) = \frac{1}{2}\left[\cos^2 2\theta_i + \cos^2 2\theta_j + (\cos 2\theta_m - \cos 2\theta_i \cos 2\theta_j)^2\right] . \qquad (3.22)$$

Equation (3.21) gives the familiar expression for a single reflection with a monochromator:

$$C_{0,i} = \frac{1}{2}(1 + \cos^2 2\alpha \cos 2\theta_i) , \qquad (3.23)$$

where ϕ is set to be $\pi/2$.

Special care needs to be taken for the Lorentz factor when a double-crystal monochromator (DCM) is used as the beam conditioner. For example, if a $(+,-)$ arrangement for the DCM is adopted, whether the occurrence of a multiple-wave diffraction forms a $(+,-,+)$ or $(+,-,-)$ arrangement will determine the value of the Lorentz factor. In other words, the FWHM of the diffraction profile could be narrower or wider, depending on the overall crystal configuration. This point will be further discussed in Chap. 9. With all the Lorentz and polarization factors derived, one can calculate the Lorentz-polarization factor of a given multiple-wave diffraction simply by obtaining the product of the Lorentz and the polarization factors.

3.4 Multiple-Wave Diffraction in Multi-Crystal Systems

When multi-crystals are involved, the diffraction geometry gets more complicated than in a plane parallel crystal. Usually multi-crystal systems can take the form of a double-slab monolithic crystal, a multi-layered crystal, a multi-block monolithic crystal, two or more separate single-crystals, and many other combinations. The diffraction geometry of these systems is described below:

(1) Double-slab monolithic crystal: The crystal consists of two crystal slabs of the same material with an air gap in between, as shown in Fig. 3.6. Consider a three-wave diffraction (O, G, L), where $\boldsymbol{K}_O$, $\boldsymbol{K}_G$, and $\boldsymbol{K}_L$ represent the incident, the Bragg reflected, and the Laue transmitted waves, respectively. The incident wave $\boldsymbol{K}_O$ is first diffracted by the top crystal slab to the transmission $\boldsymbol{K}_O$ and $\boldsymbol{K}_L$ and the reflected $\boldsymbol{K}_G$ directions. After propagating through the air gap, the transmitted $\boldsymbol{K}_O$ and $\boldsymbol{K}_L$ waves are then diffracted from the lower crystal slab simultaneously, thus generating multiple-waves, including the two transmitted $\boldsymbol{K}_O$ and $\boldsymbol{K}_L$ waves and the back-scattered $\boldsymbol{K}_G$ waves. The $\boldsymbol{K}_G$ waves are again diffracted by the top slab after going through the gap and give rise to multiple-wave diffraction. The diffracted $\boldsymbol{K}_O$ and $\boldsymbol{K}_L$ waves generated, in turn, become the incident waves to the lower slab and again create multiple-waves. These processes keep going on as illustrated in Fig. 3.7. Therefore, multiple-wave diffraction in the double-slab monolithic crystal is the superposition of an infinite number of reflections and transmissions as just described. This diffraction picture could also extend to a sample system consisting of a thin film deposited on a single-crystal substrate, provided that the difference in the Bragg angle due to lattice mis-

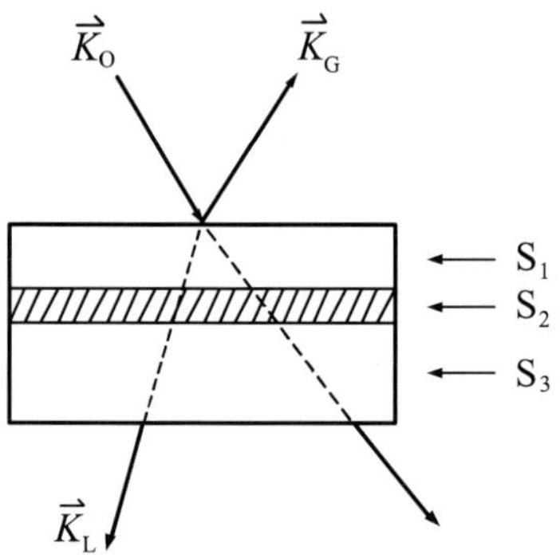

Fig. 3.6. Three-wave diffraction in a double-slab crystal

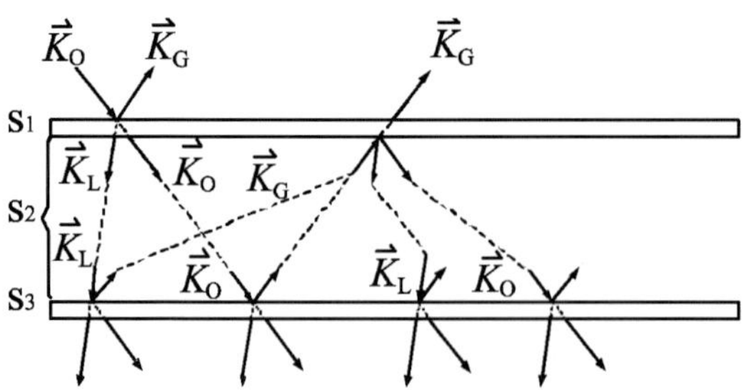

Fig. 3.7. Multiple-wave diffraction processes in a double-slab crystal

match between the film and substrate is relatively small compared to the incident beam divergence.

(2) Double- and multi-layered crystals: A double-layered crystal is similar to a double-slab monolithic crystal, except that the air gap in the former is filled with a different material or crystal. The diffraction geometry of this case is the same as that in a double-slab crystal. Only the absorption is different as far as the diffraction intensity is concerned. If, however, a multi-layered crystal like a multi-layered superlattice is involved, the diffractions from both the crystal and the superperiods, i.e., the principal and satellite reflections under multiple-wave conditions, need to be considered.

(3) Multi-block monolithic crystals: Wavevector-coplanar multiple-wave diffraction can take place in a multi-block monolithic crystal, given that the crystal is properly prepared and the wavelength is correctly chosen. As described in Sect. 3.1.2, all the diffracted waves, including the incident wave, lie in one plane for a given specific wavelength. This makes the preparation of a monolithic crystal with several blocks for coplanar multiple-wave diffraction possible. In other words, once one of the crystal blocks satisfies the diffraction condition, the other blocks are automatically in position to diffract multiply. Figure 3.8 shows the geometry of the (000)(440)(404) wavevector-coplanar three-wave diffraction in silicon for $\lambda = 1.66276$ Å, as proposed by Graeff and Bonse [375] for a three-wave X-ray interferometer. The incident wave of the wavevector $\boldsymbol{K}_O$ is first diffracted by the crystal block A and generates the (440) and (404) reflections along $\boldsymbol{K}_G$ and $\boldsymbol{K}_L$, respectively. The reflected wave $\boldsymbol{K}_G$ is then diffracted successively by blocks B and E, and diffracted back to block A. Similarly, the reflected wave $\boldsymbol{K}_L$ goes through the other path, i.e., first blocks C and D and then back to block A. It should be noted that three-wave coplanar diffraction occurs in each block. The resultant diffracted wave along $\boldsymbol{K}_O$ is thus the superposition of the waves going through the two different paths.

(4) Two or more separate single-crystals: A multi-crystal system can also be composed of several single-crystals with each of the crystals satisfying its own multiple-wave diffraction conditions. The material and crystal orientation of the crystals can be different from one another. Figure 3.9 shows an example.

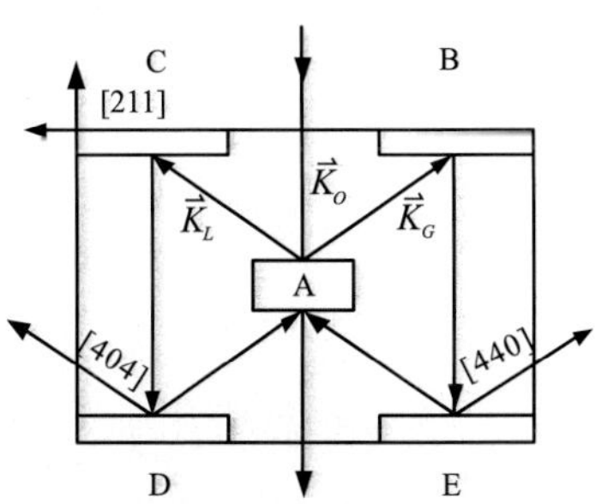

Fig. 3.8. Multiple-wave diffraction processes in a multi-block monolithic crystal [375]

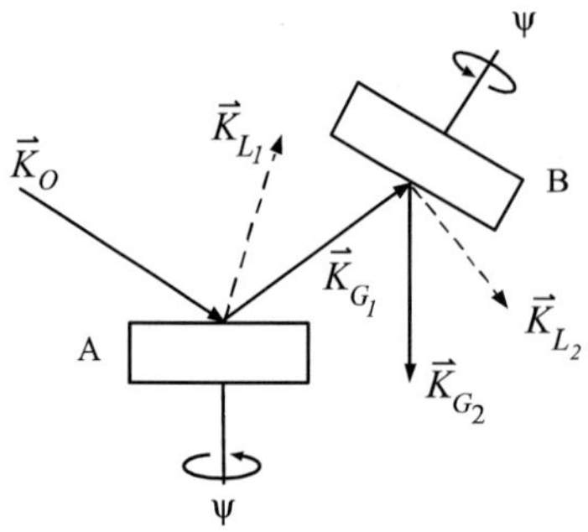

Fig. 3.9. Multiple-wave diffraction processes in two separate crystals

The crystal A is first rotated to satisfy the multiple-wave diffraction condition for the incident wave $\boldsymbol{K}_O$, thus generating, say, two diffracted waves along $\boldsymbol{K}_{G_1}$ and $\boldsymbol{K}_{L_1}$. The $\boldsymbol{K}_{G_1}$ wave is then diffracted by the second crystal B and produces multiply diffracted waves along $\boldsymbol{K}_{G_2}$ and $\boldsymbol{K}_{L_2}$. Although the $(+,-)$ arrangement is adopted here to show the crystal configuration, there could be $(+,+)$ arrangement for two separate crystals. If several crystals are involved, various configurations, such as $(+,-,+,-,\cdots)$, $(+,+,+,+,\cdots)$, and many other combinations, are possible. Under this circumstance, care must be taken in considering the Lorentz-polarization factor as far as the diffraction intensity is concerned.

There are of course many other crystal arrangements which can produce multiple-wave diffraction. However, the general geometrical feature is similar to what has been described in this chapter.

4. Experimental Techniques

The occurrence of multiple-wave diffraction, as described in Chap. 3, depends on the lattice parameters of the crystal, the wavelength of the X-ray radiation, and the lattice symmetry of the space group. Experimentally, a multiple-wave diffraction is generated by rotating the crystal to satisfy the diffraction conditions. As a matter of fact, there are several ways of rotating the crystal to generate multiple-wave diffraction, which include the collimated-beam method, the divergent-beam method, and the rotation-crystal method. In this chapter, the details about these methods and the detectors and diffractometers employed for multiple-wave diffraction experiments are described. Indexing the multiple-wave diffraction patterns obtained by these methods is also presented.

4.1 Collimated-Beam Method – Renninger Scan

One of the most frequently used methods to generate multiple-wave diffraction is the collimated-beam method, proposed by Renninger [4], where a well-collimated incident beam is employed. Figure 4.1 shows the schematic of the experimental setup, where the incident beam is collimated by an incident-beam conditioner and then impinges on a crystal. The crystal is aligned with respect to the incident beam by first tilting the θ angle for a given primary reflection, say, G-reflection. Without disturbing the θ setting for the primary reflection, the crystal is then rotated around the reciprocal lattice vector $\boldsymbol{g}$

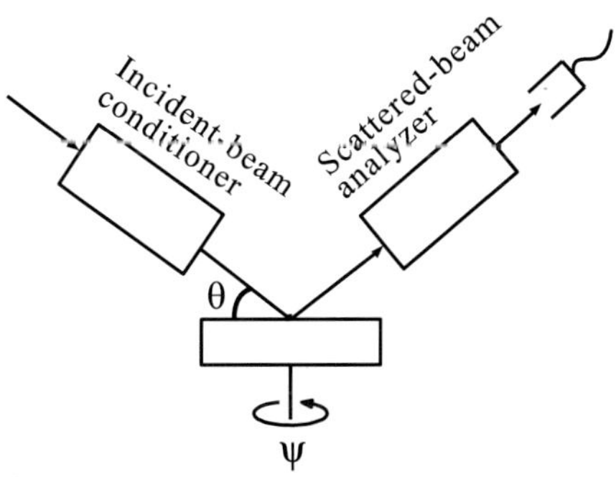

Fig. 4.1. Schematic representation of a collimated-beam arrangement

of the G reflection, the so-called ψ scan, to bring additional sets of atomic planes into position to diffract the incident beam, thus generating a secondary reflection, say, L reflection.

The interaction among the diffracted beams inside the crystal results in intensity modification on the primary reflection. This intensity-modified diffracted beam after passing through the scattered-beam analyzer is monitored by a point detector, such as an ion chamber and scintillation counter. The intensity modification versus ψ is the multiple-wave diffraction pattern, the so-called Renninger scan. Both the incident-beam conditioner and the scattered-beam analyzer are employed to minimize the beam divergence and hence to improve the angular and spectral resolutions. Both the conditioner and analyzer could be a single crystal, double crystal, and channel-cut crystal used as as the monochromator or analyzer [412]. A beam-path with pinholes at its both ends, or a vertical and horizontal slit system, or a Soller slit assembly for beam collimation could be added as part of the beam conditioner and analyzer systems [413]. For a synchrotron radiation source, a double-crystal monochromator is usually employed to select a proper X-ray wavelength. The typical beam divergence is of the order of a few arc minutes using a 1-2 meter long collimator for a conventional laboratory source. While a beam divergence of a few arc seconds can be achieved with synchrotron radiation, ultra-high angular resolution is also attainable with a specially designed multi-crystal monochromator and appropriate beam optics.

Figure 4.2 shows the Renninger scan of silicon (222) for CuKα_1 radiation, where (222), the symmetric primary reflection, is a nearly forbidden reflec-

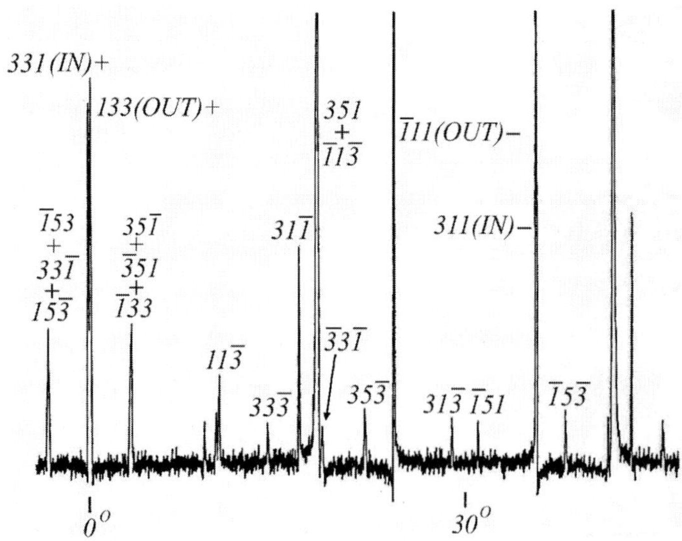

Fig. 4.2. Multiple-wave diffraction pattern of silicon (222) with CuKα_1 radiation [306]

tion. The beam divergence is two arc minutes. The vertical line profiles are the multiple-wave diffraction peaks and the horizontal positions are the ψ angles. The ψ angles equal to 0 and 30 degrees correspond to the positions at which the directions $[\bar{1}2\bar{1}]$ and $[01\bar{1}]$ lie in the plane of incidence for the primary reflection. Most multiple-wave diffractions are three-wave cases, except for the four-wave $(000)(222)(\bar{1}1\bar{3})(351)$ case. In Fig. 4.2, only the Miller indices of the secondary reflections involved in the multiple-wave diffractions are used to label those diffraction peaks. The two peaks labelled with three sets of indices, $\bar{1}53$+$33\bar{1}$+$15\bar{3}$ and $35\bar{1}$+$\bar{3}51$+$\bar{1}33$, are actually closely positioned three-wave diffractions, which could not be resolved due to the insufficient angular resolution of the incident beam. The assignment of Miller indices to diffraction peaks, i.e., the indexing of the diffraction profiles, is described below in detail.

4.1.1 Reference-Vector Method for Indexing Renninger Scan

A reference-vector is employed to define the origin of the ψ angle involved in a Renninger scan, from which the relative angular positions of multiple-wave diffraction peaks (intensity increase) and dips (intensity decrease) can be identified [413]. Figure 4.3(a) shows an overall geometry of a three-wave (O, G, L) diffraction in the reciprocal space. The points O, G, and L are the reciprocal lattice points of the direct, the primary, and the secondary reflection, respectively. $\boldsymbol{g}$ and $\boldsymbol{l}$ are the reciprocal lattice vectors of the G and L reflections. During the ψ rotation of the crystal, the secondary reciprocal lattice point L enters and then leaves the Ewald sphere, or vice versa, depending on the direction of the ψ rotation, which can be either right-handed or left-handed. Therefore, there are two ψ positions – entering and leaving – at which the three-wave (O, G, L) diffraction occurs. These two positions are denoted by 'IN' and 'OUT' for a right-handed ψ rotation as shown in Fig. 4.3(b), which is the projection of Fig. 4.3(a) onto the equatorial plane perpendicular to $\boldsymbol{g}$.

In Fig. 4.3(b), the rotation axis $\boldsymbol{g}$ at point O' is normal to this page. The plane of incidence for the primary reflection, containing the incident and the G diffracted waves, is represented as a horizontal line through point O'. The 'IN' and 'OUT' positions are symmetric about this plane of incidence. $\boldsymbol{l}_n$ and $\boldsymbol{l}_p$ are the vector components of the secondary reciprocal lattice vector $\boldsymbol{l}$ normal and parallel to $\boldsymbol{g}$. Suppose that initially the secondary reciprocal lattice point is not situated on the surface of the Ewald sphere. The projection of its initial position shown in Fig. 4.3(b) is L'. The corresponding normal component of $\boldsymbol{l}$ is $\boldsymbol{l}'_n$. At this moment, a reference unit vector $\hat{\boldsymbol{v}}'$ is chosen to lie in the plane of incidence of the G reflection and be perpendicular to $\boldsymbol{g}$. To bring the secondary reciprocal lattice point from point L' to L, the crystal needs to be rotated by an angle ψ around $\boldsymbol{g}$. Then the reference vector moves from $\hat{\boldsymbol{v}}'$ to $\hat{\boldsymbol{v}}$. Referring to Fig. 4.3(b), the angle ψ_o between $\hat{\boldsymbol{v}}'$ and $\boldsymbol{l}'_n$ and β between $\hat{\boldsymbol{v}}'$ and $\boldsymbol{l}_n$ are defined as

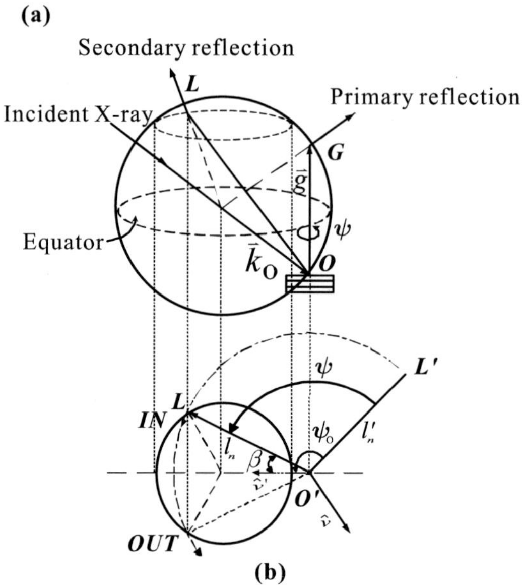

Fig. 4.3. (**a**) Geometry of three-wave diffraction in reciprocal space. (**b**) Bottom view of (**a**) [53]

$$\psi_0 = \arccos(\boldsymbol{\ell}'_n \cdot \hat{\boldsymbol{v}}') \,, \tag{4.1}$$

$$\beta = \arccos(\boldsymbol{\ell}_n \cdot \hat{\boldsymbol{v}}') \,, \tag{4.2}$$

where the reference unit vector is

$$\hat{\boldsymbol{v}}' = \frac{\boldsymbol{k}_O + \dfrac{\boldsymbol{g}}{2}}{\left|\boldsymbol{k}_O + \dfrac{\boldsymbol{g}}{2}\right|} \,. \tag{4.3}$$

The vector $\boldsymbol{k}_O$ is the wavevector of the incident beam. From the geometry, the ψ angles for the 'IN' and 'OUT' positions are

$$\psi = \psi_0 \mp \beta \,, \tag{4.4}$$

respectively. From (4.1) and (4.2) the angle β can be calculated as

$$\cos\beta = \frac{\ell^2 - \ell_p g}{2\ell_n \sqrt{\left(\dfrac{1}{\lambda^2}\right) + \left(\dfrac{g^2}{4}\right)}} \,. \tag{4.5}$$

According to (4.5), $1/\lambda$ must be greater than or equal to $g/2$. When β is $0°$ or $180°$, the reciprocal lattice point L lies in the plane of incidence of the primary reflection. Hence the multiple-wave diffraction is coplanar and coincidental. If $1/\lambda \geq g/2$ and $\beta = 90°$, then $l^2 - l_p g = 0$. This implies that point L lies on a circle of diameter g perpendicular to the plane of incidence.

Referring to Fig. 4.2, we can see that mirror symmetries appear at $\psi = 0°$ and 30°. This kind of symmetry is often encountered in multiple-wave diffraction experiments, especially for crystal systems with high symmetry. Moreover, the mirror symmetry relies directly on the symmetry of the rotation vector. Usually, the number of mirror symmetries, N_M, is equal to $2n$, where n is the n-fold rotation axis used as the reciprocal lattice vector of the primary reflection and 2 is due to the two positions, 'IN' and 'OUT', of a given multiple-wave diffraction. Hence the multiple-wave diffraction patterns generated by two-, three-, and four-fold rotation axes exhibit four-, six-, and eight-fold mirror symmetries. However, care must be taken for cases involving a three-fold rotation axis, because the symmetry of the projected plane normal to a three-fold axis may be six-fold. For example, the rotation around the [111] direction of a diamond structure, like silicon and germanium, shows twelve-fold mirror symmetries on multiple-wave diffraction patterns. That is, a mirror symmetry appears every 30° but the two adjacent ones are not equivalent to but symmetric with each other. This implies that the twelve mirror symmetries divide a 360° diffraction pattern into twelve asymmetric portions, and the symmetric portion formed by the two adjacent asymmetric portions appears only every 60°.

For an unambiguous indexing of multiple-wave diffraction pattern, one needs to identify the direction of the ψ rotation and the orientation of the crystal. Sometimes ambiguity arises when the crystal involves high symmetry, especially when the crystal orientation is not precisely known. For example, there are many ways to index the diffraction pattern shown in Fig. 4.2 when the direction of the reference vector is not quite sure. One can use either $[1\bar{1}0]$ or any other equivalent direction as the reference vector. The results are of course different. For example, the three-wave (000)(222)(311) at $\psi = 35.724°$ is a Bragg-Bragg case with $[\bar{1}2\bar{1}]$ being the reference vector, because (222) is a symmetric Bragg and (311) is an asymmetric Bragg reflection with respect to the (222) plane. Namely, $\boldsymbol{l} \cdot \boldsymbol{g} > g/2$. The peak can also be indexed as $(000)(222)(1\bar{1}1)$ if $[11\bar{2}]$ is the reference vector. This is a Bragg-Laue diffraction, because $(1\bar{1}1)$ is a transmission (Laue) reflection. That is, $\boldsymbol{l} \cdot \boldsymbol{g} < g/2$. The Bragg-Bragg case can be easily identified by detecting the presence of the secondary Bragg reflected beam for (311). If no diffracted intensity is detected at this ψ position, then this three-wave case is a Bragg-Laue diffraction, because the secondary reflected beam transmits through the crystal. The angular position at which the detector should be put can be easily determined from the diffraction geometry shown in Fig. 4.3 [414]. If the reciprocal lattice point L of the secondary reflection lies in the equatorial circle of the Ewald sphere (see Fig. 4.3), then the three-wave diffraction is a Bragg-surface case because $\boldsymbol{l} \cdot \boldsymbol{g} = g/2$.

4.2 Divergent-Beam Method – Kossel Diagram

An incident beam with a wide angular divergence, usually a few degrees to 90 degrees in solid angle, is utilized in the divergent-beam method. Owing to the large beam divergence, the diffracted beams from a crystal form a cone. This is equivalent to rotating an X-ray source to form a diffraction conic. The intersection of this diffraction cone by a film or an emulsion shows a conic image. For a wide beam divergence, many diffraction conics intersect one another. The intersection point, at which two or more reflections occur simultaneously, indicates a multiple-wave diffraction.

Historically, the divergence-beam method was employed in many diffraction studies for crystal analysis in the early days [415–427]. In those studies, the X-ray source was usually placed outside the crystal. However, according to Kossel [8–11], the X-ray source was generated inside the crystal investigated and the divergent beam was then diffracted by various sets of atomic planes before emerging from the crystal. Both diffraction conics and absorption conics were observed. The deficient images of the absorption conics compared to the background is caused by the intensity loss of the incident beam due to crystal absorption. Alternatively, internal X-ray sources can be generated by fluorescence excitation of the constituent atoms of the crystal using high-energy photons [428, 429] or by focused electron-beam bombardment on the crystal surface [430]. This technique is sometimes called the 'Kossel method'. In contrast, the technique with an external X-ray source is referred to as 'pseudo-Kossel'. Pseudo-Kossel diagrams can be obtained by using microfocus X-ray sources from rotating-anode X-ray generators with fine focus filaments or the X-ray radiation generated by electron bombardment on a thin metal foil which covers the crystal studied [431–433]. Therefore, various X-ray wavelengths can be used to produce Kossel diagrams. Owing to the small source size or incident beam size, of the order of a few microns, both Kossel and pseudo-Kossel diagrams show high-resolution diffraction conics with clear contrast.

The formation of Kossel conics is schematically shown in Fig. 4.4 for transmission and reflection geometries. For the transmission case (Fig. 4.4(a)), the incident X-rays are reflected by the (hkl) and $(\bar{h}\bar{k}\bar{l})$ planes simultaneously due to the beam divergence. The diffracted waves form a pair of diffraction conics on an X-ray film or a two-dimensional detector, while the transmitted waves suffer crystal absorption and form deficient absorption conics which lie outside the diffraction conics. For the reflection case (Fig. 4.4(b)), the back-reflected cone by the (hkl) planes is intercepted by the upper film, thus forming a diffraction conic. The transmitted cone through the crystal again suffers absorption and exhibits a deficient absorption conic on the lower film. The semi-epical angle of the cone is θ_G for the transmission and $90° - \theta_G$ for the reflection geometry, respectively, where θ_G is the Bragg angle of the (hkl) reflection. The visibility of absorption conics depends on the linear absorption coefficient μ of the X-ray used and the crystal thickness T. Hence, deficient

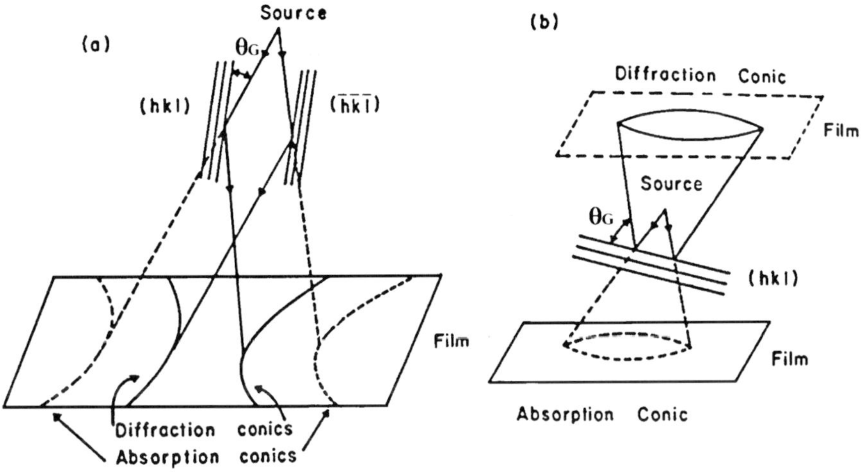

Fig. 4.4. Formation of Kossel conics: (**a**) transmission, (**b**) back reflection [53]

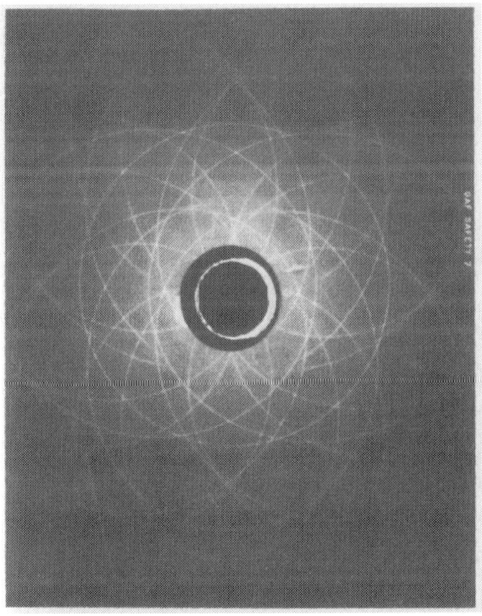

Fig. 4.5. Back-reflection Kossel diagram of KCl (100) for Cu radiation [53]

conics are not always detectable. The optimal condition for observing a clear image of absorption conics is $\mu T \simeq 1$ [340, 434]. Sometimes a slight increase in the background can help to enhance the contrast of deficient lines. Figure 4.5 shows a back-reflection pseudo-Kossel diagram of a (100) KCl crystal for Cu radiation, obtained with a microfocus X-ray generator. The incident beam size is about 10 μm in diameter and the distance between the crystal and the film is 3 mm. The (100) planes are parallel to the film. Fine Kossel lines are observed. The intersection indicates the occurrence of multiple-wave diffraction. It should be noted that pseudo-Kossel lines are not exactly conic but ellipse-like [435], and hyperbolae are usually observed in back-reflection Kossel diagrams [436, 437]. Theoretically the double wavevector method can be used to describe the geometric conditions for the formation of Kossel lines and their intersections [438]. We shall come back to this point in Chap. 9.

The details at and near an intersection of Kossel lines give the intensity distribution of a multiple-wave diffraction, which is important for the understanding of the kinematical and dynamical aspects of X-ray interaction inside a crystal. To observe the details, a Kossel diagram needs to be zoomed in to a smaller field of vision, which can be achieved by taking a section pseudo-Kossel diagram of the area interested. Figures 4.6(a) and (b) are the schematic representations of the experimental arrangements for a reflection and a transmission geometry, respectively. In Fig. 4.6(a), the incident beam divergence is limited to 5–10 degrees so as to narrow down the field of vision. The distance between the source and the crystal can be as small as 5–10 cm, depending on the source size and the desired beam divergence. The film is placed about 1–2 m away from the crystal to record part of the reflected Kossel lines at or near an intersection. The image of the small area at the intersection of Kossel lines is therefore amplified by the long film-to-crystal distance. Figure 4.7(a) shows a section pseudo-Kossel diagram of the germanium four-wave $(000)(004)(\bar{1}\bar{1}1)(\bar{1}\bar{1}3)$ diffraction for CuKα radiation. This diagram is obtained with a microfocus generator. The focal size is $20 \times 50\,\mu\text{m}^2$ and the distance from the crystal to the film is 1 m. Due to the small angular divergence of the incident beam, the Kossel conics become lines. The two horizontal lines are the images of the (004) reflection for CuKα_1 and CuKα_2 and the two arrows labelled $(\bar{1}\bar{1}1)$ and $(\bar{1}\bar{1}3)$ indicate the direction of the $(\bar{1}\bar{1}1)$ and $(\bar{1}\bar{1}3)$ reflections. The intersections of the $(\bar{1}\bar{1}1)$ and $(\bar{1}\bar{1}3)$ lines with the (004) images are the locations at which the four-wave diffractions of CuKα_1 and CuKα_2 take place. Intensity modification on the (004) is clearly seen in the vicinity of the four-wave diffraction points.

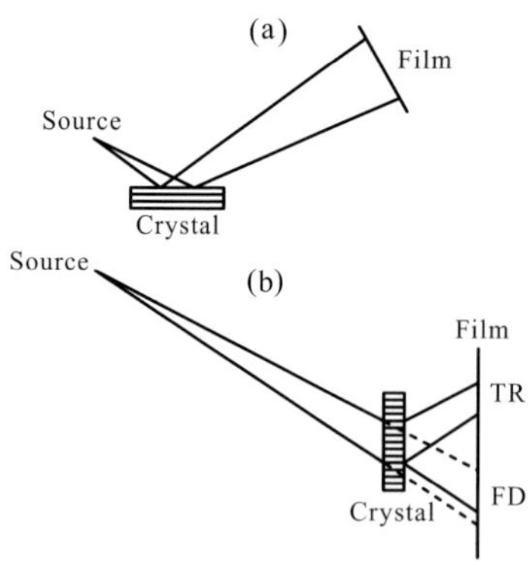

Fig. 4.6. Divergent-beam experimental setups: (**a**) reflection, (**b**) transmission [53]

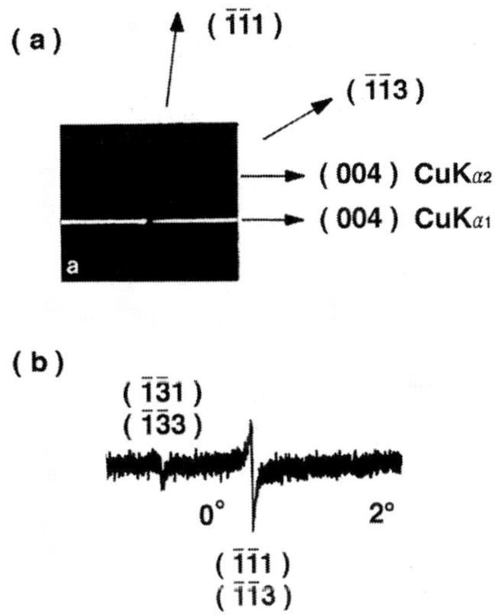

Fig. 4.7. (**a**) Section Kossel diagram of germanium (004) in the four-wave $(000)(004)(\bar{1}\bar{1}1)(\bar{1}\bar{1}3)$ case. (**b**) Intensity distribution of (**a**) along the (004) $CuK\alpha_1$ line obtained with a collimated incident beam [306]

Similar experimental setup to that in Fig. 4.6(a) can be adopted for a transmission geometry. If both transmitted and diffracted images are desired, then the film-to-crystal distance needs to be shortened, while the source-to-crystal distance has to be increased to ensure high resolution for the incident beam. Figure 4.6(b) is such an arrangement for a transmission case. The source-to-crystal distance and the crystal-to-film distance are about 1–2 m and a few centimeters, respectively. Both forward diffraction (FD) and transmitted reflection (TR) are detected by the film.

The intensity distribution of a pseudo-Kossel line near a multiple-wave diffraction position is very similar to that of a Renninger scan. For example, the intensity measured from left to right along the (004) Kossel line of $CuK\alpha_1$ shown in Fig. 4.7(a) first increases and then decreases, which resembles the intensity profile of the Renninger scan shown in Fig. 4.7(b). In other words, the scan along a Kossel line of a (hkl) reflection is equivalent to the ψ scan around the $[hkl]$ reciprocal lattice vector.

4.2.1 Indexing Kossel Diagrams

Indexing a Kossel diagram is similar to indexing a diffraction diagram in the Laue method [439], although the former involves characteristic radiation and the latter a white beam. To begin with, the symmetry elements of the crystal, such as rotation and mirror, need to be identified from the symmetry of the Kossel diagram. If the Kossel diagram shows four-, three-, and two-fold symmetries for a cubic crystal, then the principal pole axes along the incident beam direction are [001], [111], and [011], respectively. After the principal pole axis is identified, the Kossel lines can be indexed by several methods, such as gnomonic, Kossel, cylindrical, or stereographic projection [340, 343, 345, 440–444]. The easiest way seems to be using the stereographic projection to construct a Kossel diagram and then comparing it with the experimentally obtained diagram [340, 440]. Since stereographic projection is a commonly used graphics technique and the details about which can be found in many textbooks [439,445], only the essential procedure is described below. First, the Bragg angles θ of all possible reflections with $2d \leq \lambda$ are calculated, where d is the interatomic spacing of a reflection and λ the X-ray wavelength used. Then the position of the pole at which the normal of each diffraction plane intercepts the reference sphere is calculated based on the geometric relationship between the reciprocal lattice vector of a given reflection and the principal pole axis identified previously. The projected pole onto the stereographic plane can then be established. A circle of the radius corresponding to $90° - \theta$ is inscribed on the stereographic projection plane about each individual pole. Thus, a complete stereographic projection can be plotted with the help of numerical calculations [345, 441, 442]. Figure 4.8 is the calculated stereographic projection of Kossel lines for Fig. 4.5. Only a few strong reflections are shown. A four-fold symmetry due to the principal pole axis [100] is evident. Once the Kossel lines are indexed, the multiple-wave diffractions occurring at the intersections of the Kossel lines are identified as well.

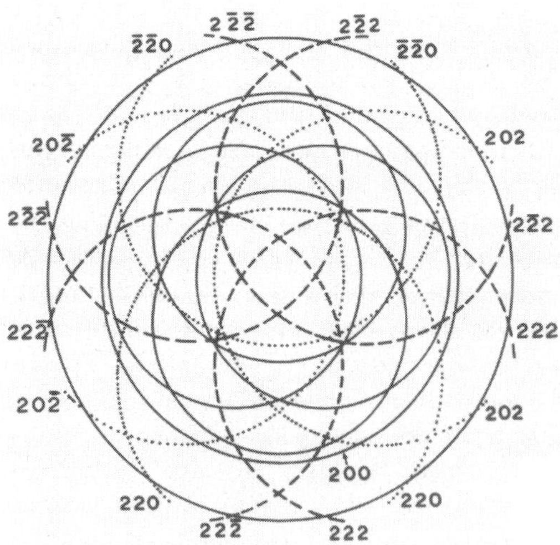

Fig. 4.8. Stereographic projection of Fig. 4.5 [53]

The Kossel lines described above are similar to the Kikuchi lines observed in electron diffraction experiments [446]. Electron multiple-wave diffraction also takes place at the intersection of Kikuchi lines.

4.3 Rotation-Crystal Methods – Reference-Beam/Stereoscopic Imaging

Rotation-crystal methods, such as rotation, Weissenberg oscillation, and precession, are commonly used photographic methods for crystallography studies [439, 445, 447, 448]. The Weissenberg camera, Burgers precession camera, and rotation camera are the hardware for collecting crystal diffraction data for crystal structure determination. Even though the automatic four-circle diffractometer has become an essential apparatus for the same purpose nowadays, crystal rotation is still a necessary step to orient a crystal for diffraction experiments, and the ψ scan is usually employed for extinction estimation and correction. Moreover, rotational photographic cameras equipped with area detectors, like image plates (IP) and charge coupled devices (CCD), are powerful tools for collecting the diffraction data of small and macromolecular crystals. In the Renninger scan, the crystal is set to rotate around a reciprocal lattice vector to generate multiple-wave diffraction. While in the data collection for structure analysis, multiple-wave diffraction is purposely avoided so as to have better accuracy in intensity measurement for two-wave diffraction. However, the occurrence of multiple-wave diffraction is sometimes unavoidable during the crystal rotation in Weissenberg, precession, and diffractometer measurements [230, 412, 449, 450]. The conditions for generating multiple-wave diffraction during the rotations about some of the major crystallographic axes in monoclinic, orthorhombic, tetragonal, hexagonal and cubic crystal systems in the Weissenberg methods have been discussed in [53, 230, 451]. Similar argument for crystal rotation in the precession methods has also been given in the literature [53, 410].

In contrast to the discussion about the undesired multiple-wave diffraction effects on the two-wave intensity measurements, multiple-wave diffraction has been proven to be capable of providing the phase information of X-ray reflection from intensity measurements [257, 259, 261, 262, 279, 287, 330, 452, 453]. Recent progress indicates that the physical measurement of reflection phases for macromolecular crystals on a qualitative and a quantitative basis is feasible [279, 292, 293, 454] without invoking multi-wavelength [455] and heavy-atom derivatives [301,305,331]. To generate a large number of multiple-wave diffraction for collecting more reflection phases is most desired for solving the X-ray phase problem, especially for macromolecular crystals. Unfortunately, the intensity profiles of multiple-wave diffraction used in phase determination are usually obtained one at a time, just like in the case of the Renninger scans. The detection of a large collection of diffraction profiles is very difficult because sometimes the macromolecular crystal deteriorates quickly under X-ray exposure. This fact has seriously hindered the practicability of this diffraction technique. To overcome this difficulty, Shen [309,318,319,456] first proposed the so-called reference-beam technique and Chang et al. [310,316,326,457–459] then suggested the so-called stereoscopic multiple-wave diffraction imaging technique. The former combined the multiple-wave diffraction with the X-ray standing wave technique [309]. The latter was derived from the grazing incidence multiple-wave diffraction experiments where a Bragg-angle scan, the so-called ω scan, was used for two symmetry-related secondary reflections [276]. Both the proposed techniques are essentially the same and involve an additional ω scan during the intensity measurement, compared with the Renninger azimuth scan. In fact, using ω scan to investigate the intensity distribution of multiple-wave diffraction has been reported earlier [460].

The idea of the reference-beam/stereoscopic imaging is simple. As stated in Chap. 3, the interaction among the diffracted waves in a multiple-wave diffraction gives rise to intensity variation on each of the diffracted waves. In other words, each diffracted wave can be treated as a reference beam for other diffracted waves, because the intensity change can be detected only in the presence of a given involved reflection. In the Renninger scans, the intensity variation of the primary reflection G is revealed by the two-wave intensity of the G reflection when the secondary reflection L comes into play in the diffraction process. Therefore, the primary reflection G provides a reference background for the intensity variation, which is monitored by a point detector. In this way, only one multiple-wave diffraction can be detected each time. Now, if the intensity variations of all possible secondary reflections can be recorded on an area detector during the crystal oscillation with the primary reflection as the reference, then many multiple-wave cases can be measured. This situation is similar to that in the usual crystal oscillation photographic method, where diffraction spots of various reflections, L's, are recorded on an area detector or X-ray film (Fig. 4.9(a)), except that now the primary reflection G should be present to fulfill the multiple-wave diffraction conditions (Fig. 4.9(b)). Hence the individual two-wave diffractions L_1, L_2, L_3, ... with the wavevectors $\boldsymbol{K}_{L_1}$, $\boldsymbol{K}_{L_2}$, $\boldsymbol{K}_{L_3}$, ... (see Fig. 4.9(a)), are modified by the presence of the primary reflection G as schematically shown in Fig. 4.9(b). Moreover, in order to reveal the intensity variations of the secondary reflections due to multiple-wave interaction, the primary reflecting planes should be initially completely off the Bragg position, i.e., $\theta < \theta_G$, and then move stepwise across the Bragg diffraction position, i.e., $\theta > \theta_G$. For each step, namely, at each θ position, the crystal needs to be oscillated around the reciprocal lattice vector $\boldsymbol{g}$, i.e., the ψ oscillation, to ensure that the secondary reflections actually take place and their diffracted intensities are recorded on an area detector, such as an image plate (IP) or a CCD detector. If an IP is used, the plate needs to be dislocated slightly from its

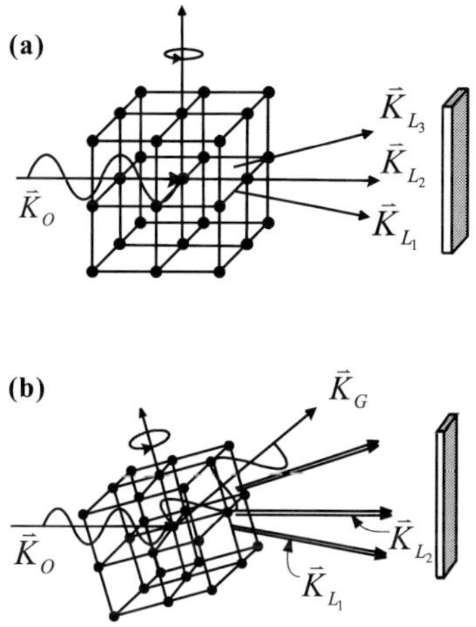

Fig. 4.9. Schematics of (**a**) conventional crystal-oscillation geometry, and (**b**) reference-beam diffraction in Bragg-inclined geometry in real space [314]

previous position for each θ angle to prevent image overlapping. If a CCD is used, the intensity can be retrieved instantaneously for a given θ setting and the CCD is ready for data acquisition for the next θ setting. The intensity versus θ for each secondary reflection forms the multiple-wave diffraction profile for the (O, G, L) situation. Figure 4.10(a) shows the oscillation images of a GaAs (004) multiple-wave diffractions recorded in an IP at sixteen θ positions [309]. The primary reflection, (004), is the reference for secondary reflections L. The diffraction images of secondary reflections are labelled with Miller indices. The sixteen recorded intensities of the (317) secondary reflection are plotted against $\Delta\theta$ in Fig. 4.10(b), where the profile asymmetry results from the effective phase u, defined in (9.7), is shown in Fig. 4.10(c). u is closely related to the triplet phase δ_3 of the involved structure factors. The details of the effects of the reflection phase on the intensity distribution of multiple-wave diffraction will be given in Chap. 9. Figure 4.10(d) showing the rocking curve of the primary (004) reference-beam indicates the participation of the reference reflection (004) in the multiple-wave diffraction processes.

Since a pair of inversion-symmetry related (ISR) multiple-wave diffraction profiles are required for the quantitative estimation of phases for non-centrosymmetric crystals like macromolecules [279, 291], both (O, G, L) and $(O, -G, -L)$ multiple-wave diffractions need to be considered. Figure 4.11 is the layout of an experimental setup for the multiple-wave stereoscopic imaging, where two ISR three-wave diffractions (O, G, L) and $(O, -G, -L)$ are considered. For the three-wave (O, G, L) diffraction, the same ω rotation for θ and the same ψ scan around the reciprocal lattice vector $\boldsymbol{g}$ of the primary G reflection as for the reference-beam technique are performed. The intensity variation on the reference background of the L reflection for a given $\Delta\theta$ is

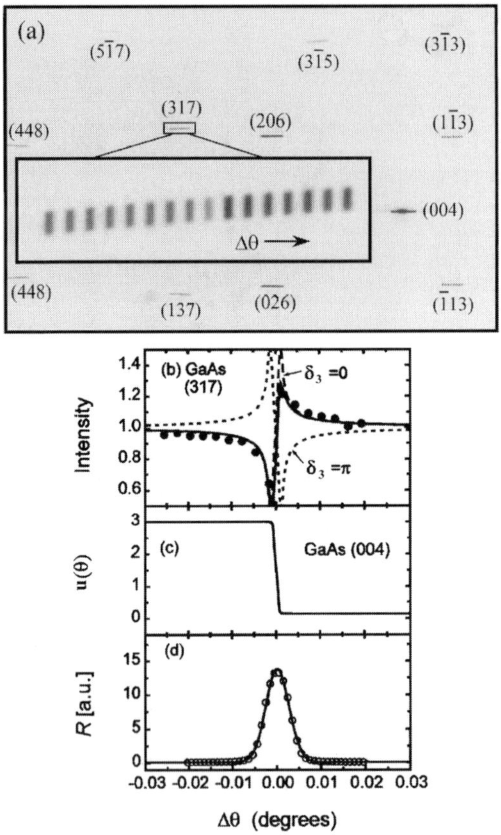

Fig. 4.10. (**a**) Oscillation images of GaAs around the (004) reference reflection at $\lambda = 0.9185\,\text{Å}$. The inset is the enlarged view of the (317) reflection images at 16 θ settings. (**b**) Integrated intensities of (317) versus θ: measurements (*full circles*), dynamical calculation (*solid lines*) and (*dashed lines*) for the triplet phase $\delta_3 = 0$, and π, respectively. (**c**) Effective phase shift u of (004). (**d**) Rocking curve of (004) [309]

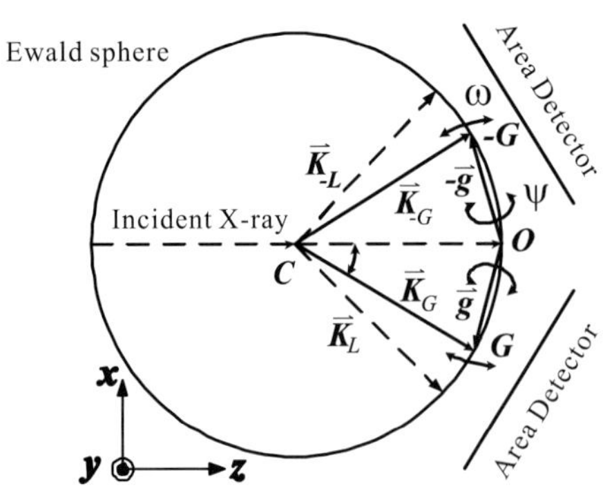

Fig. 4.11. Schematic of the experimental setup for the stereoscopic multiple-wave diffraction imaging in reciprocal space [310]

recorded on a two-dimensional detector. The intensity I_L of the L-reflection versus $\Delta\theta$ (or the ω) is also obtained.

For the ISR $(O,-G,-L)$ case, the crystal is rotated to a geometry and crystallography equivalent position so that the pathway of the reciprocal lattice point $-L$ crossing the surface of the Ewald sphere is kept the same as that of the reciprocal lattice point L in the (O,G,L) case [310, 318, 326]. Figure 4.12 shows the geometric relation between the two ISR three-wave diffractions. The angular position for the $(O,-G,-L)$ case is ($\theta=\theta_G$, $2\theta=2\theta_G$, $\psi=\psi+\pi$, $\chi=\chi+\pi$), which is equivalent to ($\theta=\theta_G$, $2\theta=2\theta_G$, $\psi=\psi$, $\chi=\chi$) for the (O,G,L) case [318]. As a matter of fact, the equivalent positions can be derived, in general, from the crystal orientation matrix for a given detector position [326]. The procedure of determining the equivalent angular positions is described below.

Let's consider an eight-circle Huber diffractometer with a ϕ-χ-ψ-ω goniostat, where the stacking of the crystal-rotating circles is in the order of ϕ, χ, ψ, and ω. An area detector is placed on the detector arm, which can be rotated along the 2θ and γ circles in the vertical and the horizontal plane, respectively. The conventional ϕ and χ circles are used to align the reciprocal lattice vector $\boldsymbol{g}$ along the ψ axis. The origin, $\chi=0$, of the χ-circle is defined as the ϕ-axis lying in the horizontal plane along the $-x$ direction. The ω circle is utilized to vary the Bragg angle of the G reflection. The ψ- and ω-axis lie in the horizontal and the vertical plane, respectively.

A laboratory coordinate system (x,y,z) is defined as follows. The incident-beam is along the z-axis and the upward vertical direction perpendicular to the incident beam is the $+y$ direction. If we face the incident beam, the x-axis perpendicular to y- and z-axis lies in the horizontal plane and points towards the right-hand side. Suppose that the three-beam (O,G,L) case occurs at ($2\theta=2\theta_G$, $\phi=\phi_O$, $\chi=\chi_O$). The position of the reciprocal lattice point G is at $(-\cos\theta_G, 0, -\sin\theta_G)$ in the laboratory coordinate system. In order to

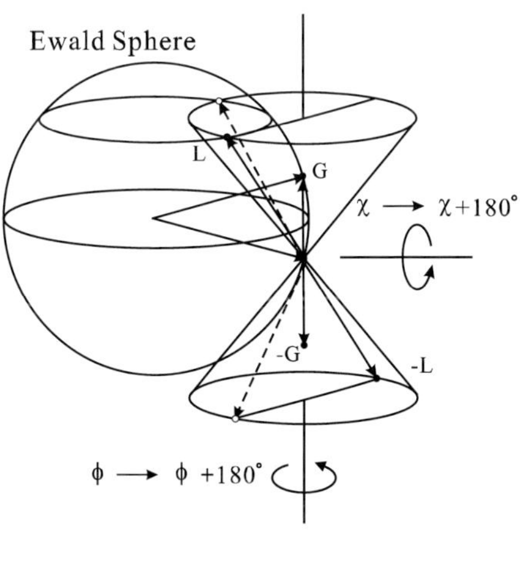

Fig. 4.12. Geometric relationship between two ISR three-wave diffractions [318]

find the angular position $(2\theta, \theta, \phi, \chi)$ for the $(O, -G, -L)$ case, we first move the crystal to the position at which the reciprocal lattice vector $\boldsymbol{g}$ of the primary reflection G is along the $-x$ direction, represented by $[-1, 0, 0]$, and the secondary reciprocal lattice vector is assumed to be $\boldsymbol{l} = [x, y, z]$ in the laboratory system. According to the inversion symmetry between the (O, G, L) and $(O, -G, -L)$, there are only two ways to find the equivalent position for the $(O, -G, -L)$ case to occur:

1) The detector position is kept unchanged. Move the reciprocal lattice vectors $-\boldsymbol{g}$ and $-\boldsymbol{l}$ to the original positions of $\boldsymbol{g} = [-1, 0, 0]$ and $\boldsymbol{l} = [x, y, z]$ or to the equivalent positions at which the (O, G, L) case occurs. In other words, $\boldsymbol{g}$ and $\boldsymbol{l}$ need to be moved to $[1, 0, 0]$ and one of the equivalent positions $[\pm x, \pm y, \pm z]$, respectively. Considering the fact that the vectors $\boldsymbol{g}$ and $\boldsymbol{l}$ are changed simultaneously, only $[-x, y, -z]$, $[x, -y, -z]$, and $[-x, -y, z]$ with two 'minus' coordinates are the possible positions for $\boldsymbol{l}$. Among these three positions, only $[-x, y, -z]$ is the correct equivalent vector position for $\boldsymbol{l}$ to move to, because the position $[-x, -y, z]$ leads to no equivalent position for $-\boldsymbol{l}$ relative to $\boldsymbol{g}$. The position $[x, -y, -z]$ is the correct choice for $\boldsymbol{l}$ only when the detector position is changed from 2θ to -2θ. The rotation operations leading $\boldsymbol{g}$ and $\boldsymbol{l}$ to change to $[100]$ and $[-x, y, -z]$, respectively, are

$$\mathbf{X}(\chi) \cdot \Phi(\phi) \cdot \mathbf{X}(-\chi_O) \begin{pmatrix} -1 \\ 0 \\ 0 \end{pmatrix} = \begin{pmatrix} 1 \\ 0 \\ 0 \end{pmatrix}, \tag{4.6}$$

$$\mathbf{X}(\chi) \cdot \Phi(\phi) \cdot \mathbf{X}(-\chi_O) \begin{pmatrix} x \\ y \\ z \end{pmatrix} = \begin{pmatrix} -x \\ y \\ -z \end{pmatrix}, \tag{4.7}$$

where the χ-circle is rotated by $-\chi_O$ around the z-axis back to its origin, then the ϕ-circle is rotated by ϕ from ϕ_O around the $-x$ direction, and subsequently the χ-circle is rotated by χ from the origin around the z-axis. The directions of rotation of the circles refer to those defined for the eight-circle diffractometer used. Equation (4.6) gives

$$\begin{pmatrix} -\cos\chi_O \cos\chi - \cos\phi \sin\chi_O \sin\chi \\ \cos\phi \cos\chi \sin\chi_O - \cos\chi_O \sin\chi \\ -\sin\phi \sin\chi_O \end{pmatrix} = \begin{pmatrix} 1 \\ 0 \\ 0 \end{pmatrix}. \tag{4.8}$$

This leads to

$$\phi = 0\,, \qquad \chi = \pm\pi + \chi_O \quad \text{or} \quad -\chi_O\,, \tag{4.9a}$$

$$\phi = \pi\,, \qquad \chi = \pi - \chi_O \quad \text{or} \quad \chi_O\,. \tag{4.9b}$$

Only $\phi = \pi$, $\chi = \pi - \chi_O$ in (4.9b) satisfies the condition imposed by (4.7). This means that the crystal needs to be rotated to $\phi = \pi + \phi_O$ and $\chi = \pi - \chi$ while the same detector position is remained as that for the (O, G, L) case. In other words, during the X-ray exposure, the direction of the ψ-rotation around $\boldsymbol{g}$ should be opposite to that of the rotation around $\boldsymbol{g}$ in the (O, G, L) case. This means that if the images of the secondary reflections L in the (O, G, L) case originally appear in the upper half of the 2-d detector above the $\boldsymbol{g}$ vector, then the images of their counterparts $-L$ in the $(O, -G, -L)$ case appear in the lower part, and vice versa. Note that the projection of the $\boldsymbol{g}$ vector on the detector lies horizontally and is nearly at the center of the detector. Under this situation, the pathway of the secondary reciprocal lattices in both cases are the same, except that one enters (or leaves) the Ewald sphere from the upper hemisphere and the other leaves (or enters) the sphere from the lower hemisphere.

2) The detector position is changed from 2θ to -2θ. Move $\boldsymbol{l}$ from $[x, y, z]$ to $[x, -y, -z]$ while keeping $\boldsymbol{g} = [-1, 0, 0]$ unchanged. The corresponding rotation operations are

$$\mathbf{X}(\chi) \cdot \Phi(\phi) \cdot \mathbf{X}(-\chi_O) \begin{pmatrix} -1 \\ 0 \\ 0 \end{pmatrix} = \begin{pmatrix} -1 \\ 0 \\ 0 \end{pmatrix} \tag{4.10}$$

and

$$\mathbf{X}(\chi) \cdot \Phi(\phi) \cdot \mathbf{X}(-\chi_O) \begin{pmatrix} x \\ y \\ z \end{pmatrix} = \begin{pmatrix} x \\ -y \\ -z \end{pmatrix} . \tag{4.11}$$

Equation (4.11) gives

$$\begin{pmatrix} -\cos\chi_O \cos\chi - \cos\phi \sin\chi_O \sin\chi \\ \cos\phi \cos\chi \sin\chi_O - \cos\chi_O \sin\chi \\ -\sin\phi \sin\chi_O \end{pmatrix} = \begin{pmatrix} -1 \\ 0 \\ 0 \end{pmatrix} . \tag{4.12}$$

This leads to

$$\phi = 0 \,, \qquad \chi = \chi_O \quad \text{or} \quad \pi - \chi_O \,, \tag{4.13a}$$

$$\phi = \pi \,, \qquad \chi = -\chi_O \quad \text{or} \quad \pm\pi + \chi_O \,. \tag{4.13b}$$

However, only $\phi = \pi, \chi = -\chi_O$ in (4.13b) satisfies the condition of (4.11). This implies that we need to rotate the crystal to $\phi = \pi + \phi_O$ and $\chi = -\chi_O$ while moving the detector from 2θ to -2θ. The direction of the ψ-rotation around $-\boldsymbol{g}$ is the same as that of the rotation around $\boldsymbol{g}$ in the (O, G, L) case. This means that if the images of the secondary reflections, L, in the (O, G, L) case appear on the left hemisphere, then the images of their counterparts, $-L$, in the $(O, -G, -L)$ case will appear on the right hemisphere.

Similar positions for a conventional four-circle diffractometer can be found by using an appropriate orientation matrix.

Figure 4.13 shows the stereoscopic images of the ISR pair with (18, 12, 1) and $(\overline{18}, \overline{12}, \overline{1})$ as the primary reflections, namely, the reference reflections, for a hexagonal lysozyme crystal (unit cell dimensions: $a = b = 78.9$ Å, $c = 38.1$ Å, space group $P4_32_12$, protein data bank (PDB) number 1LYZ) at $\Delta\theta = 0$. The π-polarized incident beam of $\lambda = 1.54228$ Å is used. The beam divergences are 0.01° in both the vertical and the horizontal directions and the energy resolution $\Delta E/E$ is about 2×10^{-4}. The experimental setup similar to that reported in [276] is equipped with an eight-circle diffractometer and an image plate. The image plate sits on the 2θ arm of the diffractometer. These oscillation images are obtained during the oscillation range of $\Delta\psi = 2°$ around the [18, 12, 1] and $[\bar{18}, \bar{12}, \bar{1}]$ directions (i.e., G = (18, 12, 1) and $(\bar{18}, \bar{12}, \bar{1})$) at the peak positions $\theta_G = 12.263°$ and $-12.263°$, respectively. The speed of oscillation is 0.125°/sec and the total exposure time for each image is 102 seconds. The distance between the crystal and the image plate is 10 cm.

In Figs. 4.13(a) and 4.13(b), all the points except (18, 12, 1) and $(\bar{18}, \bar{12}, \bar{1})$ are the diffraction spots of the secondary reflections partially or fully satisfying the multiple diffraction conditions. Compared with full diffraction, partial diffraction means the reciprocal lattice point L does not completely traverse the surface of the Ewald sphere during the oscillation. Comparing Fig. 4.13(a) with Fig. 4.13(b), it is clear that the intensities of some of the secondary reflection spots increase or decrease (see the hkl and $\bar{h}\bar{k}\bar{l}$ spots). With the help of the software DENZO [461], the integrated intensity of each full secondary reflection spot can be read and normalized by

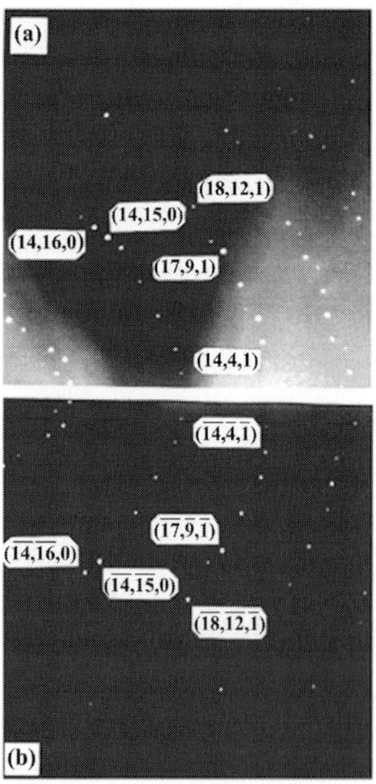

Fig. 4.13. Stereoscopic images of tetragonal lysozyme oscillating around (**a**) $(18, 12, 1)$ and (**b**) $(\bar{18}, \bar{12}, \bar{1})$ at $\Delta\theta = 0$ and $\lambda = 1,54228$ Å [310]

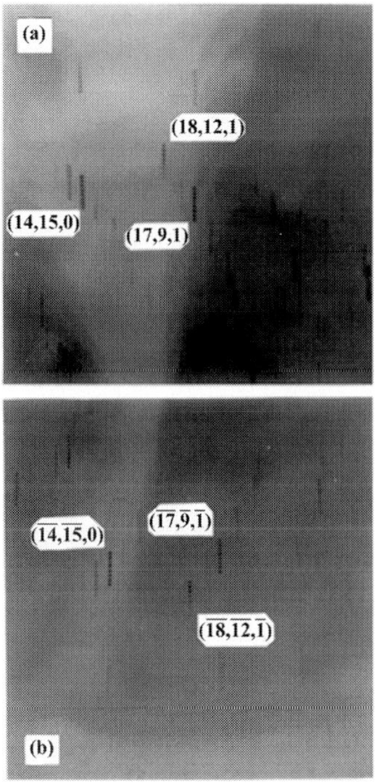

Fig. 4.14. Stereoscopic images of Fig. 4.13 at various tilt angles $\Delta\theta$ of (**a**) (18, 12, 1) and (**b**) ($\overline{18}$, $\overline{12}$, $\overline{1}$) reflections [310]

the intensity of the incident beam. Moreover, the two images, Figs. 4.13(a) and 4.13(b), show a mirror symmetry to each other, which is due to the geometry of the equivalent positions of (O, G, L) and $(O, -G, -L)$ shown in Fig. 4.12.

Figure 4.14 shows the diffracted intensities of the secondary reflections of Fig. 4.13 at various tilt angles, $\Delta\theta$, in the vicinity of the primary reflections (18, 12, 1) and ($\overline{18}$, $\overline{12}$, $\overline{1}$). During the X-ray exposure for each tilt angle, the image plate is shifted by a small fixed $2\Delta\theta$. The trace of the intensity versus $\Delta\theta$ gives the intensity profile of a given secondary reflection. Figure 4.15 shows the three-wave diffraction profiles of (000)(18, 12, 1) (14, 16, 0) and (000)($\overline{18}$, $\overline{12}$, $\overline{1}$) ($\overline{14}$, $\overline{16}$, 0), labelled as (14, 16, 0)/(18, 12, 1) and ($\overline{14}$, $\overline{16}$, 0)/($\overline{18}$, $\overline{12}$, $\overline{1}$). The identification of multiple-wave diffraction relies on the indexing of the stereosopic images, which can be done as usual for oscillation photograph by using the existing software packages, such as DENZO. In general, the crystal orientation matrix is employed in the indexing software. The details are given in Sect. 4.3.1.

Stereoscopic images with a better resolution can be obtained by using a CCD detector as demonstrated by Shen, Kycia, and Dobrianov [318, 319], which will be discussed in Sect. 4.4.

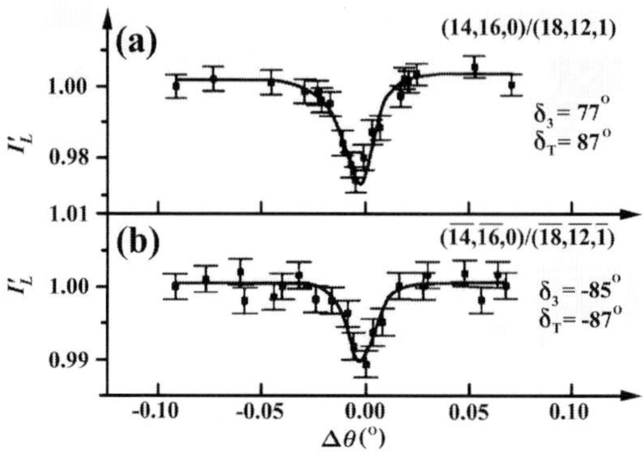

Fig. 4.15. Three-wave diffraction profiles of tetragonal lysozyme obtained from stereoscopic images: (**a**) I(14, 16, 0) vs. $\Delta\theta$ of (000)(18, 12, 1) (14, 16, 0), and (**b**) I($\overline{14}$, $\overline{16}$, 0) vs. $\Delta\theta$ of (000)($\overline{18}$, $\overline{12}$, $\overline{1}$)($\overline{14}$, $\overline{16}$, 0) at $\lambda = 1,54228$ Å [310]

4.3.1 Indexing Multiple-Wave Diffraction Diagram by Crystal Orientation Matrix

The most precise way to index multiple-wave diffraction diagrams is using the crystal orientation matrix to determine the initial positions of possible secondary reciprocal lattice points in the laboratory orthogonal coordinate system, (x, y, z). For a given crystal belonging to a crystal system with the reciprocal lattice basis vectors (a^*, b^*, c^*), the orientation matrix is defined as

$$M_{\text{or}} = \begin{pmatrix} \hat{x} \cdot \boldsymbol{a}^* & \hat{x} \cdot \boldsymbol{b}^* & \hat{x} \cdot \boldsymbol{c}^* \\ \hat{y} \cdot \boldsymbol{a}^* & \hat{y} \cdot \boldsymbol{b}^* & \hat{y} \cdot \boldsymbol{c}^* \\ \hat{z} \cdot \boldsymbol{a}^* & \hat{z} \cdot \boldsymbol{b}^* & \hat{z} \cdot \boldsymbol{c}^* \end{pmatrix} , \tag{4.14}$$

where $\hat{x}$, $\hat{y}$, $\hat{z}$ are the unit vectors along the x-, y-, and z-axis, respectively. The coordinates, (x, y, z), of the reciprocal lattice point for (h, k, l) can be calculated from the orientation matrix as

$$\begin{pmatrix} x \\ y \\ z \end{pmatrix} = M_{\text{or}} \begin{pmatrix} h \\ k \\ \ell \end{pmatrix} . \tag{4.15}$$

This matrix can be determined, as usual, from the crystal-oscillation photographs recorded on X-ray films or area detectors, or from the intensity measurements of a few reflections with an X-ray diffractometer [447]. Assume that a four-circle $(2\theta, \omega, \phi, \chi)$ diffractometer is used to orient the crystal and the ω, χ, and ϕ axes are along the $-y$-, $+z$-, and $+y$-direction, respectively. The orientation matrix determined experimentally is M_{or} when ω, χ, and ϕ are all equal to 0°. Initially, the reciprocal lattice point G of the primary reflection is brought onto the surface of the Ewald sphere by a combined rotation of the three circles of the diffractometer, i.e., ϕ, χ, and ω, and lies on the equatorial circle, i.e., the xz-plane in Fig. 4.16. This motion can be represented by the product of three rotation matrices $\Phi(\phi)$, $X(\chi)$, and $\Omega(\omega)$ around the ϕ, χ, and ω axes:

$$M_{\mathrm{r}} = \Omega(\omega)X(\chi)\Phi(\phi)$$
$$= \begin{pmatrix} \cos\omega\cos\chi\cos\phi & \cos\omega\sin\chi & -\cos\omega\cos\chi\sin\phi \\ +\sin\omega\sin\phi & & +\sin\omega\cos\phi \\ -\sin\chi\cos\phi & \cos\chi & \sin\chi\sin\phi \\ -\sin\omega\cos\chi\cos\phi & -\sin\omega\sin\chi & \sin\omega\cos\chi\sin\phi \\ +\cos\omega\sin\phi & & +\cos\omega\cos\phi \end{pmatrix} . \quad (4.16)$$

The new orientation matrix after this combined rotation becomes

$$M'_{\mathrm{or}} = M_{\mathrm{r}}M_{\mathrm{or}} . \quad (4.17)$$

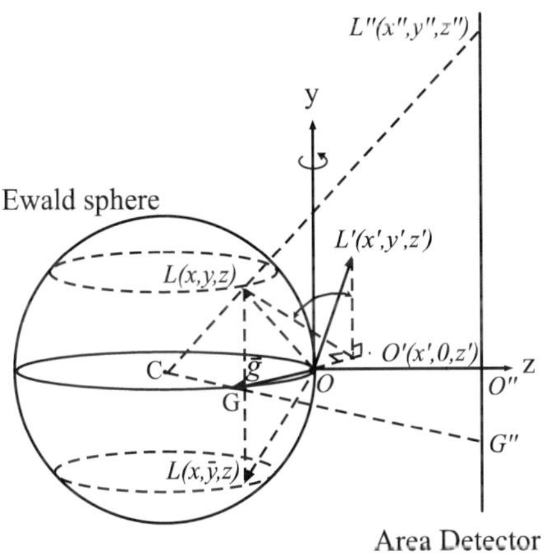

Fig. 4.16. Coordinate system used for orientation matrix

The initial position of the reciprocal lattice point L' of the secondary reflection in the laboratory system is (x', y', z'), which can be determined by using (4.15) and the new orientation matrix M'_{or}. In Fig. 4.16, the incident beam $\boldsymbol{CO}$ is along the z-direction. The ψ-rotation around the $\boldsymbol{g}$ brings the L' of the secondary reflection at $\psi = 0°$ to the position L on the surface of the Ewald sphere, where L corresponds to the ψ angular position at which the multiple-wave diffraction (O, G, L) takes place. There are two positions, (x, y, z), of the L point at which point L touches the Ewald sphere. Moreover, the two positions are symmetric about the xz plane. This can be easily proved from the following relations:

$$x^2 + y^2 + z^2 = \ell^2 ,$$
$$x^2 + y^2 + \left(z + \frac{1}{\lambda}\right)^2 = \frac{1}{\lambda^2} ,$$
$$\frac{x - x'}{z - x'} = \tan\omega , \quad (4.18)$$

where ℓ is the modulus of the reciprocal lattice vector $\boldsymbol{l} = \boldsymbol{OL}$ of the L reflection. These lead to double solutions to (4.18):

$$x = x' + (z - z')\tan\omega\ ,$$
$$y = \pm\sqrt{\ell^2 - x^2 - z^2}\ ,$$
$$z = -\frac{\lambda\ell^2}{2}\ , \tag{4.19}$$

where

$$x' = \ell\cos\alpha\cos\omega\ ,$$
$$y' = \ell\sin\alpha\ ,$$
$$z' = -\ell\cos\alpha\sin\omega\ . \tag{4.20}$$

α is the angle between $\boldsymbol{g}$ and $\boldsymbol{l}$, namely between the $\boldsymbol{OL'}$ and $\boldsymbol{OG}$ and between $\boldsymbol{OL}$ and $\boldsymbol{OG}$. As shown in Fig. 4.16, $\boldsymbol{OO'}$ is the component of $\boldsymbol{OL'}$ parallel to $\boldsymbol{g}$ and the coordinates of O' are $(x', 0, z')$. The components of $\boldsymbol{OL'}$ and $\boldsymbol{OL}$ normal to $\boldsymbol{g}$ are $\boldsymbol{O'L'}$ and $\boldsymbol{O'L}$, respectively, with the same magnitude of $l\sin\alpha$. The angle between the two vectors is the ψ angle, which can be calculated as

$$\cos\psi = \frac{y}{y'}\ . \tag{4.21}$$

Thus the angular distance in ψ between the initial position and the multiple-wave position can be determined. A computing program can be written to calculate the ψ positions of all possible secondary reflections and list the calculated positions in increasing order of ψ. This list can be used to index multiple-wave diffraction pattern obtained with the Renninger scan. By the same token, the two-dimensional multiple-wave diffraction image generated from the reference-beam/stereoscopic imaging can also be indexed unambiguously. The procedure is described below:

1. Determine the crystal orientation matrix M'_{or} when the crystal is aligned for the primary reflection G.
2. List the calculated (x, y, z) and ψ angles of all the possible secondary reflections for the range $\Delta\psi$ of oscillation around the $\boldsymbol{g}$.
3. Calculate the position (x', y', z') of the projection of (x, y, z) on the area detector according to the projection geometry (see, for example, Fig. 4.16).
4. Plot the (x', y', z') for all the possible secondary reflections within the angular range $\Delta\psi$ on the xy-plane and compare it with the multiple-wave image recorded on the detector.

Through this procedure, all the diffraction spots appearing on the detector can be indexed. The commercially available software packages, such as DENZO [461], can be used to facilitate the indexing. However, care must be taken in considering the crystal-oscillation around an inclined axis, i.e., the $\boldsymbol{g}$

of the primary reflection, with respect to the incident beam, because most of the software packages are written for a rotation axis normal to the incident beam. Figure 4.13 shows the indexed diffraction patterns of the stereoscopic multiple-wave images using the DENZO program.

4.4 Diffractometer and Detector

Generating multiple-wave diffraction in the Renninger scan and crystal oscillation, as stated in the previous sections, involves the ψ rotation around the reciprocal lattice vector of a given reflection which is chosen as the primary reflection or reference reflection. To carry out a ψ scan, an appropriate mechanism and apparatus are needed to rotate the crystal to satisfy Bragg's law for the primary reflection and then align the reciprocal lattice vector of this reflection along a rotation axis of a diffractometer. During the ψ rotation, a detector is required to register the diffraction intensity. Therefore, the minimum number of rotation axes necessary for performing a ψ scan is three, i.e., an ω for changing Bragg angle, a 2θ for the detector movement, and a ψ axis for the rotation around the reciprocal lattice vector, provided that a goniometric head is employed to orient the reciprocal lattice vector to lie in the plane of incidence of the primary reflection. Figure 4.17 shows, as an example, a three-circle diffractometer [462] suitable for multiple-wave diffraction experiments.

Precise alignment of a reciprocal lattice vector with a rotation axis is sometimes difficult, especially when the Darwin width of the reflection is very narrow and the crystal size is small. An alternative is to use a four-circle diffractometer with 2θ, ω, ϕ, and χ as the rotation axes. The reciprocal lattice vector can be accurately moved to lie in the plane of incidence by adjusting the ϕ and χ circles. The ψ scan can be carried out by the combined motion of ω, ϕ, and χ axes. However, exact ψ scans are also difficult due to the gear backlashes of the three circles. To overcome this difficulty, one

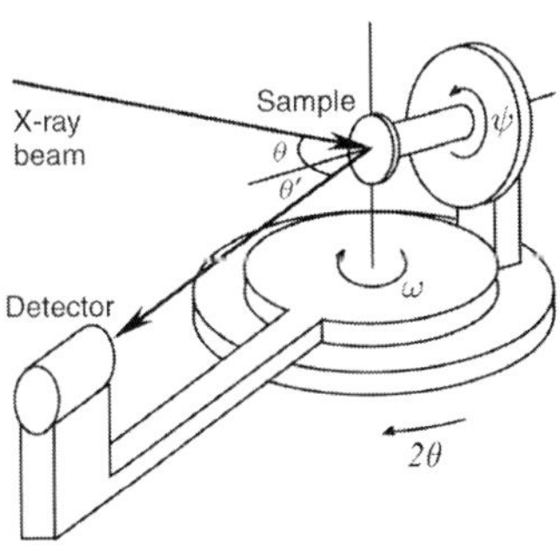

Fig. 4.17. Three-circle diffractometer [462]

can first determine the optimal four angles at two adjacent ψ positions, separated by $\Delta\psi$, by the usual crystal centering process and then calculate the four angles for each ψ position within the interval $\Delta\psi$ and drive the crystal to the precalculated four-angle positions. If the interval $\Delta\psi$ is kept small, fairly linear motion around the four axes can be achieved. If, however, the Darwin width of the primary reflection is yet very narrow, say, of the order of a few thousandths of a degree, then the combined rotation of three axes is not sufficient to provide a steady and exact ψ scan [287,305]. Thus an additional axis, ψ-axis, is required. Figure 4.18 shows the so-called ψ-circle diffractometer [305,328], where the $\phi - \chi$ circle first sits on the ψ circle and then lies on the ω circle. The detector can be rotated vertically around the horizontal ν axis which is situated on the 2θ arm. Therefore, the reciprocal lattice vector is first aligned along the ψ axis by the ϕ and χ circles and is then moved to the Bragg condition by the ω circle. The detector is rotated to the horizontal plane with an appropriate 2θ setting. In this situation, the rotation around the ψ-axis generates multiple-wave diffraction, which is monitored by the detector. If an analyzing crystal is used for the diffracted beam, then the commercially available eight-circle diffractometer can serve this purpose. The two additional axes are employed to rotate the analyzer and the detector.

The presence of a χ circle sometimes blocks either the incident beam or the diffracted beam, which limits the number of detectable reflections in the multiple-wave diffraction experiments. Hence the adoption of the so-called κ goniostat has been proposed [463–465]. Figure 4.19 shows schematically a six-circle κ diffractometer [463]. The κ axis is used to replace the conventional χ circle. The rest of the axes are similar to those mentioned for the ψ-circle diffractometer [463]. An alternative six-circle κ diffractometer is shown in Fig. 4.20, where the ψ axis is perpendicular to the ω axis because the plane of incidence is the vertical plane normal to the page of Fig. 4.20 [465]. A point detector attached to the δ circle is in the same plane. Both the ω and 2θ are

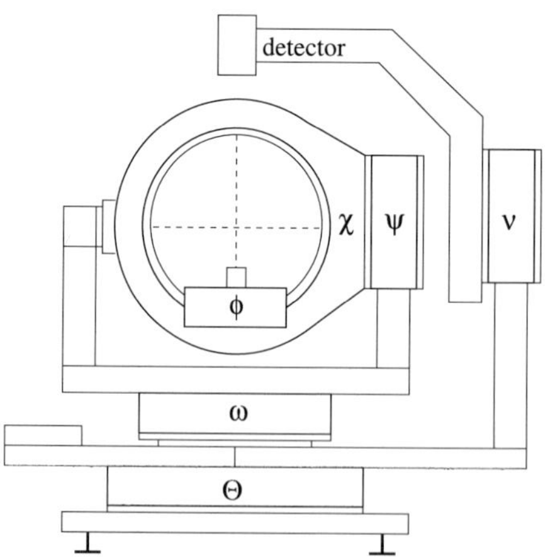

Fig. 4.18. ψ-circle diffractometer [305]

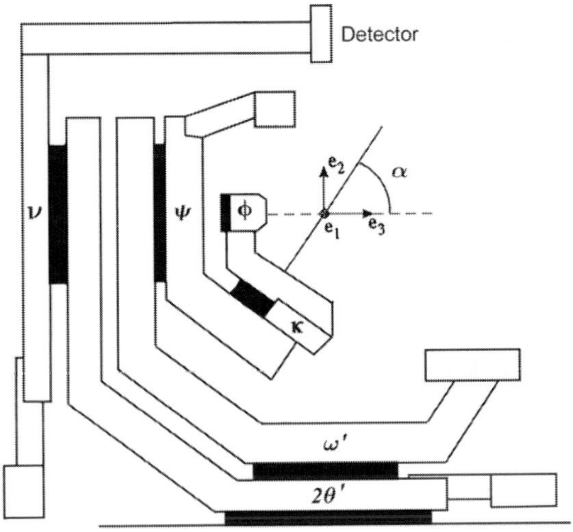

Fig. 4.19. Six-circle κ diffractometer [463]

along the horizontal line passing through the center of the diffractometer. All the six-circle and eight-circle diffractometers have a 2θ arm to accommodate a point detector, such as the scintillation counter, proportional counter, and semiconductor pin diode, which is used to record diffraction patterns of Renninger scans. For reference-beam/stereoscopic imaging, a two-dimensional detector can be placed on the 2θ circle. In the literature, another type of diffractometer with a CCD detector, shown in Fig. 4.21, has been reported. A five-circle κ diffractometer is employed with ϕ, κ, ψ, ω, and 2θ in the se-

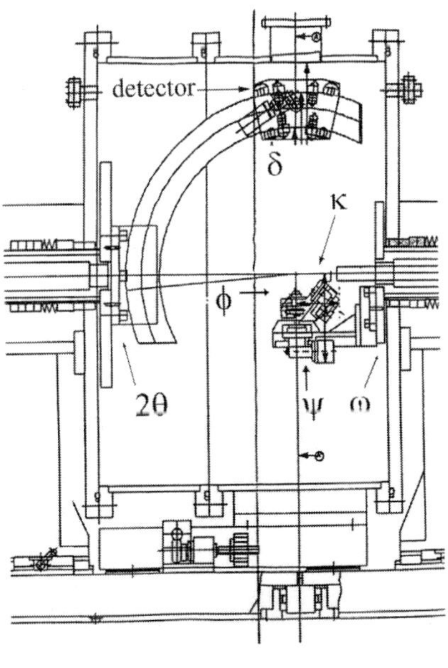

Fig. 4.20. UV compatible six-circle κ diffractometer [465]

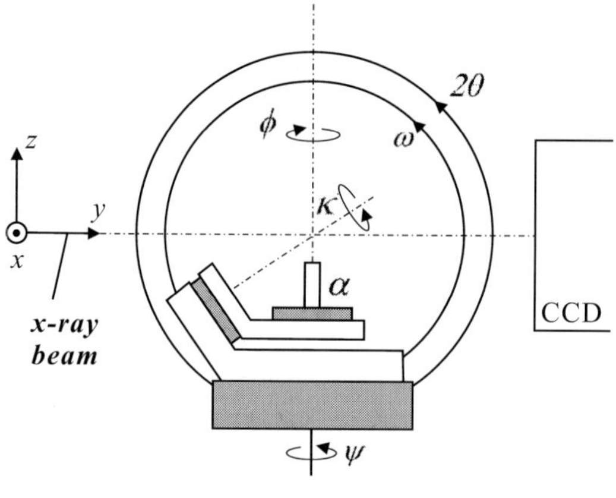

Fig. 4.21. Five-circle κ diffractometer [681]

quence of stacking as the rotation axes. The arrangement of ϕ, κ, ψ, and ω is similar to that shown in Fig. 4.20, except that the ω circle is used to rotate the crystal to the inversion-symmetry related diffraction position as discussed earlier in the previous section.

All the ψ-circle diffractometer, eight-circle diffractometer, and six-circle κ diffratometer can also be used in grazing incidence multiple-wave diffraction experiments. In this case, the plane of incidence lies in the vertical plane and the grazing incidence angle is set by tilting the crystal sample around a vertical rotation axis. The surface specularly reflected beam is monitored by a detector placed slightly off the plane of incidence. The details about the experimental setup will be discussed in Chap. 9 for multiple-wave in-plane diffraction.

As to the area detector used in the reference-beam/stereoscopic imaging, a CCD detector usually provides a larger dynamic range than an image

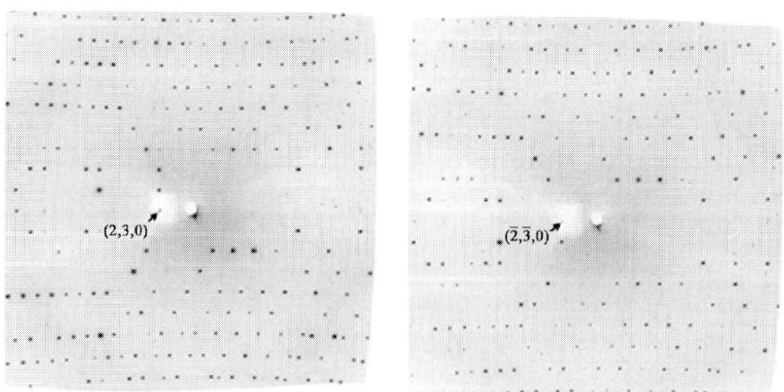

(**a** (**b**)

Fig. 4.22. Stereoscopic images of tetragonal lysozyme using a CCD detector with the reference reflections (**a**) $(2,3,0)$ and (**b**) $(\bar{2},\bar{3},0)$ [319]

plate does. Moreover, multiple exposures on an image plate give an accumulated higher background, while the exposures on a CCD are controlled electronically and each exposure can be retrieved from the CCD instantly. Therefore, it is advantageous to use a CCD detector to record multiple-wave diffraction images. Figure 4.22 shows the stereoscopic pair of multiple-wave diffraction images of $\lambda = 1.28\,\text{Å}$ for the tetragonal lysozyme with $(2, 3, 0)$ and $(\bar{2}, \bar{3}, 0)$ as the reference-beam reflections [319]. As can be seen, the quality of the images is much improved in this figure compared with Fig. 4.13, especially the background. The multiple-wave diffraction profiles extracted from the CCD images are displayed in Fig. 4.23 for the two three-wave cases (000)(2, 3, 0)($\bar{8}$, $\overline{10}$, $\bar{5}$) and (000)(2, 3, 0)(1, 11, 7). The rocking curve of the primary (230) reflection is also shown in the same figure for comparison [319].

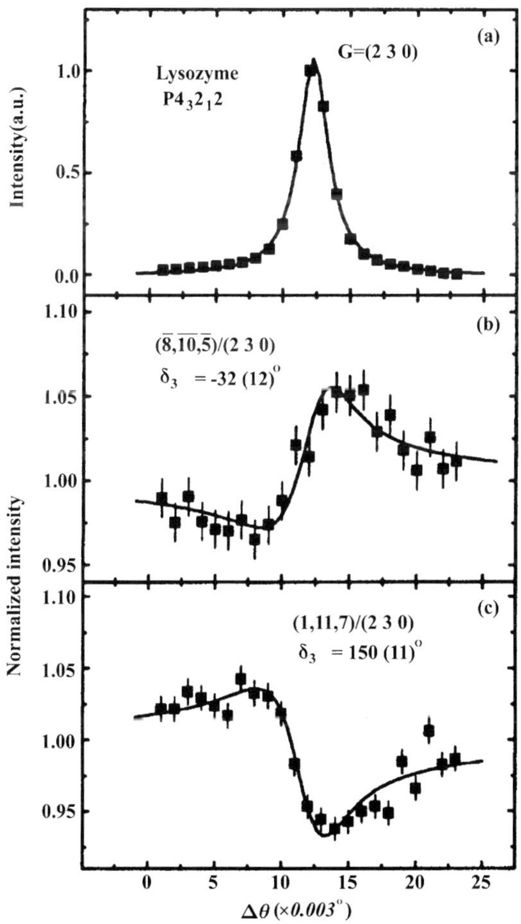

Fig. 4.23. (**a**) Two-wave (230) diffraction profile, (**b**) and (**c**) three-wave diffraction profiles of tetragonal lysozyme using a CCD detector with the reference reflection (2, 3, 0) [319]

5. Kinematical Theory of X-Ray Diffraction

Multiple-wave diffraction can be understood and predicted according to the diffraction geometry. The diffraction experiments can be realized with the techniques and methods mentioned in the previous chapters. When a multiple-wave diffraction intensity distribution or pattern is acquired, how to interpret it is an important step in understanding the multiple-wave interaction in a crystal. Since multiple-wave diffraction is a combination of many two-wave diffractions, the intensity aspect of multiply diffracted waves could be, in principle, handled by the usual ways adopted for two-wave diffraction.

Historically, the so-called kinematical theory, evolved from the geometrical theory [61, 466], was developed to account for diffracted intensity from a real crystal under the following assumptions: (1) The X-ray scattering by the atoms and lattice points of the crystal is coherent and elastic; (2) the interaction between the scattered X-ray photons and the nuclei is ignored; (3) the scattering is treated as a far-field phenomenon so that both the incident and the scattered waves are considered as plane waves; (4) multiple scattering effects are negligibly small and the scattered waves are not subject to rescattering from other atoms and lattice points. This implies that the interaction between the incident and diffracted waves is not considered. Under these assumptions, the integrated intensity, as described in Chap. 2, could not give a satisfactory account of the measured intensity. In the so-called mosaic crystal model proposed by Darwin [70], a real crystal is considered to be composed of a large number of small perfect crystal blocks slightly misaligned with one another. A mathematical function, the so-called mosaic spread distribution, is often used to describe quantitatively the distribution of mosaic blocks as a function of the angle ϕ between two adjacent blocks. Usually Gaussian and Lorentzian distribution functions are employed to describe the mosaic structure in real crystals. For simplicity, the mosaic spread $W(\phi)$ is assumed to be Gaussian:

$$W(\phi) = \frac{1}{\sqrt{2\pi}\eta} \exp\left(-\frac{\phi^2}{2\eta^2}\right) , \tag{5.1}$$

where η is the standard deviation. Equation (5.1) represents the distribution of the perfect blocks with the angular deviation ϕ from the mean value of the mosaic spread.

The assumption that the interaction between the incident and the diffracted waves is negligibly small implies that only intensities or reflecting powers need to be considered in the kinematical approach. In the mosaic crystal model, the incident and the diffracted powers are attenuated due to ordinary absorption and diffraction. The attenuation caused by absorption is related to the usual exponentially decayed function of the product of the linear absorption coefficient and the crystal thickness. The attenuation resulted from diffraction is due to extinctions. The improved kinematical approach for two-wave diffraction based on the mosaic model, proposed by Hamilton [467] and Zachariasen [48, 468, 469], deals with the power-transfer equations, the so-called Hamilton-Darwin transfer equations, to take care of the change in reflecting powers between the transmitted and the diffracted waves in the irradiated crystal blocks. In fact, a similar consideration for multiple-wave diffraction was first proposed by Renninger in 1937 [4]. In his 1937 paper, Renninger discussed two kinematical approaches, the simplest 'Einfachster Ansatz' and the sophisticated 'Verfeinerter Ansatz' [245]. The latter resembles the treatment of power-transfer for diffraction processes. In the following, we shall consider the power-transfer equation for the two-wave and multiple-wave diffraction cases.

5.1 Power-Transfer Equation

Consider first a simple two-wave (O, G) Laue diffraction from a plane-parallel crystal plate, where O and G stand for the incident and the Laue diffracted wave, respectively (see Fig. 5.1(a)). The crystal is composed of many plate-like ideal crystal blocks. Each individual block diffracts X-rays independently due to the misalignment among the blocks. In other words, there is no definite phase relationship among the scattered waves from different blocks. If one considers the diffraction from a crystal layer of the thickness $\mathrm{d}x$ located in the depth x from the crystal surface along the inward surface normal $\hat{\boldsymbol{n}}_e$, the coupling in reflecting power between the O and G waves is governed by the differential equations [48]:

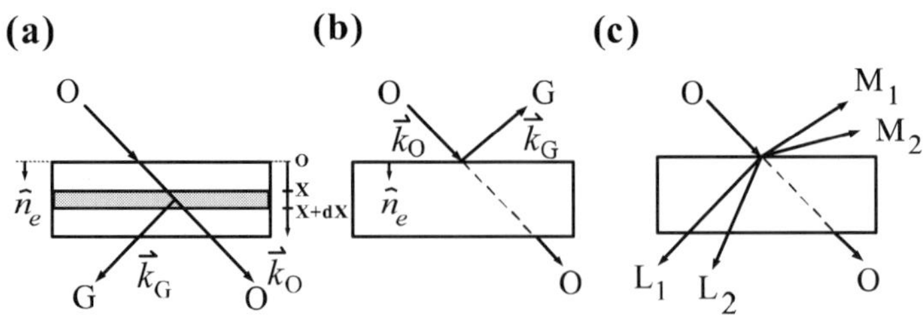

Fig. 5.1. Schematic of (**a**) two-wave Laue diffraction, (**b**) two-wave Bragg reflection, and (**c**) multiple-wave diffraction

$$\frac{\mathrm{d}P_O}{\mathrm{d}x} = -\mu_0 \frac{P_O}{\gamma_O} - \frac{Q_{OG}P_O}{\gamma_O} + \frac{Q_{GO}P_G}{\gamma_G} , \tag{5.2a}$$

$$+\frac{\mathrm{d}P_G}{\mathrm{d}x} = -\mu_0 \frac{P_G}{\gamma_G} - \frac{Q_{GO}P_G}{\gamma_G} + \frac{Q_{OG}P_O}{\gamma_O} , \tag{5.2b}$$

where P_O and P_G are the reflecting powers of the O and G waves, μ_0 is the linear absorption coefficient, and γ_O and γ_G are the absolute values of the direction cosines of the wavevectors $\boldsymbol{k}_O$ and $\boldsymbol{k}_G$ of the incident O wave and the diffracted G wave with respect to the surface normal $\hat{\boldsymbol{n}}_e$. Q_{OG} and Q_{GO} are the linear reflection coefficients of the reflection from the incident O wave to the reflected G wave and vice versa. Q_{OG} is defined as [48]

$$Q_{OG} = W(\theta - \theta_G) R_{OG} n_0 , \tag{5.3}$$

where W is the mosaic spread distribution given in (5.1) and θ_G the Bragg angle of the G reflection. R_{OG} is the integrated reflecting power defined as the product of the reflecting power and the angular deviation of a single block. n_0 is the number of single crystal blocks per unit length in depth.

Equation (5.2a) indicates that the power transfer of the O wave per unit length along the inward surface normal is equal to the sum of the power loss and gain due to absorption and diffraction. The power loss is related to the normal absorption μ_0 (the first term) and the extinction associated with the reflection from the O wave to the G wave (the second term). The power gain results from the power increase via the reflection from the G wave to the O wave (the third term). It is similar for the power variation of the Laue diffracted G wave given in the second equation, (5.2b). The 'plus' sign on the left side of the equation indicates that the reflection is of a transmission type. For the diffraction of a Bragg reflection type (see Fig. 5.1(b)), a 'minus' is attached to the first term of the equation, because the direction cosine, γ_G, of the wavevector $\boldsymbol{k}_G$ of the diffracted wave G with respect to $\hat{\boldsymbol{n}}_e$ is negative. Once the 'minus' sign is adopted, all the direction cosines given in (5.2a) and (5.2b) are taken as their absolute values.

Equation (5.2a) and (5.2b) can be generalized for a multiple-wave case involving many Bragg type reflections M_1, M_2, ... and Laue transmissions L_1, L_2, ... (see Fig. 5.1(c)) as [4, 26, 53, 230–245, 470]

$$\pm\frac{\mathrm{d}P_i}{\mathrm{d}x} = -\frac{\mu_0 P_i}{\gamma_i} + \sum_j \left(\frac{Q_{ji}P_j}{\gamma_j} - \frac{Q_{ij}P_i}{\gamma_i} \right) \tag{5.4}$$

for an involved wave i, where the sign of the first term is positive for Laue transmission and negative for Bragg reflection. The summation of the second term on the right side of the equation is taken over all the involved waves j. Q_{ij} is the linear reflecting power of the $j - i$ reflection. For an N-wave diffraction, there are N linear differential equations of the following simple form [230, 231]:

$$\frac{\mathrm{d}P_i}{\mathrm{d}x} = -\sum_j a_{ij} P_j , \tag{5.5}$$

with the solution for the reflecting power as

$$P_i = \sum_j A_{ij} \exp(Z_j x) . \tag{5.6}$$

By substituting (5.6) into (5.5), the secular equation consisting of a system of N linear equations of the following form

$$(a_{ii} - Z_m) A_{im} + \sum_{j \neq i} a_{ij} A_{jm} = 0 \tag{5.7}$$

is obtained. The coefficients a_{ii} and a_{ij} are expressed in terms of the coefficients of (5.4) as

$$a_{ii} = -\frac{\mu_0 + \sum_j Q_{ij}}{\gamma_i} , \qquad a_{ij} = \frac{Q_{ji}}{\gamma_j} . \tag{5.8}$$

The secular equation (5.7) can be solved as an eigenvalue equation. As usual, the eigenvector yields the ratio between A_{im} and A_{jm} for each eigenvalue Z_m. The absolute values of A_{im} and A_{jm} can be determined from the two boundary conditions:

At the entrance surface, $x = 0$:

$$P_O(0) = P_O^e \qquad \text{and} \qquad P_i(0) = 0 \tag{5.9}$$

for any involved Laue transmission i except for the incident O reflection, where P_O^e, the given incident power, is usually set to be unity.

At the exit surface, $x = T$:

$$P_i(T) = 0 \tag{5.10}$$

for any involved Bragg type reflection i, where T is the thickness of the crystal plate. Once the coefficients A_{ij}'s are known, the reflecting powers P's can be calculated accordingly.

5.1.1 Two-Wave Laue Diffraction and Bragg Reflection

Consider a general two-wave (O, G) Laue (transmission) diffraction, in which $\gamma_G \neq \gamma_O$ and $Q_{OG} \neq Q_{GO}$. The power-transfer equations, (5.2a) and (5.2b), hold. According to (5.7), the eigenvalues Z_j and the eigenvectors related to the coefficients A's are

$$Z_1 = -\frac{\mu_0}{\gamma_O} , \tag{5.11a}$$

$$Z_2 = -\frac{\mu_0}{\gamma_O} - \left(\frac{Q_{GO}}{\gamma_G} + \frac{Q_{OG}}{\gamma_O} \right) , \tag{5.11b}$$

and

$$[A_O(1) \quad A_G(1)] = \left[\frac{Q_{GO}}{\gamma_G} \quad \frac{Q_{OG}}{\gamma_O} \right] , \tag{5.12a}$$

$$[A_O(2) \quad A_G(2)] = [1 \quad -1] . \tag{5.12b}$$

Here the eigenvectors are expressed in a horizontal (transpose) form. After employing the boundary condition, (5.9), the corresponding reflecting powers are obtained:

$$P_O(x) = P_O^e \left\{ \left[\frac{\frac{Q_{GO}}{\gamma_G}}{\frac{Q_{GO}}{\gamma_G} + \frac{Q_{OG}}{\gamma_O}} \right] \exp\left[-\frac{\mu_0}{\gamma_O} x \right] + \left[\frac{\frac{Q_{OG}}{\gamma_O}}{\frac{Q_{GO}}{\gamma_G} + \frac{Q_{OG}}{\gamma_O}} \right] \right.$$
$$\left. \times \exp\left(\left[-\frac{\mu_0}{\gamma_O} - \left(\frac{Q_{GO}}{\gamma_G} + \frac{Q_{OG}}{\gamma_O} \right) \right] x \right) \right\} , \qquad (5.13)$$

$$P_G(x) = P_O^e \left\{ 2 \left(\frac{\frac{Q_{OG}}{\gamma_O}}{\frac{Q_{GO}}{\gamma_G} + \frac{Q_{OG}}{\gamma_O}} \right) \sinh \left[\left(\frac{\frac{Q_{GO}}{\gamma_G} + \frac{Q_{OG}}{\gamma_O}}{2} \right) x \right] \right.$$
$$\left. \times \exp\left(- \left[\frac{\mu_0}{\gamma_O} + \frac{\frac{Q_{GO}}{\gamma_G} + \frac{Q_{OG}}{\gamma_O}}{2} \right] x \right) \right\} . \qquad (5.14)$$

If this Laue diffraction is symmetric, i.e., $\gamma_G = \gamma_O$, and if Friedel's law holds, i.e., $Q_{OG} = Q_{GO}$, then (5.13) and (5.14) become

$$P_O(x) = P_O^e \cosh(Q_{OG} x) \exp\left[- \left(\frac{\mu_0 + Q_{OG}}{\gamma_O} \right) x \right] , \qquad (5.15)$$

$$P_G(x) = P_O^e \sinh(Q_{OG} x) \exp\left[- \left(\frac{\mu_0 + Q_{OG}}{\gamma_O} \right) x \right] . \qquad (5.16)$$

Consider now a two-wave (O, G) Bragg reflection. The power-transfer equations, (5.2a) and (5.2b), in this case still hold, except that a 'minus' sign needs to be added to the first term of (5.2b). The eigenvalues and eigenvectors of the power-transfer equations are

$$Z_{1,2} = \frac{1}{2} \left(\frac{\mu_0 + Q_{GO}}{\gamma_G} - \frac{\mu_0 + Q_{OG}}{\gamma_O} \right) \pm U \qquad (5.17)$$

and

$$[A_O(j) \;\; A_G(j)] = \left[\frac{Q_{GO}}{\gamma_G} \;\; \frac{1}{2} \left(\frac{\mu_0 + Q_{GO}}{\gamma_G} + \frac{\mu_0 + Q_{OG}}{\gamma_O} \right) \pm U \right] \qquad (5.18)$$

for $j = 1$ and 2. With appropriate boundary conditions, (5.10), the transmitted power of the O wave at the depth x and the reflecting power of the G wave from the crystal surface take the following forms

$$P_O(x) = P_O^e \exp[(d_A - U)x]$$
$$\times \left\{ 1 + \frac{(d_B - U) \exp[-U(3x + T)] \sinh(Ux)}{d_B \sinh(UT) + U \cosh(UT)} \right\} , \qquad (5.19)$$

$$P_G(0) = P_O^e \left(\frac{\frac{Q_{OG}}{\gamma_O}}{\frac{Q_{GO}}{\gamma_G}} \right) \left(\frac{d_B - U}{\frac{Q_{OG}}{\gamma_O}} \right) \left\{ 1 - \frac{U \exp(-UT)}{d_B \sinh(UT) + U \cosh(UT)} \right\} , \qquad (5.20)$$

where

$$d_A = \frac{1}{2}\left(\frac{\mu_0 + Q_{GO}}{\gamma_G} - \frac{\mu_0 + Q_{OG}}{\gamma_O}\right) , \tag{5.21}$$

$$d_B = \frac{1}{2}\left(\frac{\mu_0 + Q_{GO}}{\gamma_G} + \frac{\mu_0 + Q_{OG}}{\gamma_O}\right) , \tag{5.22}$$

$$U = \sqrt{d_B^2 - \frac{Q_{OG}Q_{GO}}{\gamma_O\gamma_G}} . \tag{5.23}$$

For a symmetric Bragg reflection, $Q_{OG} = Q_{GO}, \gamma_G = \gamma_O, d_A = 0$. In addition, d_B and U need to be replaced by d_o and U_o defined as

$$d_o = \frac{\mu_0 + Q_{OG}}{\gamma_O} , \tag{5.24}$$

$$U_o = \sqrt{d_o^2 - \frac{Q_{OG}^2}{\gamma_O^2}} . \tag{5.25}$$

5.1.2 Multiple-Wave Diffraction

The solutions to the power-transfer equations for multiple-wave diffraction, (5.4), can be found by treating them as being involved in an eigenvalue problem as stated and demonstrated for two-wave cases. However, analytical expressions for eigenvalues Z's cannot always be derived for a general multiple-wave situation. Zachariasen [230, 231] demonstrated that analytical expressions could be obtained for symmetric transmission diffractions, where the involved reciprocal lattice points form a symmetric polygon. For example, let us consider a symmetric three-wave (O, G, L) diffraction, where the moduli of the reciprocal lattice vectors of G, L, and $G-L$ are equal, i.e., $|g| = |\ell| = |g-\ell|$, and $\gamma_O = \gamma_G = \gamma_L$. In addition, the reflection coefficients, or reflectivities, of the reflections G, L, $L-G$, and $G-L$ are Q_1, Q_2, Q_3, and Q_3. The secular equation for a non-trivial solution of (5.4) is then expressed as

$$\begin{vmatrix} -\mu_0 - 2Q_1 - Z & Q_1 & Q_1 \\ Q_1 & -\mu_0 - Q_1 - Q_3 - Z & Q_3 \\ Q_1 & Q_3 & -\mu_0 - Q_1 - Q_3 - Z \end{vmatrix} = 0 . \tag{5.26}$$

That the determinant in (5.26) is symmetric provides the condition to find the eigenvalues analytically. Following the same derivation as in two-wave cases, the reflecting powers at the exit surface of the crystal, i.e., $x = T$, are

$$P_O(T) = \frac{1}{3}\left[1 + \exp(-3Q_1T)\right]\exp(-\mu_0 T) , \tag{5.27}$$

$$P_G(T) = P_L(T) = \frac{1}{3}\left[1 - \exp(-3Q_1T)\right]\exp(-\mu_0 T) . \tag{5.28}$$

The integrated reflections can be calculated accordingly by considering crystal mosaic spread.

The power-transfer equations for symmetric four- and six-wave diffractions can be solved analytically by the same token [230, 231]. For a general multiple-wave diffraction situation, numerical methods are usually employed to find correct eigenvalues and eigenvectors as well as the reflecting powers.

All the reflecting powers described in (5.2a), (5.2b), (5.4), (5.13), (5.14), (5.19), (5.20), and (5.26) indicate that the attenuation in power results from the normal absorption and the loss due to diffraction. Therefore, the effective linear absorption coefficient μ_e consists of the normal absorption μ_0 and the extinction proportional to the reflectivities Q's:

$$\mu_e = \mu_0 + \sum Q_{ij} \ . \tag{5.29}$$

The sum taken over the involved reflections is nothing more than the power loss per unit path length in the multiple-wave diffraction. It is also called the secondary extinction coefficient when the power attenuation among the mosaic blocks is considered. A similar effect taking place within a single perfect crystal block is the primary extinction [70], which depends on the thickness of the block and the structure factors of the involved reflections. The secondary extinction is related to the diffraction power loss among blocks; therefore it depends on the primary extinction and the mosaic distribution. Based on this discussion, the formulation dealing with power transfer in multiple-wave diffraction processes is within the framework of the secondary extinction theory. In other words, extinction effects are taken into account in the kinematical approach for multiple-wave diffraction. In the literature, there have been many theoretical approaches devoted to the extinction effects in two-wave cases [144, 145, 468, 471–477]. The same formulation can be adopted and generalized for multiple-wave cases, provided that the crystals dealt with fall into the category assumed by the theories.

5.2 Approximate Solutions to the Power-Transfer Equation

The exact solution to (5.4) for a general multiple-wave diffraction cannot always be found analytically. An alternative is to obtain a proper approximate solution according to the experimental situations. If the divergence of the incident wave is much larger than the diffraction width of a perfect crystal and much smaller than the mosaic spread, then the secondary extinction is small. Namely, $\mu_e \ell \ll 1$, or

$$\mu_0 \ell_i \ll 1 \qquad \text{and} \qquad Q_{ij} \ell_i \ll 1 \ , \tag{5.30}$$

where ℓ_i is the path length of the diffracted wave i in the crystal. For a plane parallel to the crystal plate of thickness T,

$$\ell_i = \frac{T}{\gamma_i} . \tag{5.31}$$

By considering the mosaic distribution, the reflectivity Q_{ij} of (5.4) can be expressed as [26]

$$Q_{ij} = W(\Delta\theta_{ij}) \left(\frac{\lambda^3 r_e^2 |F_{ij}|^2}{V_c^2} \right) \frac{1}{\sin 2\theta_{ij}} , \tag{5.32}$$

where $\Delta\theta_{ij}$ is the deviation from the exact Bragg angle θ_{ij} of the $(j-i)$ reflection, λ the X-ray wavelength, V_c the volume of the unit cell, r_e the classic radius of electron, and F the structure factor. The subscript ij indicates the $j-i$ reflection with the incident wave i and the reflected wave j.

The approximate solution of (5.4) is assumed to be a Taylor series expanded about $x=0$ at the entrance surface of the crystal plate,

$$P_i(T) = \sum_{n=0}^{\infty} \frac{T^n}{n!} P_i^{(n)}(0) \tag{5.33}$$

for all the diffracted waves i, where $P_i^{(n)}$ is the n-th derivative of P_i with respect to x. By substituting (5.33) into (5.4) and employing the appropriate boundary conditions like (5.9) for Laue transmission and (5.10) for Bragg reflection, the power-transfer equations can be solved as a system of linear equations with the coefficients as functions of the reflectivities and the crystal thickness.

For illustration, we consider the following three-wave (O, G, L) cases with the Taylor series up to the second-order derivatives:

(A) Three transmission reflections
The boundary conditions, $P_O(0) = 1$ and $P_G(0) = P_L(0) = 0$, lead the diffracted power of wave M to be

$$\begin{aligned} P_M(T) = Q_{OM}\ell_O - \frac{1}{2} Q_{OM}\ell_O \big[&\mu_0\ell_O + \mu_0\ell_M + Q_{OM}\ell_O + Q_{MO}\ell_M \\ &+ \sum_j (Q_{Oj}\ell_O + Q_{Mj}\ell_M) \big] + \frac{1}{2}\sum_j Q_{Oj}\ell_O Q_{jM}\ell_j , \end{aligned} \tag{5.34}$$

where $j \neq O, M$, and $M = G, L$. If those terms involving j are set to zero, then the reflecting power of the two-wave M reflection is obtained:

$$P'_M(T) = Q_{OM}\ell_O - \frac{1}{2} Q_{OM}\ell_O \left[\mu_0\ell_O + \mu_0\ell_M + Q_{OM}\ell_O + Q_{MO}\ell_M \right] . \tag{5.35}$$

This result can also be obtained from (5.14) with the assumption of small extinction. For simplicity, we assume reflection G is the primary reflection and reflection L the secondary reflection. The difference between P_G and P'_G is the variation in the reflecting power of the primary reflection G in the presence of the secondary reflection L, i.e.,

$$\Delta P_G(T) = \frac{1}{2}\sum_L (-Q_{OG}\ell_O Q_{OL}\ell_O - Q_{OG}\ell_O Q_{GL}\ell_G + Q_{OL}\ell_O Q_{LG}\ell_L) \ . \tag{5.36}$$

This is the same as that derived from Renninger's simplest approach [4,245]. If $\Delta P_G > 0$, the diffraction is of the Umweg type. If $\Delta P_G < 0$, the diffraction is of the Aufhellung type. In other words, a peak and a dip with respect to the two-wave power P'_G are expected to be observed respectively. The three terms on the right of (5.36) indicate three successive reflections. The first term implies that during the diffraction some of the incident power goes to the L diffracted wave. This power transfer is a loss for the G reflection, thus leading to the minus sign. The second term involves the successive reflections of first the G and then the L reflection. For the G wave, power loss takes place because there is a power transfer from the G wave to the L wave through the coupling $(L-G)$ reflection. In the third term, the incident wave is first diffracted into the L wave, and then the L wave is diffracted into the G wave; hence there is a power gain for the G wave. In addition, the simultaneous presence of the G–L and L–G reflections in (5.36) signifies that the multiple scattering involving both the forward G–L reflection and the backward L–G reflection has been considered. Moreover, (5.36) involves the G, L, and G–L reflections. If two of the three reflections are forbidden, then $\Delta P_G = 0$. This means that no three-wave diffraction takes place. On the other hand, if only the primary G reflection is forbidden, then the diffraction peaks are always observable because $\Delta P_G > 0$ due to the only surviving last term.

(B) Two transmission reflections (O, L) and one Bragg reflection (G)
The boundary conditions, $P_O(0) = 1$, $P_G(0) \neq 0$, $P_L(0) = 0$, $P_G(T) = 0$, and the Taylor series up to the second-order terms give the reflecting power of the reflection G as

$$P_G(0) = \frac{N_2}{D_2} \ , \tag{5.37}$$

where

$$N_2 = Q_{OG}\ell_O + \frac{1}{2}Q_{OG}\ell_O\,(A_G\ell_G - A_O\ell_O) + \frac{1}{2}Q_{OL}Q_{LG}\ell_O\ell_L \ , \tag{5.38a}$$

$$D_2 = 1 + A_G\ell_G + \frac{1}{2}\left(A_G^2\ell_G^2 + Q_{GO}^2\ell_O\ell_G - Q_{GL}^2\ell_G\ell_L\right) \ , \tag{5.38b}$$

and

$$A_O = \mu_0 + Q_{OG} + Q_{OL} \ , \tag{5.39a}$$

$$A_G = \mu_0 + Q_{GO} + Q_{GL} \ . \tag{5.39b}$$

Similarly, the reflecting power of the two-wave diffraction G is

$$P'_G(0) = \frac{N'_2}{D'_2} \ , \tag{5.40}$$

where

$$N_2' = Q_{OG}\ell_O + \frac{1}{2}Q_{OG}\ell_O\left(\mu_0\ell_G + Q_{GO}\ell_G - \mu_0\ell_O + Q_{OG}\ell_O\right) , \quad (5.41a)$$

$$D_2' = 1 + (\mu_0\ell_G + Q_{GO}\ell_G) + \frac{1}{2}\left[(\mu_0 + Q_{GO})^2 \ell_G^2 - Q_{GO}^2\ell_O\ell_G\right]$$
$$\sim 1 + \mu_0\ell_G + Q_{GO}\ell_G . \quad (5.41b)$$

The difference ΔP_G between $P_G(0)$ and $P_G'(0)$ can be calculated according to (5.37) and (5.40).

(C) One transmission reflection (O) and two Bragg reflections (G, L)
The boundary conditions, $P_O(0) = 1$, $P_G(0) \neq 0$, $P_L(0) \neq 0$, $P_G(T) = 0$, $P_L(T) = 0$, and the Taylor series up to the second-order terms lead the reflecting powers of the G and L reflections to be

$$P_G(0) = \frac{a_G b_G + b_G c_G}{\Delta} , \qquad P_L(0) = \frac{a_L b_G + b_L c_G}{\Delta} , \quad (5.42)$$

where

$$\Delta = c_G c_L - a_G a_L ,$$
$$a_G = Q_{GL}\ell_L + \left(Q_{OG}Q_{OL}\ell_O\ell_L + A_G Q_{LG}\ell_G\ell_L + A_L Q_{LG}\ell_L^2\right) ,$$
$$b_G = Q_{OG}\ell_O\left[1 + \frac{1}{2}(A_G\ell_G - A_O\ell_O)\right] - \frac{1}{2}Q_{LO}Q_{LG}\ell_O\ell_L ,$$
$$c_G = 1 + A_G\ell_G + \left(A_G^2\ell_G^2 - Q_{GO}^2\ell_O\ell_G\right) + \frac{1}{2}Q_{GL}^2\ell_G\ell_L , \quad (5.43)$$

and a_L, b_L and c_L have the same forms as a_G, b_G and c_G except that the subscripts G and L need to be interchanged.

The reflecting powers of two-wave cases and their differences from the three-wave cases can be derived in a similar way as in case B.

It should be noted that under the assumption of small extinction, i.e., $\mu_0\ell \ll 1$ and $Q\ell \ll 1$, the second-order terms in the denominators of (5.37), (5.40), and (5.42) can be neglected. Thus (5.37) and (5.42) reduce to (5.34). Therefore, with the second-order approximation, (5.36) for $\Delta P_G(T)$ is valid for all multiple-wave diffractions in a crystal plate, irrespective of the diffraction type (Bragg or Laue) and the boundary conditions. In other words, with the second-order approximation, the three successive reflections mentioned are the dominant processes that cause the variation of reflecting power in multiple-wave diffractions.

5.3 Higher-Order Approximation and Iterative Calculation

The assumption that $\mu_0\ell \ll 1$ and $Q\ell \ll 1$ is valid mostly for neutrons but partially for X-rays. The condition of $Q\ell \ll 1$ is usually fulfilled for X-ray and neutron diffractions if weak reflections are involved. However, the assumption of $\mu_0\ell \ll 1$ is not valid for X-rays. For example, when a highly absorbing crystal is considered for multiple-wave diffraction, $\mu_0\ell$ can be close to unity, which invalidates the assumption. In this situation, higher-order approximations in terms of $\mu_0\ell$ and $Q\ell$ need to be taken into account in solving the power-transfer equations. In the following, instead of mentioning third-, fourth-, and higher-order approximations one by one, a general procedure involving n-th-order derivatives and an iterative calculation scheme are presented briefly.

The assumed solution for the reflecting power $P_i(d_1)$ at depth d_1 in the crystal plate is again a Taylor series expansion about $x = 0$ given in (5.33). From (5.4), the n-th-order derivatives with respect to x can be expressed as functions of the powers of all the participating reflections [235]:

$$P_i^{(n)}(x) = \sum_{j_1}\sum_{j_2}\cdots\sum_{j_n} Y_{ij_1}Y_{j_1j_2}\cdots Y_{j_{n-1}j_n}P_{j_n}(x) , \tag{5.44}$$

where the j's are the involved reflections and

$$\begin{aligned} Y_{j_pj_q} &= \frac{S_{j_q}Q_{j_pj_q}}{\gamma_{j_q}} \quad \text{for} \quad j_p \neq j_q , \\ Y_{j_pj_p} &= -\frac{S_{j_p}A_{j_p}}{\gamma_{j_p}} \quad \text{for} \quad j_p = j_q , \end{aligned} \tag{5.45}$$

with

$$A_{j_p} = \mu_0 + \sum_{j_r \neq j_p} Q_{j_pj_r} , \tag{5.46}$$

and

$$S_{j_p} = \begin{cases} + & \text{for transmitted wave} , \\ - & \text{for reflected wave} . \end{cases} \tag{5.47}$$

From (5.33) and (5.44), the reflecting power at depth $x = d_1$ can be written as

$$P_i(d_1) = \sum_j a_{ij}(d_1)P_j(0) , \tag{5.48}$$

where

$$a_{ij}(d_1) = \sum_{n=0}^{\infty}\frac{d_1^n}{n!}\sum_{j_1}\sum_{j_2}\cdots\sum_{j_{n-1}} Y_{ij_1}Y_{j_1j_2}\cdots Y_{j_{n-1}j} . \tag{5.49}$$

By employing the boundary conditions for the incident wave O, the Laue transmissions L's, and the Bragg reflections M's:

$$P_O(0) = 1 , \qquad P_L(0) = 0 , \qquad P_M(d_1) = 0 , \tag{5.50}$$

where L and M cover all the transmissions and reflections involved in the multiple-wave diffraction, the approximate solutions can be found as usual by solving a linear equation system [260]. Since the procedure for solving the linear equations is straightforward, we shall skip the derivation here.

In the iterative calculation, the appropriate Lorentz-polarization factors for the n-th-order successive reflections involved in (5.44) need to be included. This means that in each calculation cycle, a new reflection has to be added to the successive reflections and the polarization factor has to be calculated by following the ordering of successive reflections, i.e., $i-j_1$, j_1-j_2, j_2-j_3, ... The details about the polarization factor for a successive reflection can be found in Sect. 3.3. Although the general description of Lorentz factor is also given in that section, the resultant Lorentz factor for multiple-wave diffraction will be discussed next.

5.4 Integrated Intensity and X-Ray Path Length in Crystals

The reflecting powers can be calculated based on those approaches described in the previous sections, where the reflectivity Q_{ij} defined in (5.32) as a function of $\Delta\theta_{ij}$ is used. If $\Delta\theta_{ij}$ is set to zero, then the calculation leads to the so-called peak intensity, which can, in principle, be compared with the measured peak intensity. In reality, it would be more appropriate to compare the calculated integrated intensity over $\Delta\theta_{ij}$ with the experimental measured intensity integrated over the beam divergence and crystal rotation angle. Because multiple-wave diffraction involves two angular rotations, the Bragg angle θ and the azimuth angle ψ, intensity calculations integrated over θ and ψ are necessary. For the θ integration, the θ_{ij} and the quantity $1/\sin\theta_{ij}$ given in (5.32) are the Bragg angle and the Lorentz factor for the Bragg angle rotation, respectively. For the azimuth rotation, the reflectivity takes the form similar to (5.32) as

$$Q'_{ij} = W\left(\Delta\theta_{ij}\right)\left(\frac{\lambda^2 r_e^2 |F|^2}{V_c^2}\right)\left(\frac{1}{\sin\psi\cos\chi\cos\xi}\right) , \tag{5.51}$$

where the term inside the last parentheses is the Lorentz factor L_F of the ψ rotation given in (3.10). The angles ε, ψ, χ, and ξ, are defined in Fig. 3.4, where ε is related to ψ. Referring to Fig. 3.4 and replacing G and L by i and j respectively, we obtain the following relation [26, 48]:

$$\Delta\theta_{ij} = \left(\frac{\sin\psi\cos\chi\cos\xi}{\sin 2\theta_{ij}}\right)\Delta\varepsilon = K_{ij}^{\varepsilon}\Delta\varepsilon , \tag{5.52}$$

where θ_{ij} is the Bragg angle of the $(j-i)$ reflection and $\Delta\varepsilon = \Delta\psi/2$ is the deviation from the mean value of ψ. Giving a normalized mosaic distribution in terms of $\Delta\varepsilon$

$$W(\Delta\varepsilon) = \frac{K_{ij}^{\varepsilon}}{\sqrt{2\pi}\eta}\exp\left[-\frac{\left(K_{ij}^{\varepsilon}\Delta\varepsilon\right)^2}{2\eta^2}\right] , \tag{5.53}$$

the reflectivity can be expressed as

$$Q'_{ij} = \frac{Q_{ij}}{K^{\varepsilon}_{ij}} , \tag{5.54}$$

where Q_{ij} is the reflectivity defined in (5.32). For a reflection $(j-i)$ with the reciprocal lattice vector $\boldsymbol{g}_{j-i}(=\boldsymbol{g}_j - \boldsymbol{g}_i)$ parallel to the rotation axis, the expression for K^{ε}_{ij} given in (5.52) is not valid, because $\Delta\theta_{ij}$ tends to approach zero due to $\psi = 0$. In reality, $\Delta\theta_{ij}$ is not zero. Under this circumstance, the following equation is useful to give a correct K^{ε}_{ij} value:

$$K^{\varepsilon}_{ij} = \frac{1}{\sin\theta_{ij}\cos\theta_{ij}} \left[K^{\varepsilon}_{iO}\left(\sin\theta_i\cos\theta_i - \cos\theta_i\sin\theta_j\cos\alpha_{ij}\right) \right.$$
$$\left. + K^{\varepsilon}_{jO}\left(\sin\theta_j\cos\theta_j - \sin\theta_i\cos\theta_j\cos\alpha_{ij}\right)\right] . \tag{5.55}$$

The Bragg law:

$$2\sin\theta_{ij} = \lambda\left|\boldsymbol{g_j} - \boldsymbol{g_i}\right| , \qquad 2\sin\theta_j = \lambda\left|\boldsymbol{g_j}\right| , \tag{5.56}$$

and the relation

$$\sin^2\theta_{ij} = \sin^2\theta_i + \sin^2\theta_j - 2\sin\theta_i\sin\theta_j\cos\alpha_{ij} \tag{5.57}$$

have been employed for the derivation of (5.55). α_{ij} is the angle between $\boldsymbol{g}_i$ and $\boldsymbol{g}_j$.

The differences in reflection power between the multiple-wave peak (or dip) and the background of a given primary reflection G integrated with respect to θ and ψ yield the integrated intensities R_ψ and R_θ [26],

$$R_\psi = \frac{\int \Delta P_G\left(\theta_G, \psi\right) \mathrm{d}\psi}{\int P_G\left(\theta, \psi_A\right) \mathrm{d}\theta} \tag{5.58}$$

and

$$R_\theta = \frac{\int \Delta P_G\left(\theta, \psi_o\right) \mathrm{d}\theta}{\int P_G\left(\theta, \psi_A\right) \mathrm{d}\theta} , \tag{5.59}$$

where the common denominator, representing the background of the primary reflection, is the intensity of the G reflection integrated over θ at ψ_A off the exact multiple-wave diffraction position ψ_o. The numerator of (5.58) is the integrated intensity at the two-wave Bragg position θ_G over ψ while that of (5.59) is the integrated intensity at the exact multiple-wave position ψ_o over θ. With the consideration of the Lorentz-polarization factors for the successive reflections involved in ΔP_G given in (5.36), the integrated intensities become

$$R_\psi = \frac{1}{2}\frac{p_G Q_{OG}\ell_O}{\sqrt{2\pi}\eta} \sum_L \left(- \frac{C_{GL}(L-G)}{K^{\psi}_{OL}} \frac{Q_{OL}}{Q_{OG}} \right.$$
$$- \frac{C_{G,G\text{-}L}[(G-L)-G]Q_{GL}\ell_G}{K^{\psi}_{GL}Q_{OG}\ell_O}$$
$$\left. + \frac{C_{L,L\text{-}G}[(L-G)-L]Q_{OL}Q_{LG}\ell_L}{\sqrt{\left(K^{\psi}_{OL}\right)^2 + \left(K^{\psi}_{LG}\right)^2} Q^2_{OG}\ell_O} \right) \tag{5.60}$$

and

$$R_\theta = \frac{1}{2}\frac{p_G Q_{OG}\ell_O}{\sqrt{2\pi}\eta}\sum_L \left(-\frac{C_{GL}(L-G)}{\sqrt{1+\left(K_O^\theta\right)^2}}\frac{Q_{OL}}{Q_{OG}} \right.$$
$$- \frac{C_{G,G\text{–}L}[(G-L)-G]Q_{GL}\ell_G}{\sqrt{1+\left(K_{GL}^\theta\right)^2}Q_{OG}\ell_O}$$
$$\left. + \frac{C_{L,L\text{–}G}[(L-G)-L]Q_{OL}Q_{LG}\ell_L}{\sqrt{\left(K_{OL}^\theta\right)^2+\left(K_{LG}^\theta\right)^2}Q_{OG}^2\ell_O} \right) , \tag{5.61}$$

where $K_{OG}^\psi = 0$ and $K_{OG}^\theta = 1$ have been used in the derivation. p_G is the two-wave polarization factor of the G reflection, defined in (3.14), and $C_{ij}[j-i]$ with $i, j = G, L, G\text{–}L$, and $L\text{–}G$, defined in (3.21), is the polarization factor for the corresponding successive reflection given in Sect. 3.3. It should be noted that for the higher-order approximation and the iterative calculation for ΔP_G, the resultant Lorentz factor for an n-th order successive reflection for the ψ and θ rotations

$$L_n^{\psi,\theta} = \left[\sum_{i=1}^{n} \left(K_{i-1,i}^{\psi,\theta} \right)^2 \right]^{-\frac{1}{2}} \tag{5.62}$$

needs to be considered. When $\Delta\varepsilon = 0$ and all the Lorentz factors are set to equal to zero, (5.60) leads to the peak intensity ratio

$$R_P = \frac{\Delta P_G\left(\theta_G, \psi_o\right)}{P_G\left(\theta_G, \psi_A\right)} . \tag{5.63}$$

In addition to the Lorentz-polarization factors, the X-ray path lengths ℓ's also play a very important role in affecting the intensities, as shown in (5.60), (5.61), and (5.63). For example, neutron multiple-wave diffraction experiments have shown that appreciable intensity changes result from the modification of the neutron-beam path lengths [26]. According to (5.31), the path length depends on the thickness T for a thin plate-like crystal provided that the neutron or the X-ray beam can penetrate through the crystal. Usually neutrons are more penetrating than X-rays. Equation (5.31) may not be valid for highly absorbing crystals for X-rays. In other words, absorption is a key factor influencing the X-ray path length. If the crystal is of any arbitrary shape with an absorption function denoted by A, then the average path length could be defined as [230, 231]

$$\langle \ell \rangle = \left\langle \frac{x}{\gamma} \right\rangle = A\frac{\mathrm{d}A^*}{\mathrm{d}\mu} , \tag{5.64}$$

where μ is the linear absorption coefficient and $A^* = A^{-1}$. For example, for a crystal plate of thickness T, the absorption function could take the form

$$A = \exp(-\mu_0 T) \tag{5.65}$$

for a transmitted wave [230, 231], and

$$A = \frac{\gamma}{\mu_0 T}\left[1 - \exp\left(-\frac{\mu_0 T}{\gamma}\right)\right] \tag{5.66}$$

for a reflected wave [232], where the averaged $1/\gamma = 1/\gamma_O + 1/\gamma_i$. For a highly absorbing plate-like crystal, the average path length can be considered as the reciprocal of μ_0, i.e.,

$$\langle \ell \rangle = \frac{1}{\mu_0}, \tag{5.67}$$

then the path length for each reflected wave is

$$\ell_i = \frac{1}{\mu_0}\frac{\gamma}{\gamma_i}. \tag{5.68}$$

From (5.67), it is not surprising that one of the assumptions for the small secondary extinction, $\mu_0 \ell \ll 1$, is marginally satisfied by X-rays.

5.5 Multiple-Wave Diffraction Profile

The multiple-wave diffraction profile, i.e., the intensity distribution versus the rotation angles, like those of two-wave diffractions, depends on the incident beam divergences and the spectral dispersion, the crystal mosaic spread, and the Lorentz factor. As indicated in (5.60) and (5.61), of which the first two negative terms are related to the 'Aufhellung' effect and the third positive term is due to the 'Umweganregung'. The former involves a single Gaussian distribution function and the latter a product of two Gaussians. This implies that the profile width of the positive term is narrower than that of the negative terms. In both the Umweganregung and the Aufhellung situations, symmetric distributions will result. That the overall profile is symmetric with respect to the angles of rotation is expected. This disagrees, however, with the experimentally measured diffraction profiles. For example, in Fig. 4.2, some of the multiple-wave diffractions show very asymmetric characters. This disagreement is due to that the effect of phase on the multiply diffracted intensity [257, 259, 261] is not included in the kinematical approaches. Attempts have been made recently to consider this effect in the kinematical theory. In the literature, the intensity resulting from the interference between the diffracted G wave and the detoured diffracted waves of the successive reflections of L and $G-L$ is added to the total intensity of the primary reflection G at the exact multiple-wave position [245], i.e.,

$$I_G(\theta,\psi) = \left[I_G^{(2)}(\theta,\psi) - I_{\mathrm{Aufh}}(\theta,\psi)\right] + \left[I_{\mathrm{Umweg}}(\theta,\psi)\right] + I_{\mathrm{interf}}(\theta,\psi), \tag{5.69}$$

where $I_G^{(2)}$ is the two-wave intensity, I_{Aufh} is related to the two negative terms of (5.60), and I_{Umweg} is proportional to the positive term of (5.60). The first bracket on the right side of the equality represents the resultant intensity of the G wave subject to the changes due to Aufhellung. The second bracket stands for the intensity of the detoured (Umweganregung) wave via the successive reflections mentioned. The interference between the two should, in principle, depend on the cross term of the two brackets. Therefore, the intensity of interference, I_{interf}, takes the form

$$I_{\mathrm{interf}}(\theta,\psi) = 2\left\{\left[I_G^{(2)}(\theta,\psi) - I_{\mathrm{Aufh}}(\theta,\psi)\right]\left[I_{\mathrm{Umweg}}(\theta,\psi)\right]\right\}^{\frac{1}{2}} \times \cos u(\theta,\psi) , \tag{5.70}$$

where the effective phase sum $u(\theta,\psi)$ is defined as [243, 478]

$$u(\theta,\psi) = \delta_3 + \Delta(\theta,\psi) . \tag{5.71}$$

The phase δ_3 is the triplet phase given as

$$\delta_3 = \delta_{OL} + \delta_{LG} - \delta_{OG} , \tag{5.72}$$

where the δ's are the phases of the structure factors involved. $\Delta(\theta,\psi)$ is the scattering phase shift due to the rotation of the reciprocal lattice point of the L reflection relative to the Ewald sphere. The phase Δ is π for the reciprocal lattice point L lying outside the Ewald sphere and is 0 for the L inside the Ewald sphere. The phase changes drastically from π to 0 when the reciprocal lattice point L crosses the surface of the Ewald sphere from outside to inside [279, 478]. This phase shift can be easily derived from the wavevector defined in the dynamical theory (see Chaps. 7 and 9). Based on (5.69) with appropriate distribution functions, normalized Gaussian, Lorentzian, or pseudo-Voigt, for the two-wave, Aufhellung, Umweganregung, and interference profiles, a computer code *UMWEG*99 was developed by Rossmanith [243–245, 470] to calculate the intensity distribution of multiple-wave diffractions. Figure 5.2 shows the measured and calculated diffraction patterns of germanium (222) for $\lambda = 1.540562$ Å, where the primary reflection (222) is a nearly forbidden reflection [470]. Figure 5.2(a) is the experimental diffraction pattern obtained with a conventional Cu X-ray source [479]. Figure 5.2(b) is the calculated diffraction pattern based on the Bethe approximation for the dynamical theory (see Chap. 7) [270]. Figure 5.2(c) is the pattern simulated according to (5.69) [470]. Satisfactory qualitative and even quantitative agreement between the kinematical and dynamical calculations and the measured profiles is clearly seen in this figure. If the interference intensity is not included, the multiple-wave profile asymmetry disappears in the calculated pattern. It should be noted that appreciable interference effect, especially the profile asymmetry, can be observed when a perfect crystal or strong reflections are involved. This effect is therefore dynamical rather than kinematical.

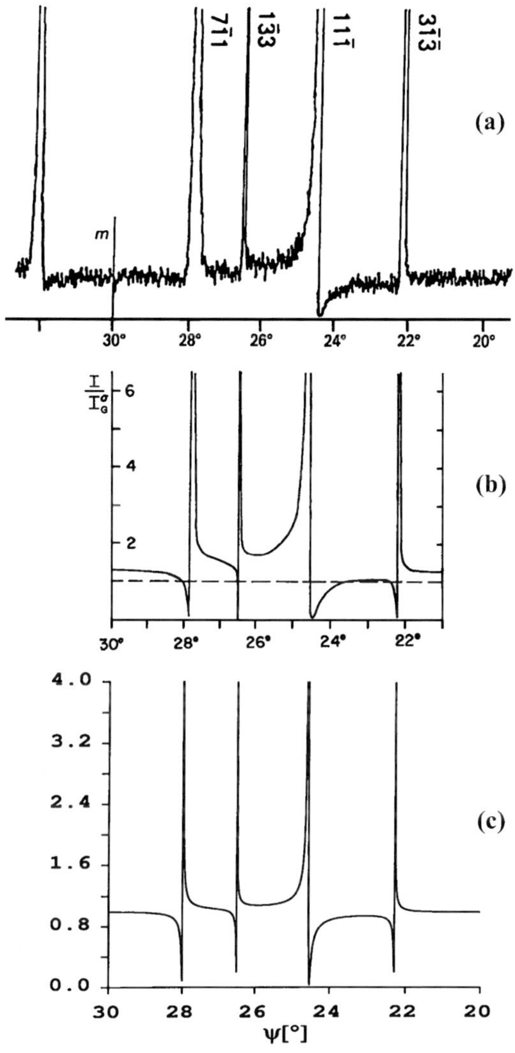

Fig. 5.2. Multiple-wave diffraction pattern of Ge (222) for 1.540562 Å from (**a**) experiments, (**b**) dynamical calculation, and (**c**) kinematical calculation [470]

The addition of the interference term to the expression for integrated intensity is reasonable from the physics point of view. However, it is not mathematically rigorous, because such a term cannot be obtained directly from the power-transfer equations governing the multiple-wave diffraction processes. In fact, to reveal the phase effect, the EM wavefields rather than the reflection powers need to be considered in describing the multiple-wave diffraction behavior. The Takagi-Taupin equations, where the X-ray wavefields in an imperfect crystal lattice are considered [327], can serve this purpose. This will be discussed in detail in Chap. 7.

6. Dynamical Theory of X-Ray Diffraction

The kinematical theory of X-ray diffraction follows closely the secondary extinction theory, which can be applied to situations involving very weak or no interaction between the incident wave and the diffracted wave. The theory is therefore suited for diffractions from small and imperfect crystals. If the diffraction from a large perfect crystal is concerned, the primary extinction becomes important. This is because the atomic planes in a perfect crystal are well ordered. Once the atomic planes are set to the exact Bragg diffraction position for an incident wave, they are also perfectly aligned for the reflected wave to be diffracted back into the direction of the incident wave. In analogy with the optics of the visible spectrum, a phase lag of about $\pi/2$ of the reflected wave in the incident-wave direction relative to the incident wave results. Namely, the phase difference between the twice-reflected wave and the incident wave leads to the modification of the amplitude of the incident wave. Similar modification takes place for the diffracted wave. Thus, the interaction between the incident and the diffracted waves via multiple reflection dominates the diffraction process. If a relatively strong reflection is involved, this multiple reflection effect cannot be ignored. Moreover, the phase lag of the reflected wave occurs for each plane of atoms inside the crystal. The resultant phase lags of the incident and diffracted waves modify their phase velocities in the crystal, thus changing the index of refraction. In other words, the dispersion relation describing the wavevectors of the incident and diffracted waves inside a crystal is also modified.

The interaction of the incident wave with the diffracted wave through multiple reflection was first considered by Darwin [66,67] in 1914 and then modified by taking into account crystal absorption by Prins [68] and Kohler [69]. A different dynamical theoretical approach dealing with the interaction of the electromagnetic fields with a three-dimensional array of electric dipoles, which has a more fundamental view and wider scope than Darwin's, was investigated by Ewald [57–59] even earlier than the discovery of X-ray diffraction by Laue, Friedrich, and Knipping [61, 466]. In 1931, Laue reformulated the dynamical theory, based on the solutions to Maxwell's equations, for the interaction of X-rays with a crystal by treating the crystal as a medium with a periodic complex dielectric constant [30, 50]. Laue's treatment has been more frequently used than Ewald's formulation in the literature due to its simple mathematical form and being easy to be understood. In this chapter, the fundamentals of the dynamical theory suited for multiple-wave diffraction will be described. The formulations will be discussed in Chap. 8 as special cases for the two-wave Bragg reflection, Laue transmission, grazing incident in-plane diffraction, and back reflection with a Bragg angle equal to $\pi/2$. Since many excellent review articles and books have been written for these two-wave diffractions [32–37,39,40,48–52,55], we will only mention key points and equations here. Interested readers are suggested to refer to the recently published book by Authier [55]. In the following, Laue's dynamical theory and the Gaussian unit system will be adopted.

6.1 Fundamental Equation of Wavefield

X-ray diffraction from a crystal is considered by Laue [30] the interaction of EM waves with a continuous distribution of negative charges with the positive charges concentrated at the nuclei. Unless under nuclear resonance conditions [81, 480], the contribution of X-ray scattering from the nuclei is very small compared with that from the negative charges. Only the X-ray scattering from the electron distribution is considered here. The formation of X-ray wavefields in the periodic arrays of atoms during the X-ray diffraction can be described by Maxwell's equations:

$$\boldsymbol{\nabla} \times \boldsymbol{H} = \frac{1}{c}\left(\frac{\partial \boldsymbol{D}}{\partial t} + 4\pi \boldsymbol{J}_t\right) , \tag{6.1a}$$

$$\boldsymbol{\nabla} \times \boldsymbol{E} = -\frac{1}{c}\frac{\partial \boldsymbol{B}}{\partial t} , \tag{6.1b}$$

$$\boldsymbol{\nabla} \cdot \boldsymbol{D} = 4\pi \varrho_t , \tag{6.1c}$$

$$\boldsymbol{\nabla} \cdot \boldsymbol{B} = 0 , \tag{6.1d}$$

where the vector fields $\boldsymbol{E}$, $\boldsymbol{D}$, $\boldsymbol{H}$, and $\boldsymbol{B}$ are the electric field, electric displacement, magnetic field, and magnetic induction, respectively, for a given net charge density ρ_t and current density $\boldsymbol{J}_t$. The quantity c is the speed of light in vacuum. Because the crystal is neutral, there is neither a net charge nor a net current.

The charge distribution of the crystal is polarized by the electromagnetic field of the X-rays, which can be described by

$$\boldsymbol{D} = \boldsymbol{E} + 4\pi \boldsymbol{P} , \tag{6.2}$$

provided that the X-ray frequency is far from the absorption edges of the constituent atoms of the crystal. The vector $\boldsymbol{P}$ is the electric polarization. Similarly, for the magnetization $\boldsymbol{M}$ due to the magnetic field $\boldsymbol{H}$, the magnetic induction is expressed as

$$\boldsymbol{B} = \boldsymbol{H} + 4\pi \boldsymbol{M} . \tag{6.3}$$

Since X-rays are high-frequency EM waves ($\sim 10^{19}$ Hertz), the linear relation between $\boldsymbol{E}$ and $\boldsymbol{D}$ and between $\boldsymbol{H}$ and $\boldsymbol{B}$ are

$$\boldsymbol{D} = \varepsilon \boldsymbol{E} , \qquad \boldsymbol{B} = \mu_n \boldsymbol{H} , \tag{6.4}$$

where the electric conductivity is assumed to be zero and the magnetic permeability $\mu_n \simeq 1$. The electric polarization then takes the form

$$\boldsymbol{P} = \frac{\chi}{4\pi} \boldsymbol{E} , \tag{6.5}$$

and

$$\chi = \varepsilon - 1 , \tag{6.6}$$

where $\chi/4\pi$ is the electric susceptibility and ε the dielectric constant for the X-ray used.

The polarization $\boldsymbol{P}$ can be easily obtained from the equation of motion of the induced dipoles by the electric field $\boldsymbol{E}$ of the X-rays. In general, the dipoles undergo a forced oscillation. For simplicity, let's assume that the angular frequency ω of the X-ray used is far greater than a natural frequency ω_o, namely, an absorption edge, then the contribution of damping is negligibly small. Under such a circumstance, the dipole moment $\boldsymbol{p}$ can be expressed as

$$\boldsymbol{p} = e\boldsymbol{x} = -\frac{e^2}{m\omega^2} \boldsymbol{E} , \tag{6.7}$$

where $\boldsymbol{x}$ is the displacement of the negative charge with respect to the positive charge. The polarization then takes the form

$$\boldsymbol{P} = \varrho(\boldsymbol{r}) \boldsymbol{p} = -\frac{e^2}{m\omega^2} \varrho(\boldsymbol{r}) \boldsymbol{E} , \tag{6.8}$$

where $\varrho(\boldsymbol{r})$ is the electron density. By substituting (6.8) into (6.5), the χ of (6.6) is given by

$$\chi = -\frac{r_e \lambda^2}{\pi} \varrho(\boldsymbol{r}) , \tag{6.9}$$

where r_e is the classic radius of electron. The order of magnitude of χ is about 10^{-5}–10^{-7} for X-rays. Since the crystal is treated, according to Laue, as a medium with a periodic complex dielectric constant, the χ is a periodic function, which can be expressed as a Fourier series:

$$\chi = \sum_G \chi_G \exp\left[-2\pi \mathrm{i} \left(\boldsymbol{g}_G \cdot \boldsymbol{r}\right)\right] . \tag{6.10}$$

χ_G, the Fourier component for a given G reflection, is

$$\chi_G = -\frac{r_e \lambda^2 F_G}{\pi V_c} . \tag{6.11}$$

F_G is the structure factor of the G reflection, which is a Fourier transform of the electron density distribution $\varrho(\boldsymbol{r})$, i.e., (2.29a):

$$\varrho(\boldsymbol{r}) = \frac{1}{V_c} \sum_G F_G \exp\left[-2\pi\mathrm{i}\left(\boldsymbol{g}_G \cdot \boldsymbol{r}\right)\right] , \tag{6.12}$$

where $\boldsymbol{g}_G$ is the reciprocal lattice vector of the G reflection and V_c the volume of the crystal unit cell.

The electric susceptibility $\chi/4\pi$ is usually isotropic for $\omega \gg \omega_o$. However, it becomes anisotropic due to the local anisotropic atomic arrangement caused by spin and orbital ordering, especially when the energies of the X-ray photons are near absorption edges. That means under resonance scattering conditions, anisotropic polarization also takes place when the contribution of the interaction between an electric dipole and a quadrupole and the interactions among higher-order multipoles are appreciable [97, 173, 174, 481–488].

Equations (6.1a) and (6.1b) with $\boldsymbol{J}_t = 0$, together with (6.4), lead to the following wave equation:

$$\nabla^2 \boldsymbol{D}(\boldsymbol{r},t) - \frac{1}{c^2}\frac{\partial^2 \boldsymbol{D}(\boldsymbol{r},t)}{\partial t^2} = -\boldsymbol{\nabla} \times \boldsymbol{\nabla} \times \left(\frac{\chi}{1+\chi}\boldsymbol{D}(\boldsymbol{r},t)\right) . \tag{6.13}$$

In the case of X-rays, for a very small χ, (6.13) can be approximated in a differential form as

$$\nabla^2 \boldsymbol{D}(\boldsymbol{r},t) - \frac{1}{c^2}\frac{\partial^2 \boldsymbol{D}(\boldsymbol{r},t)}{\partial t^2} \simeq -\boldsymbol{\nabla} \times \boldsymbol{\nabla} \times \left[\chi \boldsymbol{D}(\boldsymbol{r},t)\right] , \tag{6.14}$$

or, in an integral form as

$$\begin{aligned} \boldsymbol{D}(\boldsymbol{r},t) = \boldsymbol{D}_e(\boldsymbol{r},t) + \frac{1}{4\pi}\int \frac{1}{|\boldsymbol{r}-\boldsymbol{r}'|}\boldsymbol{\nabla}' \times \boldsymbol{\nabla}' \\ \times \left[\chi(\boldsymbol{r}')\boldsymbol{D}\left(\boldsymbol{r}', t - \frac{|\boldsymbol{r}-\boldsymbol{r}'|}{c}\right)\right] \mathrm{d}\boldsymbol{r}' , \end{aligned} \tag{6.15}$$

which is the solution to the inhomogeneous wave equation, (6.14). $\boldsymbol{D}_e$ is the solution to the homogeneous equation of (6.14) with $\chi = 0$. Similar to $\boldsymbol{D}(\boldsymbol{r},t)$ in (6.14), the wave equation in terms of the electric field $\boldsymbol{E}(\boldsymbol{r},t)$ takes the following differential form:

$$\boldsymbol{\nabla} \times \boldsymbol{\nabla} \times \boldsymbol{E}(\boldsymbol{r},t) - 4\pi k^2[1+\chi]\boldsymbol{E}(\boldsymbol{r},t) = 0 . \tag{6.16}$$

Equations (6.1c) and (6.1d) with $\varrho_t = 0$, i.e.,

$$\boldsymbol{\nabla} \cdot \boldsymbol{D} = 0 , \qquad \boldsymbol{\nabla} \cdot \boldsymbol{B} = 0 , \tag{6.17}$$

give, as usual, the transversality of the electric displacement and the magnetic induction. By considering this characteristic and the periodic nature of the crystal, the solution to the inhomogeneous wave equation (6.14) is assumed to be a Bloch function, i.e., a superposition of an infinite number of plane waves due to infinite number of reflections:

$$\boldsymbol{D}(\boldsymbol{r},t) = \sum_G \boldsymbol{D}_G \exp\left(\mathrm{i}\omega t - 2\pi\mathrm{i}\boldsymbol{K}_G \cdot \boldsymbol{r}\right) , \tag{6.18a}$$

$$\boldsymbol{E}(\boldsymbol{r},t) = \sum_G \boldsymbol{E}_G \exp\left(\mathrm{i}\omega t - 2\pi\mathrm{i}\boldsymbol{K}_G \cdot \boldsymbol{r}\right) , \tag{6.18b}$$

$$\boldsymbol{H}(\boldsymbol{r},t) = \sum_G \boldsymbol{H}_G \exp\left(\mathrm{i}\omega t - 2\pi\mathrm{i}\boldsymbol{K}_G \cdot \boldsymbol{r}\right) , \tag{6.18c}$$

$$\boldsymbol{B}(\boldsymbol{r},t) = \sum_G \boldsymbol{B}_G \exp\left(\mathrm{i}\omega t - 2\pi\mathrm{i}\boldsymbol{K}_G \cdot \boldsymbol{r}\right) , \tag{6.18d}$$

where $\boldsymbol{K}_G$, the wavevector of the G reflection, satisfies Bragg's law, i.e.,

$$\boldsymbol{K}_G = \boldsymbol{K}_O + \boldsymbol{g}_G . \tag{6.19}$$

$\boldsymbol{K}_O$ is the wavevector of the incident wave. All the $\boldsymbol{K}$'s denote the wavevectors inside the crystal and all the $\boldsymbol{k}$'s the wavevectors outside the crystal. The magnitude of the incident wavevector is k, which is equal to $1/\lambda$ in vacuum. By substituting the Bloch function and Bragg's law into (6.14), the wave equation becomes

$$(k^2 - K_G^2)\boldsymbol{D}_G - \sum_L \chi_{G\text{–}L}\left[\boldsymbol{K}_G \times (\boldsymbol{K}_G \times \boldsymbol{D}_L)\right] = 0 , \tag{6.20}$$

or,

$$\frac{(K_G^2 - k^2)}{K_G^2}\boldsymbol{D}_G = \sum_L \chi_{G\text{–}L}\boldsymbol{D}_{L[G]} , \tag{6.21}$$

for all L reflections, where the relation $\boldsymbol{A}\times(\boldsymbol{B}\times\boldsymbol{C}) = \boldsymbol{B}(\boldsymbol{A}\cdot\boldsymbol{C}) - \boldsymbol{C}(\boldsymbol{A}\cdot\boldsymbol{B})$ is employed. $\boldsymbol{D}_{L[G]}$ is the vector component of $\boldsymbol{D}_L$ perpendicular to $\boldsymbol{K}_G$, i.e.,

$$\boldsymbol{D}_{L[G]} = -\frac{\boldsymbol{K}_G \times (\boldsymbol{K}_G \times \boldsymbol{D}_L)}{K_G^2} . \tag{6.22}$$

Equation (6.20), or (6.21), is the fundamental equation of wavefield. Equations (6.14) and (6.15) are sometimes called the fundamental equations of wavefield of differential and integral forms, respectively. The Fourier components of the polarization $\boldsymbol{P}$ and the electric field $\boldsymbol{E}$ in terms of the electric displacement $\boldsymbol{D}$ are

$$\boldsymbol{P}_G = \frac{\sum_L \chi_{G\text{–}L}\boldsymbol{D}_L}{4\pi} , \tag{6.23}$$

$$\boldsymbol{E}_G = \boldsymbol{D}_G - \sum_L \chi_{G\text{–}L}\boldsymbol{D}_L , \tag{6.24}$$

and

$$\boldsymbol{D}_G = \boldsymbol{E}_G + \sum_L \chi_{G\text{–}L}\boldsymbol{E}_L . \tag{6.25}$$

With the Fourier representations of $\boldsymbol{D}$, $\boldsymbol{H}$, $\boldsymbol{E}$, and $\boldsymbol{B}$, Maxwell's equations, (4.1), lead to

$$
\begin{aligned}
\boldsymbol{D}_G &= -\frac{\boldsymbol{K}_G \times \boldsymbol{H}_G}{|\boldsymbol{K}_G|} , \\
\boldsymbol{B}_G &= \frac{\boldsymbol{K}_G \times \boldsymbol{E}_G}{|\boldsymbol{K}_G|} , \\
\boldsymbol{K}_G &\cdot \boldsymbol{D}_G = 0 , \\
\boldsymbol{K}_G &\cdot \boldsymbol{H}_G = 0 .
\end{aligned} \tag{6.26}
$$

Therefore $\boldsymbol{K}_G$, $\boldsymbol{D}$, and $\boldsymbol{H}$ are mutually orthogonal and the electric field $\boldsymbol{E}$ lies in the $\boldsymbol{K}_G$-$\boldsymbol{D}$ plane. Since the difference in magnitude between $\boldsymbol{E}$ and $\boldsymbol{D}$ is of the order of χ, $\boldsymbol{E}$ and $\boldsymbol{D}$ are nearly parallel to each other. Yet, if we consider the physical difference between $\boldsymbol{E}$ and $\boldsymbol{D}$, then the fundamental equation in terms of $\boldsymbol{E}$ can be written [489] as

$$
\frac{K_G^2 \boldsymbol{E}_{G[G]} - k^2 \boldsymbol{E}_G}{k^2} = \sum_L \chi_{G\text{–}L} \boldsymbol{E}_L , \tag{6.27}
$$

where $\boldsymbol{E}_{G[G]} = \boldsymbol{E}_G$ cannot be a null vector because $\boldsymbol{\nabla} \cdot \boldsymbol{E} \neq 0$. Compared with (6.27), the fundamental equation in terms of $\boldsymbol{D}$ is written in the following alternative form:

$$
\boldsymbol{D}_G = \frac{K_G^2}{K_G^2 - k^2} \sum_L \chi_{G\text{–}L} \boldsymbol{D}_{L[G]} , \tag{6.28}
$$

where $K_G^2/(K_G^2 - k^2)$ is the resonance function. This implies that the wavefield $\boldsymbol{D}_G$ has an appreciable value when the magnitude of $\boldsymbol{K}_G$ is very close to k. In other words, the crystal is excited for a Bragg reflection when the reciprocal lattice point G is situated very close to the surface of the Ewald sphere.

Both (6.27) and (6.28) can be used for dynamical calculations. The difference in the calculations between the two is less than $10^{-3}\%$. Moreover, the number of fundamental equations depends on the number of reciprocal lattice points considered; thus for an N wave diffraction, there are N-vector equations.

6.2 Formalism of Photonic Band Theory

There is an infinite number of reciprocal lattice points for a given crystal. If a great many of reciprocal lattice points are considered, the fundamental equations for a range of angular frequencies ω could, in principle, map out the so-called photonic band structure, an analogy to the electronic band structure.

The idea of photonic band was first proposed and demonstrated experimentally with microwaves by Yablonovitch in 1987 [490–493]. The so-called photonic bandgap material, or, photonic crystal, is composed of periodic dielectric spheres with a dielectric constant different from that in the surroundings [494]. It is a new type of periodic dielectric medium with respect to photons, like semiconductors with respect to electrons. It is known that the electronic energy band structure, the most important characteristic of a semiconductor, results from the periodicity of the atomic arrangement on an Angstrom scale. In contrast, the photonic crystal has a periodicity comparable with a photon wavelength of interest, which is closely related to the large period of the dielectrics. In a photonic crystal, the EM waves are forbidden in a photonic bandgap, irrespective of their propagation direction.

The scattering of EM waves from a photonic crystal with a periodic dielectric constant $\varepsilon(\boldsymbol{r})$ can be described by the wave equation directly derived from (6.14):

$$-\nabla^2 \boldsymbol{D} = \left(\frac{\omega^2}{c^2}\right) \boldsymbol{D} + \boldsymbol{\nabla} \times \boldsymbol{\nabla} \times [V(\boldsymbol{r})\boldsymbol{D}] \,, \tag{6.29}$$

where the 'potential' $V(\boldsymbol{r})$, replacing the χ, is expressed as

$$V(\boldsymbol{r}) = 1 - \frac{1}{\varepsilon(\boldsymbol{r})} \,. \tag{6.30}$$

By considering the Bloch waves as the solution to (6.29), the following fundamental equation is obtained:

$$\alpha_G \boldsymbol{D}_G + \sum_L V_{G\text{–}L} \left\{ (\boldsymbol{K}_G \cdot \boldsymbol{D}_L)\, \boldsymbol{K}_G - K_G^2 \boldsymbol{D}_L \right\} = 0 \,, \tag{6.31}$$

where

$$\alpha_G = K_G^2 - \left(\frac{\omega}{2\pi c}\right)^2 = K_G^2 - k^2 \,. \tag{6.32}$$

Equation (6.31) is exactly the same as (6.20) with $V_{G\text{–}L}$ being equivalent to $\chi_{G\text{–}L}$. Alternative forms in terms of the electric field $\boldsymbol{E}$ and magnetic field $\boldsymbol{H}$ can be found in the literature [495–497]. If scalar waves, instead of vector waves, are considered for electrons in solids, the corresponding fundamental equation gives the familiar electronic band structure. In this case, the χ for X-rays is replaced by the ratio of the periodic potential energy to the total energy in a crystal.

Similar to X-ray diffraction and photonic crystals, the equivalent fundamental equation of wavefield for electron diffraction derived from Schrödinger equation was reported by Bethe in 1928 [12].

6.3 Polarization of Wavefields

There are N vector equations of wavefield in the form of (6.20) for an N-wave diffraction. For a three-dimensional periodic structure, like a crystal, these N vector equations can be decomposed into $3N$ scalar equations by properly defined polarizations for wavefileds. For simplicity, if a Cartesian coordinate system is used to define the three components of the X-ray wavefield – one longitudinal and two transverse with respect to the directions of wave propagation, i.e., $\boldsymbol{K}$'s – for a general situation, then there will be $3N$ fundamental equations to be dealt with. However, as long as the transversality of the EM waves is concerned, two mutually orthogonal polarization unit vectors, denoted by $\hat{\boldsymbol{\sigma}}$ and $\hat{\boldsymbol{\pi}}$, perpendicular to the corresponding wavevector $\boldsymbol{K}$ are usually adopted to define the electric displacements, $\boldsymbol{D}$'s, inside a crystal. With this choice of polarization, the number of scalar equations reduces to $2N$. Thus the conventional σ-π representation has the advantage of solving only $2N$ equations. But it cannot be applied to cases involving grazing incidence diffraction due to the difficulties in finding appropriate solutions for X-ray wavefields. Here we shall follow the conventional consideration for polarization first. The representation of wavefield in the Cartesian coordinate system will be discussed in Chap 7.

As stated in Sect. 2.4.2 and Sect. 3.3, the electric displacement $\boldsymbol{D}_M$ of a given diffracted wave M can be written as

$$\boldsymbol{D}_M = \hat{\boldsymbol{\sigma}}_M D_{\sigma M} + \hat{\boldsymbol{\pi}}_M D_{\pi M} \ , \tag{6.33}$$

where $\hat{\boldsymbol{\sigma}}_M$ and $\hat{\boldsymbol{\pi}}_M$ are the σ- and π-polarization unit vectors. Both $\hat{\boldsymbol{\sigma}}_M$ and $\hat{\boldsymbol{\pi}}_M$ are perpendicular to the wavevector $\boldsymbol{K}_M$. For a two-wave (O, G) diffraction the σ- and π-polarized unit vectors of the incident wave O and the diffracted wave G are defined according to the plane of incidence containing the two involved wavevectors $\boldsymbol{K}_O$ and $\boldsymbol{K}_G$. The $\hat{\boldsymbol{\sigma}}_M$, $\hat{\boldsymbol{\pi}}_M$, and the unit vector $\hat{\boldsymbol{K}}_M$ of $\boldsymbol{K}_M$ for $M = O, G$ forms a mutually orthonormal system, so that

$$\hat{\boldsymbol{\sigma}}_O = \hat{\boldsymbol{\sigma}}_G = -\frac{\hat{\boldsymbol{K}}_O \times \hat{\boldsymbol{K}}_G}{\left|\hat{\boldsymbol{K}}_O \times \hat{\boldsymbol{K}}_G\right|} \ , \tag{6.34a}$$

$$\hat{\boldsymbol{\pi}}_O = \hat{\boldsymbol{K}}_O \times \hat{\boldsymbol{\sigma}}_O \ , \qquad \hat{\boldsymbol{\pi}}_G = \hat{\boldsymbol{K}}_G \times \hat{\boldsymbol{\sigma}}_G \ . \tag{6.34b}$$

In other words, all the σ's are perpendicular to the plane of incidence of the G reflection and all the π's lie in that plane. At the exact Bragg diffraction position, i.e., when $\theta = \theta_G$,

$$\hat{\boldsymbol{\sigma}}_O \cdot \hat{\boldsymbol{\sigma}}_G = 1 \ , \quad \hat{\boldsymbol{\pi}}_O \cdot \hat{\boldsymbol{\pi}}_G = \cos 2\theta_G \ , \quad \hat{\boldsymbol{\sigma}}_O \cdot \hat{\boldsymbol{\pi}}_G = \hat{\boldsymbol{\sigma}}_G \cdot \hat{\boldsymbol{\pi}}_O = 0 \ . \tag{6.35}$$

For an N-wave (O, G_1, G_2, $\ldots$, G_{N-1}) diffraction, the polarization relations (6.34a) and (6.34b) hold, i.e.,

$$\hat{\boldsymbol{\sigma}}_{G_j} = -\frac{\hat{\boldsymbol{K}}_O \times \hat{\boldsymbol{K}}_{G_j}}{\left|\hat{\boldsymbol{K}}_O \times \hat{\boldsymbol{K}}_{G_j}\right|} \ , \tag{6.36a}$$

$$\hat{\boldsymbol{\pi}}_{G_j} = \hat{\boldsymbol{K}}_{G_j} \times \hat{\boldsymbol{\sigma}}_{G_j} \ , \tag{6.36b}$$

for $j = 1, 2, \ldots, N-1$. Unless the N-wave diffraction is a wavevector-coplanar case, $\hat{\boldsymbol{\sigma}}_M$ and $\hat{\boldsymbol{\pi}}_L$ for any two different reflections M and L are not necessarily perpendicular to each other. The vector form of the fundamental equations of wavefield, (6.21), is expressed as

$$2\varepsilon_O \boldsymbol{D}_O = \chi_O \boldsymbol{D}_{O[O]} + \sum_{j=1}^{N-1} \chi_{O\text{-}G_j} \boldsymbol{D}_{G_j[O]} \ , \tag{6.37a}$$

$$2\varepsilon_{G_j}\boldsymbol{D}_{G_j} = \chi_{G_j-O}\boldsymbol{D}_{O[G_j]} + \chi_O\boldsymbol{D}_{G_j[G_j]} + \sum_{\ell \neq j}^{N-1} \chi_{G_j-G_\ell}\boldsymbol{D}_{G_\ell[G_j]} , \quad (6.37b)$$

for $j = 1, 2, \ldots, N-1$, where

$$2\varepsilon_M = \frac{K_M^2 - k^2}{K_M^2} \quad (6.38)$$

for $M = O, G_1, G_2, \ldots, G_{N-1}$, and

$$\boldsymbol{D}_{G_i[G_j]} = -\left[\hat{\boldsymbol{K}}_{G_j} \times \left(\hat{\boldsymbol{K}}_{G_j} \times \boldsymbol{D}_{G_i}\right)\right] . \quad (6.39)$$

By carrying out the scalar products of $\hat{\boldsymbol{\sigma}}_M$ and $\hat{\boldsymbol{\pi}}_M$ on the M-th vector equation, respectively, (6.37a) and (6.37b) can be decomposed into the following $2N$ scalar equations:

$$(2\varepsilon_O - \chi_O)\, D_{\sigma O} = \sum_{j=1}^{N-1} \chi_{O\;G_j}\left[D_{\sigma G_j} p_{\sigma O,\sigma G_j} + D_{\pi G_j} p_{\sigma O,\pi G_j}\right] , \quad (6.40a)$$

$$(2\varepsilon_O - \chi_O)\, D_{\pi O} = \sum_{j=1}^{N-1} \chi_{O-G_j}\left[D_{\sigma G_j} p_{\pi O,\sigma G_j} + D_{\pi G_j} p_{\pi O,\pi G_j}\right] , \quad (6.40b)$$

$$\left(2\varepsilon_{G_j} - \chi_O\right) D_{\sigma G_j} = \chi_{G_j-O}\left[D_{\sigma O} p_{\sigma G_j,\sigma O} + D_{\pi O} p_{\sigma G_j,\pi O}\right] + \sum_{j \neq \ell}^{N-1} \chi_{G_j-G_\ell} \times \left[D_{\sigma G_\ell} p_{\sigma G_j,\sigma G_\ell} + D_{\pi G_\ell} p_{\sigma G_j,\pi G_\ell}\right] , \quad (6.40c)$$

$$\left(2\varepsilon_{G_j} - \chi_O\right) D_{\pi G_j} = \chi_{G_j-O}\left[D_{\sigma O} p_{\pi G_j,\sigma O} + D_{\pi O} p_{\pi G_j,\pi O}\right] + \sum_{j \neq \ell}^{N-1} \chi_{G_j-G_\ell} \times \left[D_{\sigma G_\ell} p_{\pi G_j,\sigma G_\ell} + D_{\pi G_\ell} p_{\pi G_j,\pi G_\ell}\right] , \quad (6.40d)$$

where the polarization factors are defined as

$$p_{\sigma O,\pi G_j} = \hat{\boldsymbol{\sigma}}_O \cdot \hat{\boldsymbol{\pi}}_{G_j} , \qquad p_{\sigma G_j,\pi G_\ell} = \hat{\boldsymbol{\sigma}}_{G_j} \cdot \hat{\boldsymbol{\pi}}_{G_\ell} , \quad (6.41)$$

and $\chi_O = \chi_{O-O}$. In short, (6.40a)–(6.40d) can be expressed in the following simple form:

$$(2\varepsilon_M - \chi_O)\, D_{\alpha M} = \sum_{\substack{L=O \\ L \neq M}}^{G_{N-1}} \chi_{M-L}\left[D_{\sigma L} p_{\alpha M,\sigma L} + D_{\pi L} p_{\alpha M,\pi L}\right] , \quad (6.42)$$

where $\alpha = \sigma$ and π, $M = O, G_1, G_2, \ldots, G_{N-1}$, and the summation is taken over all L except M.

The polarization vectors in a multiple-wave diffraction can be considered as references for wavevectors and wavefields. There are many ways to define the polarization vectors, as long as they are convenient to use and can provide a simpler form for the fundamental equations of wavefield. And most importantly, they should be able to lead to correct solutions to the fundamental equations. For example, for a systematic multiple-wave diffraction, sometimes it is convenient to define all the σ unit vectors to lie in the plane containing all the reciprocal lattice points [184].

6.4 Equation of Dispersion and Excitation of Dispersion Surface

The fundamental equation, (6.40a)–(6.40d), is a system of $2N$ homogeneous complex linear equations with $2N$ unknown $2\varepsilon_M$ and $2N$ unknown wavefield amplitudes $D_{\sigma M}$ and $D_{\pi M}$. It can be represented in a matrix form as

$$\tilde{\Phi}[\boldsymbol{D}] = [0] \,, \tag{6.43}$$

where $\tilde{\Phi}$ is a $2N \times 2N$ matrix defined as

$$\tilde{\Phi} = \begin{bmatrix} 2\varepsilon_O - \chi_O & 0 & -\chi_{O-G_1} p_{\sigma O,\sigma G_1} & -\chi_{O-G_1} p_{\sigma O,\pi G_1} & \cdots & -\chi_{O-G_{N-1}} p_{\sigma O,\pi G_{N-1}} \\ 0 & 2\varepsilon_O - \chi_O & -\chi_{O-G_1} p_{\pi O,\sigma G_1} & -\chi_{O-G_1} p_{\pi O,\pi G_1} & \cdots & -\chi_{O-G_{N-1}} p_{\pi O,\pi G_{N-1}} \\ -\chi_{G_1-O} p_{\sigma G_1,\sigma O} & -\chi_{G_1-O} p_{\sigma G_1,\pi O} & 2\varepsilon_{G_1} - \chi_O & 0 & \cdots & \cdot \\ -\chi_{G_1-O} p_{\pi G_1,\sigma O} & -\chi_{G_1-O} p_{\pi G_1,\pi O} & 0 & 2\varepsilon_{G_1} - \chi_O & \cdots & \cdot \\ \vdots & \vdots & \vdots & \vdots & \vdots\vdots\vdots & \vdots \\ \cdot & \cdot & \cdot & \cdot & \cdots & 0 \\ -\chi_{G_{N-1}-O} p_{\pi G_{N-1},\sigma O} & -\chi_{G_{N-1}-O} p_{\pi G_{N-1},\pi O} & \cdot & -\chi_{G_{N-1}-O} p_{\pi G_{N-1},\pi G_1} & \cdots 0 & 2\varepsilon_{G_{N-1}} - \chi_O \end{bmatrix} , \tag{6.44}$$

the transpose of the $2N \times 1$ vector, $[\boldsymbol{D}]$, is

$$[\boldsymbol{D}]^T = \begin{bmatrix} D_{\sigma O} & D_{\pi O} & D_{\sigma G_1} & D_{\pi G_1} & \cdots & D_{\sigma G_{N-1}} & D_{\pi G_{N-1}} \end{bmatrix} . \tag{6.45}$$

For (6.43) to have nontrivial solutions, the determinant of $\tilde{\Phi}$ is set to be zero:

$$\det \tilde{\Phi} = 0 \,. \tag{6.46}$$

Equation (6.46) is known as the equation of dispersion, which describes the relationship between the wavevector $\boldsymbol{K}_M$ inside the crystal and the crystal setting for a given photon energy or X-ray wavelength. In reciprocal space, a surface, called dispersion surface, can be constructed based on the solutions of (6.46). Since the wavevector $\boldsymbol{K}_M$ is an unknown quantity in $2\varepsilon_M$ and the polarization factors p's depend on $\boldsymbol{K}_M$ according to (6.36a) and (6.36b), almost every term in (6.40a)–(6.40d) and (6.44) is related to the unknown $\boldsymbol{K}_M$.

It is therefore not easy to find solutions to (6.43). However, to the first-order approximation of χ, the polarization factors p's can be considered constant. This approximation is valid when the angular position (θ, ψ) of the crystal is not far from the exact multiple-wave diffraction position and the diffraction involves neither grazing incidence and emergence nor back diffraction. Under such a condition, (6.43) can be solved as a complex eigenvalue problem. From (6.46), at a given crystal setting, the real part of the eigenvalue $\boldsymbol{K}_M$, or, $2\varepsilon_M$, determines the coordinates of the starting point of the wavevectors $\boldsymbol{K}_M$ for all the involved reflections on the dispersion surface. This starting point is called the tie point, named by Ewald [57–59], or sometimes called the wavepoint. The loci of the tie point, as the crystal setting varies with respect to the incident wave, form the dispersion surface. The imaginary part of the eigenvalue yields the linear absorption coefficient. The corresponding eigenvector $[\boldsymbol{D}]$ gives the ratio among the wavefield amplitudes, which can be represented by the cofactors involved in the determinant $\det\tilde{\Phi}$ [498]. For example, for the j-th eigenvalue $\boldsymbol{K}_M(j)$, or, $2\varepsilon_M(j)$, the j-th eigenvector gives the amplitude ratio of any pair of diffracted waves:

$$\frac{E_{\sigma G_l}(j)}{E_{\pi G_m}(j)} = \frac{\mathrm{Cof}\,(\sigma G_l, \sigma G_l)_j}{\mathrm{Cof}\,(\pi G_m, \sigma G_l)_j},$$

$$\frac{E_{\sigma G_l}(j)}{E_{\pi G_m}(j)} = \frac{\mathrm{Cof}\,(\sigma G_l, \pi G_m)_j}{\mathrm{Cof}\,(\pi G_m, \pi G_m)_j}, \tag{6.47}$$

where $\mathrm{Cof}(\sigma G_l, \pi G_m)$ is the cofactor of the element $\chi_{G_m - G_l} p_{\pi G_m, \sigma G_l}$ in the determinant $\det\tilde{\Phi}$.

The dispersion surface, as defined by $2\varepsilon_M$, depends on the difference between the wavevectors inside and outside the crystal, namely, the variation in the refractive index of X-rays in the crystal along a certain direction. This implies that the refractive index is modified by the presence of multiple-wave diffraction. It is therefore reasonable to state that the dispersion surface is nothing but a surface of refractive index, representing the surface of constant index of refraction. Since it is direction-dependent, i.e., related to the wavevector $\boldsymbol{K}_M$, the refractive index of X-rays is anisotropic when multiple-wave diffraction takes place. Since $2\varepsilon_M$ is complex, the dispersion surface, strictly speaking, should be described in complex reciprocal space or complex momentum space. The real and imaginary parts of $2\varepsilon_M$ are related to the real and imaginary parts of the index of refraction. Because an N-wave diffraction is the combination of N one-wave or $N-1$ two-wave diffractions, we shall discuss the dispersion surface in one-, two-, three-, and N-wave cases following the complex equation of dispersion, (6.46).

i) One-wave case

A one-wave (O) diffraction in real space is shown in Fig. 6.1(a). The incident wave of the wavevector $\boldsymbol{k}_O$ impinges on a crystal at an incident angle θ' and is refracted along the wavevector $\boldsymbol{K}'_O$ without any reflection (see Fig. 6.1(a)).

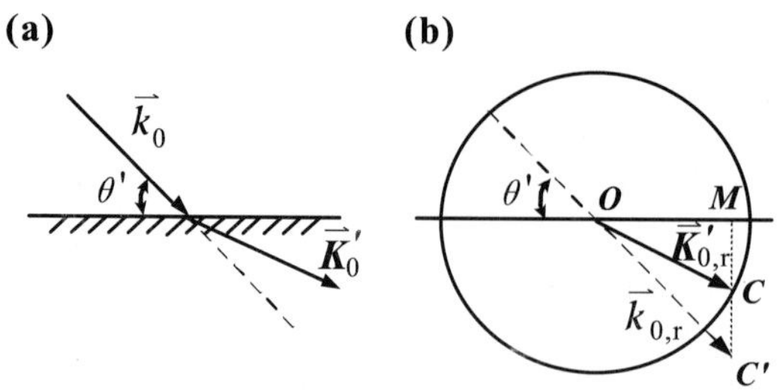

Fig. 6.1. Schematic of one-wave diffraction in (**a**) real space, and (**b**) reciprocal space

The equation of dispersion for this one-wave (O) diffraction, actually, a refraction, is

$$2\varepsilon_O - \chi_O = 0 \,. \tag{6.48}$$

Namely,

$$K_O'^2 = k^2 \left(1 - \chi_O\right)^{-1} \cong k^2 \left(1 + \chi_O\right) \,. \tag{6.49}$$

The approximation appearing in (6.49) is due to the small magnitude of χ_O. According to (6.49), the dispersion surface is spherical. In terms of the complex index of refraction $n = n_r + in_i$, where the subscript r means real part and subscript i stands for imaginary part, the equation of dispersion takes the form

$$K_O' = nk \,, \tag{6.50}$$

where n is related to χ_O by

$$n^2 = 1 + \chi_O \,. \tag{6.51}$$

Referring to (6.18a), (6.49) and (6.51), we obtain the real and imaginary parts of the wavevector $\boldsymbol{K}_O'$ inside the crystal as

$$K_{O,r}' = k\left(1 + \frac{\chi_{O,r}}{2}\right) \,, \tag{6.52a}$$

$$K_{O,i}' = -\frac{k\chi_{O,i}}{2} \,, \tag{6.52b}$$

and the corresponding real and imaginary parts of the index of refraction as

$$n_r \cong 1 + \frac{\chi_{O,r}}{2} = 1 - \Delta \,, \tag{6.53a}$$

$$n_i \cong \frac{\chi_{O,i}}{2} \,, \tag{6.53b}$$

where the small correction Δ in the index of refraction is

$$\Delta = -\frac{\chi_{O,r}}{2} \cong \frac{r_e \lambda^2}{2\pi} \varrho \,. \tag{6.54}$$

The magnitude of Δ is of the order of χ_O, which is about 10^{-5}. Equation (6.53b) yields the linear absorption coefficient

$$\mu = 4\pi K'_{O,i} = 4\pi k n_i = -2\pi k \chi_{O,i} \,. \tag{6.55}$$

Note that χ is a negative quantity. For simplicity, if only the real part of the wavevector is concerned, then the dispersion surface of one-wave diffraction is a sphere of radius $n_r k$ with the center at the reciprocal lattice point O. Figure 6.1(b) shows the relationship between the refracted wavevector $\boldsymbol{K}'_{O,r}(=\boldsymbol{OC})$ and the incident $\boldsymbol{k}_{O,r}(=\boldsymbol{OC'})$ for a given incident angle θ'. The line connecting points C and C' along $\boldsymbol{MC}$ is normal to the crystal surface so that the tangential components of $\boldsymbol{K}'_{O,r}$ and $\boldsymbol{k}_{O,r}$ with respect to the crystal surface at the crystal boundary are equal. Namely, Snell's law holds. When the incident angle θ' changes, the direction of the $\boldsymbol{K}'_{O,r}$ is varied by moving point C on the surface of the dispersion sphere shown in the figure. Although a crystal boundary is introduced in Fig. 6.1(a) to reveal the relation between the two wavevectors inside and outside the crystal, the dispersion surface shown in Fig. 6.1(b) is independent of the crystal boundary. Hence, the dispersion surface is intrinsic for a given diffraction situation.

If both the σ- and π-polarizations are considered for the incident wave, there will be two dispersion sheets satisfying the equation of dispersion, (6.48).

ii) Two-wave case
Two reciprocal lattice points, O and G, and two wavevectors, $\boldsymbol{K}_O$ and $\boldsymbol{K}_G$, are involved in a two-wave (O, G) diffraction. The corresponding dispersion surface is constructed in the same way as for the one-wave case. Two spheres of the radius nk centered at points O and G are drawn with the distance between O and G equal to the magnitude of the reciprocal lattice vector of the G reflection, i.e., $OG = g$. Figure 6.2 shows the dispersion surface in the plane of incidence of the G reflection.

The intersection point Lo is the so-called Lorentz point. For simplicity, the wavevectors $\boldsymbol{K}_O$ and $\boldsymbol{K}_G$, instead of their real parts, are shown in the figure. As mentioned in the one-wave case, the spheres represent the surfaces of constant index of refraction n. At the intersection of the two spheres,

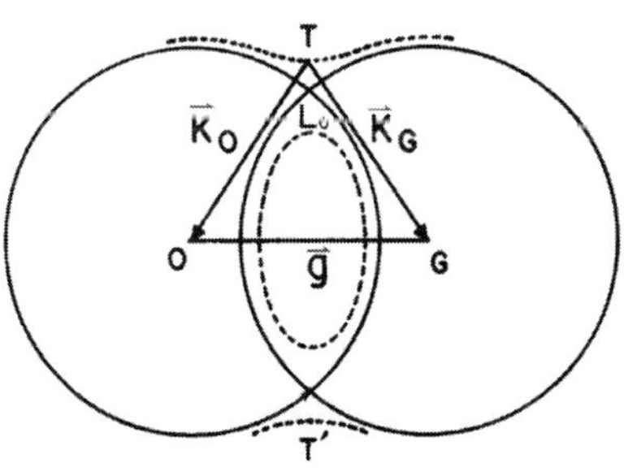

Fig. 6.2. Schematic of the dispersion surface for a two-wave diffraction

Bragg's law, $\boldsymbol{K}'_G = \boldsymbol{K}'_O + \boldsymbol{g}$, holds, where $\boldsymbol{K}'_O$ and $\boldsymbol{K}'_G$ are the one-wave wavevectors of the incident and diffracted waves inside the crystal. The interaction between these two waves modifies further the index of refraction. This gives rise to the dispersion surface, labelled with broken dashed lines, as schematically shown in Fig. 6.2. This dashed surface, formed by the loci of tie points, can be described, according to (6.46), by the following equation of dispersion

$$\begin{vmatrix} \dfrac{K_O^2 - k^2}{K_O^2} - \chi_O & -C\chi_{O\text{-}G} \\ -C\chi_{G\text{-}O} & \dfrac{K_G^2 - k^2}{K_G^2} - \chi_O \end{vmatrix} = 0 \,. \tag{6.56}$$

The polarization factor C of either the σ or the π polarization is defined as

$$C = \begin{cases} p_{\sigma O,\sigma G} = p_{\sigma G,\sigma O} = 1 \,, \\ p_{\pi O,\pi G} = p_{\pi G,\pi O} = \cos 2\theta_G \,. \end{cases} \tag{6.57}$$

With a first-order approximation in χ, (6.56) can be written as

$$\left(K_O^2 - K_O'^2\right)\left(K_G^2 - K_G'^2\right) = C^2 k^4 \chi_{O\text{-}G}\chi_{G\text{-}O} \,, \tag{6.58}$$

where the K'_O and K'_G are the moduli of $\boldsymbol{K}'_O$ and $\boldsymbol{K}'_G$, which are respectively the wavevectors of the one-wave diffractions O and G with an equal magnitude of nk. Assume that the wavevectors $\boldsymbol{K}_O$ and $\boldsymbol{K}_G$ of the two-wave case differ from the one-wave wavevectors $\boldsymbol{K}'_O$ and $\boldsymbol{K}'_G$ by $\triangle \boldsymbol{K}$, i.e.,

$$\boldsymbol{K}_O = \boldsymbol{K}'_O + \Delta\boldsymbol{K} \,, \qquad \boldsymbol{K}_G = \boldsymbol{K}'_G + \Delta\boldsymbol{K} \,. \tag{6.59}$$

The equation of dispersion, (6.58), becomes

$$y_O y_G = \frac{1}{4} k^2 C^2 \chi_{O\text{-}G}\chi_{G\text{-}O} \,, \tag{6.60}$$

where

$$y_O = \frac{\boldsymbol{K}'_O \cdot \Delta\boldsymbol{K}}{k} \,, \qquad y_G = \frac{\boldsymbol{K}'_G \cdot \Delta\boldsymbol{K}}{k} \,. \tag{6.61}$$

Since ΔK is of the order of χ, only the first-order terms are considered in deriving (6.60). This gives y_O and y_G the order of 10^{-5} compared with k. The dispersion surface in the plane of incidence is shown in Fig. 6.3.

The portion in the vicinity of the Lorentz point Lo is magnified, not in proportion to the real scale, to show the details. According to (6.58), the dispersion curve is a hyperbola. The intersection of the two wavefronts represented by straight lines of the incident wave $\boldsymbol{k}_O$ and the diffracted wave $\boldsymbol{k}_G$ in vacuum is the Laue point, denoted by La. The two dashed lines labelled as Σ_O and Σ_G respectively are the dispersion surface for the one-wave case. These two lines (or surface) serve as the asymptotic curves for the dispersion surface. The distances from La and from Lo to the reciprocal lattice points O and G are LaO = LaG = k and LoO = LoG = nk. Point E is an entrance point, arbitrarily chosen such that $\boldsymbol{EO}$, the wavevector of an incident

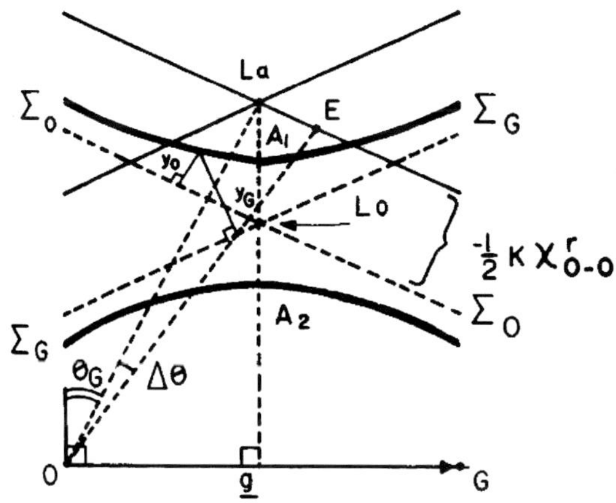

Fig. 6.3. Dispersion sheets in the vicinity of Lorentz point of Fig. 6.2 for a two-wave diffraction

wave, deviates from the Bragg angle θ_G by $\Delta\theta$. From the geometry shown in Fig. 6.3,

$$\mathrm{LaLo} = \frac{kC\chi_{G\text{-}O}}{2\cos\theta_G}, \tag{6.62a}$$

$$\mathrm{A_1A_2} = \frac{kC\sqrt{\chi_{O\text{-}G}\chi_{G\text{-}O}}}{\cos\theta_G}, \tag{6.62b}$$

where A_1 and A_2 are the tie points at the vertices of the hyperbola.

Referring to Fig. 6.2, if both σ and π polarizations are considered, then there are four continuous dispersion sheets for the two-wave diffraction. If, however, the intersection of the dispersion surface with the cross section cutting through points T and T' and perpendicular to $\boldsymbol{g}$ is considered, then the number of discrete dispersion sheets associated with the intersection points (the tie points) is eight.

The dispersion surface discussed here is valid for conventional two-wave diffractions. If an extreme diffraction geometry is involved, such as in the extremely asymmetric diffraction, the grazing-incidence diffraction, and the cases with $\pi/2$ as the Bragg angle, then modification on the equation of dispersion results. This aspect will be discussed in Chap. 8.

iii) Three-wave case

The dispersion surface of a three-wave (O, G_1, G_2) diffraction can be constructed by the same token as for the two-wave case. Three one-wave dispersion spheres Σ_O, Σ_{G_1} and Σ_{G_2} for a given polarization, centered at the reciprocal lattice points O, G_1, and G_2, respectively, can be drawn in the reciprocal space. The radii of the spheres are nk. The vectors connecting these three points, O, G_1, and G_2, are the reciprocal lattice vectors $\boldsymbol{g_1}$, $\boldsymbol{g_2}$ and $\boldsymbol{g_2} - \boldsymbol{g_1}$ of the primary reflection G_1, the secondary reflection G_2 and the coupling reflection $G_2 - G_1$, respectively. Figure 6.4 shows this construction in the vicinity of the Lorentz point Lo.

As discussed for the two-wave case, the one-wave dispersion spheres are modified due to the interaction among the diffracted waves, and the modified dispersion surface is asymptotic to the one-wave dispersion surface. This statement can be equally applied to three-wave cases. As an example, the three one-wave spheres in Fig. 6.4 in a small region near point Lo can be represented as three planes as shown in Fig. 6.5(a). The asymptotic component surface and the actual dispersion sheets are shown in Figs. 6.5(b) and (c) [19]. In these figures, the three surface components are dislocated purposely for clarity. This three-dimensional feature is best viewed from properly chosen two-dimensional cross sections. Figure 6.6(a) is a section at point Lo and perpendicular to $\boldsymbol{g}_1$. The intercept of the dispersion surface with the section C-plane is shown in Fig 6.6(b) as three dispersion curves. Points La and Lo

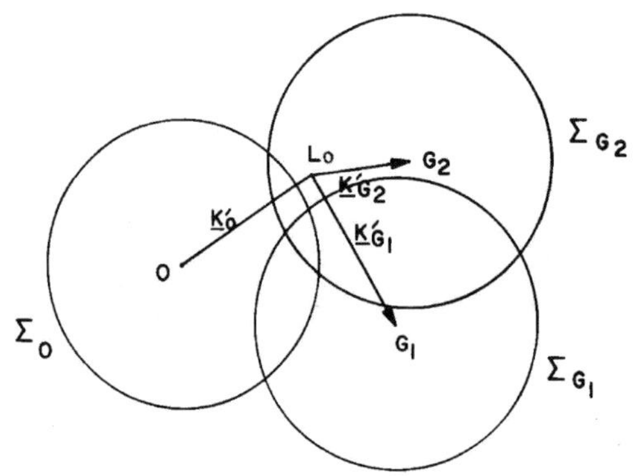

Fig. 6.4. Schematic of the dispersion surface for a three-wave diffraction

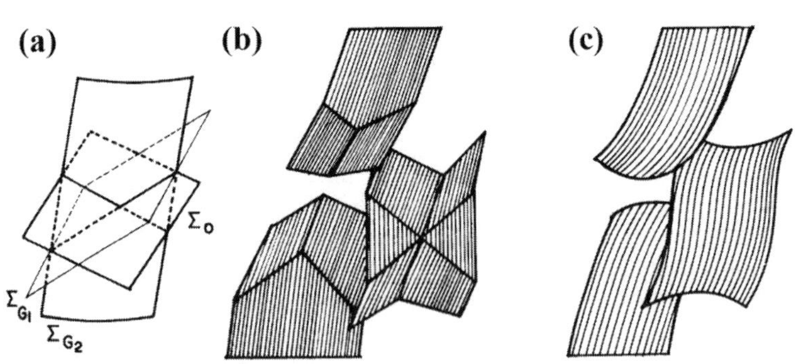

Fig. 6.5. Schematic representation of (**a**) sections of spheres of Fig. 6.4 near Lorentz point Lo, (**b**) the approximate planes for the dispersion sheets, (**c**) the dispersion sheets

are the three-wave Laue point and Lorentz point. The distances from La and Lo to the three reciprocal lattice points are k and nk, respectively. The quantities y_O, y_{G_1}, and y_{G_2}, similar to those defined in (6.61), are the distances from the tie point T to the respective asymptotic lines denoted by Σ_O, Σ_{G_1} and Σ_{G_2}. The horizontal coordinate is proportional to the azimuth angle ψ of rotation around $\boldsymbol{g}_1$. The coordinates, y_O, y_{G_1} and y_{G_2}, of the tie point T are governed by the equation of dispersion, (6.46), for the involvement of three reciprocal lattice points and a single polarization. For simplicity, for the σ polarization and for the cross section at $\Delta\theta = 0$, the equation of dispersion takes the form

$$\begin{vmatrix} 2y_O & kC_1\chi_{O-G_1} & kC_2\chi_{O-G_2} \\ kC_1\chi_{G_1-O} & 2y_{G_1} & kC_3\chi_{G_1-G_2} \\ kC_2\chi_{G_2-O} & kC_3\chi_{G_2-G_1} & 2y_{G_2} \end{vmatrix} = 0\,, \tag{6.63}$$

where the polarization factors C's are

$$C_1 = p_{\sigma O,\sigma G_1}\,, \qquad C_2 = p_{\sigma O,\sigma G_2}\,, \qquad C_3 = p_{\sigma G_1,\sigma G_2}\,. \tag{6.64}$$

For two polarizations, there are 6 continuous dispersion sheets and 12 discrete dispersion sheets.

iv) N-wave case
The dispersion surface of an N-wave diffraction, with $N > 3$, is more complicated than that of a three-wave case, because there are so many dispersion sheets stacked together in the reciprocal space. That is, there are $2N$ continuous and $4N$ discrete dispersion sheets for two polarizations. Hence it is not easy to visualize the three-dimensional structure of the dispersion surface. Yet, with the aid of numerical calculation based on the equation of

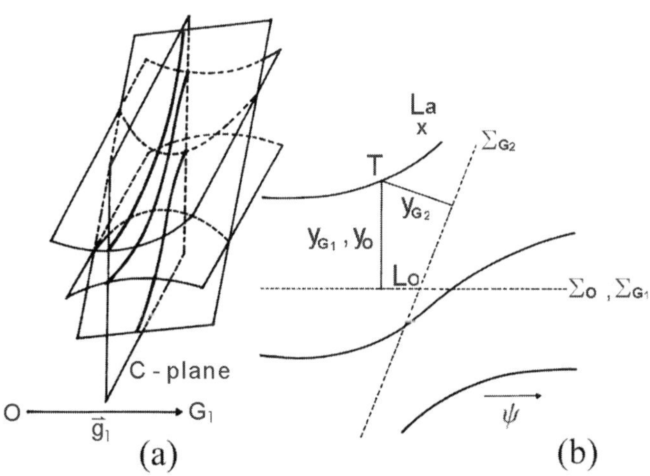

Fig. 6.6. (**a**) A close view of the dispersion sheets of Fig. 6.5(c); (**b**) the projection of the dispersion sheets onto the C plane normal to the reciprocal lattice vector $\boldsymbol{g}_1$

dispersion, (6.46), the cross-section view described for the three-wave case can be adopted to reveal the dispersion surface. However, complications may arise when two dispersion sheets cross over. In other words, the solution to the equation of dispersion is degenerate. Finding non-degenerate solutions is necessary for constructing a correct dispersion surface.

The dispersion surface, as has been described, represents the loci of the tie point. The surface is an intrinsic feature of the dynamical diffraction, which is independent of the crystal boundaries. Moreover, the tie point on the dispersion surface determines the wavevectors of all the reflections involved in a multiple-wave diffraction inside the crystal. Now a question arises. How are tie points related to a given diffraction condition, for example, an incident wavevector? More precisely, what is the relationship between a tie point and a given entrance point E? According to Laue [30], the tie point is also called the excitation point . Then the question becomes how an incident wave excites the tie point of the dispersion surface and generates diffracted waves inside the crystal. The answer has to rely on the condition of phase matching between the incident wave and the diffracted wave at the crystal boundary.

Refer to Fig. 6.7. For a three-wave (O, G_1, G_2) case with a single polarization, the incident wave with the wavevector $\boldsymbol{EO} = \boldsymbol{k}_O$ impinges on the crystal at an angular position specified as (θ, ψ). Points O, G_1 and G_2 are the reciprocal lattice points and point E is the entrance point on the wavefront LaE of the incident wave O. Point La is the three-wave Laue point, at which the crystal is set at the exact three-wave diffraction position, say, (θ_o, ψ_o). The incident wave excites the dispersion surface at the tie points T_1, T_2, and T_3, thus generating diffracted waves along the wavevectors $\boldsymbol{T_jO} = \boldsymbol{K}_O(j)$, $\boldsymbol{T_jG_1} = \boldsymbol{K}_{G_1}(j)$ and $\boldsymbol{T_jG_2} = \boldsymbol{K}_{G_2}(j)$ with j =1, 2, and 3. This excitation satisfies the phase matching condition, i.e., the continuity of the tangential

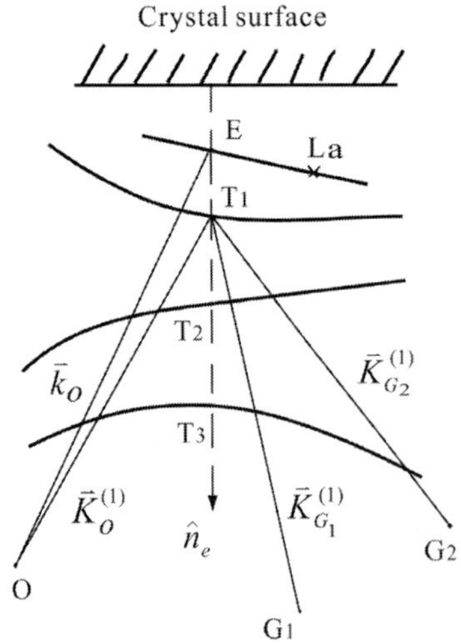

Fig. 6.7. Geometric relationship between tie points and the entrance point E in the reciprocal space

components of the wavevectors inside and outside the crystal at the crystal boundary:

$$k_{O,t} = K_{O,t}(j) \,, \tag{6.65a}$$

$$K_{G_1,t}(j) = K_{G_1,t}(m) \,, \qquad j \neq m \tag{6.65b}$$

$$K_{G_2,t}(j) = K_{G_2,t}(m) \,, \qquad j \neq m \tag{6.65c}$$

for j, m =1, 2, 3. The subscript t stands for the tangential component. Equations (6.65a)–(6.65c) imply that points E, T_1, T_2 and T_3 are lined up along the inward surface normal $\hat{\boldsymbol{n}}_e$. In other words, for a given entrance point E on the incident wavefront, the excited tie points T_j can be found by drawing a line perpendicular to the crystal surface from the entrance point E. The intersections of this line with the dispersion surface are the tie points T_j. The distance between the entrance point E and the tie point T_j is expressed as

$$ET_j = k\zeta(j) \,, \tag{6.66}$$

where $\zeta(j)$ is the 'accommodation' (Anpassung) according to Ewald [57–59]. The wavevectors of the excited waves at the tie points inside the crystal then take the form

$$\boldsymbol{K}_O(j) = \boldsymbol{k}_O - k\zeta(j)\hat{\boldsymbol{n}}_e \,, \tag{6.67a}$$

$$\boldsymbol{K}_{G_m}(j) = \boldsymbol{k}_{G_m} - k\zeta(j)\hat{\boldsymbol{n}}_e \,, \tag{6.67b}$$

where $m = 1, \ldots, N-1$ for an N-wave diffraction. The $\zeta(j)$ is determined by the j-th eigenvalue of (6.46). In terms of the real part, ζ_r, and the imaginary part, ζ_i, of ζ, both $\boldsymbol{K}_O$ and $\boldsymbol{K}_G$ can be expressed as

$$\boldsymbol{K}_O(j) = \boldsymbol{K}_{O,r}(j) - \mathrm{i}\boldsymbol{K}_{O,i}(j) \,, \tag{6.68a}$$

$$\boldsymbol{K}_{G_m}(j) = \boldsymbol{K}_{G_m,r}(j) - \mathrm{i}\boldsymbol{K}_{G_m,i}(j) \,, \tag{6.68b}$$

where

$$\boldsymbol{K}_{O,r}(j) = \boldsymbol{k}_O - k\zeta_r(j)\hat{\boldsymbol{n}}_e \,, \tag{6.69a}$$

$$\boldsymbol{K}_{G_m,r}(j) = \boldsymbol{k}_{G_m,r} - k\zeta_r(j)\hat{\boldsymbol{n}}_e \,, \tag{6.69b}$$

$$\boldsymbol{K}_{O,i}(j) = \boldsymbol{K}_{G_m,i}(j) = k\zeta_i(j)\hat{\boldsymbol{n}}_e \,. \tag{6.69c}$$

Equation (6.69c) indicates that the imaginary parts of the wavevectors inside the crystal are all equal and are normal to the crystal surface.

The $2\varepsilon_{G_m}$ defined in (6.38) can be written as a quadratic function of ζ as

$$2\varepsilon_{G_m} = \zeta^2(j) - 2\zeta(j)\gamma_{G_m} + \alpha_{G_m} \,, \tag{6.70}$$

where the direction cosines γ's and the geometric factors α's are

$$\gamma_{G_m} = \hat{\boldsymbol{k}}_{G_m} \cdot \hat{\boldsymbol{n}}_e \,, \tag{6.71}$$

$$\alpha_{G_m} = \frac{g_m^2 + 2\boldsymbol{k}_O \cdot \boldsymbol{g}_m}{k^2} \,. \tag{6.72}$$

At the exact diffraction condition for the G_m reflection,

$$g_m^2 + 2\boldsymbol{k}_O \cdot \boldsymbol{g}_m = 0 \,, \tag{6.73}$$

which leads to

$$\alpha_{G_m} = 0 \,. \tag{6.74}$$

For an N-wave diffraction with two polarizations, there are $2N$ continuous dispersion sheets, which can be excited at $4N$ possible tie points by a given incident wave. This fact is illustrated schematically in Figs. 6.8(a) and 6.8(b) for a two-wave and a three-wave case with a single polarization. As shown in Fig. 6.8(a), there are two continuous dispersion sheets with closed surfaces. T_1, T_2, T_1' and T_2' are the four excited tie points, defined by the inward surface normal $\hat{\boldsymbol{n}}_e$. Points T_1 and T_2 are the excitation points close to the entrance point E on the upper intersection of the two spheres (not shown) of the radius $1/\lambda$ centered at O and G. Points T_1' and T_2' are associated with the lower intersection far from point E. Similarly, there are three continuous dispersion sheets and six tie points (see Fig. 6.8(b)) with three in the upper and the other three in the lower intersection of three spheres centered at the three reciprocal lattice points (Only two spheres are partially shown in the figure.).

In contrast to the dispersion surface, the location of tie points and the value of the accommodation ζ rely on the crystal boundaries. Hence the excitations of the dispersion surface for a Bragg reflection and for a Laue transmission are quite different. The details of the dynamical behavior of these two cases will be discussed in Chap. 8.

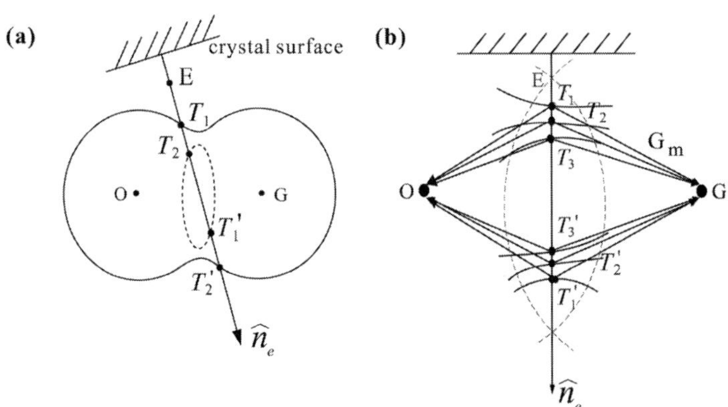

Fig. 6.8. Geometric relation between the continuous dispersion sheets and the tie points in (**a**) two-wave diffraction and (**b**) three-wave diffraction

6.5 Energy Flow and Group Velocity

The energy flow of the wavefield can be described, as usual, by the Poynting vector

$$\boldsymbol{S} = \frac{c}{4\pi}(\boldsymbol{E} \times \boldsymbol{H}) \,. \tag{6.75}$$

The electric field $\boldsymbol{E}$ and the magnetic field $\boldsymbol{H}$ are expressed in terms of the Bloch functions similar to those given in (6.18a)–(6.18d) as

$$\begin{aligned} \boldsymbol{E}(\boldsymbol{r},t) &= \exp(\mathrm{i}\omega t)\sum_j \sum_{\sigma G} E_{\sigma G}(j)\exp\left[-2\pi\mathrm{i}\boldsymbol{K}_G(j)\cdot\boldsymbol{r}\right] \,, \\ \boldsymbol{H}(\boldsymbol{r},t) &= \exp(\mathrm{i}\omega t)\sum_j \sum_{\sigma G} H_{\sigma G}(j)\exp\left[-2\pi\mathrm{i}\boldsymbol{K}_G(j)\cdot\boldsymbol{r}\right] \,, \end{aligned} \tag{6.76}$$

where $\boldsymbol{E}_{\sigma G}(j)$ and $\boldsymbol{H}_{\sigma G}(j)$ are related to the eigenvector $\boldsymbol{D}_{\sigma G}(j)$ of (6.45) and the boundary conditions. For the j-th dispersion sheet, the Poynting vector $\boldsymbol{S}(j)$ is

$$\begin{aligned} \boldsymbol{S}(j) = &\frac{c}{4\pi}\exp\left(-4\pi\boldsymbol{K}_{O,i}(j)\cdot\boldsymbol{r}\right)\sum_{\sigma G}\sum_{\sigma G'}\left[\boldsymbol{E}_{\sigma G}(j)\times\boldsymbol{H}_{\sigma G'}(j)\right] \\ &\times\exp\left[2\pi\mathrm{i}\left(\boldsymbol{K}_{G,r}(j)-\boldsymbol{K}_{G',r}(j)\right)\cdot\boldsymbol{r}\right] \,, \end{aligned} \tag{6.77}$$

where

$$\boldsymbol{E}_{\sigma G}(j)\times\boldsymbol{H}_{\sigma G'}(j) = \boldsymbol{E}_{\sigma G}(j)\times\frac{\hat{\boldsymbol{K}}_{G'}(j)\times\boldsymbol{E}_{\sigma G'}(j)}{\mu_n} \,. \tag{6.78}$$

For a nonmagnetic material, μ_n is assumed to be approximately unity. By averaging over time and space, the Poynting vector, which is physically measurable, becomes

$$\langle\boldsymbol{S}(j)\rangle = \frac{c}{8\pi}\exp\left(-4\pi\boldsymbol{K}_i(j)\cdot\boldsymbol{r}\right)\sum_{\sigma G}\hat{\boldsymbol{K}}_G\left(E_{\sigma G}E^*_{\sigma G}\right) \,. \tag{6.79}$$

The relation

$$\langle\exp\left[2\pi\mathrm{i}\left(\boldsymbol{K}_{G,r}(j)-\boldsymbol{K}_{G',r}(j)\right)\cdot\boldsymbol{r}\right]\rangle = \frac{1}{V_o}\delta\left(\boldsymbol{K}_G-\boldsymbol{K}_{G'}\right) \tag{6.80}$$

has been used for deriving (6.80), where V_o is the volume of the crystal. Based on (6.25), the Fourier component $E_{\sigma G}$ can be replaced by $D_{\sigma G}$. With this approximation, the averaged Poynting vector can be expressed in terms of D's as

$$\langle\boldsymbol{S}(j)\rangle \sim \frac{c}{8\pi}\exp\left(-4\pi\boldsymbol{K}_{O,i}(j)\cdot\boldsymbol{r}\right)\sum_{\sigma G}\hat{\boldsymbol{K}}_G\left|D_{\sigma G}\right|^2 \,. \tag{6.81}$$

Equations (6.79) and (6.81) amount to that the Poynting vector of a Bloch wave is the sum of the Poynting vectors of all the component waves.

At the first sight, it seems not easy to express $|D_{\sigma G}|^2$ in terms of the matrix elements of (6.43). If, however, the crystal is assumed to be non-absorbing, then the transpose of the Fourier coefficient χ is equal to the complex conjugate of the χ itself, i.e., $\chi_{G_l,G_m} = \chi^*_{G_m,G_l}$, and the corresponding cofactors satisfy the condition

$$\mathrm{Cof}\,(\sigma G_l, \pi G_m)_j = \left[\mathrm{Cof}\,(\pi G_m, \sigma G_l)_j\right]^* . \tag{6.82}$$

Now by multiplying the first equation of (6.47) by the complex conjugate of the second equation, the ratio of the square of the wavefield amplitude among the diffracted waves associated with the j-th eigenvalue can be expressed as either

$$\frac{|D_{\sigma G_l}(j)|^2}{|D_{\pi G_m}(j)|^2} = \frac{\mathrm{Cof}\,(\sigma G_l, \sigma G_l)_j}{\mathrm{Cof}\,(\pi G_m, \pi G_m)_j} \tag{6.83}$$

or

$$\frac{|D_{\sigma G_l}(j)|^2}{\mathrm{Cof}\,(\sigma G_l, \sigma G_l)_j} = \frac{|D_{\pi G_m}(j)|^2}{\mathrm{Cof}\,(\pi G_m, \pi G_m)_j} = \cdots = \text{constant} . \tag{6.84}$$

From (6.81) and (6.84), the Poyning vector becomes

$$\langle \boldsymbol{S}(j) \rangle \propto \frac{c}{8\pi} \exp\left[-4\pi \boldsymbol{K}_i(j) \cdot \boldsymbol{r}\right] \sum_{\sigma G} \hat{\boldsymbol{K}}_G \mathrm{Cof}(\sigma G, \sigma G)_j . \tag{6.85}$$

The determinant of the matrix $\tilde{\Phi}$ given in (6.44) for a given mode j can be expanded in terms of the matrix elements of the σG_1 and the cofactors as

$$\begin{aligned} \det \tilde{\Phi} = {} & (2\varepsilon_{G_1} - \chi_O)\,\mathrm{Cof}\,(\sigma G_1, \sigma G_1) \\ & + \sum_{\sigma M} \chi_{G_1 - M} p_{\sigma M, \sigma G_1} \mathrm{Cof}\,(\sigma G_1, \sigma M) , \end{aligned} \tag{6.86}$$

where the summation is taken over all the σ's and π's. Now if we consider the positional variation of the tie point T on a given j-th dispersion sheet, there are two possible ways to shift the tie point T. First, the tie point T can be shifted to T' by an infinitesimal variation $\boldsymbol{\Delta\tau}$ tangent to the dispersion sheet due to momentum transfer, while the angular frequency ω being kept constant (Fig. 6.9(a)). Second, the tie point T can be shifted to T'' by $\boldsymbol{\Delta K}_\nu$ normal to the dispersion sheet due to an infinitesimal frequency variation $\Delta\omega$ (Fig. 6.9(b)). Both ways are schematically shown in Fig. 6.9, where $\boldsymbol{\Delta\tau} = \Delta\tau\hat{\boldsymbol{\tau}}$ and $\boldsymbol{\Delta K}_\nu = \Delta K_\nu \hat{\boldsymbol{\nu}}$ are along the unit vectors $\hat{\boldsymbol{\tau}}$ and $\hat{\boldsymbol{\nu}}$ tangent and normal to the j-th dispersion sheet, respectively. The two possible ways are discussed below.

(i) Variation due to momentum transfer
All the wavevectors $\boldsymbol{K}_M$ become $\boldsymbol{K}_M + \boldsymbol{\Delta\tau}$ and satisfy the dispersion relation

$$\Delta^k[\det \tilde{\Phi}] = 0 . \tag{6.87}$$

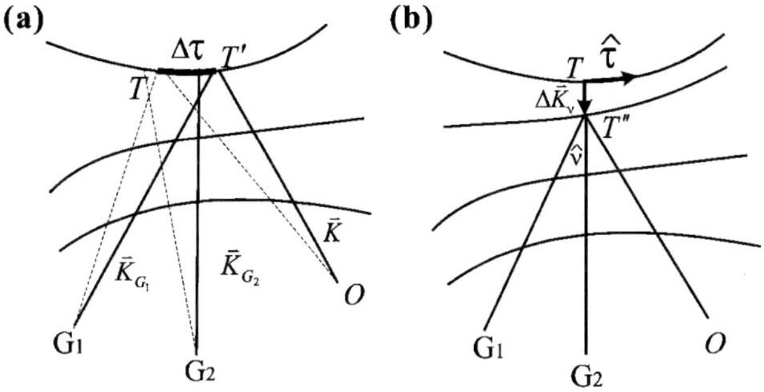

Fig. 6.9. Definitions of the variations on the dispersion sheets: (**a**) $\boldsymbol{\Delta\tau}$, and (**b**) $\boldsymbol{\nu}$

Δ^k is the variational operator with respect to K due to this infinitesimal $\boldsymbol{\Delta\tau}$. Since $2\varepsilon_{G_1}$ depends on the wavevector $\boldsymbol{K}_{G_1}$, the derivative of the matrix $\tilde{\Phi}$ with respect to $\boldsymbol{K}_{G_1}$ leads to

$$\begin{aligned}\Delta^k[\det\tilde{\Phi}] &= (\boldsymbol{\Delta\tau}\cdot\boldsymbol{K}_{G_1})\,\frac{2\mathrm{Cof}\,(\sigma G_1,\sigma G_1)}{k^2}\\ &\quad+(2\varepsilon_{G_1}-\chi_O)\,\Delta^k\,[\mathrm{Cof}\,(\sigma G_1,\sigma G_1)]\\ &\quad+\sum_{\sigma M}\chi_{G_1-M}\,p_{\sigma M,\sigma G_1}\,\Delta^k\,[\mathrm{Cof}\,(\sigma G_1,\sigma M)]\\ &= -\,(\boldsymbol{\Delta\tau}\cdot\boldsymbol{K}_{G_1})\,\frac{2\mathrm{Cof}\,(\sigma G_1,\sigma G_1)}{k^2}+\Delta^k_{\sigma G_1}[\det\tilde{\Phi}]\,,\end{aligned}\tag{6.88}$$

where $\Delta^k_{\sigma G_1}$ is the variational operator acting on all the matrix elements except the elements of the σG_1 column and the σG_1 row. This operator acting on the matrix $\tilde{\Phi}$ yields

$$\Delta^k_{\sigma G_1}[\det\tilde{\Phi}] = (\boldsymbol{\Delta\tau}\cdot\boldsymbol{K}_{G_2})\,\frac{2\mathrm{Cof}\,(\pi G_2,\pi G_2)}{k^2}+\Delta^k_{\sigma G_1,\pi G_2}[\det\tilde{\Phi}]\,. \tag{6.89}$$

Continuing this variational operation over all the σM's gives

$$\Delta^k[\det\tilde{\Phi}] = \boldsymbol{\Delta\tau}\cdot\sum_{\sigma M}\boldsymbol{K}_M\frac{2\mathrm{Cof}\,(\sigma M,\sigma M)}{k^2}=0\,, \tag{6.90}$$

since

$$\Delta^k_{\sigma O,\pi O,\sigma G_1,\pi G_1,\cdots,\sigma G_{N-1},\pi G_{N-1}}[\det\tilde{\Phi}]=0\,. \tag{6.91}$$

From (6.85) and (6.91), the Poynting vector associated with the j-th dispersion sheet is normal to the tangential plane of the sheet, so that

$$\langle\boldsymbol{S}(j)\rangle\cdot\boldsymbol{\Delta\tau}=0\,. \tag{6.92}$$

The same conclusion can also be drawn from the consideration of wave packet [498]. This orthogonality of the energy flow with respect to the dispersion surface is analogous to that ray trajectories are always normal to the surface of refractive index in the crystal optics of the visible spectrum.

(ii) Variation due to frequency change
The current frequency can be expressed as

$$\omega = \omega_0 + \boldsymbol{v}_g \cdot \Delta \boldsymbol{K}_\nu \, , \tag{6.93}$$

where $\boldsymbol{v}_g$ is the group velocity defined as

$$\boldsymbol{v}_g = \left(\frac{\partial \omega}{\partial K_\nu}\right)_\tau \hat{\boldsymbol{\nu}} = \mathrm{grad}_{K_\nu} \omega \, . \tag{6.94}$$

The variation of the dispersion surface with respect to the changes ΔK_ν and $\Delta\omega$ yields

$$\Delta[\det \tilde{\Phi}] = \frac{\partial[\det \tilde{\Phi}]}{\partial K_\nu} \Delta K_\nu + \frac{\partial[\det \tilde{\Phi}]}{\partial \omega} \Delta\omega = 0 \, , \tag{6.95}$$

where

$$\frac{\partial[\det \tilde{\Phi}]}{\partial K_\nu} = \hat{\boldsymbol{\nu}} \cdot \sum_{\sigma M} \frac{2\mathrm{Cof}(\sigma M, \sigma M)}{k^2} \, , \tag{6.96}$$

$$\frac{\partial[\det \tilde{\Phi}]}{\partial \omega} \simeq -\frac{\omega}{c^2} \sum_{\sigma M} \frac{2\mathrm{Cof}(\sigma M, \sigma M)}{k^2} \, , \tag{6.97}$$

following the same procedure as for $\Delta^k[\det \tilde{\Phi}]$ of (6.88). The relations $k = (\omega/c)/(2\pi)$ and $K_M \simeq k$ have been employed for deriving (6.97). Equation (6.95) then gives

$$\frac{1}{c}\left(\frac{\partial \omega}{\partial K_\nu}\right)_\tau \sum_{\sigma M} \mathrm{Cof}(\sigma M, \sigma M) = \hat{\boldsymbol{\nu}} \cdot \sum_{\sigma M} \hat{\boldsymbol{K}}_M \mathrm{Cof}(\sigma M, \sigma M) \, . \tag{6.98}$$

This implies that

$$(\langle W_\mathrm{e} \rangle + \langle W_\mathrm{m} \rangle) \, \boldsymbol{v}_g = \langle \boldsymbol{S} \rangle \, , \tag{6.99}$$

because the average electric and magnetic energy densities associated with each dispersion sheet are

$$\begin{aligned} \langle W_\mathrm{e} \rangle &= \langle W_\mathrm{m} \rangle \\ &\propto \frac{1}{16\pi} \exp\left(-4\pi \boldsymbol{K}_i \cdot \boldsymbol{r}\right) \sum_{\sigma M} D^*_{\sigma M} D_{\sigma M} \\ &= \frac{1}{16\pi} \exp\left(-4\pi \boldsymbol{K}_i \cdot \boldsymbol{r}\right) \sum_{\sigma M} \mathrm{Cof}(\sigma M, \sigma M) \, . \end{aligned} \tag{6.100}$$

6.6 Mode of Wave-Propagation

The dispersion equation defines not only the direction of energy flow but also the type of wave propagation in the crystal. The type of wave propagation, sometimes called the mode of propagation [37, 177, 178], in analog to the normal modes in the classical vibration of continuum, is specified by the eigenvalue $\boldsymbol{K}_G(j)$ and the corresponding eigenvector related to the ratio of wavefield amplitudes $D(j)$'s. Hence, the j-th mode of propagation is defined by and associated with the j-th dispersion sheet. In other words, the X-ray wavefields in the crystal are described as the normal modes of vibration of the electric displacements, each of which not only propagates along the direction specified by the eigenvalue but forms a standing wave normal to that direction.

The modes of wave-propagation are illustrated below for symmetric two-wave, three-wave, and N-wave diffractions.

6.6.1 Wavefields of Modes for Symmetric Two-Wave Transmission Diffraction

The fundamental equation of wavefields for a two-wave (O, G) Laue diffraction, according to (6.56), is

$$\begin{bmatrix} \dfrac{K_O^2 - k^2}{K_O^2} - \chi_O & -C\chi_{O\text{-}G} \\ -C\chi_{G\text{-}O} & \dfrac{K_G^2 - k^2}{K_G^2} - \chi_O \end{bmatrix} \begin{bmatrix} D_O \\ D_G \end{bmatrix} = 0 \,, \tag{6.101}$$

where C is the polarization factor; $C = 1$ for the σ polarization and $C = \cos 2\theta_G$ for the π polarization (see Fig. 6.10(a)). This fundamental equation can be simplified as

$$\begin{pmatrix} 2y_O & kC\chi_{O\text{-}G} \\ kC\chi_{G\text{-}O} & 2y_G \end{pmatrix} \begin{pmatrix} D_O \\ D_G \end{pmatrix} = 0 \,, \tag{6.102}$$

whcre

$$y_O = k\left(\chi_O - \frac{K_O^2 - k^2}{K_O^2}\right) \,, \tag{6.103a}$$

$$y_G = k\left(\chi_O - \frac{K_G^2 - k^2}{K_G^2}\right) \,. \tag{6.103b}$$

Equations (6.103a) and (6.103b) are equivalent to (6.61). The eigenvector of (6.102) gives the wavefield-amplitude ratio

$$\frac{D_G}{D_O} = -\frac{kC\chi_{G\text{-}O}}{2y_G} = -\frac{2y_O}{kC\chi_{O\text{-}G}} \,. \tag{6.104}$$

At the exact two-wave diffraction position, according to (6.60),

$$y_O = y_G = y = \pm\frac{kC}{2}\sqrt{\chi_{O\text{-}G}\chi_{G\text{-}O}} \,. \tag{6.105}$$

Therefore, there are four eigenvalues:

$$\left.\begin{matrix} y(1) \\ y(4) \end{matrix}\right\} = \mp\frac{k}{2}\sqrt{\chi_{O\text{-}G}\chi_{G\text{-}O}} \tag{6.106}$$

for the σ polarization, and

$$\left.\begin{array}{l} y(2) \\ y(3) \end{array}\right\} = \mp\frac{k}{2}\sqrt{\chi_{O\text{-}G}\chi_{G\text{-}O}}\cos 2\theta_G \tag{6.107}$$

for the π polarization. These four eigenvalues define four modes of propagation. Referring to (6.104), the eigenvectors are related to the ratios for all the modes j:

$$\xi(j) = \frac{D_G(j)}{D_O(j)} , \tag{6.108}$$

which leads to

$$\begin{aligned} \xi(1) = \xi(2) &= \frac{\chi_{G\text{-}O}}{\sqrt{\chi_{O\text{-}G}\chi_{G\text{-}O}}} , \\ \xi(3) = \xi(4) &= -\frac{\chi_{G\text{-}O}}{\sqrt{\chi_{O\text{-}G}\chi_{G\text{-}O}}} . \end{aligned} \tag{6.109}$$

For a non-absorbing crystal, $\chi_{O\text{-}G} = \chi^*_{G\text{-}O}$. Thus (6.109) becomes

$$\begin{aligned} \xi(1) = \xi(2) &= \exp\left[\mathrm{i}\left(\pi + \delta_{G\text{-}O}\right)\right] , \\ \xi(3) = \xi(4) &= \exp\left[\mathrm{i}\delta_{G\text{-}O}\right] , \end{aligned} \tag{6.110}$$

where $\delta_{G\text{-}O}$ is the crystallographic phase of the $G-O$ reflection. Namely, the magnitudes of ξ aside from the crystallographic phase can be expressed as

$$\begin{aligned} |\xi(1)| = |\xi(2)| &= -1 , \\ |\xi(3)| = |\xi(4)| &= 1 . \end{aligned} \tag{6.111}$$

Note that $\chi_{G\text{-}O}$ is negative. Equations (6.110) and (6.111) indicate that the diffracted waves of modes 1 and 2 are out of phase with respect to the refracted waves, whereas the diffracted waves of modes 3 and 4 are in phase with the refracted waves. Because modes 1 and 4 are associated with the σ-polarization and modes 2 and 3 with the π-polarization, the electric displacements of modes 1 and 4 are perpendicular to the plane of incidence, while those of modes 2 and 3 lie in that plane. Hence, there is no correlation between the σ- and the π-polarized wavefields. These four modes of propagation are depicted schematically in Fig. 6.10(b–e), which are the side-views of Fig. 6.10(a) for the polarized components of the wavefields.

The wavefield $\boldsymbol{D}(j)$ of mode j is the sum of the wavefields of all the diffracted waves M ($M = O,\ G$) associated with the j-th dispersion sheet, i.e.,

$$\begin{aligned} \boldsymbol{D}(j) &= \sum_M \boldsymbol{D}_M(j)\exp\left[-2\pi\mathrm{i}\boldsymbol{K}_M(j)\cdot\boldsymbol{r}\right] \\ &= \boldsymbol{D}_O(j)\exp\left[-2\pi\mathrm{i}\boldsymbol{K}_O(j)\cdot\boldsymbol{r}\right] + \boldsymbol{D}_G(j)\exp\left[-2\pi\mathrm{i}\boldsymbol{K}_G(j)\cdot\boldsymbol{r}\right] . \end{aligned} \tag{6.112}$$

The intensity of this wavefield can be expressed in terms of the amplitude ratios ξ's and the polarization factors C's as

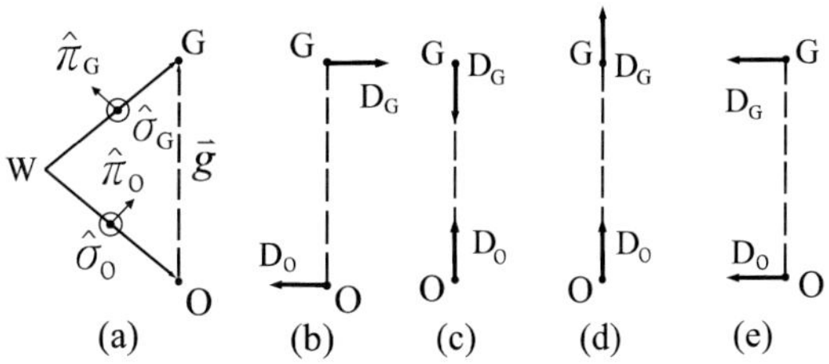

Fig. 6.10. (**a**) Definitions of polarization vectors for a symmetric two-wave Laue diffraction at the exact diffraction position; (**b-e**) Schematic of the projected wavefields on the plane perpendicular to OGW for modes 1, 2, 3 and 4, respectively

$$
\begin{aligned}
I(j) &= \boldsymbol{D}^*(j) \cdot \boldsymbol{D}(j) \\
&= |\boldsymbol{D}_O(j)|^2 \left\{1 + |\xi(j)|^2 + 2C|\xi(j)| \cos\left[2\pi(\boldsymbol{g} \cdot \boldsymbol{r}) + \delta_{G\text{-}O}\right]\right\} . \quad (6.113)
\end{aligned}
$$

The Bragg diffraction condition, $\boldsymbol{K}_G(j) - \boldsymbol{K}_O(j) + \boldsymbol{g}$, has been used in deriving (6.111), where $\boldsymbol{g}$ is the reciprocal lattice vector of the G reflection. The intensity maxima of $I(1)$ of mode 1 occur when

$$\boldsymbol{g} \cdot \boldsymbol{r} = N + \frac{1}{2} - \frac{\delta_{G\text{-}O}}{2\pi} , \tag{6.114}$$

because $|\xi(1)| = -1$ and $C = 1$. On the contrary, the maxima of $I(4)$ of mode 4 take place when

$$\boldsymbol{g} \cdot \boldsymbol{r} = N - \frac{\delta_{G\text{-}O}}{2\pi} , \tag{6.115}$$

because $|\xi(1)| = 1$ and $C = 1$. This means that the standing wave of mode 1 is formed with antinodes in the middle of the atomic planes, while that of mode 4 with antinodes coincident with the atomic planes for the G reflection (see Fig. 6.11). It should be noted that although (6.113) is valid for the π polarized modes 2 and 3, these two modes do not form standing waves in the crystal. This is because the wavefield components lie in the plane normal to the atomic planes.

6.6.2 Wavefields of Modes for Symmetric Three-Wave Transmission Diffraction

The fundamental equation of wavefield, (6.43), for a three-wave Laue diffraction involves at least a 6×6 matrix equation for the two polarizations, σ and π. To find an analytical solution to the dispersion equation, (6.46), is difficult for a general three-wave case. In order to illustrate the modes of propagation, a symmetric three-wave (O, G, L) Laue diffraction at the exact three-wave diffraction position is considered. The involved reciprocal lattice vectors $\boldsymbol{g}$, $\boldsymbol{l}$ and $\boldsymbol{g}-\boldsymbol{l}$ of the primary reflection G, the secondary L, and the

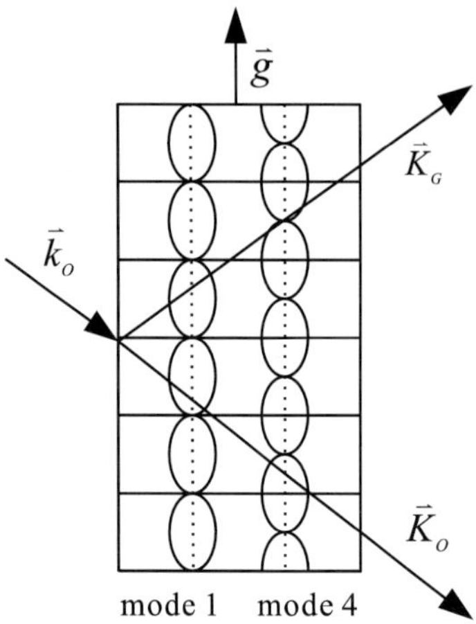

Fig. 6.11. Standing-wave patterns of modes 1 and 4

coupling G–L, respectively, form an equilateral triangle OGL (See Fig. 6.12), so that

$$\chi_{O\text{–}G} = \chi_{O\text{–}L} = -\chi_{G\text{–}L} = \chi_{L\text{–}G} = \chi_{G\text{–}O} = \chi_{L\text{–}O} = a\ . \tag{6.116}$$

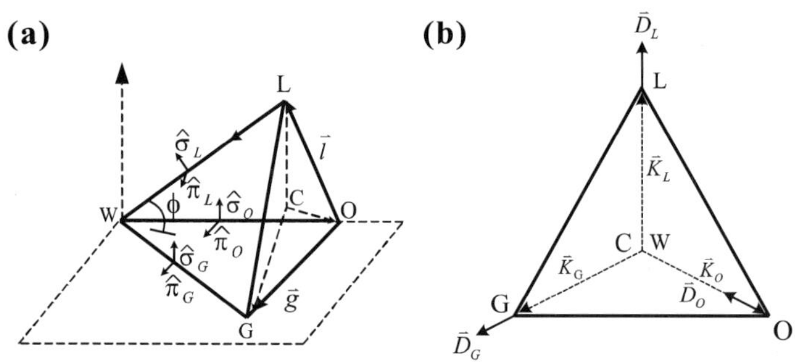

Fig. 6.12. (**a**) Definitions of polarization vectors for a symmetric three-wave Laue diffraction at the exact diffraction position; (**b**) schematic of the projected wavefields of the non-degenerate mode

In Fig. 6.12(a), the polarized unit vectors $\hat{\boldsymbol{\sigma}}_O$ and $\hat{\boldsymbol{\sigma}}_G$ are perpendicular to the plane of incidence, WOG, of the G reflection and the $\hat{\boldsymbol{\pi}}_O$ and $\hat{\boldsymbol{\pi}}_G$ are chosen to lie in that plane. Vectors $\boldsymbol{WO}$, $\boldsymbol{WG}$ and $\boldsymbol{WL}$ are the wavevectors $\boldsymbol{K}_O$, $\boldsymbol{K}_G$ and $\boldsymbol{K}_L$. Points W and C are the center of the Ewald sphere and the center of the circle circumscribing O, G and L, respectively. The angle between $\boldsymbol{K}_L$ and the plane of incidence of the G reflection is ϕ. The unit vector $\hat{\boldsymbol{\sigma}}_L$ is chosen to be perpendicular to both $\boldsymbol{g}$ and $\boldsymbol{K}_L$. All the $\hat{\boldsymbol{\sigma}}$'s, $\hat{\boldsymbol{\pi}}$'s and $\hat{\boldsymbol{K}}$'s follow the orthogonal conditions given in (6.34a) and (6.34b). The polarization factors are the products among these polarization unit vectors:

$$
\begin{aligned}
&\hat{\boldsymbol{\sigma}}_O \cdot \hat{\boldsymbol{\pi}}_G = \hat{\boldsymbol{\sigma}}_G \cdot \hat{\boldsymbol{\pi}}_O = \hat{\boldsymbol{\sigma}}_O \cdot \hat{\boldsymbol{\pi}}_L = \hat{\boldsymbol{\sigma}}_G \cdot \hat{\boldsymbol{\pi}}_L \,, \\
&\hat{\boldsymbol{\sigma}}_L \cdot \hat{\boldsymbol{\pi}}_O = -\hat{\boldsymbol{\sigma}}_L \cdot \hat{\boldsymbol{\pi}}_G = \sin\theta_G \sin\psi = C_1 \,, \\
&\hat{\boldsymbol{\sigma}}_O \cdot \hat{\boldsymbol{\sigma}}_L = \hat{\boldsymbol{\sigma}}_G \cdot \hat{\boldsymbol{\sigma}}_L = \cos\psi = C_2 \,, \\
&\hat{\boldsymbol{\pi}}_O \cdot \hat{\boldsymbol{\pi}}_L = \hat{\boldsymbol{\pi}}_G \cdot \hat{\boldsymbol{\pi}}_L = \cos\theta_G = C_3 \,, \\
&\hat{\boldsymbol{\pi}}_O \cdot \hat{\boldsymbol{\pi}}_G = \cos 2\theta_G = C \,,
\end{aligned}
\tag{6.117}
$$

where θ_G is the Bragg angle of the G reflection and the factors C_1, C_2 and C_3 are related to C as

$$
C_1^2 = \frac{(1+2C)(1-C)^2}{2(1+C)} \,, \quad C_2^2 = \frac{2C^2}{1+C} \,, \quad C_3^2 = \frac{1+C}{2} \,. \tag{6.118}
$$

Thus, the fundamental equation of wavefield, (6.43), takes the following simple form

$$
\begin{bmatrix}
y'_O & 0 & a & 0 & C_2 a & 0 \\
0 & y'_O & 0 & Ca & C_1 a & C_3 a \\
a & 0 & y'_G & 0 & -C_2 a & 0 \\
0 & Ca & 0 & y'_G & C_1 a & -C_3 a \\
C_2 a & C_1 a & -C_2 a & C_1 a & y'_L & 0 \\
0 & C_3 a & 0 & -C_3 a & 0 & y'_L
\end{bmatrix}
\begin{bmatrix}
D_{\sigma O} \\ D_{\pi O} \\ D_{\sigma G} \\ D_{\pi G} \\ D_{\sigma L} \\ D_{\pi L}
\end{bmatrix}
= 0 \,, \tag{6.119}
$$

where $y'_M = 2y_M/k$ for $M = O$, G and L. For this symmetric three-wave case, $y'_O = y'_G = y'_L = y'$ at the exact three-wave diffraction point. The solutions to the equation of dispersion are

$$
\begin{aligned}
&y'(1) = 1 - 2C \,, \\
&y'(2) = y'(3) = 1 + C \,, \\
&y'(4) = y'(5) = y'(6) = -1 \,,
\end{aligned}
\tag{6.120}
$$

which indicate that the eigenvalues $y'(1)$, $y'(2)$ and $y'(4)$ are singly, doubly and triply degenerate, respectively. Namely, mode 1 is the singly degenerate mode, modes 2 and 3 are the doubly degenerate modes, and modes 4, 5 and 6 are the triply degenerate modes. The corresponding eigenvectors derived from (6.119) are discussed below.

(i) The singly degenerate mode (mode 1)
The wavefield-amplitude relations for the eigenvalue $y'(1)$ are

$$
\begin{aligned}
&D_{\sigma O}(1) : D_{\pi O}(1) : D_{\sigma G}(1) : D_{\pi G}(1) : D_{\sigma L}(1) = -\frac{1}{\beta} : 1 : \frac{1}{\beta} : 1 : -\frac{\beta_1}{\beta} \,, \\
&D_{\pi L}(1) = 0 \,,
\end{aligned}
\tag{6.121}
$$

where $\beta = \sqrt{1+2C}$, $\beta_1 = \sqrt{2(1+C)}$. The resultant wavefield amplitudes of the reflections are given by

$$
|D_O| = |D_G| = |D_L| = \sqrt{\frac{2(1+C)}{1+2C}}\,|D_{\pi O}| \,. \tag{6.122}
$$

The projection of these wavefields onto the triangle plane OGL is schematically shown in Fig. 6.12(b). The wavefields $\boldsymbol{D}_O$, $\boldsymbol{D}_G$, and $\boldsymbol{D}_L$ perpendicular to the reciprocal lattice vectors $\boldsymbol{g}-\boldsymbol{l}$, $\boldsymbol{l}$, and $\boldsymbol{g}$ lie in the planes CWO, CWG and CWL, respectively. This implies that $\boldsymbol{D}_O$, $\boldsymbol{D}_G$ and $\boldsymbol{D}_L$ lie in the atomic planes of the $G-L$, L, and G, respectively.

(ii) The triply degenerate modes (modes 4, 5 and 6)
The corresponding eigenvector of the eigenvalue $y'(4) = -1$ of mode 4 gives

$$D_{\pi O} = D_{\pi G} = 0\,, \qquad D_{\sigma L} = D_{\pi L} = 0\,, \qquad D_{\sigma O} = D_{\sigma G}\,. \qquad (6.123)$$

Only the σ components of the $\boldsymbol{D}_O$ and $\boldsymbol{D}_G$ survive. The schematic representation of this mode is shown in Fig. 6.13(a). Since the degeneracy of this eigenvalue $y' = -1$ is associated with the symmetry of the equilateral triangle formed by the involved reciprocal lattice vectors, the other two modes, modes 5 and 6, can be determined by permuting the O, G and L in (6.119). This is equivalent to redefine the plane of incidence to which the two associated σ-polarization vectors are perpendicular. For mode 5, the plane of incidence is redefined by the $\boldsymbol{K}_O$ and $\boldsymbol{K}_L$, i.e., the WOL plane, so that $\hat{\boldsymbol{\sigma}}_O$ and $\hat{\boldsymbol{\sigma}}_L$ are perpendicular to the WOL plane. Similarly, for mode 6, $\hat{\boldsymbol{\sigma}}_G$ and $\hat{\boldsymbol{\sigma}}_L$ are perpendicular to the plane of incidence WGL. The eigenvectors of modes 5 and 6 give the following two relations

$$D_{\pi O} = D_{\pi G} = 0\,, \qquad D_{\pi L} = D_{\sigma G} = 0\,, \qquad D_{\sigma O} = D_{\sigma L}\,, \qquad (6.124)$$

and

$$D_{\pi O} = D_{\pi G} = 0\,, \qquad D_{\sigma O} = D_{\pi L} = 0\,, \qquad D_{\sigma G} = -D_{\sigma L}\,, \qquad (6.125)$$

respectively. The minus sign appearing in the last equation of (6.125) is due to $\chi_{G-L} = -\chi_{L-G}$. The corresponding wavevectors of these two modes are shown in Figs. 6.13(b) and (c). Clearly, these three degenerate modes are σ polarized with only two waves being effective in each mode. In comparison with the wavevectors of the two-wave diffraction shown in Fig. 6.10, the triply degenerate modes act as if they were the σ mode in the corresponding two-wave diffraction, i.e., mode 4 for the G reflection, mode 5 for the L reflection and mode 6 for the G–L reflection. This is equivalent to the situation in that a two-wave diffraction with a σ-polarized incident wave takes place when the crystal undergoes a 120-degree rotation around the vector $\boldsymbol{WC}$. The only difference between this three-wave diffraction and the two-wave case is in the phases of the reflections involved. As indicated in Fig. 6.13(c), the directions of the wavefields depend on the signs of the related Fourier components χ's. In other words, the three-wave case is sensitive to the phases of the G, L and $G-L$ reflections, whereas the two-wave case is phase-insensitive.

(iii) The doubly degenerate modes (modes 2 and 3)
The wavefield-amplitude relations of the degenerate eigenvalue $y' = 1 + C$ are

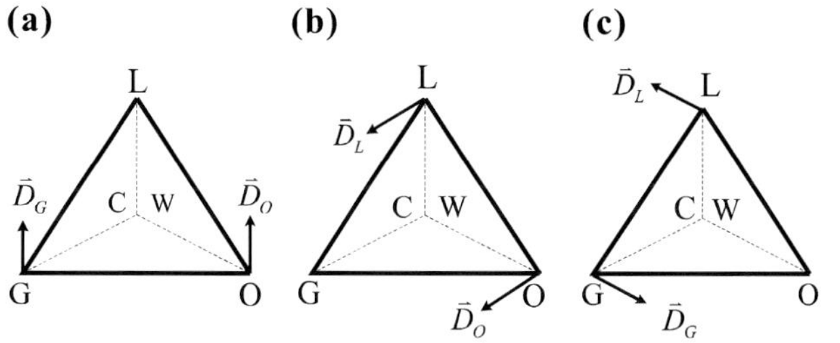

Fig. 6.13. Wavefields of the triply degenerate mode for a symmetric three-wave Laue diffraction

$$D_{\sigma O} : D_{\pi O} : D_{\sigma G} : D_{\pi G} : D_{\sigma L} = 1 : \beta_2 : -1 : \beta_2 : -\frac{C}{C_2} ,$$
$$D_{\pi L} = 0 , \tag{6.126}$$

where $\beta_2 = (CC_1)/[C_2(1+2C)]$. These relations imply that

$$|D_G| = |D_O| , \qquad |D_L| = \sqrt{\frac{2(1+2C)}{5+C}}\,|D_O| . \tag{6.127}$$

The angle between $\boldsymbol{D}_O$ and $\hat{\boldsymbol{\sigma}}_O$ is α, and the angle between $\boldsymbol{D}_G$ and $\hat{\boldsymbol{\sigma}}_G$ is -α, where α is defined as

$$\tan\alpha = \frac{1-C}{2\sqrt{1+2C}} . \tag{6.128}$$

The wavefields of mode 2 are shown in Fig. 6.14(a). By following the same procedure of permuting O, G and L in Fig. 6.14 as for the triply degenerate modes, two additional modes are generated which are shown in Figs. (6.14)(b) and (c). However, the last one is merely a linear combination of mode 1 and mode 6. It is therefore not a normal mode. The total number of normal mode for this particular three-wave diffraction is six. The resultant wavefield of a given reflection G is the vector sum of the G diffracted waves over all the normal modes, i.e.,

$$D_G = \sum_j [\hat{\boldsymbol{\sigma}}_G D_{\sigma G}(j) + \hat{\boldsymbol{\pi}}_G D_{\pi G}(j)] , \tag{6.129}$$

where only the wavefield-amplitude ratios are known. To know the absolute wavefields, appropriate boundary conditions have to be employed.

6.6.3 Wavefields of Modes for Symmetric N-Wave Diffraction

The polarization vectors serve as references to define the direction of a wavefield. A proper choice of polarization vectors facilitates the description of a mode of wave-propagation, especially when the number of waves becomes larger. In Fig. 6.15, for a symmetric N-wave $(O, G_1, G_2, \ldots, G_{N-1})$ systematic diffraction, the reciprocal lattice vectors, $\boldsymbol{g}_1$, $\boldsymbol{g}_2$, $\boldsymbol{g}_3$, $\ldots$, $\boldsymbol{g}_{N-1}$, lie in the

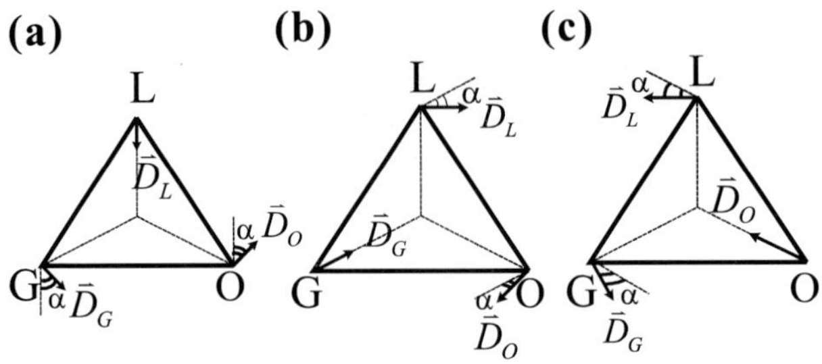

Fig. 6.14. Wavefields of the doubly degenerate mode for a symmetric three-wave Laue diffraction

plane $OG_1G_2G_3 \dots G_{N-1}$ (see Fig. 6.15). Points O, G_1, G_2, G_3, …, G_{N-1} are the reciprocal lattice points, which lie on the surface of the Ewald sphere with the center at the point W. The vectors from W to the reciprocal lattice points are the wavevectors, $\boldsymbol{K}$'s. According to Joko and Fukuhara [184], all the $\hat{\boldsymbol{\sigma}}$'s are defined to lie in the $OG_1G_2G_3 \dots G_{N-1}$ plane and are respectively perpendicular to the line connecting the center C to each reciprocal lattice point. Hence $\hat{\boldsymbol{\sigma}}_M$ is normal to $\boldsymbol{K}_M$ for $M = O$, G_1, G_2, …, G_{N-1}. The corresponding $\hat{\boldsymbol{\pi}}_M$ is defined by following (6.36b), whose projection onto the $OG_1G_2G_3 \cdots G_{N-1}$ plane is also indicated in the same figure.

Let us rewrite the wavefield amplitude of a given reflection G_m as

$$\boldsymbol{D}_{G_m} = \hat{\boldsymbol{\sigma}}_m D_{\sigma G_m} + \hat{\boldsymbol{\pi}}_m D_{\pi G_m} \,, \tag{6.130}$$

where

$$\begin{aligned} D_{\sigma G_m} &= (\boldsymbol{E}^e_{\sigma O} + \boldsymbol{E}^e_{\pi O}) \cdot \hat{\boldsymbol{\sigma}}_m \,, \\ D_{\pi G_m} &= (\boldsymbol{E}^e_{\sigma O} + \boldsymbol{E}^e_{\pi O}) \cdot \hat{\boldsymbol{\pi}}_m \,. \end{aligned} \tag{6.131}$$

Both $D_{\sigma G_m}$ and $D_{\pi G_m}$ are contributed from the σ- and π-polarized components of the incident wave, $\boldsymbol{E}^e_{\sigma O}$ and $\boldsymbol{E}^e_{\pi O}$. In terms of the polarization

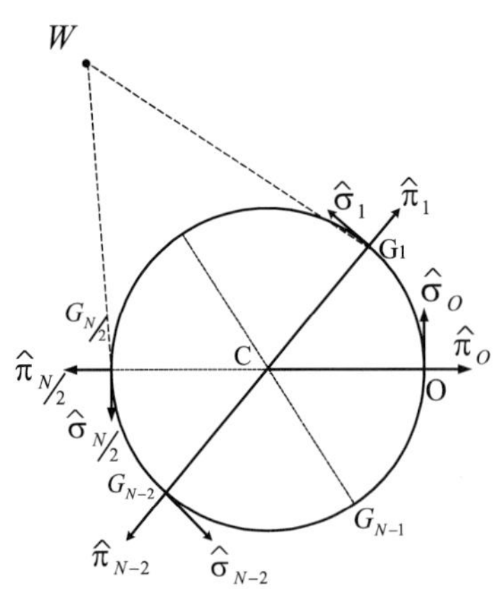

Fig. 6.15. Polarization vectors in a general N-wave symmetric diffraction

factors defined according to Fig. 6.15 as

$$\begin{aligned}
\hat{\boldsymbol{\sigma}}_O \cdot \hat{\boldsymbol{\sigma}}_m &= \cos\left(\frac{2\pi m}{N}\right) , \\
\hat{\boldsymbol{\pi}}_O \cdot \hat{\boldsymbol{\sigma}}_m &= -\sin\left(\frac{2\pi m}{N}\right) , \\
\hat{\boldsymbol{\sigma}}_O \cdot \hat{\boldsymbol{\pi}}_m &= \sin\left(\frac{2\pi m}{N}\right) , \\
\hat{\boldsymbol{\pi}}_O \cdot \hat{\boldsymbol{\pi}}_m &= \cos\left(\frac{2\pi m}{N}\right) ,
\end{aligned} \tag{6.132}$$

(6.130) can be expressed as

$$\begin{aligned}
\boldsymbol{D}_{G_m} &= \hat{\boldsymbol{\sigma}}_m \left[A_m \cos\left(\frac{2\pi m}{N}\right) + B_m \sin\left(\frac{2\pi m}{N}\right)\right] \\
&\quad + \hat{\boldsymbol{\pi}}_m \left[A'_m \cos\left(\frac{2\pi m}{N}\right) + B'_m \sin\left(\frac{2\pi m}{N}\right)\right] .
\end{aligned} \tag{6.133}$$

For an N-wave diffraction, there are normally $2N$ modes of wave-propagation. Usually, a mode j with an odd number j, i.e., $j = 2n + 1$, is associated with the σ component, while those with even j's are related to the π component. Moreover, for a symmetric multiple-wave case, if the number j is properly chosen, then the wavefield amplitude of mode j depends on the integer multiple n of the angle $2\pi m/N$, with $n = 0, 1, \ldots, N$. By taking this fact into account, the wavefield amplitude $\boldsymbol{D}_{G_m}(j)$ can be written as

$$\begin{aligned}
\boldsymbol{D}_{G_m}(j) &= \hat{\boldsymbol{\sigma}}_m \left[A_m \cos\left(\frac{2\pi nm}{N}\right) + B_m \sin\left(\frac{2\pi nm}{N}\right)\right] \\
&\quad + \hat{\boldsymbol{\pi}}_m \left[A'_m \cos\left(\frac{2\pi nm}{N}\right) + B'_m \sin\left(\frac{2\pi nm}{N}\right)\right] ,
\end{aligned} \tag{6.134}$$

where $n = (j - 1)/2$ for an odd j and $n = j/2$ for an even j. If the total number of waves, N, is even as shown in Fig. 6.15, then

$$\hat{\boldsymbol{\sigma}}_m = -\hat{\boldsymbol{\sigma}}_{m'} , \qquad \hat{\boldsymbol{\pi}}_m = -\hat{\boldsymbol{\pi}}_{m'} , \tag{6.135}$$

for $m' = N - m - 1$. Consequently,

$$D_{G_m}(j = 2n - 1) = D_{\sigma G_m} = -D_{\sigma G_{N-m'-1}} , \tag{6.136a}$$

$$D_{G_m}(j = 2n) = D_{\pi G_m} = -D_{\pi G_{N-m'-1}} . \tag{6.136b}$$

As a result, (6.134) is simplified as

$$\begin{aligned}
\boldsymbol{D}_{G_m}(j = 2n - 1) &= \hat{\boldsymbol{\sigma}}_m A_m \cos\left(\frac{2\pi nm}{N}\right) + \hat{\boldsymbol{\pi}}_m B'_m \sin\left(\frac{2\pi nm}{N}\right) , \\
\boldsymbol{D}_{G_m}(j = 2n) &= \hat{\boldsymbol{\sigma}}_m B_m \sin\left(\frac{2\pi nm}{N}\right) + \hat{\boldsymbol{\pi}}_m A'_m \cos\left(\frac{2\pi nm}{N}\right) ,
\end{aligned} \tag{6.137}$$

where the normalization condition

$$\sum_{m=0}^{N-1} \left(A_m^2 + B_m'^{\,2}\right) = \sum_{m=0}^{N-1} \left(B_m^2 + A_m'^{\,2}\right) = 1 \tag{6.138}$$

holds. When $n = N$,

$$\boldsymbol{D}_{G_m}(j = 2N-1) = \hat{\boldsymbol{\sigma}}_m A_m \,, \quad \boldsymbol{D}_{G_m}(j = 2N) = \hat{\boldsymbol{\pi}}_m A_m' \tag{6.139}$$

for $m = 0, 1, 2, \ldots, N-1$, where

$$A_m = A_m' = \frac{1}{\sqrt{N}} \,. \tag{6.140}$$

When $n = N/2$, if N is even,

$$\begin{aligned} \boldsymbol{D}_{G_m}(j = N-1) &= (-1)^m \hat{\boldsymbol{\pi}}_m B_m' \,, \\ \boldsymbol{D}_{G_m}(j = N) &= (-1)^m \hat{\boldsymbol{\sigma}}_m B_m \,, \end{aligned} \tag{6.141}$$

for all m, where

$$B_m' = B_m = \frac{1}{\sqrt{N}} \,. \tag{6.142}$$

These two modes with $n = N$ and $n = N/2$ are non-degenerate. The eigenvalues and eigenvectors for a given n can be calculated by substituting (6.137) into the fundamental equation, (6.43). When $n \neq N$ and $n \neq N/2$, two sets of A, B' and A', B of the forms [190,193]

$$\begin{aligned} A_m &= -B_m' = \sqrt{\frac{2}{N}} \frac{1}{\sqrt{1+Z_\pm^2}} \,, \\ A_m' &= B_m = -\sqrt{\frac{2}{N}} \frac{Z_\pm}{\sqrt{1+Z_\pm^2}} \,, \end{aligned} \tag{6.143}$$

can be obtained for each eigenvalue, where

$$Z_\pm = \xi \pm \sqrt{1+\xi^2} \,. \tag{6.144}$$

The quantity ξ is the solution of the dispersion equation. Therefore, the eigenvalue is doubly degenerate.

6.7 Number of Modes and Number of Permitted Modes

The dispersion surface as a whole consists of $2N$ closed curved sheets for two polarizations, as discussed in Sect. 6.4. Since the eigenvalue 2ε is a quadratic function of the accommodation ζ according to (6.70), the number of the solutions of the dispersion equation (6.46) is $4N$. Half of the total modes are associated with the upper intersection ($\zeta > 0$) and the other half with the lower intersection ($\zeta < 0$) of the dispersion sheets with respect to $\zeta = 0$. The total number of modes is therefore also $4N$. According to their corresponding eigenvectors, not all the modes have an appreciable wavefield-amplitude ratio. Under certain circumstances, a mode may not be allowed to participate in a given diffraction process. For example, those modes with the Poynting vector pointing out of the crystal with respect to the incident wave usually have a negative absorption. More rigorously speaking, they are related, according to Kato et al. [499], to positive Riemann sheets. When the crystal thickness is semi-infinite, the energy flow of these modes blows up and the conservation of total energy is violated. These modes are therefore not permitted in the diffraction process. The permitted modes are those with the Poynting vector pointing towards the interior of the crystal. In this case the total number of permitted modes is not $4N$. It is well known that the number of permitted modes for a symmetric two-wave Laue diffraction in an infinitely thick crystal is $2N = 4$. While there are only $N = 2$ permitted modes for a symmetric Bragg reflection in a very thick crystal. This fact is depicted in Fig 6.16, where the dispersion surface (Fig 6.16(a)) of a single polarization for the two-wave symmetric Laue (Fig 6.16(b)) and symmetric Bragg diffractions (Fig 6.16(c)) with respect to the crystal surfaces NL and MR is shown. At the tie points T_1' and T_2' in the Laue geometry and T_1 and T_2 in the Bragg geometry excited by the incident wave of the wavevector $\boldsymbol{EO}$, the Poynting vectors are $\boldsymbol{S}_1'$, $\boldsymbol{S}_2'$, $\boldsymbol{S}_1$, and $\boldsymbol{S}_2$, respectively. The point E is the entrance point. Referring to the surface NL and Fig 6.16(b), $\boldsymbol{S}_1'$ and $\boldsymbol{S}_2'$ are pointing towards the interior of the crystal. Hence the two modes associated with T_1' and T_2' are permitted modes. For the Bragg case (Fig 6.16(c)), the Poynting vector $\boldsymbol{S}_1$ is in the direction away from the crystal, whereas $\boldsymbol{S}_2$ is towards the crystal. Hence only the mode with the tie point T_2 is a permitted mode. Moreover, from Fig 6.16(a), it is clear that the dispersion sheets lying below the Laue point La with respect to the crystal surface at the exact diffraction position always have the Poynting vectors towards the interior of the crystal. Therefore, the number of permitted modes equals the number of the accommodations with a positive real-part ζ_r in (6.66). For the two polarizations, σ and π, the number of permitted modes is doubled.

The number of permitted modes in an N-wave diffraction can be derived from the dispersion equation at the exact N-wave diffraction position together with the known properties of the dynamical diffraction [53, 249]. For simplicity, let us consider a general three-wave (O, G, L) diffraction. The dispersion equation is given by

$$\begin{vmatrix} 2\varepsilon_O - \chi_O & -C_1\chi_{O\text{-}G} & -C_2\chi_{O\text{-}L} \\ -C_1\chi_{G\text{-}O} & 2\varepsilon_G - \chi_O & -C_3\chi_{G\text{-}L} \\ -C_2\chi_{L\text{-}O} & -C_3\chi_{L\text{-}G} & 2\varepsilon_L - \chi_O \end{vmatrix} = 0 \,, \tag{6.145}$$

where the coefficients C's are the polarization factors defined in (6.64). The second order terms of ζ in the eigenvalue 2ε given in (6.70) is neglected. 2ε for the situation at the exact three-wave diffraction then takes the form

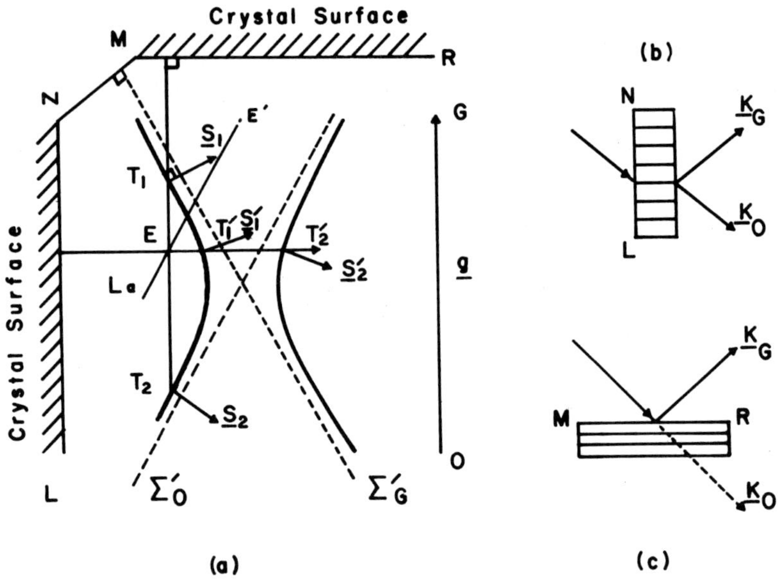

Fig. 6.16. (**a**) Geometric relation between the dispersion surface and the crystal surface for (**b**) two-wave Laue and (**c**) two-wave Bragg diffraction

$$2\varepsilon_M \cong -2\zeta\gamma_M \tag{6.146}$$

for $M = O, G$, and L, where the relation (6.74) has been used. γ_M is the direction cosine of the wavevector $\boldsymbol{K}_M$ with respect to the inward surface normal $\hat{\boldsymbol{n}}_e$. By the approximation of the small imaginary part, (6.145) can be expressed in terms of the real parts of χ's and ζ's as

$$\begin{vmatrix} \zeta_r + \dfrac{\chi_{O,r}}{2\gamma_O} & \dfrac{C_1\chi_{O\text{–}G,r}}{2\gamma_O} & \dfrac{C_2\chi_{O\text{–}L,r}}{2\gamma_O} \\ \dfrac{C_1\chi_{G\text{–}O,r}}{2\gamma_G} & \zeta_r + \dfrac{\chi_{O,r}}{2\gamma_G} & \dfrac{C_3\chi_{G\text{–}L,r}}{2\gamma_G} \\ \dfrac{C_2\chi_{L\text{–}O,r}}{2\gamma_L} & \dfrac{C_3\chi_{L\text{–}G,r}}{2\gamma_L} & \zeta_r + \dfrac{\chi_{O,r}}{2\gamma_L} \end{vmatrix} = 0\,. \tag{6.147}$$

The sign of ζ_r is independent of the off-diagonal elements of the secular determinant of (6.147), because the absolute value of the real part of the Fourier component of the incident reflection is always greater than that of the diffracted ones, i.e., $|\chi_{O,r}| > |\chi_{O\text{–}G,r}|$, $|\chi_{O,r}| > |\chi_{O\text{–}L,r}|$, and $|\chi_{O,r}| > |\chi_{G\text{–}L,r}|$. By neglecting the off-diagonal elements, (6.147) becomes

$$(\zeta_r - a_{0,r})\,(\zeta_r - a_{1,r})\,(\zeta_r - a_{2,r}) = 0\,, \tag{6.148}$$

where

$$a_{0,r} = -\frac{\chi_{O,r}}{2\gamma_O}\,, \quad a_{1,r} = -\frac{\chi_{O,r}}{2\gamma_G}\,, \quad a_{2,r} = -\frac{\chi_{O,r}}{2\gamma_L}\,. \tag{6.149}$$

Equation (6.147) can be generalized for an N-wave case as

$$\Pi_{j=0}^{N-1}\left(\zeta_r - a_{j,r}\right) = 0\ , \tag{6.150}$$

with

$$a_{j,r} = -\frac{\chi_{O,r}}{2\gamma_j} \tag{6.151}$$

for all the reflections j (= 0, 1, 2, ..., $N-1$). Note that $\chi_{O,r} < 0$.

The sign of ζ_r depends therefore on the sign of the direction cosine γ_j. Since γ_j is positive for a Laue transmission and negative for a Bragg reflection, the number of permitted modes is then equal to the number of Laue diffractions, including the incident reflection, or equal to $N - N_R$, where N_R is the number of Bragg reflections involved. If the crystal is semi-infinite, then the number of permitted modes is

$$N_p = 2\left(N - N_R\right) \tag{6.152}$$

for two polarizations. If the crystal is so thin that the product of the linear absorption coefficient μ and the crystal thickness T is much less than unity, or the angle of incidence is of the order of the critical angle of total reflection, i.e, a few arc minutes, then the number of permitted modes becomes

$$N_P = 4N\ . \tag{6.153}$$

6.8 Absorption

The absorption of X-rays in crystals, as described in Chap. 2, results mainly from the inelastic scattering of X-rays by atoms, which includes the photoelectric absorption, the Compton effect, and the thermal diffuse scattering due to phonons. In the dynamical diffraction, however, the effect due to the photoelectric absorption is about two order of magnitude higher than other inelastic scattering. Moreover, in terms of photoelectric absorption, only the contribution from dipoles is important. The contribution from quadrupoles and higher-order multipoles are negligibly small. The corresponding absorption coefficient due to the dipole contribution is described based on the fundamental equation of wavefield.

Now let us rewrite (6.40a)–(6.40d) as

$$\left(2\varepsilon_M - \chi_O\right) D_{\sigma M} = \sum_{\substack{\sigma L \neq \sigma M \\ L \neq M}} p_{\sigma M, \sigma L} \chi_{M-L} D_{\sigma L} \tag{6.154}$$

for all the M reflections, where σM and σL denote all σ's and π's except that $\sigma L \neq \sigma M$. This equation is valid for all the modes j involved in the diffraction process. For simplicity, the indices referred to j are omitted here. Multiplying (6.154) by $D^*_{\sigma M}$, the complex conjugate of $D_{\sigma M}$, leads to

$$\sum_{\sigma M}\left(2\varepsilon_M - \chi_O\right)\left|D_{\sigma M}\right|^2 = \sum_{\sigma M}\sum_{\substack{\sigma L \neq \sigma M \\ L \neq M}} p_{\sigma M, \sigma L} \chi_{M-L}\left(D^*_{\sigma M} D_{\sigma L}\right)\ . \tag{6.155}$$

That only the imaginary parts of this equation are related to absorption gives

$$\sum_{\sigma M} (2\varepsilon_{M,i} - \chi_{O,i}) |D_{\sigma M}|^2 = \sum_{\sigma M} \sum_{\substack{\sigma L \neq \sigma M \\ L \neq M}} p_{\sigma M, \sigma L} \chi_{M-L,i} (D^*_{\sigma M} D_{\sigma L}) . \tag{6.156}$$

In (6.70), the second-order term of ζ_i is so small that can be neglected and the imaginary part of the geometric factor α_M is zero. The $2\varepsilon_i$ then takes the form

$$2\varepsilon_{M,i} \cong -2\zeta_i \gamma_M . \tag{6.157}$$

The linear absorption coefficient μ along the inward normal $\hat{\boldsymbol{n}}_e$ of the crystal surface, proportional to the imaginary part of the wavevector $\boldsymbol{K}_i$ of any reflection involved, can be expressed as

$$\begin{aligned} \mu &= 4\pi K_i = 4\pi k \zeta_i \\ &= \mu_0 \frac{\sum_{\sigma M} |D_{\sigma M}|^2}{\sum_{\sigma M} \gamma_M |D_{\sigma M}|^2} \\ &\quad \times \left[1 + \frac{\sum_{\sigma M} \sum_{\substack{\sigma L \neq \sigma M \\ L \neq M}} p_{\sigma M, \sigma L} \left(\frac{\chi_{M-L,i}}{\chi_{O,i}} \right) D^*_{\sigma M} D_{\sigma L}}{\sum_{\sigma M} |D_{\sigma M}|^2} \right] , \end{aligned} \tag{6.158}$$

where $\mu_0 = -2\pi\chi_{O,i}$ is the normal linear absorption coefficient. An alternative expression of (6.158) is

$$\mu = \mu_0 \frac{c}{(\boldsymbol{v}_e \cdot \hat{\boldsymbol{n}}_e)} \left[1 + \frac{\sum_{\sigma M} \sum_{\substack{\sigma L \neq \sigma M \\ L \neq M}} p_{\sigma M, \sigma L} \left(\frac{\chi_{M-L,i}}{\chi_{O,i}} \right) D^*_{\sigma M} D_{\sigma L}}{\sum_{\sigma M} |D_{\sigma M}|^2} \right] , \tag{6.159}$$

where the velocity of energy transport, $\boldsymbol{v}_e$, is defined as

$$\boldsymbol{v}_e = \frac{c \sum_{\sigma M} \hat{\boldsymbol{K}}_M |D_{\sigma M}|^2}{\sum_{\sigma M} |D_{\sigma M}|^2} . \tag{6.160}$$

The absorption coefficient along $\boldsymbol{v}_e$ is simply the same equation as (6.159), except that the denominator $(\boldsymbol{v}_e \cdot \hat{\boldsymbol{n}}_e)$ is replaced by $|v_e|$.

Equation (6.159) is valid for non-degenerate multiple-wave cases, in which the eigenvectors are orthogonal to one another. For a degenerate case, the eigenvectors of degenerate modes should be expressed as a linear combination of the eigenvectors satisfying the fundamental equation of wavefield. Under such a circumstance, (6.159) is valid again for the modified eigenvectors.

The linear absorption coefficient given in (6.159) for mode j can be considered as the sum of the normal absorption μ_0 and the correction $\Delta\mu$ due to abnormal absorption, i.e.,

$$\mu(j) = \mu_0 + \Delta\mu_j , \tag{6.161}$$

where

$$\Delta\mu_j = 2\pi k \left[2\varepsilon_i(j) - \chi_{O,i}\right] . \tag{6.162}$$

That the trace of the matrix $\tilde{\Phi}$ in (6.44) is zero, i.e.,

$$\sum_j \left[2\varepsilon_i(j) - \chi_{O,i}\right] = 0 , \tag{6.163}$$

implies that

$$\sum_j \Delta\mu_j = 0 . \tag{6.164}$$

Hence, the normal absorption coefficient μ_0 is nothing but the average linear absorption coefficient over the total number of permitted modes:

$$\mu_0 = -\frac{2\pi k \sum_{j=1}^{N_P} 2\varepsilon_i(j)}{N_P} . \tag{6.165}$$

The minimum linear absorption coefficient $\mu_{\min}$ defined as

$$\mu_{\min} = -2\pi k \min\left[2\varepsilon_i(1), 2\varepsilon_i(2), \cdots, 2\varepsilon_i(N_P)\right] \tag{6.166}$$

is of particular interest in dynamical diffraction. For example, for a two-wave symmetric Laue diffraction, the minimum linear absorption coefficient is dependent on the Debye-Waller factor M as [75, 76, 500], i.e.,

$$\mu_{\min} = \frac{\mu_0}{\cos\theta_G}(1 - \Omega) , \tag{6.167}$$

where

$$\Omega = \frac{\chi_{G-O,i}}{\chi_{O,i}} \cong \exp(-M) . \tag{6.168}$$

6.9 Boundary Conditions

The boundary conditions for the EM waves inside and outside a crystal at the crystal boundaries involve phase matching and amplitude matching of the waves in question. The continuity of the tangential components of the wavevectors inside and outside the crystal ensures phase matching at the crystal boundaries, which has been discussed in relation to the excitation of the dispersion surface for the entrance crystal surface. For amplitude matching, the continuities of the tangential components of the electric and magnetic fields, $\boldsymbol{E}$ and $\boldsymbol{H}$, and of the normal components of the electric displacements $\boldsymbol{D}$ and magnetic inductions $\boldsymbol{B}$ at the boundaries are usually employed. That is,

$$\hat{\boldsymbol{n}}_e \cdot (\boldsymbol{D}_{\text{in}} - \boldsymbol{D}_{\text{out}}) = 0 , \tag{6.169a}$$

$$\hat{\boldsymbol{n}}_e \cdot (\boldsymbol{B}_{\text{in}} - \boldsymbol{B}_{\text{out}}) = 0 , \tag{6.169b}$$

$$\hat{\boldsymbol{n}}_e \times (\boldsymbol{E}_{\text{in}} - \boldsymbol{E}_{\text{out}}) = 0 , \tag{6.169c}$$

$$\hat{\boldsymbol{n}}_e \times (\boldsymbol{H}_{\text{in}} - \boldsymbol{H}_{\text{out}}) = 0 , \tag{6.169d}$$

according to Maxwell's equations given in (6.1a)–(6.1d). The subscripts, 'in' and 'out', indicate the fields inside and outside the crystal. $\hat{\boldsymbol{n}}_e$ is the unit vector normal to the boundary considered. Unless the polarization $\boldsymbol{P}$ involves higher order terms in $\boldsymbol{E}$ or the diffraction geometry is either grazing incidence/emergence or normal incidence, the four equations given above can be reduced to two equations, because $\boldsymbol{D}$'s, $\boldsymbol{E}$'s and also $\boldsymbol{H}$'s, $\boldsymbol{B}$'s are linearly proportional to each other.

6.9.1 Boundary Conditions at an Entrance Surface

Suppose that an N-wave diffraction takes place in a plane-parallel plate of thickness T (Fig. 6.17). There are N_L transmission-type Laue diffractions and N_R reflection-type Bragg diffractions. The transmission diffraction includes the refraction of the incident wave. Diffraction along the crystal surface and at grazing incidence will be considered in Chap. 8. The boundary conditions for the N-wave diffraction with the incident fields $\boldsymbol{D}^e$, $\boldsymbol{E}^e$, $\boldsymbol{B}^e$, and $\boldsymbol{H}^e$ at an entrance surface defined by the position vector $\boldsymbol{r}_e$ are

$$\sum_j c_j D_{\sigma O}(j) \exp\left[-2\pi \mathrm{i} \boldsymbol{K}_O(j) \cdot \boldsymbol{r}_e\right] = D^e_{\sigma O} \exp\left(-2\pi \mathrm{i} \boldsymbol{k}^e_O \cdot \boldsymbol{r}_e\right) , \tag{6.170a}$$

$$\sum_j c_j D_{\sigma G}(j) \exp\left[-2\pi \mathrm{i} \boldsymbol{K}_G(j) \cdot \boldsymbol{r}_e\right] = D^e_{\sigma G} \exp\left(-2\pi \mathrm{i} \boldsymbol{k}^e_G \cdot \boldsymbol{r}_e\right) , \tag{6.170b}$$

$$\sum_j c_j D_{\sigma L}(j) \exp\left[-2\pi \mathrm{i} \boldsymbol{K}_L(j) \cdot \boldsymbol{r}_e\right] = 0 , \tag{6.170c}$$

for the σ- and π-components of all the fields $\boldsymbol{D}^e$, $\boldsymbol{E}^e$, $\boldsymbol{B}^e$, and $\boldsymbol{H}^e$. c_j is the proportionality constant of mode j, which depends on the percentage of the

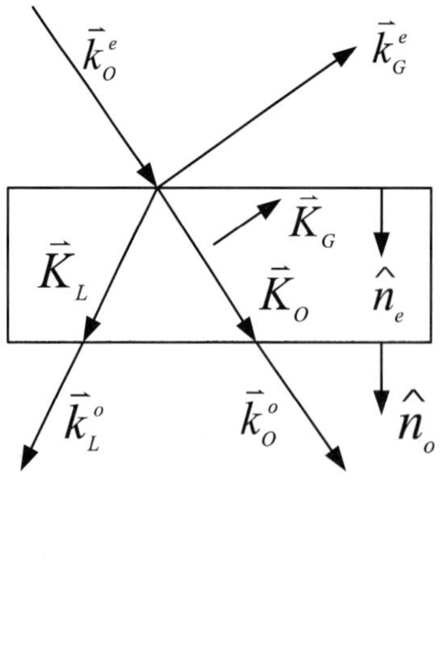

Fig. 6.17. Schematic of N-wave diffraction in a crystal plate (Only the incident O, Bragg reflected G and transmitted L waves are shown.)

incident wavefield amplitude being allocated to mode j. For simplicity, only the conditions for the σ-components of $\boldsymbol{D}$ are given here. The reflection O is the incident diffraction and reflections G and L represent a Bragg reflection and a Laue transmission, respectively, of the N-wave case. $\boldsymbol{D}_G^e$ is the wavefield reflected from the crystal. The summation is taken over all the permitted modes. $\boldsymbol{K}$'s and $\boldsymbol{k}$'s are the wavevectors inside and outside the crystal as shown schematically in Fig. 6.17. $\boldsymbol{k}^e$'s are the wavevectors above the entrance surface and $\boldsymbol{k}^o$'s the wavevectors below the exit surface. $\hat{\boldsymbol{n}}_e$ and $\hat{\boldsymbol{n}}_o$ are the unit vectors of the inward surface normals perpendicular to the entrance and the exit surface, respectively. Phase matching at the entrance surface $\boldsymbol{r}_e$, as discussed in previous sections, requires that

$$\boldsymbol{k}_O \cdot \boldsymbol{r}_e = \boldsymbol{k}_G^e \cdot \boldsymbol{r}_e = \boldsymbol{K}_M(j) \cdot \boldsymbol{r}_e \tag{6.171}$$

for all modes j and reflections M. On this condition, (6.170a)–(6.170c) become

$$\sum_j c_j D_{\sigma O}(j) = D_{\sigma O}^e \,, \tag{6.172a}$$

$$\sum_j c_j D_{\sigma G}(j) = D_{\sigma G}^e \,, \tag{6.172b}$$

$$\sum_j c_j D_{\sigma L}(j) = 0 \,. \tag{6.172c}$$

From the wavefield-amplitude ratios of (6.47), the following relations hold for all modes j:

$$\begin{aligned} \frac{c_j D_{\sigma O}(j)}{\mathrm{Cof}(\sigma O, \sigma O)_j} &= \frac{c_j D_{\pi O}(j)}{\mathrm{Cof}(\pi O, \sigma O)_j} = \frac{c_j D_{\sigma G}(j)}{\mathrm{Cof}(\sigma G, \sigma O)_j} = \cdots \\ &= \frac{c_j D_{\sigma M}(j)}{\mathrm{Cof}(\sigma M, \sigma O)_j} = C_{\sigma j} \end{aligned} \tag{6.173}$$

with respect to $\boldsymbol{D}_{\sigma O}$, and

$$\begin{aligned} \frac{c_j D_{\pi O}(j)}{\mathrm{Cof}(\pi O, \pi O)_j} &= \frac{c_j D_{\sigma O}(j)}{\mathrm{Cof}(\sigma O, \pi O)_j} = \frac{c_j D_{\sigma G}(j)}{\mathrm{Cof}(\sigma G, \pi O)_j} = \cdots \\ &= \frac{c_j D_{\sigma M}(j)}{\mathrm{Cof}(\sigma M, \pi O)_j} = C_{\pi j} \end{aligned} \tag{6.174}$$

with respect to $\boldsymbol{D}_{\pi O}$. The quantities $C_{\sigma j}$ and $C_{\pi j}$ are not independent variables and are related to each other as

$$\frac{C_{\sigma j}}{C_{\pi j}} = \frac{\mathrm{Cof}(\pi O, \pi O)_j}{\mathrm{Cof}(\sigma O, \sigma O)_j} \,. \tag{6.175}$$

Since not all the σ-polarization vectors can be chosen to be perpendicular to all the π-polarization vectors in a general N-wave case, the correlation between the σ and π components of wavefields play an important role in affecting diffraction intensities. In other words, the incident $\boldsymbol{D}_{\sigma O}^e$ can excite

$\boldsymbol{D}_{\sigma G}$ and $\boldsymbol{D}_{\pi G}$ and $\boldsymbol{D}^e_{\pi O}$ can also contribute to $\boldsymbol{D}_{\sigma G}$ and $\boldsymbol{D}_{\pi G}$. Taking this correlation into account, we can write the boundary conditions in terms of the cofactors as

$$\begin{aligned}
&\sum_j C_{\sigma j}\mathrm{Cof}(\sigma O, \sigma O)_j = D^e_{\sigma O} \, , \\
&\sum_j C_{\pi j}\mathrm{Cof}(\sigma O, \pi O)_j = 0 \, , \\
&\sum_j C_{\sigma j}\mathrm{Cof}(\pi O, \sigma O)_j = 0 \, , \\
&\sum_j C_{\pi j}\mathrm{Cof}(\pi O, \pi O)_j = D^e_{\pi O}
\end{aligned} \tag{6.176}$$

for $\boldsymbol{D}_O$,

$$\begin{aligned}
&\sum_j C_{\sigma j}\mathrm{Cof}(\sigma G, \sigma O)_j = D^e_{\sigma G}(\sigma) \, , \\
&\sum_j C_{\pi j}\mathrm{Cof}(\sigma G, \sigma O)_j = D^e_{\sigma G}(\pi) \, , \\
&\sum_j C_{\sigma j}\mathrm{Cof}(\pi G, \pi O)_j = D^e_{\pi G}(\sigma) \, , \\
&\sum_j C_{\pi j}\mathrm{Cof}(\pi G, \pi O)_j = D^e_{\pi G}(\pi)
\end{aligned} \tag{6.177}$$

for $\boldsymbol{D}_G$, and

$$\begin{aligned}
&\sum_j C_{\sigma j}\mathrm{Cof}(\sigma L, \sigma O)_j = 0 \, , \\
&\sum_j C_{\pi j}\mathrm{Cof}(\sigma L, \sigma O)_j = 0 \, , \\
&\sum_j C_{\sigma j}\mathrm{Cof}(\pi L, \pi O)_j = 0 \, , \\
&\sum_j C_{\pi j}\mathrm{Cof}(\pi L, \pi O)_j = 0
\end{aligned} \tag{6.178}$$

for $\boldsymbol{D}_L$. In (6.177), $D^e_{\sigma G}(\sigma)$ and $D^e_{\sigma G}(\pi)$ are the σ-components of the Bragg reflected G wave excited by the incident $D^e_{\sigma G}$ and $D^e_{\pi G}$, respectively. The resultant wavefield-amplitude $D^e_{\sigma G}$ is then the sum of the wavefield-amplitude of the two polarized components. So is the $D^e_{\pi G}$. These lead to

$$\begin{aligned}
D^e_{\sigma G} &= D^e_{\sigma G}(\sigma) + D^e_{\sigma G}(\pi) \, , \\
D^e_{\pi G} &= D^e_{\pi G}(\sigma) + D^e_{\pi G}(\pi) \, .
\end{aligned} \tag{6.179}$$

6.9.2 Boundary Conditions at an Exit Surface

The boundary conditions for wavevectors at the exit surface defined by the position vector $\boldsymbol{r}_o$ are different for the Laue transmission L and the Bragg reflection G. We shall consider a multiple-wave diffraction involving an incident wave O, a Bragg reflected wave G and a transmission reflection L.

The geometries of wavevectors in real space and reciprocal space are shown in Figs. 6.18(a) and 6.18(b), respectively. For simplicity, only the related wavevectors and the Poynting vectors with respect to the dispersion sheets of a single polarization for the transmission reflections O and L and the Bragg reflection G are shown. In Fig. 6.18(a), the incident wave with the wavevector $\boldsymbol{k}_O^e$ impinges on the entrance surface of the crystal plate at point A and generates six wavefields inside the crystal. The wavevectors of these six wavefields are $\boldsymbol{K}_O(j) = \boldsymbol{T_j O}$, $\boldsymbol{K}_G(j) = \boldsymbol{T_j G}$, and $\boldsymbol{K}_L(j) = \boldsymbol{T_j L}$ with $j = 1, 2$ (see Fig. 6.18(b)). The tie point T_1' is not excited, because the Poynting vector $\boldsymbol{S}_1''$ points out of the crystal. The excited wavefields then propagate inside the crystal in two groups along the Poynting vectors $\boldsymbol{S}_1'$ and $\boldsymbol{S}_2'$. When the wavefields of $j = 1$ associated with $\boldsymbol{S}_1'$ reach the exit surface at point A' (Fig. 6.18(a)), part of them are reflected back to the crystal along the Poynting vector $\boldsymbol{S}_1''$ and the rest produce outgoing waves of the wavevectors $\boldsymbol{k}_O^o(1)$ and $\boldsymbol{k}_L^o(1)$ emerging from the exit surface. In terms of the reciprocal space (Fig. 6.18(b)), the tie point T_1' on the dispersion sheet and

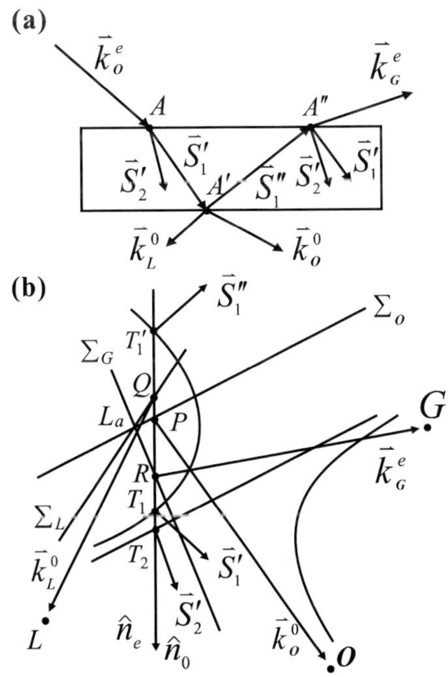

Fig. 6.18. Relations of wavevectors in (**a**) real space and (**b**) reciprocal space

points P and Q on the wavefronts Σ_O and Σ_L of the waves O and L are excited, thus generating the Poynting vector $\boldsymbol{S}_1''$ and the outgoing wavevectors $\boldsymbol{k}_O^o(1) = \boldsymbol{PO}$ and $\boldsymbol{k}_L^o(1) = \boldsymbol{QL}$. When the back reflected wavefields from the lower exit surface reach the entrance surface at point A'', the reflected wave with the wavevector $\boldsymbol{k}_G^e(1) = \boldsymbol{RG}$ emerging from the entrance surface is generated. In other words, point R on the wavefront Σ_G of the Bragg reflected wave G shown in Fig. 6.18(b) is excited. Meanwhile, the tie points T_1 and T_2 are also excited and generate wavefields propagating along $\boldsymbol{S}_1'$ and $\boldsymbol{S}_2'$. Similar situation occurs for the wavefields of $j = 2$ associated with $\boldsymbol{S}_2'$. The outgoing waves $\boldsymbol{k}_O^o(2) = \boldsymbol{PO}$ and $\boldsymbol{k}_L^o(2) = \boldsymbol{QL}$ and the reflected wave $\boldsymbol{k}_G^e(2) = \boldsymbol{RG}$ are generated. Note that $\boldsymbol{k}_O^o(1) = \boldsymbol{k}_O^o(2)$, $\boldsymbol{k}_L^o(1) = \boldsymbol{k}_L^o(2)$, and $\boldsymbol{k}_G^e(1) = \boldsymbol{k}_G^e(2)$, because the unit vectors $\hat{\boldsymbol{n}}_e$ and $\hat{\boldsymbol{n}}_o$ are in the same direction for a plane-parallel crystal plate.

This consideration on the excitation of waves and diffraction processes with respect to the entrance and the exit surfaces for the three-wave (O, G, L) case can be easily extended to a general N-wave diffraction. There the same excitation and diffraction processes take place at points A, A' and A'' for other Laue transmissions and Bragg reflections involved in the N-wave diffraction at appropriate positions at the entrance and exit surfaces.

Phase matching at the exit surface gives the relations among the wavevectors as

$$\begin{aligned} &\boldsymbol{k}_O^o(j) \cdot \boldsymbol{r}_{ot} = \boldsymbol{K}_O(j) \cdot \boldsymbol{r}_{ot} = \boldsymbol{K}_G(j) \cdot \boldsymbol{r}_{ot} \qquad \text{and} \\ &\boldsymbol{k}_L^o(j) \cdot \boldsymbol{r}_{ot} = \boldsymbol{K}_L(j) \cdot \boldsymbol{r}_{ot} \end{aligned} \tag{6.180}$$

for the refraction O, all the Bragg reflections G, and all the transmission reflections L, where $\boldsymbol{r}_{ot}$ is the vector component of $\boldsymbol{r}_o$ tangent to the exit surface. The origin of $\boldsymbol{r}_o$, the same as $\boldsymbol{r}_e$, is a point chosen at the entrance surface. $\boldsymbol{k}_O^o(j) = \boldsymbol{PO}$, $\boldsymbol{k}_L^o(j) = \boldsymbol{QL}$, $\boldsymbol{K}_O(j) = \boldsymbol{T_jO}$, $\boldsymbol{K}_G(j) = \boldsymbol{T_jG}$ and $\boldsymbol{K}_L(j) = \boldsymbol{T_jL}$ (see Fig. 6.18(b)) are the involved wavevectors.

The amplitude matching leads to the following boundary conditions

$$\begin{aligned} &\sum_j c_j D_{\sigma O}(j) \exp\left[-2\pi \mathrm{i} \boldsymbol{K}_O(j) \cdot \boldsymbol{r}_o\right] = D_{\sigma O}^o \exp\left[-2\pi \mathrm{i} \boldsymbol{k}_O^o \cdot \boldsymbol{r}_o\right] \,, \\ &\sum_j c_j D_{\sigma G}(j) \exp\left[-2\pi \mathrm{i} \boldsymbol{K}_G(j) \cdot \boldsymbol{r}_o\right] = 0 \,, \\ &\sum_j c_j D_{\sigma L}(j) \exp\left[-2\pi \mathrm{i} \boldsymbol{K}_L(j) \cdot \boldsymbol{r}_o\right] = D_{\sigma L}^o \exp\left[-2\pi \mathrm{i} \boldsymbol{k}_L^o \cdot \boldsymbol{r}_o\right] \,, \end{aligned} \tag{6.181}$$

for all the σO, πO, σG, πG, σL and πL components. From (6.180), we obtain

$$\begin{aligned} &[\boldsymbol{K}_O(j) - \boldsymbol{k}_O^o] \cdot \boldsymbol{r}_o = \overline{PT_j}T = k\zeta(j)T \,, \\ &[\boldsymbol{K}_G(j) - \boldsymbol{k}_G^o] \cdot \boldsymbol{r}_o = \overline{RT_j}T = \left[k\zeta(j) + \overline{PR}\right] T \,, \\ &[\boldsymbol{K}_L(j) - \boldsymbol{k}_L^o] \cdot \boldsymbol{r}_o = \overline{QT_j}T = \left[k\zeta(j) + \overline{PQ}\right] T \,, \end{aligned} \tag{6.182}$$

where T is the thickness of the crystal plate. With (6.182), the boundary conditions (6.181) reduce to

$$\begin{aligned}
&\sum_j c_j D_{\sigma O}(j) \exp\left[-2\pi \mathrm{i} k\zeta(j)T\right] = D^o_{\sigma O} \; , \\
&\sum_j c_j D_{\sigma G}(j) \exp\left[-2\pi \mathrm{i} \overline{RT_j} T\right] = 0 \; , \\
&\sum_j c_j D_{\sigma L}(j) \exp\left[-2\pi \mathrm{i} \overline{QT}_j T\right] = D^o_{\sigma L} \; .
\end{aligned} \tag{6.183}$$

Again, each equation can be decomposed into four equations due to the correlation between the σ and π components of the wavefields, i.e.,

$$\begin{aligned}
&\sum_j C_{\sigma j} \mathrm{Cof}(\sigma O, \sigma O)_j V_{Oj} = D^o_{\sigma O}(\sigma) \; , \\
&\sum_j C_{\pi j} \mathrm{Cof}(\sigma O, \pi O)_j V_{Oj} = D^o_{\sigma O}(\pi) \; , \\
&\sum_j C_{\sigma j} \mathrm{Cof}(\pi O, \sigma O)_j V_{Oj} = D^o_{\pi O}(\sigma) \; , \\
&\sum_j C_{\pi j} \mathrm{Cof}(\pi O, \pi O)_j V_{Oj} = D^o_{\pi O}(\pi) \; ,
\end{aligned} \tag{6.184}$$

for the outgoing wave O, where

$$V_{Oj} = \exp[-2\pi \mathrm{i} k\zeta(j)T] \; . \tag{6.185}$$

$D^o_{\sigma O}(\sigma)$ and $D^o_{\sigma O}(\pi)$ are the σ components of the wave O excited by the σ and π components of the incident wave. Similarly, $D^o_{\pi O}(\sigma)$ and $D^o_{\pi O}(\pi)$ are the π components of the wave O contributed by the σ and π components of the incident wave. For the waves G and L,

$$\begin{aligned}
&\sum_j C_{\sigma j} \mathrm{Cof}(\sigma G, \sigma O)_j V_{Gj} = 0 \; , \\
&\sum_j C_{\pi j} \mathrm{Cof}(\sigma G, \sigma O)_j V_{Gj} = 0 \; , \\
&\sum_j C_{\sigma j} \mathrm{Cof}(\pi G, \pi O)_j V_{Gj} = 0 \; , \\
&\sum_j C_{\pi j} \mathrm{Cof}(\pi G, \pi O)_j V_{Gj} = 0 \; ,
\end{aligned} \tag{6.186}$$

and

$$\begin{aligned}
&\sum_j C_{\sigma j} \mathrm{Cof}(\sigma L, \sigma O)_j V_{Lj} = D^o_{\sigma L}(\sigma) \; , \\
&\sum_j C_{\pi j} \mathrm{Cof}(\sigma L, \sigma O)_j V_{Lj} = D^o_{\sigma L}(\pi) \; ,
\end{aligned}$$

$$\sum_j C_{\sigma j}\mathrm{Cof}(\pi L,\pi O)_j V_{Lj} = D^o_{\pi L}(\sigma) ,$$
$$\sum_j C_{\pi j}\mathrm{Cof}(\pi L,\pi O)_j V_{Lj} = D^o_{\pi L}(\pi) , \tag{6.187}$$

where

$$V_{Gj} = V_{Oj} = \exp[-2\pi\mathrm{i}\overline{RT_j}T] ,$$
$$V_{Lj} = \exp\left[-2\pi\mathrm{i}\overline{QT}_jT\right] . \tag{6.188}$$

Similar to (6.179), the σ and π components of the outgoing waves are defined as

$$D^o_{\sigma M} = D^o_{\sigma M}(\sigma) + D^o_{\sigma M}(\pi) ,$$
$$D^o_{\pi M} = D^o_{\pi M}(\sigma) + D^o_{\pi M}(\pi) , \tag{6.189}$$

for $M = O$ and L.

For an N-wave diffraction consisting of N_R Bragg reflections G and $N - N_R$ Laue transmissions L, including the incident wave O, there are $4N$ unknown C_j's, $4N_R$ unknown D^e_G's, and $4(N-N_R)$ unknown D^o_L's. The total number of unknowns is $8N$. The total number of equations for the boundary conditions is also $8N$, i.e., the $4N$ equations given in (6.176)–(6.178) and the $4N$ equations (6.184), (6.186), and (6.187). Therefore, the equations for the boundary conditions are solvable. The solutions lead to

$$\begin{aligned}\boldsymbol{D}_M(\boldsymbol{r}) &= \sum_{j=1}^{2N}\{\hat{\boldsymbol{\sigma}}_M\left[C_{\sigma j}\mathrm{Cof}(\sigma M,\sigma O)_j + C_{\pi j}\mathrm{Cof}(\sigma M,\pi O)_j\right] \\ &\quad + \hat{\boldsymbol{\pi}}_M\left[C_{\sigma j}\mathrm{Cof}(\pi M,\sigma O)_j + C_{\pi j}\mathrm{Cof}(\pi M,\pi O)_j\right]\} \\ &\quad \times\exp\left[-2\pi\mathrm{i}\boldsymbol{K}_M(j)\cdot\boldsymbol{r}\right] \\ &= \sum_{j=1}^{2N}\boldsymbol{D}_M(j)\exp\left[-2\pi\mathrm{i}\boldsymbol{K}_M(j)\cdot\boldsymbol{r}\right]\end{aligned} \tag{6.190}$$

for all the M reflections inside the crystal,

$$\boldsymbol{D}^e_G(\boldsymbol{r}) = \exp\left(-2\pi\mathrm{i}\boldsymbol{k}^e_G\cdot\boldsymbol{r}\right)\sum_{j=1}^{2N} D_G(j) \tag{6.191}$$

for all the Bragg reflected waves G from the entrance surface into the empty space, and

$$\boldsymbol{D}^o_L(\boldsymbol{r}) = \exp\left(-2\pi\mathrm{i}\boldsymbol{k}^o_L\cdot\boldsymbol{r}\right)\left(\hat{\boldsymbol{\sigma}}_L D^o_{\sigma L} + \hat{\boldsymbol{\pi}}_L D^o_{\pi L}\right) \tag{6.192}$$

for all the outgoing waves L, including O, emerging from the exit surface. The transmission coefficient T_L and the reflection coefficient R_G are then defined as

$$T_L = \frac{\langle \boldsymbol{S}^o \cdot \hat{\boldsymbol{K}}_L \rangle}{\langle \boldsymbol{S}_O \cdot \hat{\boldsymbol{k}}_O \rangle_{T=0}} = \frac{\gamma_L}{\gamma_O} \frac{\left| \sum_j \boldsymbol{D}_L(j) \right|^2}{\left| D^e_{\sigma O} \right|^2 + \left| D^e_{\pi O} \right|^2} ,$$

$$R_G = \frac{\langle \boldsymbol{S}^e \cdot \hat{\boldsymbol{k}}_G \rangle}{\langle \boldsymbol{S}_O \cdot \hat{\boldsymbol{k}}_O \rangle_{T=0}} = \frac{\gamma_G}{\gamma_O} \frac{\left| \sum_j \boldsymbol{D}_G(j) \right|^2}{\left| D^e_{\sigma O} \right|^2 + \left| D^e_{\pi O} \right|^2} , \quad (6.193)$$

where $\boldsymbol{S}^o$ and $\boldsymbol{S}^e$ are the Poynting vectors below the exit surface and above the entrance surface, respectively. The reflected intensity is

$$I_G = \frac{\left| \sum_j \boldsymbol{D}_G(j) \right|^2}{\left| D^e_{\sigma O} \right|^2 + \left| D^e_{\pi O} \right|^2} . \quad (6.194)$$

The boundary conditions for an infinitely thick crystal are (6.176)–(6.178), where the number of permitted modes $N_P = 2(N - N_R)$. The sum is taken over j from $j = 1$ to N_P. There are $2(N - N_R)$ $C_{\sigma j}$'s, $2(N - N_R)$ $C_{\pi j}$'s and $4N_R$ D^e_G's, which are the unknowns. These unknowns can be determined because there are $4N$ equations. The corresponding wavefields for reflected waves and outgoing waves, the coefficients of transmission and reflection, and the diffracted intensities are the same as that in (6.190)–(6.194), except that the summation is taken over the permitted modes.

6.10 Fundamental Differences Between Laue Transmission and Bragg Reflection: Scattering Matrix

The geometrical difference between the Laue transmission and the Bragg reflection involved in an N-wave diffraction leads to different boundary conditions as described in the previous section. As a matter of fact, the basic difference between the two types of diffraction can be clearly revealed from the comparison of the fundamental equation of wavefields between an N-wave Borrmann transmission and an N-wave diffraction involving Bragg reflection.

The fundamental equations of wavefield, (6.154), for mode j_p and of the complex-conjugate wavefield for mode j_q involved in an N-wave Borrmann transmission in a non-absorbing crystal can be written as

$$2\varepsilon_M (j_p) D_{\sigma M} (j_p) = \sum_{\sigma L} p_{\sigma M, \sigma L} \chi_{M-L} D_{\sigma L} (j_p) , \quad (6.195a)$$

$$2\varepsilon^*_M (j_q) D^*_{\sigma M} (j_q) = \sum_{\sigma L} p_{\sigma M, \sigma L} \chi^*_{M-L} D^*_{\sigma L} (j_q) , \quad (6.195b)$$

where the summation is taken over all σ- and π-polarized components and the relations

$$p_{\sigma M, \sigma M} = 1 \qquad \text{and} \qquad \chi_{M-M} = \chi_O \quad (6.196)$$

have been employed. For non-absorbing crystals, $\chi^*_{M-L} = \chi_{M-L}$. By multiplying (6.195a) and (6.195b) with $D^*_{\sigma M}(j_q)$ and $D_{\sigma M}(j_p)$, respectively, and by summing up the two equations over all the σM, the following expression

$$\sum_{\sigma M} \left[2\varepsilon_M(j_p) - 2\varepsilon^*_M(j_q)\right] D^*_{\sigma M}(j_q) D_{\sigma M}(j_p) = 0 \tag{6.197}$$

is obtained if

$$\sum_{\sigma M}\sum_{\sigma L} p_{\sigma M,\sigma L}\chi_{M-L} D^*_{\sigma M}(j_q) D_{\sigma L}(j_p) = \sum_{\sigma M}\sum_{\sigma L} p_{\sigma M,\sigma L}\chi_{M-L} D_{\sigma M}(j_p) D^*_{\sigma L}(j_q) \ . \tag{6.198}$$

This implies that (6.197) and (6.198) are valid when σM and σL are interchangeable. With a first-order approximation in ζ for 2ε's given in (6.70), (6.197) becomes

$$\sum_{\sigma M} \gamma_M D^*_{\sigma M}(j_q) D_{\sigma M}(j_p) = 0 \tag{6.199}$$

for $\zeta^*(j_q) \neq \zeta(j_p)$. This equation leads to the hermitian condition for the eigenvectors $\boldsymbol{A}_{\sigma M}$:

$$\sum_{\sigma M} A^*_{\sigma M}(j_q) A_{\sigma M}(j_p) = 0 \ , \tag{6.200}$$

where $A_{\sigma M}$ is the magnitude of $\boldsymbol{A}_{\sigma M}$. The $\boldsymbol{A}_{\sigma M}$ is defined as

$$\boldsymbol{A}_{\sigma M} = \sqrt{\gamma_M}\boldsymbol{D}_{\sigma M} \ . \tag{6.201}$$

With these eigenvectors, the corresponding eigenvalue equation can be expressed in terms of the scattering matrix $\tilde{S}$ as

$$\tilde{S}\boldsymbol{A} = \tilde{\zeta}\boldsymbol{A} \ , \tag{6.202}$$

where $\tilde{\zeta}$ is a diagonal matrix with the eigenvalues $\zeta(j)$ as the elements. The diagonal and off-diagonal elements of $\tilde{S}$ are

$$S_{\sigma M,\sigma M} = -\frac{1}{2\gamma_M}(\alpha_M - \chi_O) \ , \quad S_{\sigma M,\sigma L} = -\frac{1}{\sqrt{\gamma_M\gamma_L}} p_{\sigma M,\sigma L}\chi_{M-L} \ , \tag{6.203}$$

where α_M is the geometric factor given in (6.72).

The scattering matrix of an N-wave Borrmann transmission is hermitian, because the direction cosines γ_M are positive for all the reflections involved. The hermitian condition (6.200) holds for all the reflections. When Bragg reflections are involved in a reflection-type multiple-wave diffraction, $\tilde{S}$ is no longer hermitian. Consequently, the wavefield amplitude $A_{\sigma G}$ is equal to $\mathrm{i}\sqrt{|\gamma_G|}D_{\sigma G}$ for a given Bragg reflection G. This describes the total reflection associated with the Bragg reflection where the reflectivity is proportional to the square of the amplitude $D_{\sigma G}$.

6.11 Excitation of Mode and Intensity of Wavefield

The energy associated with a given mode j of propagation is nothing but the amount of the energy specified by the corresponding Poynting vector $\boldsymbol{S}(j)$. This energy, compared with the incident energy, is called the excitation of mode, defined as

$$\mathrm{Ex}(j) = \frac{|\boldsymbol{S}(j)\cdot\hat{\boldsymbol{n}}_e|}{|\boldsymbol{S}_O\cdot\hat{\boldsymbol{n}}_e|} = \frac{\sum_M \gamma_M \left[|D_{\sigma M}(j)|^2 + |D_{\pi M}(j)|^2\right]}{\gamma_O\left[|D^e_{\sigma O}|^2 + |D^e_{\pi O}|^2\right]}\ , \tag{6.204}$$

where $\boldsymbol{S}_O$ is the Poynting vector of the incident wave O. Similarly, the excitation of wave proportional to the energy associated with a given wave M can be defined as

$$\mathrm{Ex}(M) = \left(\frac{\gamma_M}{\gamma_O}\right)\frac{\sum_j \left[|D_{\sigma M}(j)|^2 + |D_{\pi M}(j)|^2\right]}{\left[|D^e_{\sigma O}|^2 + |D^e_{\pi O}|^2\right]}\ . \tag{6.205}$$

The excitations of mode and of wave usually do not exceed unity for a transmission diffraction, because all the involved Poynting vectors are towards the interior of the crystal studied and the total energy is conserved. However, the excitation can be more than unity, but less than or equal to two, when a Bragg reflection is involved in a multiple-wave diffraction. The reason is due to the occurrence of total reflection for the Bragg reflection.

Although the excitations of mode and wave can be calculated from (6.204) and (6.205), they are physically immeasurable. An alternative quantity which can reflect the amount of the incident energy going to a certain mode is the measure of the intensity of the total wavefield of mode j normalized to the intensity of the incident wave, i.e.,

$$I_F(j) = \frac{|\boldsymbol{D}(\boldsymbol{r})_j|^2}{|\boldsymbol{D}^e_O(\boldsymbol{r})|^2}\ , \tag{6.206}$$

where the total wavefield $\boldsymbol{D}(\boldsymbol{r})_j$ of mode j and the incident wave $\boldsymbol{D}^e_O(\boldsymbol{r})$ are given by

$$\boldsymbol{D}(\boldsymbol{r})_j = \sum_{\sigma M} D_{\sigma M}(j)\exp\left[-2\pi\mathrm{i}\boldsymbol{K}_M(j)\cdot\boldsymbol{r}\right] \tag{6.207}$$

and

$$\boldsymbol{D}^e_O(\boldsymbol{r}) = (\hat{\boldsymbol{\sigma}}_O D^e_{\sigma O} + \hat{\boldsymbol{\pi}}_O D^e_{\pi O})\exp\left[-2\pi\mathrm{i}\boldsymbol{k}_O\cdot\boldsymbol{r}\right]\ . \tag{6.208}$$

The former is usually called the standing-wave of mode j. Substituting (6.207) and (6.208) into (6.206) yields the normalized intensity of the standing-wave of mode j,

$$I_F(j) = \frac{\exp\left[-4\pi k\zeta_i(j)\hat{\boldsymbol{n}}_e \cdot \boldsymbol{r}\right]}{\left|D^e_{\sigma O}\right|^2 + \left|D^e_{\pi O}\right|^2}\Big\{\sum_{\sigma M}|D_{\sigma M}(j)|^2$$
$$+ \sum_{\substack{\sigma M \\ M \neq L}} \sum_{\sigma L} D^*_{\sigma L}(j) D_{\sigma M}(j) \exp\left[-2\pi \mathrm{i}\left(\boldsymbol{g}_M - \boldsymbol{g}_L\right) \cdot \boldsymbol{r}\right]\Big\} . \quad (6.209)$$

This intensity can be considered as the sum of the contributions from the σ-, π- and mixed-polarized wavefield components:

$$I_F(j) = I_{F,\sigma}(j) + I_{F,\pi}(j) + I_{F,\sigma\pi}(j) \,, \quad (6.210)$$

where the intensities of the component waves in terms of the wavefield-amplitude ratios are

$$I_{F,\sigma}(j) = \frac{|D_{\sigma O}(j)|^2}{\left|D^e_O\right|^2}\left\{1 + \sum_M \left|\frac{D_{\sigma M}(j)}{D_{\sigma O}(j)}\right|^2 + 2\mathrm{Re}\left[\sum_{M \neq O}\left(\frac{D^*_{\sigma M}(j)}{D^*_{\sigma O}(j)} p_{\sigma M,\sigma O}\right.\right.\right.$$
$$\left.\left. + \frac{D^*_{\pi M}(j)}{D^*_{\sigma O}(j)} p_{\pi M,\sigma O}\right) \exp\left(2\pi \mathrm{i}\boldsymbol{g}_M \cdot \boldsymbol{r}\right)\right]$$
$$+ \frac{|D_{\sigma O}(j)|^2}{\left|D^e_O\right|^2} \sum_{M \neq 0} \sum_{L \neq 0} \frac{D_{\sigma M}(j)}{D_{\sigma O}(j)} \frac{D^*_{\sigma L}(j)}{D^*_{\sigma O}(j)}$$
$$\left. \times \exp\left[-2\pi \mathrm{i}\left(\boldsymbol{g}_M - \boldsymbol{g}_L\right) \cdot \boldsymbol{r}\right] \exp\left[-4\pi k\zeta_i(j)\hat{\boldsymbol{n}}_e \cdot \boldsymbol{r}\right]\right\} \quad (6.211)$$

for the σ polarization, and

$$I_{F,\sigma\pi}(j) = 2\mathrm{Re}\left\{\frac{|D_{\sigma O}(j)|^2}{\left|D^e_O\right|^2} \sum_M \sum_L \frac{D_{\sigma M}(j) D^*_{\pi L}(j)}{D_{\sigma O}(j) D^*_{\pi O}(j)}\right.$$
$$\left. \times \exp\left[-2\pi \mathrm{i}\left(\boldsymbol{g}_M - \boldsymbol{g}_L\right) \cdot \boldsymbol{r}\right]\right\} \exp\left[-4\pi k\zeta_i(j)\hat{\boldsymbol{n}}_e \cdot \boldsymbol{r}\right] \quad (6.212)$$

for the mixed polarization. The intensity for the π-polarized wavefield has the same expression as (6.211), except that the σ's and π's are interchanged.

The intensity of the total wavefield of a given mode j is a function of the angular positions θ and ψ of the crystal setting and the photon energy, or wavelength, used. If the photon energies are close to the absorption edges of the constituent atoms of the crystal, then appreciable fluorescence intensities from excited atoms can be detected. This is because under the diffraction conditions standing-waves are formed inside the crystal and in the vicinity of the crystal surface. These standing-waves then excite the constituent atoms at proper photon energies. If impurity atoms are the main concern for a crystal sample, fluorescence at various θ and ψ from the impurity atoms excited by the standing waves of proper energy can be measured. It is then possible to unambiguously identify the positions of impurity atoms in the crystal lattice from the measured fluorescence profiles as a function of the angular position, namely, the intensity of wavefield versus the θ and ψ.

6.12 Spherical-Wave Nature of X-Rays

The dynamical theory discussed so far has been devised for plane-wave X-rays. The incident waves are assumed to be extremely parallel so that the beam divergences are null or very small. In fact, the incident waves possess a spherical coherent wavefront, therefore their beam divergences are of a finite angular width. Usually, the plane-wave characteristics for incident waves are not always attainable in diffraction experiments with conventional X-ray sources. But they could be obtained with synchrotron radiation using appropriate optical components. The condition under which the plane-wave theory is valid is that the divergence Δ of the incident wave is much less than the intrinsic width W_θ of a dynamical diffraction, which is proportional to the width of the gap of the dispersion surface, i.e.,

$$\Delta \ll W_\theta \, , \tag{6.213}$$

where

$$W_\theta = \frac{p_G \mathrm{Re}\left[\chi_{G-O}\chi_{O-G}\right]\sqrt{\dfrac{\gamma_G}{\gamma_O}}}{\sin 2\theta_G} \tag{6.214}$$

can be derived from a two-wave diffraction in an asymmetric geometry. p_G and θ_G are respectively the polarization factor and the Bragg angle of the G reflection. The width W_θ of an ordinary reflection is of the order of 10^{-6}, while the usual beam divergence ranges from 10^{-2} to 10^{-4} for conventional X-ray sources and can reach 10^{-7} or 10^{-8} for synchrotron radiation with special optical arrangements. Hence, under usual conditions, the incident waves may be considered as spherical waves.

A spherical wave emerging from a point source is usually expressed as

$$\phi(\boldsymbol{r}) = \frac{\phi_o \exp(-2\pi \mathrm{i} k r)}{4\pi r} \, , \tag{6.215}$$

where ϕ_o is the wave-amplitude, k is the modulus of the wavevector $\boldsymbol{k}$, and $\boldsymbol{r}$ is the vector from the source to the observation point. If the spherical symmetry of the Fourier transform of $\phi(\boldsymbol{r})$ is considered, the spherical wave propagating along $\boldsymbol{k}_z$ may be written as a superposition of plane waves [40, 55, 119, 120], i.e.,

$$\phi(\boldsymbol{r}) = \mathrm{i}\frac{\phi_o}{4\pi}\int_{-\infty}^{\infty}\int_{-\infty}^{\infty}\frac{1}{k_z}\exp(-2\pi\mathrm{i}\boldsymbol{k}\cdot\boldsymbol{r})\mathrm{d}k_x\mathrm{d}k_y \, , \tag{6.216}$$

where

$$k_z = \sqrt{k^2 - \left(k_x^2 + k_y^2\right)} \, , \tag{6.217}$$

for a monochromatic wave, namely, k being a constant. According to Kato [119], the electromagnetic waves generated by spontaneous emission of an atom can be expressed as the following spherical waves:

$$\boldsymbol{E} = -\mathrm{i}k\frac{E_O}{4\pi}\int\int \frac{\boldsymbol{J}\times\boldsymbol{k}\times\boldsymbol{k}}{k_z}\exp(-2\pi\mathrm{i}\boldsymbol{k}\cdot\boldsymbol{r})\mathrm{d}k_x\mathrm{d}k_y\ ,$$
$$\boldsymbol{H} = -\mathrm{i}k\frac{H_O}{4\pi}\int\int \frac{\boldsymbol{J}\times\boldsymbol{k}}{k_z}\exp(-2\pi\mathrm{i}\boldsymbol{k}\cdot\boldsymbol{r})\mathrm{d}k_x\mathrm{d}k_y\ , \quad (6.218)$$

where $\boldsymbol{J}$ is the current density induced by the spontaneous emission. The sphericity of the wavefront is neglected, which means the wavefront of the incident wave is a plane. Because the incident wave is usually confined in a small solid angle, the polarization factors associated with the vector products of $\boldsymbol{J}$ and $\boldsymbol{k}$ can be considered as constant. Therefore the polarization unit vectors, $\hat{\boldsymbol{\sigma}}$'s and $\hat{\boldsymbol{\pi}}$'s, can be adopted as reference directions for the wavefields, and each wave in a reference direction is assumed to be a scalar wave with the amplitude proportional to $\boldsymbol{J}\times\boldsymbol{k}$ or $\boldsymbol{J}\times\boldsymbol{k}\times\boldsymbol{k}$. Under this condition, the wavefield amplitude of each component wave for a multiple-wave diffraction is

$$D_{\sigma M}(\boldsymbol{r}) = \frac{\mathrm{i}}{4\pi}\int_{-\infty}^{\infty}\int_{-\infty}^{\infty}\frac{1}{k_z}d_{\sigma M}(\boldsymbol{k};\boldsymbol{r})\exp(-2\pi\mathrm{i}\boldsymbol{k}\cdot\boldsymbol{r})\mathrm{d}k_x\mathrm{d}k_y\ , \quad (6.219)$$

where

$$d_{\sigma M}(\boldsymbol{k};\boldsymbol{r}) = \sum_{j=1}^{N_P} D_{\sigma M}(j)\exp\left[-2\pi\mathrm{i}\left(\boldsymbol{K}_M(j)-\boldsymbol{k}\right)\cdot(\boldsymbol{r}-\boldsymbol{r}_e)\right]\ , \quad (6.220)$$

for all σ- and π-polarized components and for all reflections M. $\boldsymbol{r}_e$ is the position vector of the crystal surface. All the $D_{\sigma M}(j)$ and $D_{\pi M}(j)$ can be determined from the dispersion equation and the boundary conditions. In other words, an incident beam with a plane wavefront can excite the whole dispersion surface near the Lorentz point owing to the relatively large beam divergence. The excited wavefields are the superposition of the plane waves generated at the tie points on the excited portion of the dispersion surface. According to Kato, only those waves excited at different branches of the dispersion surface with Poynting vectors in the same direction can be superposed. In other words, only the conjugated waves are considered.

The double integral in (6.219) can be calculated using Kevin's stationary-phase method [501,502], if the integrand is independent of either of the variables k_x and k_y. For a two-wave (O,G) diffraction, if the incident wave $\boldsymbol{k}_O$ is along $\boldsymbol{k}_z$, $\boldsymbol{k}_x$ lies in the plane of incidence of the G reflection, and $\boldsymbol{k}_y$ is normal to the $k_x - k_z$ plane, then $D_{\sigma M}(j)$ and $D_{\pi M}(j)$ are independent of k_y. Hence, the integration over k_y can be obtained with Kevin's stationary-phase method. In terms of a two-dimensional oblique coordinate system with $\boldsymbol{K}_O$ and $\boldsymbol{K}_G$ as the axes, the integration leads to Bessel functions. For a general multiple-wave case, the wavefield amplitudes always depend on both k_x and k_y due to the correlation between the σ- and π-polarization components. Moreover, it is difficult to find an analytical expression for the wavefield. In such a case, the integral can be calculated by numerical methods.

6.13 Dynamical Diffraction in Distorted Crystals

The dynamical theories discussed in the previous sections are not generally applicable to distorted crystals of any kind of deformation. A generalized theory was developed by Takagi [128, 129], and similar treatments were given by Taupin [130] and Schlengenotto [503]. In these generalized theories, wavefield amplitudes are considered as a slowly varying function of position so that

$$\boldsymbol{D}(\boldsymbol{r},t) = \exp(\mathrm{i}\omega t)\sum_{G}\boldsymbol{D}_G(\boldsymbol{r})\exp\left(-2\pi\mathrm{i}\boldsymbol{K}_G\cdot\boldsymbol{r}\right) . \qquad (6.221)$$

By substituting (6.221) into (6.14) and neglecting the second-order terms, the fundamental equation of wavefield can be expressed as

$$\mathrm{i}\left(\boldsymbol{K}_G\cdot\boldsymbol{\nabla}\right)\boldsymbol{D}_G = \pi K_G^2\sum_{L}\chi_{G\text{–}L}\boldsymbol{D}_{L[G]} - \pi\left(k_G^2-k^2\right)\boldsymbol{D}_G , \qquad (6.222)$$

which can be approximated as

$$\frac{\partial\boldsymbol{D}_G}{\partial x_G} = -\mathrm{i}\pi k\sum_{L\neq G}\chi_{G\text{–}L}\boldsymbol{D}_{L[G]} + \mathrm{i}\pi k\left(2\beta_G\right)\boldsymbol{D}_G , \qquad (6.223)$$

where

$$\frac{\partial\boldsymbol{D}_G}{\partial x_G} = \left(\boldsymbol{K}_G\cdot\boldsymbol{\nabla}\right)\boldsymbol{D}_G , \qquad (6.224)$$

$$2\beta_G = \frac{K_G^2-k^2}{k^2} - \chi_O = \frac{K_G^2-K_O'^2}{k^2} , \qquad (6.225)$$

with $K_O'^2 = k^2(1+\chi_O)$, and $K_O \simeq K_G \simeq k$. The parameter 2β depends on the coupling between $\boldsymbol{K}_G$ and $\boldsymbol{k}_O$. Equation (6.223) is called Takagi's equation, or Takagi-Taupin's equation. For a two wave (O, G) case, the Takagi-Taupin equations consist of two sets of equations for two polarizations, i.e.,

$$\begin{aligned}\frac{\partial D_O(\boldsymbol{r})}{\partial x_O} &= -\mathrm{i}\pi k C\chi_{O\text{–}G}D_G(\boldsymbol{r}) ,\\ \frac{\partial D_G(\boldsymbol{r})}{\partial x_G} &= -\mathrm{i}\pi k C\chi_{G\text{–}O}D_O(\boldsymbol{r}) + \mathrm{i}\pi k\left(2\beta_G\right)D_G(\boldsymbol{r}) ,\end{aligned} \qquad (6.226)$$

where the polarization factor $C = 1$ for the σ polarization and $C = \cos 2\theta_G$ for the π polarization. x_O and x_G are the coordinates along $\boldsymbol{K}_O$ and $\boldsymbol{K}_G$ respectively. Note that $\beta_O = 0$ for the O wave. These generalized fundamental equations have been extensively used to deal with diffractions from crystals with small lattice distortion as well as with large deformation. The former involves migration of tie points on the dispersion sheet of the same branch. The latter is related to the jump of tie points from one branch of the dispersion surface to another branch. Interested readers are suggested to refer to the recent book by Authier [55] for detailed updated information.

For a three-wave (O, G, L) wavevector-coplanar diffraction, where all the σ components are perpendicular to the π components, the Takagi-Taupin equations can be written as

$$\begin{aligned}\frac{\partial D_O}{\partial x_O} &= \mathrm{i}Y_{OO}D_O + \mathrm{i}Y_{OG}D_G + \mathrm{i}Y_{OL}D_L\ ,\\ \frac{\partial D_G}{\partial x_G} &= \mathrm{i}Y_{GO}D_O + \mathrm{i}Y_{GG}D_G + \mathrm{i}Y_{GL}D_L\ ,\\ \frac{\partial D_L}{\partial x_L} &= \mathrm{i}Y_{LO}D_O + \mathrm{i}Y_{LG}D_G + \mathrm{i}Y_{LL}D_L\ ,\end{aligned} \tag{6.227}$$

where

$$\begin{aligned}Y_{jj} &= \pi k\,(2\beta_j)\ ,\\ Y_{j\ell} &= -\pi k p_{j\ell}\chi_{j-\ell}\ ,\end{aligned} \tag{6.228}$$

for j and $\ell = O, G, L$. The polarization factor of the $j-\ell$ reflection, $p_{j\ell}$, is defined in (6.41), which involves scalar products of either only the σ unit vectors or only the π unit vectors.

For a general three-wave diffraction, the correlation between the σ and π components cannot be neglected. Thus, each of the equations in (6.227) has to be decomposed into two equations relating to the σ and π components of the wave considered, i.e.,

$$\begin{aligned}\frac{\partial D_{\sigma G}}{\partial x_G} &= -\mathrm{i}\pi k \sum_{L\neq G} \chi_{G-L}\left(p_{\sigma G,\sigma L}D_{\sigma L} + p_{\sigma G,\pi L}D_{\pi L}\right) + \mathrm{i}\pi k\,(2\beta_G)\,D_{\sigma G}\ ,\\ \frac{\partial D_{\pi G}}{\partial x_G} &= -\mathrm{i}\pi k \sum_{L\neq G} \chi_{G-L}\left(p_{\pi G,\sigma L}D_{\sigma L} + p_{\pi G,\pi L}D_{\pi L}\right) + \mathrm{i}\pi k\,(2\beta_G)\,D_{\pi G}\ .\end{aligned} \tag{6.229}$$

Therefore there are six equations for the wavefields.

The solutions to the Takagi-Taupin equations can be found from appropriate boundary conditions. Since the fundamental equations involve first-order derivatives, the boundary conditions include the continuity of the derivatives of the corresponding wavefield amplitudes at the crystal boundaries in addition to those regarding the amplitude matching at the boundaries. The exact solutions for the three-wave diffraction will be given in Chap. 7 for a crystal of a finite shape [327].

7. Theoretical Approaches

The fundamentals of the dynamical theories for multiple-wave diffraction have so far been described in the previous chapter. It is understood that analytical expressions for wavefields and diffracted intensities of a general multiple-wave diffraction cannot be derived from the fundamental equation of wavefield, mainly due to the complexity imposed by the involvement of equations of high-order polynomials and by the complicated coupling of different polarized wavefield components. To reveal dynamical effects and to understand the interaction of X-ray waves inside crystalline materials, numerical calculations based on the fundamentals of the theories usually give a direct estimation and exact results, on a quantitative basis, of the essential physical quantities, such as mode accommodation, wavefield amplitude, absorption coefficient, Poynting vector, excitation of mode, wavefield intensity and diffracted intensity. All these quantities, as mentioned in Chap. 6, are associated with the solutions to the equation of dispersion, which can be solved numerically as an eigenvalue problem. Therefore, the procedures for calculating eigenvalues by linearizing eigenvalue equations [246] and by using a Cartesian coordinates representation [255] will be described in the first half of this chapter. The latter, as will be seen, is considered as a general procedure for finding solutions to the eigenvalue equations dealing with situations involving wide-angle and grazing-angle incidence and scattering. Although numerical calculations are very powerful and effective to quantify those important physical items, the calculations may not always provide explicitly information about the qualitative aspects of those items. Then approximations may be needed in giving analytical expressions for those physical items in terms of important variables, such as phase, structure factor, polarization factor, angular position, etc.. For this reason, various approximations and approaches, including Born, Bethe, distorted-wave Born, resonance perturbation Bethe approximation, Takagi-Taupin equations, and quantum mechanics are presented in the second half of this chapter.

7.1 Numerical Calculation Procedures

7.1.1 Eigenvalue Problem

The fundamental equation of wavefield, (6.43), can be rewritten in a matrix form as [246, 255]

$$\tilde{\Phi}_{2N}\boldsymbol{D}_{2N} = 0 \,, \tag{7.1}$$

where $\boldsymbol{D}_{2N}$ is a column vector expressed in terms of the σ and π components of wavefields defined in (6.45):

$$\boldsymbol{D}_{2N} = \left(D_{\sigma O}\ D_{\sigma G_1}\ \cdots\ D_{\sigma G_{N-1}}\ D_{\pi O}\ D_{\pi G_1}\ \cdots\ D_{\pi G_{N-1}} \right)^T \,, \tag{7.2}$$

and $\tilde{\Phi}_{2N}$ is the scattering matrix, which is a $2N \times 2N$ complex matrix composed of the $N \times N$ matrices $\tilde{R}_N$, $\tilde{S}_N$, $\tilde{U}_n$, $\tilde{W}_N$ and $\tilde{P}_N$:

$$\tilde{\Phi}_{2N} = \begin{bmatrix} \tilde{R}_N - \tilde{S}_N & -\tilde{U}_N \\ -\tilde{W}_N & \tilde{R}_N - \tilde{P}_N \end{bmatrix} . \tag{7.3}$$

$\tilde{R}_N$ is a diagonal complex matrix represented by $\{R_{ML}\}$ with the elements

$$R_{MM} = 2\varepsilon_M \,, \qquad R_{ML} = 0 \quad (M \neq L) \,. \tag{7.4}$$

Similarly, the elements of $\tilde{S}_N$, $\tilde{U}_N$, $\tilde{W}_N$ and $\tilde{P}_N$ are defined respectively as

$$\begin{aligned} S_{ML} &= \chi_{M\text{–}L}p_{\sigma M,\sigma L} \,, \\ U_{ML} &= \chi_{M\text{–}L}p_{\sigma M,\pi L} \,, \\ W_{ML} &= \chi_{M\text{–}L}p_{\pi M,\sigma L} \,, \\ P_{ML} &= \chi_{M\text{–}L}p_{\pi M,\pi L} \,. \end{aligned} \tag{7.5}$$

According to (6.70), matrix $\tilde{R}_N$ can be expressed as

$$\tilde{R}_N = \zeta^2\tilde{I}_N - 2\zeta\tilde{\Gamma}_N + \tilde{V}_N \,, \tag{7.6}$$

where $\tilde{I}_N$ is an $N \times N$ unit matrix, and $\tilde{\Gamma}_N$ and $\tilde{V}_N$ are $N \times N$ diagonal matrices with the elements

$$\Gamma_{MM} = \gamma_M \,, \qquad V_{MM} = \alpha_M \,. \tag{7.7}$$

By combining (7.4)–(7.7), (7.3) becomes

$$\tilde{\Phi}_{2N} = \zeta^2\tilde{I}_{2N} - 2\zeta\tilde{\Gamma}_{2N} + \tilde{C}_{2N} \,, \tag{7.8}$$

where

$$\begin{aligned} \tilde{I}_{2N} &= \begin{bmatrix} \tilde{I}_N & \tilde{O}_N \\ \tilde{O}_N & \tilde{I}_N \end{bmatrix} , \\ \tilde{\Gamma}_{2N} &= \begin{bmatrix} \tilde{\Gamma}_N & \tilde{O}_N \\ \tilde{O}_N & \tilde{\Gamma}_N \end{bmatrix} , \\ \tilde{C}_{2N} &= \begin{bmatrix} \tilde{V}_N - \tilde{S}_N & \tilde{U}_N \\ \tilde{W}_N & \tilde{V}_N - \tilde{P}_N \end{bmatrix} , \end{aligned} \tag{7.9}$$

and $\tilde{O}_N$ being an $N \times N$ null matrix. According to Colella [246], the quadratic equation in ζ, (7.8), can be linearized by introducing a new eigenvector $\boldsymbol{D}'_{2N}$,

$$\boldsymbol{D}'_{2N} = \zeta \boldsymbol{D}_{2N} \; , \tag{7.10}$$

such that

$$\tilde{\Phi}_{2N} \boldsymbol{D}_{2N} = \left(\zeta \tilde{I}_{2N} - 2\tilde{\Gamma}_{2N} \right) \boldsymbol{D}'_{2N} + \tilde{C}_{2N} \boldsymbol{D}_{2N} = 0 \; . \tag{7.11}$$

The fundamental equation of wavefield, together with (7.10), takes the following $4N \times 4N$ matrix form:

$$\left(\tilde{\Gamma}_{4N} - \zeta \tilde{I}_{4N} \right) \boldsymbol{D}_{4N} = 0 \; , \tag{7.12}$$

where

$$\begin{aligned} \tilde{\Gamma}_{4N} &= \begin{bmatrix} 2\hat{\Gamma}_{2N} & -\tilde{C}_{2N} \\ \tilde{I}_{2N} & \tilde{O}_{2N} \end{bmatrix} , \\ \tilde{I}_{4N} &= \begin{bmatrix} \tilde{I}_{2N} & \tilde{O}_{2N} \\ \tilde{O}_{2N} & \tilde{I}_{2N} \end{bmatrix} , \end{aligned} \tag{7.13}$$

and

$$\boldsymbol{D}_{4N} = \begin{bmatrix} \boldsymbol{D}_{2N} \\ \boldsymbol{D}'_{2N} \end{bmatrix} . \tag{7.14}$$

Matrix $\tilde{O}_{2N}$ is an $2N \times 2N$ null matrix. Equation (7.12) can be solved numerically as a linearized eigenvalue problem. There are $4N$ solutions for the eigenvalues ζ and $4N$ solutions for the eigenvectors $\boldsymbol{D}_M(j)$, where $j = 1, 2, 3, \ldots, 4N$. Hence there are $4N$ plane waves with amplitudes $D_M(j)$ propagating along the directions of $\boldsymbol{K}_M(j)$ inside the crystal. This algorithm of linearizing the fundamental equation into a $4N \times 4N$ matrix form was first proposed by Colella [246]. In principle, this approach of solving the fundamental equation for multiple-wave diffraction is quite general. Under grazing incidence conditions, however, modification on the algorithm is necessary. This modification will be presented in Sect. 7.1.3.

7.1.2 Linear Approximation

The quadratic matrix $\tilde{\Phi}_{2N}$ given in (7.8) reduces to a linear function in ζ,

$$\tilde{\Phi}_{2N} = -\zeta \tilde{\Gamma}_{2N} + \tilde{C}_{2N} \; , \tag{7.15}$$

when the values of direction cosines γ_M are large. This implies that the solutions with large ζ's are neglected. In other words, the resonance term becomes

$$\frac{K_M^2}{K_M^2 - k^2} \cong \frac{k^2}{K_M^2 - k^2} \cong \frac{k}{2\,(K_M - k)} \; . \tag{7.16}$$

This approximation is reasonable, because those waves with a large distance between the tie point and the entrance point have negligibly small amplitudes. For small ζ of the order of $\chi^2 \sim 10^{-10}$, the following system of equations in a matrix form is obtained:

$$\left(\tilde{C}_{2N} - 2\zeta\tilde{\Gamma}_{2N}\right) \boldsymbol{D}_{2N} = 0 \,. \tag{7.17}$$

Multiplying (7.17) by the inverse matrix $\tilde{\Gamma}_{2N}^{-1}$ of $\tilde{\Gamma}_{2N}$, we obtain a standard eigenvalue equation as

$$\left(\frac{1}{2}\tilde{\Gamma}_{2N}^{-1} - \zeta\tilde{I}_{2N}\right) \boldsymbol{D}_{2N} = 0 \,. \tag{7.18}$$

This equation can be easily solved, as usual, and there are $2N$ solutions for both eigenvalues and eigenvectors.

7.1.3 Cartesian Coordinates Representation of Eigenvalue and Eigenvector

The fundamental equation (7.1), as described in Chap. 6, is a result of the transformation of a system of vector equations to that of scalar equations based on the mutually orthogonal unit vectors $\hat{\boldsymbol{\sigma}}_M$, $\hat{\boldsymbol{\pi}}_M$, and $\hat{\boldsymbol{K}}_M = \boldsymbol{K}_M/k$. The polarization factors are then defined accordingly. For wide-angle incidence and scattering, the magnitude of $\boldsymbol{K}_M$ is usually taken as the distance from the tie point in the vicinity of the multiple-wave Lorentz point to the reciprocal lattice point M, which is nearly constant. The corresponding polarization factor is considered to be constant for a given reflection M. Strictly speaking, polarization factors involved in the scattering matrix are variables and depend on the location of the tie point excited by the incident wave. Large variation in polarization factors is usually encountered when grazing incidence is involved. This is because the variation in the normal component of $\boldsymbol{K}_M$ to the crystal surface, namely, the accommodation ζ, increases dramatically as the incident or the scattered angle decreases and reaches $\sqrt{\chi_M} \simeq 10^{-2}$–$10^{-3}$ near the critical angle of total reflection. In other words, the polarization factors are related to the eigenvalue ζ. This fact complicates the eigenvalue problem. The equation of dispersion then becomes a nonlinear high-order polynomial, whose solutions are difficult to be found. To overcome this difficulty, it would be ideal to remove the polarization factors from the scattering matrix and form a linear polynomial for the equation of dispersion. It is equivalent to represent the amplitudes of the diffracted waves on the basis independent of the unknown $\boldsymbol{K}_M$. Along with this consideration, Stetsko and Chang developed an algorithm to solve this problem using a Cartesian coordinates representation for eigenvalues and eigenvectors [255]. The details are given below.

A Cartesian coordinates system is chosen such that the z-axis is normal to the crystal entrance surface, pointing towards the empty space. The

x and y axes and the origin are arbitrarily chosen, based on the convenience of representing the geometry of a particular multiple-wave diffraction. The coordinates of the involved reciprocal lattice point, M, are denoted by (X_M, Y_M, Z_M), and the coordinates of the tie point T_j of mode j, the origins of the wavevectors $\boldsymbol{K}(j)$'s inside the crystal, are denoted by (x_j, y_j, z). The wavevector $\boldsymbol{K}_M(j)$ is then

$$\boldsymbol{K}_M(j) = (X_M - x_j, Y_M - y_j, Z_M - z) = (x_M, y_M, Z_M - z) \ . \tag{7.19}$$

Here, x_j and y_j are defined by the angular and spectral parameters of the incident wave, while the unknown quantity z is obtained from the dispersion equation.

The fundamental equations of wavefield are expressed in terms of the electric field $\boldsymbol{E}$ as

$$\left(K_M^2 - k^2\right) \boldsymbol{E}_M = k^2 \left(\boldsymbol{K}_M \cdot \boldsymbol{E}_M\right) \boldsymbol{K}_M + k^2 \sum_L \chi_{M\text{–}L} \boldsymbol{E}_L \ , \tag{7.20}$$

where $M = O, G_1, G_2, \ldots, G_{N-1}$, and the summation is taken over all L. The electric field $\boldsymbol{E}_M$ is represented as

$$\boldsymbol{E}_M = \left(E_M^x,\ E_M^y,\ E_M^z\right) \ . \tag{7.21}$$

Substituting the Cartesian representations of the wavevector $\boldsymbol{K}(j)$ and wavefield $\boldsymbol{D}_M(j)$ into (7.20) yields a system of $3N$ scalar equations:

$$\begin{aligned} &\tilde{\Phi}\boldsymbol{E} \\ &= \begin{bmatrix} (\tilde{C} - z\tilde{I})^2 + \tilde{B}^2 - \tilde{G}^2 & -\tilde{A}\tilde{B} & -\tilde{A}(\tilde{C} - z\tilde{I}) \\ -\tilde{A}\tilde{B} & (\tilde{C} - z\tilde{I})^2 + \tilde{A}^2 - \tilde{G}^2 & -\tilde{B}(\tilde{C} - z\tilde{I}) \\ -\tilde{A}(\tilde{C} - z\tilde{I}) & -\tilde{B}(\tilde{C} - z\tilde{I}) & \tilde{A}^2 + \tilde{B}^2 - \tilde{G}^2 \end{bmatrix} \\ &\quad \times \begin{bmatrix} \boldsymbol{E}_x \\ \boldsymbol{E}_y \\ \boldsymbol{E}_z \end{bmatrix} = 0 \ , \end{aligned} \tag{7.22}$$

where all the matrices in the system are $N \times N$ matrices. $\tilde{I}$ is the unit matrix and $\tilde{A}$, $\tilde{B}$, and $\tilde{C}$ are diagonal matrices whose diagonal elements are

$$a_{MM} = x_M \ , \qquad b_{MM} = y_M \ , \qquad c_{MM} = z_M \ , \tag{7.23}$$

respectively. The matrix $\tilde{G}^2$ is defined as

$$\tilde{G}^2 = k^2(\tilde{I} + \tilde{F}) \ , \tag{7.24}$$

where $\tilde{F}$ is a matrix with the elements given by

$$f_{MN} = \chi_{M\text{–}N} \ . \tag{7.25}$$

The wavefields $\boldsymbol{E}$'s are column vectors defined as

$$\begin{aligned} \boldsymbol{E}_x &= \left(E_O^x \quad E_{G_1}^x \quad \cdots \quad E_{G_{N-1}}^x\right)^T \ , \\ \boldsymbol{E}_y &= \left(E_O^y \quad E_{G_1}^y \quad \cdots \quad E_{G_{N-1}}^y\right)^T \ , \\ \boldsymbol{E}_x &= \left(E_O^z \quad E_{G_1}^z \quad \cdots \quad E_{G_{N-1}}^z\right)^T \ . \end{aligned} \tag{7.26}$$

The matrices $\tilde{A}$, $\tilde{B}$, and $\tilde{C}$ are commutative. The unknown z can be obtained from the determinant of the scattering matrix, namely, $\det \tilde{\Phi} = 0$, where the determinant is a $4N$th-order polynomial.

To reduce (7.22) to an eigenvalue equation, the following new variables are introduced:

$$\begin{aligned} \boldsymbol{E}_v &= (\tilde{C} - z\tilde{I})\boldsymbol{E}_x - \tilde{A}\boldsymbol{E}_z \,, \\ \boldsymbol{E}_w &= (\tilde{C} - z\tilde{I})\boldsymbol{E}_y - \tilde{B}\boldsymbol{E}_z \,. \end{aligned} \tag{7.27}$$

This transformation makes (7.22) an extended system of size $5N \times 5N$:

$$\begin{bmatrix} \tilde{C} - z\tilde{I} & \tilde{0} & -\tilde{I} & \tilde{0} & -\tilde{A} \\ \tilde{0} & \tilde{C} - z\tilde{I} & \tilde{0} & -\tilde{I} & -\tilde{B} \\ \tilde{B}^2 - \tilde{G}^2 & -\tilde{A}\tilde{B} & \tilde{C} - z\tilde{I} & \tilde{0} & \tilde{0} \\ -\tilde{A}\tilde{B} & \tilde{A}^2 - \tilde{G}^2 & \tilde{0} & \tilde{C} - z\tilde{I} & \tilde{0} \\ -\tilde{A}(\tilde{C} - z\tilde{I}) & -\tilde{B}(\tilde{C} - z\tilde{I}) & \tilde{0} & \tilde{0} & \tilde{A}^2 + \tilde{B}^2 - \tilde{G}^2 \end{bmatrix} \begin{bmatrix} \boldsymbol{E}_x \\ \boldsymbol{E}_y \\ \boldsymbol{E}_v \\ \boldsymbol{E}_w \\ \boldsymbol{E}_z \end{bmatrix} = 0 \,, \tag{7.28}$$

where $\tilde{0}$ is a null matrix. The elements involving $\tilde{C} - z\tilde{I}$ in the fifth row of the matrix given in (7.28) can be eliminated by multiplying the first row to the left by $\tilde{A}$ and the second row by $\tilde{B}$, and then adding the result to the fifth row. Similarly, the off-diagonal elements in the fifth column can be eliminated in the same way. Equation (7.22) then becomes

$$\begin{aligned} &\left(\tilde{Q}_4 - z\tilde{I}_4\right)\boldsymbol{E}_4 = 0 \,, \\ &\boldsymbol{E}_z = -\tilde{G}^{-2}\left(\tilde{A}\boldsymbol{E}_v + \tilde{B}\boldsymbol{E}_w\right) \,, \end{aligned} \tag{7.29}$$

where

$$\tilde{Q}_4 = \begin{bmatrix} \tilde{C} & \tilde{0} & \tilde{A}\tilde{G}^{-2}\tilde{A} - \tilde{I} & \tilde{A}\tilde{G}^{-2}\tilde{B} \\ \tilde{0} & \tilde{C} & \tilde{B}\tilde{G}^{-2}\tilde{A} & \tilde{B}\tilde{G}^{-2}\tilde{B} - \tilde{I} \\ \tilde{B}^2 - \tilde{G}^2 & -\tilde{A}\tilde{B} & \tilde{C} & \tilde{0} \\ -\tilde{A}\tilde{B} & \tilde{A}^2 - \tilde{G}^2 & \tilde{0} & \tilde{C} \end{bmatrix} \tag{7.30}$$

and

$$\boldsymbol{E}_4 = \left[(E_x)^T \ (E_y)^T \ (E_v)^T \ (E_w)^T\right]^T \,. \tag{7.31}$$

$\boldsymbol{E}_4$ is independent of the component E_z and $\tilde{I}_4$ is a $4N \times 4N$ unit matrix. From the first equation of (7.29), the eigenvalue problem now reduces to determining the eigenvalues z_j and eigenvectors of the matrix $\tilde{Q}_4$, where $\tilde{Q}_4$ is independent of z. The first and second N components of the eigenvectors are related to the x and y components of the amplitudes of the diffracted waves. With the third and fourth N components of the eigenvectors, $\boldsymbol{E}_z$ can be calculated from the second equation of (7.29).

Simple solutions to the fundamental equation can be found under some special diffraction conditions. For example, for an N-wave Laue diffraction, all the reflecting planes are normal to the entrance surface, namely, all the coordinates Z_M of the reciprocal lattice points are the same for $M = O, G_1, \ldots, G_{N-1}$. For simplicity, and for not losing the generality, $Z_M = 0$ is chosen. Then the off-diagonal elements in the third row of the scattering matrix in (7.22) can be eliminated by multiplying the first row to the left by the matrix $\tilde{A}$, the second row by $\tilde{B}$, the third row by $-z\tilde{I}$, and then adding them to the third row. Similarly, the off-diagonal elements of the third column can be eliminated. This leads to

$$\left(\tilde{U} + z^2\tilde{I}_2\right)\boldsymbol{E}_2 = \tilde{0}\,,$$
$$\boldsymbol{E}_z = z^{-1}\left(\tilde{G}^{-2}\tilde{A}\tilde{G}^2\boldsymbol{E}_x + \tilde{G}^{-2}\tilde{B}\tilde{G}^2\boldsymbol{E}_y\right)\,, \tag{7.32}$$

where

$$\tilde{U} = \begin{bmatrix} \tilde{B}^2 + \tilde{A}\tilde{G}^{-2}\tilde{A}\tilde{G}^2 - \tilde{G}^2 & -\tilde{A}\tilde{B} + \tilde{A}\tilde{G}^{-2}\tilde{B}\tilde{G}^2 \\ -\tilde{A}\tilde{B} + \tilde{B}\tilde{G}^{-2}\tilde{A}\tilde{G}^2 & \tilde{A}^2 + \tilde{B}\tilde{G}^{-2}\tilde{B}\tilde{G}^2 - \tilde{G}^2 \end{bmatrix}\,. \tag{7.33}$$

$\tilde{I}_2$ is a $2N \times 2N$ unit matrix and $\boldsymbol{E}_2$ is a column vector defined as

$$\boldsymbol{E}_2 = \left[(E_x)^T \ (E_y)^T\right]^T\,. \tag{7.34}$$

Now the problem reduces to determining the eigenvalues $-z_j^2$ and the eigenvectors of the $2N \times 2N$ scattering matrix $\tilde{U}$. The solutions are expected to be symmetric with respect to $z = 0$.

Suppose that an N-wave diffraction takes place in a plane-parallel crystal plate of thickness T_o (see Fig. 7.1). The incident wave with a wavevector $\boldsymbol{k}_O$ generates surface specularly reflected waves along $\boldsymbol{k}_M^s$ with $M = O, G_1, G_2, \ldots, G_{N-1}$ above the entrance surface and waves propagating in the $\boldsymbol{K}_M(j)$ inside the crystal. At the exit surface, there are transmitted waves with wavevectors $\boldsymbol{k}_M^t$. The normal components of $\boldsymbol{k}_M^s$ and $\boldsymbol{k}_M^t$ are

$$k_M^{sz} = -k_M^z = \left[k^2 - \left(x_M^2 + y_M^2\right)\right]^{\frac{1}{2}} \qquad \text{and}$$
$$k_M^{tz} = k_M^z \tag{7.35}$$

at the entrance surface and the exit surface, respectively.

The boundary conditions for the wavefield amplitudes at the entrance and exit surfaces follow the continuity of the tangential components and normal components of the electric fields $\boldsymbol{E}^T$ and the continuity of the electric displacements $\boldsymbol{D}^z$ and the magnetic field $\boldsymbol{H}$, where $\boldsymbol{D}$, $\boldsymbol{E}$, and $\boldsymbol{H}$ are related according to (6.25) and (6.26). The x, y, and z components of the wavefield amplitudes inside and outside the crystal at the entrance crystal boundary are connected by the following relations:

$$\sum_{j=1}^{4N} c_j E_M^x(j) = E_{(e)}^x \delta_{MO} + E_M^{sx}\,,$$

$$\sum_{j=1}^{4N} c_j E_M^y(j) = E_{(e)}^y \delta_{MO} + E_M^{sy}\,,$$

$$\sum_{j=1}^{4N} c_j \left[E_M^z(j) + \sum_{L=0}^{G_{N-1}} \chi_{M-L} E_L^z(j) \right] = E_{(e)}^z \delta_{MO} + E_M^{sz} ,$$

$$\sum_{j=1}^{4N} c_j \left[z_{Mj} E_M^y(j) - y_M E_M^z(j) \right]$$
$$= \left[k_M^z E_{(e)}^y - y_M E_{(e)}^z \right] \delta_{MO} + \left[-k_M^z E_M^{sy} - y_M E_M^{sz} \right] ,$$

$$\sum_{j=1}^{4N} c_j \left[x_M E_M^z(j) - z_{Mj} E_M^x(j) \right]$$
$$= \left[x_M E_{(e)}^z - k_M^z E_{(e)}^x \right] \delta_{MO} + \left[x_M E_M^{sz} + k_M^z E_M^{sx} \right] ,$$

$$\sum_{j=1}^{4N} c_j \left[y_M E_M^x(j) - x_M E_M^y(j) \right]$$
$$= \left[y_M E_{(e)}^x - x_M E_{(e)}^y \right] \delta_{MO} + \left[y_M E_M^{sx} - x_M E_M^{sy} \right] . \tag{7.36}$$

$E_{(e)}^x$, $E_{(e)}^y$, and $E_{(e)}^z$ are the x-, y- and z-components of the electric field of the incident wave and E_M^{sx}, E_M^{sy}, and E_M^{sz} are the components of the electric field of the specularly reflected wave of the M reflection. The quantities z_{Mj} is

$$z_{Mj} = Z_M - z_j . \tag{7.37}$$

δ_{MO} is a Kronecker delta with

$$\delta_{MO} = \begin{cases} 1 & \text{for} \quad M = O , \\ 0 & \text{for} \quad M \neq O . \end{cases} \tag{7.38}$$

The boundary conditions at the exit surface are

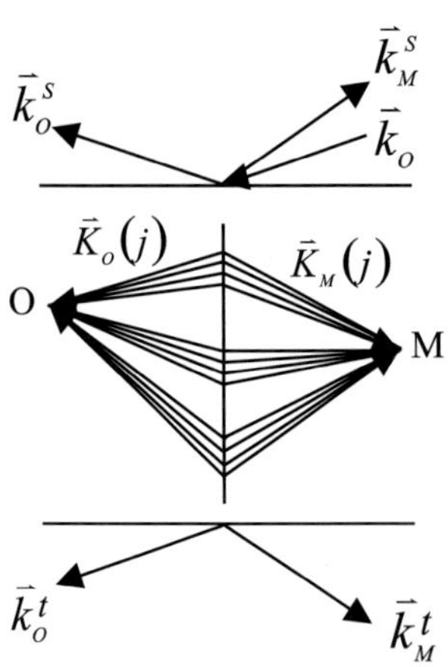

Fig. 7.1. Schematic representation of a general N-wave diffraction in a crystal plate [255]

$$
\begin{aligned}
&\sum_{j=1}^{4N} c_j E_M^x(j)\psi_j = E_M^{rx}\varphi_M \ , \\
&\sum_{j=1}^{4N} c_j E_M^y(j)\psi_j = E_M^{ry}\varphi_M \ , \\
&\sum_{j=1}^{4N} c_j \left[E_M^z(j) + \sum_{L=0}^{G_{N-1}} \chi_{M-L} E_L^z(j) \right] \psi_j = E_M^{rz}\varphi_M \ , \\
&\sum_{j=1}^{4N} c_j \left[z_{Mj} E_M^y(j) - y_M E_M^z(j) \right] \psi_j = \left[k_M^z E_M^{ry} - y_M E_M^{rz} \right] \varphi_M \ , \\
&\sum_{j=1}^{4N} c_j \left[x_M E_M^z(j) - z_{Mj} E_M^x(j) \right] \psi_j = \left[x_M E_M^{rz} - k_M^z E_M^{rx} \right] \varphi_M \ , \\
&\sum_{j=1}^{4N} c_j \left[y_M E_M^x(j) - x_M E_M^y(j) \right] \psi_j = \left[y_M E_M^{rx} - x_M E_M^{ry} \right] \varphi_M \ ,
\end{aligned}
\tag{7.39}
$$

where

$$
\begin{aligned}
\psi_j &= \exp\left(2\pi \mathrm{i} z_{Mj} T_o\right) \ , \\
\varphi_M &= \exp\left(2\pi \mathrm{i} k_M^z T_o\right) \ .
\end{aligned}
\tag{7.40}
$$

The superscript r stands for the exit surface and T_o is the thickness of the crystal plate.

The first two equations of (7.36) and (7.39) are the boundary conditions for the tangential components of $\boldsymbol{E}$, and the third equation is for the normal components of $\boldsymbol{D}$. The last three equations are associated with the tangential and normal components of $\boldsymbol{H}$. In fact, the last equation of (7.36) and (7.39) is a linear superposition of the first two equations. Thus, there are $10N$ linearly independent equations in total for both crystal surfaces. The total unknowns are also $10N$, including $4N$ unknown c_j's and $3N$ unknown components of the electric fields of the surface specularly reflected waves for the entrance surface and $3N$ unknown components of the transmitted waves for the exit surface.

The unknown c_j can be easily determined by eliminating the unknown E_M^{sx}, E_M^{sy} and E_M^{sz} in the fourth and fifth equations of (7.36) by multiplying the second equation of (7.36) by K_M^z, the third by y_M and adding them to the fourth equation. Similar procedure can be carried out for (7.39), except that the second equation of (7.39) needs to be multiplied by $-K_M^z$. Thus,

$$
\sum_{j=1}^{4N} c_j \left\{ \left[z_{Mj} + K_M^z \right] E_M^x(j) + x_M \sum_{L=O}^{G_{N-1}} \chi_{M-L} E_L^z(j) \right\} = 2K_M^z E_{(e)}^x \delta_{MO} \ ,
$$

$$\sum_{j=1}^{4N} c_j \left\{ [z_{Mj} + K_M^z] \, E_M^y(j) + y_M \sum_{L=O}^{G_{N-1}} \chi_{M-L} E_L^z(j) \right\} = 2K_M^z E_{(e)}^y \delta_{MO} \tag{7.41}$$

for the entrance surface, and

$$\sum_{j=1}^{4N} c_j \left\{ [z_{Mj} - K_M^z] \, E_M^x(j) + x_M \sum_{L=O}^{G_{N-1}} \chi_{M-L} E_L^z(j) \right\} \psi_j = 0 \, ,$$
$$\sum_{j=1}^{4N} c_j \left\{ [z_{Mj} - K_M^z] \, E_M^y(j) + y_M \sum_{L=O}^{G_{N-1}} \chi_{M-L} E_L^z(j) \right\} \psi_j = 0 \tag{7.42}$$

for the exit surface. By substituting the c_j determined from (7.41) and (7.42) into (7.36) and (7.39), respectively, the vector components of the electric fields of the diffracted waves can be obtained straightforwardly.

7.1.4 Numerical Computing Procedures

A computing program of dynamical calculation for multiple-wave diffraction can be written based on the algorithm presented in the previous sections and chapters. As usual, the program should consists of input data, calculation for the solutions to the eigenvalue equation, calculation for the boundary conditions, and output.

The input data needed are the lattice constants of the crystal studied, the X-ray wavelength used, the coordinates of the entrance point, the angular ranges interested, the Miller indices of the involved reflections and the structure factors of the reflections, including reflections M, L, $M-L$ and $L-M$ for all M and L. As described previously, multiple-wave diffraction involves the Bragg angle θ of a preselected primary reflection, say, G_1, and the azimuthal angle ψ around the reciprocal lattice vector of the primary reflection. The coordinates of the entrance points depend therefore on these two angles. Precisely speaking, the entrance point is related to the angular deviations $\Delta\theta$ and $\Delta\psi$ from the Bragg angle θ_{G_1} of the G_1 reflection and the exact multiple-wave diffraction position ψ_E in ψ, respectively. In the Cartesian coordinates system, the entrance point is at (x, y), which is related to $\Delta\theta$ and $\Delta\psi$.

The calculation procedure for eigenvalues and eigenvectors includes the construction of the dispersion equation in a matrix form, the linearization of the eigenvalue equation, and the matrix calculation for eigenvalues and eigenvectors. To form the matrix $\tilde{\Phi}$, each matrix element needs to be calculated according to (7.4) and (7.5) with correct polarization factors and structure factors if the σ- and π-polarizations are adopted. For the Cartesian coordinates representation, the polarization factors are excluded in the matrix elements of (7.22), and the calculation for matrices $\tilde{A}$, $\tilde{B}$, $\tilde{C}$, $\tilde{G}^2$ follows the definitions in (7.23), (7.24) and (7.25). The linearization of the eigenvalue equation and the determination of eigenvalues and eigenvectors can be achieved by carrying out the matrix algebra given in (7.17) and (7.29) for the conventional (σ, π) and the Cartesian coordinates (x, y, z) representations, respectively. The transformation of the scattering matrix into a Hessenberg matrix is usually adopted [504].

Wavefield-Amplitude Calculation in the Conventional (σ, π) Representation. The wavefield amplitudes in the (σ, π) representation can be obtained by solving a set of linear equations given in the boundary conditions (6.176)–(6.178) for a thick crystal and (6.184)–(6.187) for a thin crystal.

For an N-wave diffraction in a very thick crystal, if there are N_L Laue-type diffractions (O, L_1, ..., L_{N_L-1}), N_R Bragg-type reflections (G_1, G_2, ..., G_{N_R}), and no surface reflections, i.e., ($N_L = N - N_R$), then the linear equation for the boundary conditions takes the matrix form as

$$\begin{bmatrix} \tilde{S}^a & \tilde{S}^c \\ \tilde{S}^b & -\tilde{S}^d \end{bmatrix} \begin{bmatrix} \tilde{C}_a \\ \tilde{D}_R \end{bmatrix} = \tilde{D} \,, \tag{7.43}$$

where

$$\begin{aligned} \tilde{S}^a &= \begin{bmatrix} \tilde{S}_O\,(1, N_P) \\ \tilde{S}_{L_1}\,(1, N_P) \\ \vdots \\ \tilde{S}_{L_{N_L-1}}\,(1, N_P) \end{bmatrix} , \\ \tilde{S}^b &= \begin{bmatrix} \tilde{S}_{G_1}\,(1, N_P) \\ \vdots \\ \tilde{S}_{G_{N_R}}\,(1, N_P) \end{bmatrix} , \\ \tilde{S}^c &= \tilde{0} \,, \\ \tilde{S}^d &= \tilde{I} \,. \end{aligned} \tag{7.44}$$

N_P is the total number of permitted modes, which is equal to $2N_L$. The matrix element $\tilde{S}_M(1, N_P)$ for all the reflections M is given in terms of cofactors as

$$\begin{aligned} &\tilde{S}_M\,(1, N_P) \\ &= \begin{bmatrix} \mathrm{Cof}(\sigma M, \sigma O)_1 & \mathrm{Cof}(\sigma M, \sigma O)_2 & \cdots & \mathrm{Cof}(\sigma M, \sigma O)_{N_P} \\ \mathrm{Cof}(\pi M, \pi O)_1 & \mathrm{Cof}(\pi M, \pi O)_2 & \cdots & \mathrm{Cof}(\pi M, \pi O)_{N_P} \end{bmatrix} . \end{aligned} \tag{7.45}$$

$\tilde{S}^a$ is a $2N_L \times N_P$ matrix, $\tilde{S}^b$ a $2N_R \times N_P$ matrix, $\tilde{S}^c$ a $2N_L \times 2N_R$ null matrix, and $\tilde{S}^d$ a $2N_R \times 2N_R$ unit matrix. The other matrices $\tilde{C}_a$, $\tilde{D}_R$, and $\tilde{D}$, whose dimensions are $N_P \times 2$, $2N_R \times 2$, and $2N \times 2$, respectively, are defined as

$$\tilde{C}_a = \begin{bmatrix} C_{\sigma 1} & C_{\pi 1} \\ \vdots & \vdots \\ C_{\sigma N_P} & C_{\pi N_P} \end{bmatrix} ,$$

$$\tilde{D}_R = \begin{bmatrix} D_{\sigma G_1}(\sigma) & D_{\sigma G_1}(\pi) \\ D_{\pi G_1}(\sigma) & D_{\pi G_1}(\pi) \\ \vdots & \vdots \\ D_{\sigma G_{N_R}}(\sigma) & D_{\sigma G_{N_R}}(\pi) \\ D_{\pi G_{N_R}}(\sigma) & D_{\pi G_{N_R}}(\pi) \end{bmatrix},$$

$$\tilde{D} = \begin{bmatrix} D^e_{\sigma O} & 0 \\ 0 & D^e_{\pi O} \\ 0 & 0 \\ \vdots & \vdots \\ 0 & 0 \end{bmatrix}, \tag{7.46}$$

where $C_{\sigma j}$ and $C_{\pi j}$ are defined in (6.173) and (6.174). $D^e_{\sigma O}$ and $D^e_{\pi O}$ are the σ and π components of the wavefields of the incident wave, and $D_{\sigma G}(\sigma)$ and $D_{\sigma G}(\pi)$ are the σ components of the reflected wave M excited by the incident $D^e_{\sigma O}$ and $D^e_{\pi O}$, respectively.

The linear equation (7.43) can be easily solved with Gaussian reduction [505].

The boundary conditions for a plane-parallel thin crystal plate, given in (6.184)–(6.187), together with (6.176)–(6.178) can also be written in a linear matrix equation as

$$\begin{bmatrix} \tilde{S}^a & \tilde{S}^c & \tilde{S}^i \\ \tilde{S}^b & -\tilde{S}^d & \tilde{S}^j \\ \tilde{S}^e & \tilde{S}^g & -\tilde{S}^k \\ \tilde{S}^f & \tilde{S}^h & \tilde{S}^l \end{bmatrix} \begin{bmatrix} \tilde{C}_a \\ \tilde{D}_R \\ \tilde{D}_L \end{bmatrix} = \tilde{D}, \tag{7.47}$$

where $\tilde{S}^a$, $\tilde{S}^b$, $\tilde{S}^c$ and $\tilde{S}^d$ are the same as in (7.44). Matrices $\tilde{S}^e$ and $\tilde{S}^f$ are similar to $\tilde{S}^a$ and $\tilde{S}^b$, except that each matrix element is now multiplied by its corresponding V_{Mj}, which is defined either in (6.185) or in (6.188). $\tilde{S}^k$ is a $2N_L \times 2N_L$ unit matrix. $\tilde{S}^g$, $\tilde{S}^h$, $\tilde{S}^i$, $\tilde{S}^j$, and $\tilde{S}^l$ are $2N_L \times 2N_R$, $2N_R \times 2N_R$, $2N_L \times 2N_L$, $2N_R \times 2N_L$, $2N_R \times 2N_L$ null matrices, respectively. $\tilde{D}$ is the same as that in (7.46), except that its dimension is $2N_L \times 2$. $\tilde{D}_L$ is also a $2N_L \times 2$ matrix given by

$$\tilde{D}_L = \begin{bmatrix} D^o_{\sigma O}(\sigma) & D^o_{\sigma O}(\pi) \\ D^o_{\pi O}(\sigma) & D^o_{\pi O}(\pi) \\ D^o_{\sigma L_1}(\sigma) & D^o_{\sigma L_1}(\pi) \\ D^o_{\pi L_1}(\sigma) & D^o_{\pi L_1}(\pi) \\ \vdots & \vdots \\ D^o_{\sigma L_{N_L-1}}(\sigma) & D^o_{\sigma L_{N_L-1}}(\pi) \\ D^o_{\pi L_{N_L-1}}(\sigma) & D^o_{\pi L_{N_L-1}}(\pi) \end{bmatrix}. \tag{7.48}$$

The output of the calculation contains the information about both the real and imaginary parts of the eigenvalues and eigenvectors, the polarization factors if any, the linear absorption coefficient, the wavefield amplitude, the intensities of all the diffracted waves, the excitation of mode and of wave, and the intensities of wavefields inside the crystal.

Wavefield-Amplitude Calculation in the Cartesian Coordinates (x, y, z) Representation. The wavefield amplitudes in the (x, y, z) representation can be determined, according to the boundary conditions, (7.36) and (7.39), by solving a set of linear equations given below :

$$\begin{bmatrix} \tilde{S}_a & \tilde{S}_b & \tilde{0}' \\ \tilde{S}_{a,\psi} & \tilde{0}' & \tilde{S}_{b,\varphi} \end{bmatrix} \begin{bmatrix} \tilde{c} \\ \tilde{E}_s \end{bmatrix} = \tilde{E}_{(e)} \,, \tag{7.49}$$

where $\tilde{0}'$ is a null matrix and the matrices $\tilde{S}$'s are defined as

$$\tilde{S}_a = \begin{bmatrix} \tilde{S}^a_O & \tilde{S}^a_{G_1} & \cdots & \tilde{S}^a_{G_{N-1}} \end{bmatrix}^T ,$$

$$\tilde{S}_b = \begin{bmatrix} \tilde{S}^b_O & \tilde{0} & \cdot & \cdots & \tilde{0} \\ \tilde{0} & \tilde{S}^b_{G_1} & \tilde{0} & \cdots & \cdot \\ \cdot & \tilde{0} & \tilde{S}^b_{G_2} & \cdots & \cdot \\ \cdot & \cdot & \tilde{0} & \cdots & \cdot \\ \cdot & \cdot & \cdot & & \\ \cdot & \cdot & & \cdot & \\ \cdot & \cdot & & \cdot & \tilde{0} \\ \tilde{0} & \cdot & \cdot & \cdot\cdot\,\tilde{0} & \tilde{S}^b_{G_{N-1}} \end{bmatrix} ,$$

$$\tilde{S}_{a,\psi} = \begin{bmatrix} \tilde{S}^a_O \varphi_O & \tilde{S}^a_{G_1} \varphi_{G_1} & \cdots \tilde{S}^a_{G_{N-1}} \varphi_{G_{N-1}} \end{bmatrix}^T ,$$

$$\tilde{S}_{b,\varphi} = \begin{bmatrix} \tilde{S}^b_{O,\varphi} & \tilde{0} & \tilde{0} & \cdots & \tilde{0} \\ \tilde{0} & \tilde{S}^b_{G_1,\varphi} & \tilde{0} & \cdots & \cdot \\ \cdot & \tilde{0} & \tilde{S}^b_{G_2,\varphi} & \cdots & \cdot \\ \cdot & \cdot & \tilde{0} & \cdots & \cdot \\ \cdot & \cdot & \cdot & & \\ \cdot & \cdot & & \cdot & \\ \cdot & \cdot & & \cdot & \tilde{0} \\ \tilde{0} & \cdot & \cdot & \cdot\cdot\,\tilde{0} & \tilde{S}^b_{G_{N-1},\varphi} \end{bmatrix} . \tag{7.50}$$

$\tilde{S}_b$ and $\tilde{S}_{b,\varphi}$ are diagonal matrices. The matrices $\tilde{S}^a_M$, $\tilde{S}^b_M$, and $\tilde{S}^b_{M,\varphi}$ for $M = O, G_1, G_2, \cdots, G_{N-1}$ are

$$\tilde{S}^a_M = \begin{bmatrix} E^x_M(1) & E^x_M(2) & \cdots & E^x_M(N_P) \\ E^y_M(1) & E^y_M(2) & \cdots & E^y_M(N_P) \\ a^z_M(1) & a^z_M(2) & \cdots & a^z_M(N_P) \\ a^{yz}_M(1) & a^{yz}_M(2) & \cdots & a^{yz}_M(N_P) \\ a^{zx}_M(1) & a^{zx}_M(2) & \cdots & a^{zx}_M(N_P) \end{bmatrix} ,$$

$$\tilde{S}^b_M = \begin{bmatrix} -1 & 0 & 0 \\ 0 & -1 & 0 \\ 0 & 0 & -1 \\ 0 & k^z_M & y_M \\ -k^z_M & 0 & -x_M \end{bmatrix} ,$$

$$\tilde{S}^b_{M,\varphi} = \begin{bmatrix} -\varphi_M & 0 & 0 \\ 0 & -\varphi_M & 0 \\ 0 & 0 & -\varphi_M \\ 0 & -k^z_M \varphi_M & y_M \varphi_M \\ k^z_M \varphi_M & 0 & -x_M \varphi_M \end{bmatrix} . \tag{7.51}$$

The matrices $\tilde{c}$, $\tilde{E}_s$ and $\tilde{E}_{(e)}$ are expressed as

$$\begin{aligned} \tilde{c} &= \begin{bmatrix} c_1 & c_2 & \cdots & c_{N_P} \end{bmatrix}^T , \\ \tilde{E}_s &= \left[(\boldsymbol{E}^s_O)^T \; (\boldsymbol{E}^s_{G_1})^T \cdots \left(\boldsymbol{E}^s_{G_{N-1}}\right)^T \; (\boldsymbol{E}^r_O)^T \; (\boldsymbol{E}^r_{G_1})^T \cdots \left(\boldsymbol{E}^r_{G_{N-1}}\right)^T \right]^T , \\ \tilde{E}_{(e)} &= \left[E^x_{(e)} \; E^y_{(e)} \; E^z_{(e)} \quad k^z_O E^y_{(e)} - y_O E^z_{(e)} \quad x_O E^z_{(e)} - k^z_O E^x_{(e)} \right]^T , \end{aligned} \tag{7.52}$$

where

$$\begin{aligned} (\boldsymbol{E}^s_M)^T &= \left(E^{sx}_M \quad E^{sy}_M \quad E^{sz}_M \right) , \\ (\boldsymbol{E}^r_M)^T &= \left(E^{rx}_M \quad E^{ry}_M \quad E^{rz}_M \right) . \end{aligned} \tag{7.53}$$

The dimensions of the matrices are $5 \times N_P$ for $\tilde{S}^a_M$, 5×3 for the null matrix $\tilde{0}$, $5N \times N_P$ for $\tilde{S}_a$ and $\tilde{S}_{a,\psi}$, and $5N \times 3N$ for $\tilde{S}_b$, $\tilde{S}_{b,\varphi}$, and $\tilde{0}'$. The number of permitted modes N_P for this general N-wave diffraction involving specularly reflected waves is $4N$. Therefore, the dimension of the scattering matrix as a whole is $10N \times 10N$. Since there are $10N$ linear equations for $10N$ unknowns, the proportionality constants c_j and the amplitudes E^s_M at the entrance surface and E^r_M at the exit surface can be determined without difficulties.

7.2 Approximations

The two-wave approximation for three-wave and multiple-wave diffraction is usually adopted to derive analytical expressions for diffracted intensities, where the presence of additional reflections is treated as a small perturbation to the two-wave diffraction. In other words, multiple-wave diffraction is considered as a modified two-wave diffraction. Recently, several two-wave approximations have been developed, including Bethe, Born, and distorted-wave Born approximation, which will be described below.

7.2.1 Bethe Approximation

A modified two-wave approximation, the so-called pseudo-two-wave formulation [262,263,270,281,282], for a three-wave (O, G, L) diffraction is described here. The diffraction conditions are nearly satisfied by the O and G reflections. The involved L reflection is far from the exact three-wave diffraction position. The influence of the L reflection on the two-wave (O, G) diffraction is relatively small. This situation is most akin to Bethe's treatment for electron diffraction, where weak reflections are considered as a perturbation to a Bragg diffraction. Dynamical potentials are employed as a perturbation correction. In the case of X-rays, a generalized electrical susceptibility or structure factor, being analogous to the dynamical potentials, can be derived from the fundamental equations of wavefield [53, 262, 263, 267, 270, 281, 282]. In the following, the formulation given in the references [53,270,281,282] will be adopted.

The fundamental equations of a three-wave diffraction are written, according to (6.40a)–(6.40d), as

$$\begin{aligned}
&2\xi_O D_{\sigma O} + P_\sigma \chi_{O\text{–}G} D_{\sigma G} + d_1 \chi_{O\text{–}L} D_{\sigma L} + d_2 \chi_{O\text{–}L} D_{\pi L} = 0\,,\\
&2\xi_O D_{\pi O} + P_\pi \chi_{O\text{–}G} D_{\pi G} + d_3 \chi_{O\text{–}L} D_{\sigma L} + d_4 \chi_{O\text{–}L} D_{\pi L} = 0\,,\\
&P_\sigma \chi_{G\text{–}O} D_{\sigma O} + 2\xi_G D_{\sigma G} + d_1 \chi_{G\text{–}L} D_{\sigma L} + d_2 \chi_{G\text{–}L} D_{\pi L} = 0\,,\\
&P_\pi \chi_{G\text{–}O} D_{\pi O} + 2\xi_G D_{\pi G} + d_5 \chi_{G\text{–}L} D_{\sigma L} + d_6 \chi_{G\text{–}L} D_{\pi L} = 0\,,\\
&d_1 \chi_{L\text{–}O} D_{\sigma O} + d_3 \chi_{L\text{–}O} D_{\pi O} + d_1 \chi_{L\text{–}G} D_{\sigma G} + d_5 \chi_{L\text{–}G} D_{\pi G}\\
&\quad + 2\xi_L D_{\sigma L} = 0\,,\\
&d_2 \chi_{L\text{–}O} D_{\sigma O} + d_4 \chi_{L\text{–}O} D_{\pi O} + d_2 \chi_{L\text{–}G} D_{\sigma G} + d_6 \chi_{L\text{–}G} D_{\pi G}\\
&\quad + 2\xi_L D_{\pi L} = 0\,,
\end{aligned} \tag{7.54}$$

where

$$\begin{aligned}
2\xi_O &= \chi_O - 2\varepsilon_O\,,\\
2\xi_G &= \chi_O - 2\varepsilon_G\,,\\
2\xi_L &= \chi_O - 2\varepsilon_L\,.
\end{aligned} \tag{7.55}$$

The polarization factors are

$$\begin{aligned}
P_\sigma &= \hat{\boldsymbol{\sigma}}_O \cdot \hat{\boldsymbol{\sigma}}_G = 1\,,\\
P_\pi &= \hat{\boldsymbol{\pi}}_O \cdot \hat{\boldsymbol{\pi}}_G = \cos 2\theta_G\,,\\
d_1 &= \hat{\boldsymbol{\sigma}}_O \cdot \hat{\boldsymbol{\sigma}}_L = \hat{\boldsymbol{\sigma}}_G \cdot \hat{\boldsymbol{\sigma}}_L = \cos\phi_L\,,\\
d_2 &= \hat{\boldsymbol{\sigma}}_O \cdot \hat{\boldsymbol{\pi}}_L = \hat{\boldsymbol{\sigma}}_G \cdot \hat{\boldsymbol{\pi}}_L = -\sin\eta_L \sin\phi_L\,,\\
d_3 &= \hat{\boldsymbol{\pi}}_O \cdot \hat{\boldsymbol{\sigma}}_L = -\sin\theta_G \sin\phi_L\,,\\
d_4 &= \hat{\boldsymbol{\pi}}_O \cdot \hat{\boldsymbol{\pi}}_L = -\sin\theta_G \sin\eta_L \cos\phi_L + \cos\theta_G \cos\eta_L\,,\\
d_5 &= \hat{\boldsymbol{\pi}}_G \cdot \hat{\boldsymbol{\sigma}}_L = \sin\theta_G \sin\phi_L\,,\\
d_6 &= \hat{\boldsymbol{\pi}}_G \cdot \hat{\boldsymbol{\pi}}_L = \sin\theta_G \sin\eta_L \cos\phi_L + \cos\theta_G \cos\eta_L\,,
\end{aligned} \tag{7.56}$$

which are defined according to the polarization unit vectors given in a Cartesian coordinates system (see Fig. 7.2):

$$\begin{aligned}
\hat{\boldsymbol{\sigma}}_O &= (0,1,0)\,,\\
\hat{\boldsymbol{\sigma}}_G &= (0,1,0)\,,\\
\hat{\boldsymbol{\sigma}}_L &= (-\sin\phi_L, \cos\phi_L, 0)\,,\\
\hat{\boldsymbol{\pi}}_O &= (\sin\theta_G, 0, \cos\theta_G)\,,\\
\hat{\boldsymbol{\pi}}_G &= (-\sin\theta_G, 0, \cos\theta_G)\,,\\
\hat{\boldsymbol{\pi}}_L &= (-\sin\eta_L \cos\phi_L, -\sin\eta_L \sin\phi_L, \cos\eta_L)\,.
\end{aligned} \tag{7.57}$$

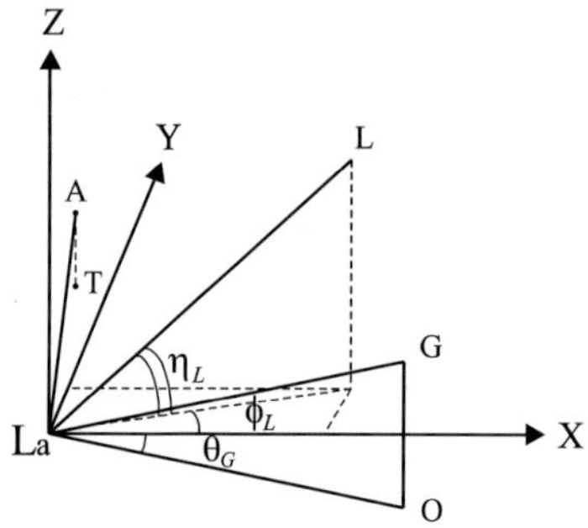

Fig. 7.2. Geometric relation among the wavevectors $\boldsymbol{K_O}, \boldsymbol{K_G}$, and $\boldsymbol{K_L}$ for a three-wave diffraction [281]

In Fig. 7.2, points O, G, and L are the reciprocal lattice points of the three reflections involved. La, A, and T are the Laue point, the entrance point, and the tie point of a given mode of wave-propagation, respectively. Assume that the entrance crystal surface is the xy-plane and the coordinates of A are (x, y, z). The accommodation vector $\boldsymbol{AT} = -k\zeta\hat{\boldsymbol{n}}_e$ is normal to the crystal surface. Since the entrance point A does not coincide with La, the incident wave with the wavevector $\boldsymbol{k}_O^e$ is slightly deviated from the exact Bragg angle θ_G of the G reflection by $\Delta\theta$ so that

$$\boldsymbol{k}_O^e = \boldsymbol{AO} = \boldsymbol{ALa} + \boldsymbol{LaO} = \boldsymbol{ALa} + \boldsymbol{k}_O \,, \tag{7.58}$$

where $\overline{ALa} = k(\Delta\theta)$ and $\boldsymbol{k}_O$ satisfies the Bragg condition for the G reflection. The corresponding wavevectors inside the crystal are

$$\begin{aligned} \boldsymbol{K}_O &= \boldsymbol{TO} = \boldsymbol{AO} - k\zeta\hat{\boldsymbol{n}}_e \,, \\ \boldsymbol{K}_G &= \boldsymbol{TG} = \boldsymbol{AG} - k\zeta\hat{\boldsymbol{n}}_e \,, \\ \boldsymbol{K}_L &= \boldsymbol{TL} = \boldsymbol{AL} - k\zeta\hat{\boldsymbol{n}}_e \,. \end{aligned} \tag{7.59}$$

The polar angle η_L and the azimuth ϕ_L define the direction of $\boldsymbol{K}_L$.

For simplicity, the crystal setting is assumed to nearly satisfy the two-wave (O, G) diffraction, but to be off the exact three-wave diffraction position for the L reflection by ψ_T. In other words, the coordinates of the entrance point A are

$$\begin{aligned} x &= k(\Delta\theta)\sin\theta_G \,, \\ y &= k\cos\theta_G \tan\psi_T \,, \\ z &= k(\Delta\theta)\cos\theta_G \,, \end{aligned} \tag{7.60}$$

where a linear approximation has been employed to define the deviation in ψ_T along the y-axis. This approximation relies on the fact that the tie point T lies on the locus of two-wave tie points of the G reflection. From the geometry shown in Fig. 7.2, it is easy to show, to the first-order approximation, that

$$\begin{aligned} 2\xi_O &= \chi_O + 2\zeta\gamma_O \,, \\ 2\xi_G &= \chi_O + 2(\Delta\theta)\sin 2\theta_G + 2\zeta\gamma_G \,, \end{aligned}$$

$$2\xi_L = \chi_O + 2\zeta\gamma_L + 2\cos\theta_G \tan\psi_T \cos\eta_L \sin\phi_L \\ +2(\Delta\theta)\sin\theta_G \cos\eta_L \cos\phi_L + 2(\Delta\theta)\cos\theta_G \sin\eta_L \,. \qquad (7.61)$$

Since the L reflection is far from the three-wave diffraction position, in (7.54) ξ_L is much larger than ξ_O and ξ_G. From (7.61), the relations between ξ_G, ξ_L, and ξ_O are

$$\xi_G = \beta_G + \frac{\gamma_G}{\gamma_O}\xi_O \,, \\ \xi_L = \beta_L + \frac{\gamma_L}{\gamma_O}\xi_O + \cos\theta_G \tan\psi_T \cos\eta_L \sin\phi_L \,, \qquad (7.62)$$

where

$$\beta_G = \frac{\chi_O}{2}\left(1 - \frac{\gamma_G}{\gamma_O}\right) + (\Delta\theta)\sin 2\theta_G \,, \\ \beta_L = \frac{\chi_O}{2}\left(1 - \frac{\gamma_L}{\gamma_O}\right) + (\Delta\theta)\left(\sin\theta_G \cos\eta_L \cos\phi_L + \cos\theta_G \sin\eta_L\right) \,. (7.63)$$

General Pseudo-Two-Wave Formulation. Equation (7.54) can be rearranged by substituting $D_{\sigma L}$ and $D_{\pi L}$ of the last two equations into the first four equations as

$$\left[2\xi_O - \frac{(d_1^2 + d_2^2)\,\chi_{O\text{–}L}\chi_{L\text{–}O}}{2\xi_L}\right] D_{\sigma O} - \left[\frac{(d_1 d_3 + d_2 d_4)\,\chi_{O\text{–}L}\chi_{L\text{–}O}}{2\xi_L}\right] D_{\pi O} \\ + \left[P_\sigma \chi_{O\text{–}G} - \frac{(d_1^2 + d_2^2)\,\chi_{O\text{–}L}\chi_{L\text{–}G}}{2\xi_L}\right] D_{\sigma G} \\ - \left[\frac{(d_1 d_5 + d_2 d_6)\,\chi_{O\text{–}L}\chi_{L\text{–}G}}{2\xi_L}\right] D_{\pi G} = 0 \,, \\ - \left[\frac{(d_1 d_3 + d_2 d_4)\,\chi_{O\text{–}L}\chi_{L\text{–}O}}{2\xi_L}\right] D_{\sigma O} + \left[2\xi_O - \frac{(d_3^2 + d_4^2)\,\chi_{O\text{–}L}\chi_{L\text{–}O}}{2\xi_L}\right] \\ \times D_{\pi O} - \left[\frac{(d_1 d_3 + d_2 d_4)\,\chi_{O\text{–}L}\chi_{L\text{–}G}}{2\xi_L}\right] D_{\sigma G} \\ + \left[P_\pi \chi_{O\text{–}G} - \frac{(d_3 d_5 + d_4 d_6)\,\chi_{O\text{–}L}\chi_{L\text{–}G}}{2\xi_L}\right] D_{\pi G} = 0 \,, \\ \left[P_\sigma \chi_{G\text{–}O} - \frac{(d_1^2 + d_2^2)\,\chi_{G\text{–}L}\chi_{L\text{–}O}}{2\xi_L}\right] D_{\sigma O} - \left[\frac{(d_1 d_3 + d_2 d_4)\,\chi_{G\text{–}L}\chi_{L\text{–}O}}{2\xi_L}\right] \\ \times D_{\pi O} + \left[2\xi_G - \frac{(d_1^2 + d_2^2)\,\chi_{G\text{–}L}\chi_{L\text{–}G}}{2\xi_L}\right] D_{\sigma G} \\ - \left[\frac{(d_1 d_5 + d_2 d_6)\,\chi_{G\text{–}L}\chi_{L\text{–}G}}{2\xi_L}\right] D_{\pi G} = 0 \,, \\ - \left[\frac{(d_1 d_5 + d_2 d_6)\,\chi_{G\text{–}L}\chi_{L\text{–}O}}{2\xi_L}\right] D_{\sigma O} + \left[P_\pi \chi_{G\text{–}O} - \frac{(d_3 d_5 + d_4 d_6)}{2\xi_L}\right.$$

$$\left.\times\chi_{G\text{–}L}\chi_{L\text{–}O}\right]D_{\pi O}-\left[\frac{(d_1d_5+d_2d_6)\,\chi_{G\text{–}L}\chi_{L\text{–}G}}{2\xi_L}\right]D_{\sigma G}$$
$$+\left[2\xi_G-\frac{\left(d_5^2+d_6^2\right)\chi_{G\text{–}L}\chi_{L\text{–}G}}{2\xi_L}\right]D_{\pi G}=0\,. \qquad (7.64)$$

By introducing the reduced variables

$$\begin{aligned}
\tilde{\xi}_{O\sigma}&=\xi_O-\frac{P_1\chi_{O\text{–}L}\chi_{L\text{–}O}}{4\xi_L}\,,\\
\tilde{\xi}_{O\pi}&=\xi_O-\frac{P_2\chi_{O\text{–}L}\chi_{L\text{–}O}}{4\xi_L}\,,\\
\tilde{\xi}_{G\sigma}&=\xi_G-\frac{P_3\chi_{G\text{–}L}\chi_{L\text{–}G}}{4\xi_L}\,,\\
\tilde{\xi}_{G\pi}&=\xi_G-\frac{P_4\chi_{G\text{–}L}\chi_{L\text{–}G}}{4\xi_L}\,,
\end{aligned} \qquad (7.65)$$

and the modified polarizabilities

$$\begin{aligned}
\chi^{\sigma}_{OG,OL}&=P_\sigma\chi_{O\text{–}G}-\frac{P_1\chi_{O\text{–}L}\chi_{L\text{–}G}}{2\xi_L}\,,\\
\chi^{\sigma}_{GO,LO}&=P_\sigma\chi_{G\text{–}O}-\frac{P_1\chi_{L\text{–}O}\chi_{G\text{–}L}}{2\xi_L}\,,\\
\chi^{\pi}_{OG,OL}&=P_\pi\chi_{O\text{–}G}-\frac{P_5\chi_{O\text{–}L}\chi_{L\text{–}G}}{2\xi_L}\,,\\
\chi^{\pi}_{GO,LO}&=P_\pi\chi_{G\text{–}O}-\frac{P_5\chi_{L\text{–}O}\chi_{G\text{–}L}}{2\xi_L}\,,
\end{aligned} \qquad (7.66)$$

(7.64) can be simplified as

$$\begin{aligned}
&2\tilde{\xi}_{O\sigma}D_{\sigma O}+\chi^{\sigma}_{OG,OL}D_{\sigma G}-\frac{P_6\chi_{O\text{–}L}\chi_{L\text{–}O}}{2\xi_L}D_{\pi O}-\frac{P_7\chi_{O\text{–}L}\chi_{L\text{–}O}}{2\xi_L}D_{\pi G}\\
&\quad=0\,,\\
&\chi^{\sigma}_{GO,LO}D_{\sigma O}+2\tilde{\xi}_{G\sigma}D_{\sigma G}-\frac{P_6\chi_{L\text{–}O}\chi_{G\text{–}L}}{2\xi_L}D_{\pi O}-\frac{P_7\chi_{G\text{–}L}\chi_{L\text{–}G}}{2\xi_L}D_{\pi G}\\
&\quad=0\,,\\
&-\frac{P_6\chi_{O\text{–}L}\chi_{L\text{–}O}}{2\xi_L}D_{\sigma O}-\frac{P_6\chi_{O\text{–}L}\chi_{L\text{–}G}}{2\xi_L}D_{\sigma G}+2\tilde{\xi}_{O\pi}D_{\pi O}+\chi^{\pi}_{OG,OL}D_{\pi G}\\
&\quad=0\,,\\
&-\frac{P_7\chi_{L\text{–}O}\chi_{G\text{–}L}}{2\xi_L}D_{\sigma O}-\frac{P_7\chi_{G\text{–}L}\chi_{L\text{–}G}}{2\xi_L}D_{\sigma G}+\chi^{\pi}_{GO,LO}D_{\pi O}+2\tilde{\xi}_{G\pi}D_{\pi G}\\
&\quad=0\,,
\end{aligned} \qquad (7.67)$$

where the polarization factors are

$$\begin{aligned}
P_1&=P_3=d_1^2+d_2^2\,,\\
P_2&=d_3^2+d_4^2\,,
\end{aligned}$$

$$
\begin{aligned}
P_4 &= d_5^2 + d_6^2 \,, \\
P_5 &= d_3 d_5 + d_4 d_6 \,, \\
P_6 &= d_1 d_3 + d_2 d_4 \,, \\
P_7 &= d_1 d_5 + d_2 d_6 \,.
\end{aligned} \tag{7.68}
$$

The equation of dispersion, derived from the secular equation of (7.67), takes the form

$$
\begin{aligned}
&\left(\tilde{\xi}_{O\sigma}\tilde{\xi}_{G\sigma} - \frac{1}{4}\chi^{\sigma}_{OG,OL}\chi^{\sigma}_{GO,LO}\right)\left(\tilde{\xi}_{O\pi}\tilde{\xi}_{G\pi} - \frac{1}{4}\chi^{\pi}_{OG,OL}\chi^{\pi}_{GO,LO}\right) \\
&= \frac{1}{16\xi_L^2}\left\{\left[\tilde{\xi}_{O\sigma}\chi_{G\text{-}L}\chi_{L\text{-}G} + \tilde{\xi}_{G\sigma}\chi_{O\text{-}L}\chi_{L\text{-}O} - \frac{1}{2}\left(\chi_{O\text{-}L}\chi_{L\text{-}G}\chi^{\sigma}_{GO,LO}\right.\right.\right. \\
&\quad \left.\left.+\chi_{L\text{-}O}\chi_{G\text{-}L}\chi^{\sigma}_{OG,OL}\right)\right] \times \left[\tilde{\xi}_{O\pi}P_7^2\chi_{G\text{-}L}\chi_{L\text{-}G} + \tilde{\xi}_{G\pi}P_6^2\chi_{O\text{-}L}\chi_{L\text{-}O}\right. \\
&\quad \left.\left.-\frac{1}{2}P_6P_7\left(\chi_{O\text{-}L}\chi_{L\text{-}G}\chi^{\pi}_{GO,LO} + \chi_{L\text{-}O}\chi_{G\text{-}L}\chi^{\pi}_{OG,OL}\right)\right]\right\} = A_o \,.
\end{aligned} \tag{7.69}
$$

The fundamental equation (7.67) and the equation of dispersion (7.69) are the exact formulations for the three-wave diffraction, which can, in general, be applied to cases involving Bragg and Laue diffraction. The reduced variables $\tilde{\xi}_{O\sigma}$, $\tilde{\xi}_{O\pi}$, $\tilde{\xi}_{G\sigma}$ and $\tilde{\xi}_{G\pi}$ are not independent of one another, as indicated in (7.65). When the three-wave case becomes a two-wave diffraction, the quantity A_o approaches zero as $1/\xi_L$ tends to zero. $1/\xi_L$ and A_o being finite implies that the normal modes of wave-propagation in the crystal are no longer plane polarized and the correlation between the σ- and π-polarized components becomes appreciable in the diffraction processes. This kind of correlation is absent for electron diffraction due to the scalar nature of electron waves.

First-Order Approximation. The two-wave approximation for three-wave diffraction requires that the reciprocal lattice point L should not be very close to the surface of the Ewald sphere, and the other points, O and G, should nearly touch the surface. This implies that ξ_L is much larger than ξ_O and ξ_G. Thus, according to (7.61), this approximation is valid as long as

$$
\tan\psi_T \gg \chi_O \,. \tag{7.70}
$$

Under this condition, (7.67) reduces to

$$
\begin{aligned}
&2\tilde{\xi}_{O\sigma}D_{\sigma O} + \chi^{\sigma}_{OG,OL}D_{\sigma G} = 0 \,, \\
&\chi^{\sigma}_{GO,LO}D_{\sigma O} + 2\tilde{\xi}_{G\sigma}D_{\sigma G} = 0 \,, \\
&2\tilde{\xi}_{O\pi}D_{\pi O} + \chi^{\pi}_{OG,OL}D_{\pi G} = 0 \,, \\
&\chi^{\pi}_{GO,LO}D_{\pi O} + 2\tilde{\xi}_{G\pi}D_{\pi G} = 0 \,,
\end{aligned} \tag{7.71}
$$

where the σ and π components are decoupled. This first-order approximation in $1/\xi_L$ is equivalent to setting A_o of (7.69) to be zero, i.e., those terms involving $1/(\xi_L)^2$ are neglected. The equations of dispersion are then

$$\tilde{\xi}_{O\sigma}\tilde{\xi}_{G\sigma} - \frac{1}{4}\chi^{\sigma}_{OG,OL}\chi^{\sigma}_{GO,LO} = 0 \, ,$$
$$\tilde{\xi}_{O\pi}\tilde{\xi}_{G\pi} - \frac{1}{4}\chi^{\pi}_{OG,OL}\chi^{\pi}_{GO,LO} = 0 \, . \tag{7.72}$$

Hence the $\tilde{\xi}$'s given in (7.65) can be considered as first-order variables, which determine the positions of tie points of the two-wave (O, G) diffraction. Similarly, the polarizabilities defined in (7.66) are also of the first-order.

The reduced variables $\tilde{\xi}$'s are expressed explicitly in terms of $\xi_{O\sigma}$ and $\xi_{G\sigma}$ for the σ components as

$$4\xi_{O\sigma}\xi_{G\sigma} - 2\xi_{O\sigma}C_{\sigma G} - 2\xi_{G\sigma}C_{\sigma O} + C_{\sigma O}C_{\sigma G} - \chi^{\sigma}_{OG,OL}\chi^{\sigma}_{GO,LO} = 0 \, , \tag{7.73}$$

where

$$C_{\sigma O} = \frac{P_1\chi_{O\text{–}L}\chi_{L\text{–}O}}{2\xi_L} \, , \qquad C_{\sigma G} = \frac{P_3\chi_{G\text{–}L}\chi_{L\text{–}G}}{2\xi_L} \, ,$$
$$C_{\pi O} = \frac{P_2\chi_{O\text{–}L}\chi_{L\text{–}O}}{2\xi_L} \, , \qquad C_{\pi G} = \frac{P_4\chi_{G\text{–}L}\chi_{L\text{–}G}}{2\xi_L} \, . \tag{7.74}$$

A similar equation for the π components can be obtained by changing σ to π in the subscripts. In contrast, the two-wave dispersion equation for the σ components is

$$4\xi_{O\sigma}\xi_{G\sigma} - P_{\sigma}^2\chi_{O\text{–}G}\chi_{G\text{–}O} = 0 \, . \tag{7.75}$$

Since the fundamental equation and the equation of dispersion have similar forms for the σ and π components, in the following, we shall only consider the relevant equations for the σ components.

Substituting (7.62) and (7.63) into the dispersion relation (7.73) yields

$$\xi^2_{O\sigma} + \xi_{O\sigma}\left(\frac{\gamma_O}{\gamma_G}\right)\left[\beta_G - \frac{1}{2}C_{\sigma G} - \frac{1}{2}C_{\sigma O}\left(\frac{\gamma_G}{\gamma_O}\right)\right] - \frac{1}{4}\left(\frac{\gamma_O}{\gamma_G}\right)$$
$$\times \left[\chi^{\sigma}_{OG,OL}\chi^{\sigma}_{GO,LO} - C_{\sigma O}C_{\sigma G} + 2C_{\sigma O}\beta_G\right] = 0 \, . \tag{7.76}$$

The solutions for the eigenvalues $\xi_{O\sigma}$ are

$$\left.\begin{matrix}\xi_{O\sigma}(1)\\ \xi_{O\sigma}(2)\end{matrix}\right\} = \frac{1}{2}\left(z \mp \sqrt{z^2+q}\right) + \frac{1}{4}C_{\sigma O} \, , \tag{7.77}$$

where

$$z = -\frac{\gamma_O}{\gamma_G}\left[\beta_G - \frac{1}{2}C_{\sigma G} + \frac{1}{2}C_{\sigma O}\left(\frac{\gamma_G}{\gamma_O}\right)\right] ,$$
$$q = \frac{\gamma_O}{\gamma_G}\left(\chi^{\sigma}_{OG,OL}\chi^{\sigma}_{GO,LO}\right) . \tag{7.78}$$

The corresponding eigenvectors for modes 1 and 2 are

$$\frac{D_{\sigma G}(1)}{D_{\sigma O}(1)} = -\frac{\left(\frac{\gamma_O}{\gamma_G}\right)\chi^{\sigma}_{GO,LO}}{z + \sqrt{z^2+q}} = B_1 \, ,$$
$$\frac{D_{\sigma G}(2)}{D_{\sigma O}(2)} = -\frac{\left(\frac{\gamma_O}{\gamma_G}\right)\chi^{\sigma}_{GO,LO}}{z - \sqrt{z^2+q}} = B_2 \, . \tag{7.79}$$

Similar wavefield-amplitude ratios can be obtained for the π components by replacing σ with π in the subscripts. Expressions for the diffraction intensities can also be derived, based on whether the reflection is strong or weak. The derivation is common for both first-order and second-order approximations. We shall discuss the intensities after presenting the second-order approximation.

Second-Order Approximation. Equation (7.67) is led to the first-order approximation, (7.71), by ignoring those terms involving explicitly $1/\xi_L$. For a second-order approximation, we shall also begin with (7.67) but retain all the terms in the equation. For the σ-polarized wavefields, $D_{\pi O}$ and $D_{\pi G}$ can be expressed in terms of $D_{\sigma O}$ and $D_{\sigma G}$ from the last two equations of (7.67). With the aid of these relations between the σ's and π's, the common term in the first two equations involving $D_{\pi O}$ and $D_{\pi G}$ then takes the form

$$\begin{aligned}
&\frac{1}{2\xi_L}\left(P_6\chi_{L\text{-}O}D_{\pi O}+P_7\chi_{L\text{-}G}D_{\pi G}\right)\\
&=\frac{1}{4\xi_L^2C_\pi}\Big\{\left(\chi_{L\text{-}O}D_{\sigma O}+\chi_{L\text{-}G}D_{\sigma G}\right)\Big[2P_7^2\chi_{L\text{-}G}\chi_{G\text{-}L}\tilde{\xi}_{O\pi}\\
&\quad+2P_6^2\chi_{O\text{-}L}\chi_{L\text{-}O}\tilde{\xi}_{G\pi}-P_6P_7f_\pi\Big]\Big\}\,, \qquad (7.80)
\end{aligned}$$

where

$$\begin{aligned}
C_\pi &= 4\tilde{\xi}_{O\pi}\tilde{\xi}_{G\pi}-\chi^{\pi}_{OG,OL}\chi^{\pi}_{GO,LO}\,,\\
f_\pi &= \chi_{O\text{-}L}\chi_{L\text{-}G}\chi^{\pi}_{GO,LO}+\chi_{L\text{-}O}\chi_{G\text{-}L}\chi^{\pi}_{OG,OL}\,. \qquad (7.81)
\end{aligned}$$

Similarly, for the π-polarized wavefields, substituting the relations between $D_{\sigma O}$, $D_{\sigma G}$ and $D_{\pi O}$, $D_{\pi G}$ obtained from the first two equations of (7.67) into the common term in the last two equations, we obtain the following relation

$$\begin{aligned}
&\frac{1}{2\xi_L}\left(\chi_{L\text{-}O}D_{\sigma O}+\chi_{L\text{-}G}D_{\sigma G}\right)\\
&=\frac{1}{4\xi_L^2C_\sigma}\Big\{\left(P_6\chi_{L\text{-}O}D_{\pi O}+P_7\chi_{L\text{-}G}D_{\pi G}\right)\Big[2\chi_{L\text{-}G}\chi_{G\text{-}L}\tilde{\xi}_{O\sigma}\\
&\quad+2\chi_{O\text{-}L}\chi_{L\text{-}O}\tilde{\xi}_{G\sigma}-f_\sigma\Big]\Big\}\,, \qquad (7.82)
\end{aligned}$$

where

$$\begin{aligned}
C_\sigma &= 4\tilde{\xi}_{O\sigma}\tilde{\xi}_{G\sigma}-\chi^{\sigma}_{OG,OL}\chi^{\sigma}_{GO,LO}\,,\\
f_\sigma &= \chi_{O\text{-}L}\chi_{L\text{-}G}\chi^{\sigma}_{GO,LO}+\chi_{L\text{-}O}\chi_{G\text{-}L}\chi^{\sigma}_{OG,OL}\,. \qquad (7.83)
\end{aligned}$$

Both (7.80) and (7.82) are proportional to the second order terms $1/(\xi_L)^2$. As given in (7.71) and (7.72), the pair of the first-order fundamental equations and of the equation of dispersion have the same mathematical forms as that for the two-wave diffraction. The only difference is in the involvement of the σ and π related components, which have different definitions. Since the behaviors of the wavefield-amplitude ratios and the dispersion surface for σ

and π-polarized wavefields are similar for dynamical diffractions, the σ and π related components, to a first-order approximation, are interchangeable, so that

$$\begin{aligned}
&\tilde{\xi}_{O\pi}\tilde{\xi}_{G\pi} - \frac{1}{4}\chi^{\pi}_{OG,OL}\chi^{\pi}_{GO,LO} \simeq \tilde{\xi}_{O\sigma}\tilde{\xi}_{G\sigma} - \frac{1}{4}\chi^{\pi}_{OG,OL}\chi^{\pi}_{GO,LO} \\
&\simeq \frac{1}{4}\chi^{\sigma}_{OG,OL}\chi^{\sigma}_{GO,LO} - \frac{1}{4}\chi^{\pi}_{OG,OL}\chi^{\pi}_{GO,LO} = \frac{C_I}{2} , \\
&\tilde{\xi}_{O\sigma}\tilde{\xi}_{G\sigma} - \frac{1}{4}\chi^{\sigma}_{OG,OL}\chi^{\sigma}_{GO,LO} \simeq \frac{1}{4}\chi^{\pi}_{OG,OL}\chi^{\pi}_{GO,LO} - \frac{1}{4}\chi^{\sigma}_{OG,OL}\chi^{\sigma}_{GO,LO} \\
&= -\frac{C_I}{2} , \\
&\tilde{\xi}_{O\pi}D_{\sigma G} \simeq -\frac{1}{2}\chi^{\sigma}_{GO,LO}D_{\sigma O} , \\
&\tilde{\xi}_{O\pi}D_{\sigma O} \simeq -\frac{1}{2}\chi^{\sigma}_{OG,OL}D_{\sigma G} .
\end{aligned} \tag{7.84}$$

Substituting (7.80), (7.82), and (7.84) into (7.67) gives the fundamental equation as

$$\begin{aligned}
&\left[2\tilde{\xi}_{O\sigma} - \frac{\chi_{O\text{–}L}}{2\xi_L^2 C_I}B_{\sigma O}\right]D_{\sigma O} + \left[\chi^{\sigma}_{OG,OL} - \frac{\chi_{O\text{–}L}}{2\xi_L^2 C_I}B_{\sigma G}\right]D_{\sigma G} = 0 , \\
&\left[\chi^{\sigma}_{GO,LO} - \frac{\chi_{G\text{–}L}}{2\xi_L^2 C_I}B_{\sigma O}\right]D_{\sigma O} + \left[2\tilde{\xi}_{G\sigma} - \frac{\chi_{G\text{–}L}}{2\xi_L^2 C_I}B_{\sigma G}\right]D_{\sigma G} = 0 , \\
&\left[2\tilde{\xi}_{O\pi} + \frac{\chi_{O\text{–}L}P_6}{2\xi_L^2 C_I}B_{\pi O}\right]D_{\pi O} + \left[\chi^{\pi}_{OG,OL} + \frac{\chi_{O\text{–}L}P_6}{2\xi_L^2 C_I}B_{\pi G}\right]D_{\pi G} = 0 , \\
&\left[\chi^{\pi}_{GO,LO} + \frac{\chi_{G\text{–}L}}{2\xi_L^2 C_I}B_{\pi O}\right]D_{\pi O} + \left[2\tilde{\xi}_{G\pi} + \frac{\chi_{G\text{–}L}P_7}{2\xi_L^2 C_I}B_{\pi G}\right]D_{\pi G} = 0 ,
\end{aligned} \tag{7.85}$$

where

$$\begin{aligned}
B_{\sigma O} &= P_7^2\left(\chi_{L\text{–}G}\right)^2\chi_{G\text{–}L}\chi^{\sigma}_{OG,OL} + P_6^2\chi_{O\text{–}L}\chi_{L\text{–}O}\chi_{L\text{–}G}\chi^{\sigma}_{GO,LO} \\
&\quad -P_6P_7\chi_{L\text{–}O}f_{\pi} , \\
B_{\sigma G} &= P_7^2\chi_{L\text{–}O}\chi_{L\text{–}G}\chi_{G\text{–}L}\chi^{\sigma}_{OG,OL} + P_6^2\chi_{O\text{–}L}\left(\chi_{L\text{–}O}\right)^2\chi^{\sigma}_{GO,LO} \\
&\quad -P_6P_7\chi_{L\text{–}G}f_{\pi} , \\
B_{\pi O} &= P_7\left(\chi_{L\text{–}G}\right)^2\chi_{G\text{–}L}\chi^{\pi}_{OG,OL} + P_6\chi_{O\text{–}L}\chi_{L\text{–}O}\chi_{L\text{–}G}\chi^{\pi}_{GO,LO} \\
&\quad -P_6\chi_{L\text{–}O}f_{\sigma} , \\
B_{\pi G} &= P_6\chi_{L\text{–}O}\chi_{L\text{–}G}\chi_{G\text{–}L}\chi^{\pi}_{OG,OL} + P_6\left(\chi_{L\text{–}O}\right)^2\chi_{O\text{–}L}\chi^{\pi}_{GO,LO} \\
&\quad -P_7\chi_{L\text{–}G}f_{\sigma} .
\end{aligned} \tag{7.86}$$

Equation (7.85) can be further reduced to the compact form

$$\begin{aligned}
&2\bar{\xi}_{O\sigma}D_{\sigma O} + \bar{\chi}^{\sigma}_{OG,OL}D_{\sigma G} = 0 , \\
&\bar{\chi}^{\sigma}_{GO,LO}D_{\sigma O} + 2\bar{\xi}_{G\sigma}D_{\sigma G} = 0 , \\
&2\bar{\xi}_{O\pi}D_{\pi O} + \bar{\chi}^{\pi}_{OG,OL}D_{\pi G} = 0 , \\
&\bar{\chi}^{\pi}_{GO,LO}D_{\pi O} + 2\bar{\xi}_{G\pi}D_{\pi G} = 0 ,
\end{aligned} \tag{7.87}$$

where the second-order modified polarizabilities are

$$\bar{\chi}^{\sigma}_{OG,OL} = \chi^{\sigma}_{OG,OL} - \frac{\chi_{O\text{–}L}}{2\xi_L^2 C_I} B_{\sigma G} ,$$
$$\bar{\chi}^{\sigma}_{GO,LO} = \chi^{\sigma}_{GO,LO} - \frac{\chi_{G\text{–}L}}{2\xi_L^2 C_I} B_{\sigma O} ,$$
$$\bar{\chi}^{\pi}_{OG,OL} = \chi^{\pi}_{OG,OL} + \frac{P_6\chi_{O\text{–}L}}{2\xi_L^2 C_I} B_{\pi G} ,$$
$$\bar{\chi}^{\pi}_{GO,LO} = \chi^{\pi}_{GO,LO} + \frac{P_7\chi_{G\text{–}L}}{2\xi_L^2 C_I} B_{\pi O} , \quad (7.88)$$

and the second-order variables $\bar{\xi}$'s are defined as

$$2\bar{\xi}_{O\sigma} = 2\tilde{\xi}_{O\sigma} - \frac{\chi_{O\text{–}L}}{2\xi_L^2 C_I} B_{\sigma O} ,$$
$$2\bar{\xi}_{G\sigma} = 2\tilde{\xi}_{G\sigma} - \frac{\chi_{G\text{–}L}}{2\xi_L^2 C_I} B_{\sigma G} ,$$
$$2\bar{\xi}_{O\pi} = 2\tilde{\xi}_{O\pi} + \frac{P_6\chi_{O\text{–}L}}{2\xi_L^2 C_I} B_{\pi O} ,$$
$$2\bar{\xi}_{G\pi} = 2\tilde{\xi}_{G\pi} + \frac{P_7\chi_{G\text{–}L}}{2\xi_L^2 C_I} B_{\pi G} . \quad (7.89)$$

$\bar{\xi}$'s can be expressed in terms of ξ's as

$$2\bar{\xi}_{O\sigma} = 2\xi_{O\sigma} - \frac{P_1\chi_{O\text{–}L}\chi_{L\text{–}O}}{2\xi_L} - \frac{\chi_{O\text{–}L}}{2\xi_L^2 C_I} B_{\sigma O} ,$$
$$2\bar{\xi}_{G\sigma} = 2\xi_{G\sigma} - \frac{P_3\chi_{G\text{–}L}\chi_{L\text{–}G}}{2\xi_L} - \frac{\chi_{G\text{–}L}}{2\xi_L^2 C_I} B_{\sigma G} ,$$
$$2\bar{\xi}_{O\pi} = 2\xi_{O\pi} - \frac{P_2\chi_{O\text{–}L}\chi_{L\text{–}O}}{2\xi_L} + \frac{P_6\chi_{O\text{–}L}}{2\xi_L^2 C_I} B_{\pi O} ,$$
$$2\bar{\xi}_{G\pi} = 2\xi_{G\pi} - \frac{P_4\chi_{G\text{–}L}\chi_{L\text{–}G}}{2\xi_L} + \frac{P_7\chi_{G\text{–}L}}{2\xi_L^2 C_I} B_{\pi G} . \quad (7.90)$$

This indicates that the ξ's are further modified due to the second-order approximation. This also implies that the occurrence of the two-wave (O, G) diffraction and of the three-wave diffraction are shifted from their first-order positions.

The solutions of (7.87) for the eigenvalues ξ's and the wavefield-amplitude ratios are

$$\left.\begin{matrix}\xi_{O\sigma}(1)\\ \xi_{O\sigma}(2)\end{matrix}\right\} = \frac{1}{2}\left(\bar{z} \mp \sqrt{\bar{z}^2 + \bar{q}}\right) + \frac{1}{4}\bar{C}_{\sigma O} , \quad (7.91)$$

and

$$\frac{D_{\sigma G}(1)}{D_{\sigma O}(1)} = -\frac{\left(\frac{\gamma_O}{\gamma_G}\right)\bar{\chi}^{\sigma}_{GO,LO}}{\bar{z} + \sqrt{\bar{z}^2 + \bar{q}}} = \bar{B}_1 ,$$

$$\frac{D_{\sigma G}(2)}{D_{\sigma O}(2)} = -\frac{\left(\frac{\gamma_O}{\gamma_G}\right)\bar{\chi}^{\sigma}_{GO,LO}}{\bar{z} - \sqrt{\bar{z}^2 + \bar{q}}} = \bar{B}_2 \,, \tag{7.92}$$

respectively, where

$$\begin{aligned}
\bar{z} &= -\left(\frac{\gamma_O}{\gamma_G}\right)\left[\beta_G - \frac{1}{2}\bar{C}_{\sigma G} + \frac{1}{2}\left(\frac{\gamma_G}{\gamma_O}\right)\bar{C}_{\sigma O}\right] \,, \\
\bar{q} &= \frac{\gamma_O}{\gamma_G}\bar{\chi}^{\sigma}_{OG,OL} \,, \\
\bar{C}_{\sigma O} &= \frac{P_1\chi_{O\text{–}L}\chi_{L\text{–}O}}{2\xi_L} + \frac{\chi_{O\text{–}L}}{2\xi_L^2 C_I}B_{\sigma O} \,, \\
\bar{C}_{\sigma G} &= \frac{P_3\chi_{G\text{–}L}\chi_{L\text{–}G}}{2\xi_L} + \frac{\chi_{G\text{–}L}}{2\xi_L^2 C_I}B_{\sigma G} \,.
\end{aligned} \tag{7.93}$$

For the π-polarized waves, the eigenvalues and eigenvectors have the same forms as that in (7.91) and (7.92), except that the subscript σ needs to be changed to π and the $\bar{C}$'s are

$$\begin{aligned}
\bar{C}_{\pi O} &= \frac{P_2\chi_{O\text{–}L}\chi_{L\text{–}O}}{2\xi_L} - \frac{P_6\chi_{O\text{–}L}}{2\xi_L^2 C_I}B_{\pi O} \,, \\
\bar{C}_{\pi G} &= \frac{P_4\chi_{G\text{–}L}\chi_{L\text{–}G}}{2\xi_L} - \frac{P_7\chi_{G\text{–}L}}{2\xi_L^2 C_I}B_{\pi G} \,.
\end{aligned} \tag{7.94}$$

Diffraction Intensities. Both the first- and second-order approximations give the same mathematical form for the wavefield-amplitude ratios, as can be seen from (7.79) and (7.92). Only the modified polarizabilities and the reduced variables are different. This facilitates the presentation of diffraction intensities for both approximations. That is, a general formulation can be adopted for both approximations. As described in Chap. 6, to derive the expressions for diffraction intensities, it is necessary to consider the boundary conditions which depend on the diffraction geometry. In the following, we shall discuss the three-wave diffraction for both Laue transmission and Bragg reflection in a plane-parallel crystal separately.

(i) Three-Wave Transmission Case

There are four modes of wave-propagation, two for σ and two for π, under a two-wave approximation. The boundary conditions at the entrance surface, i.e., at $T = 0$, for the σ-polarized fields are

$$D_{\sigma O}(1) + D_{\sigma O}(2) = D^e_O \,, \qquad D_{\sigma G}(1) + D_{\sigma G}(2) = 0 \,, \tag{7.95}$$

where D^e_O is the wavefield amplitude of the incident wave $\boldsymbol{D}^e_O(\boldsymbol{r})$,

$$\boldsymbol{D}^e_O(\boldsymbol{r}) = \boldsymbol{D}^e_O \exp\left(-2\pi \mathrm{i}\boldsymbol{k}_O \cdot \boldsymbol{r}\right) \,, \tag{7.96}$$

in empty space. Substituting the wavefield-amplitude ratios, the B_1 and B_2 of (7.79) for the first-order and the $\bar{B}_1$ and $\bar{B}_2$ of (7.92) for the second-order approximation, into (7.95) yields the following relations

$$D_O(1) = \frac{B_2}{B_2 - B_1} D_O^e , \qquad D_O(2) = -\frac{B_1}{B_2 - B_1} D_O^e ,$$
$$D_G(1) = \frac{B_1 B_2}{B_2 - B_1} D_O^e , \qquad D_G(2) = -\frac{B_1 B_2}{B_2 - B_1} D_O^e . \tag{7.97}$$

For simplicity, hereafter the subscript σ will be left out and only the first-order solutions will be considered. The second-order solutions can be easily deduced by replacing the B's and the first-order variables by the $\bar{B}$'s and the second-order variables, respectively. The boundary conditions at the exit surface of the crystal plate, i.e., $T = T_o$, are

$$D_O(T_o) = \exp(-2\pi \mathrm{i} \boldsymbol{k}_O \cdot \boldsymbol{r}) \Big[D_O(1) \exp(-\mathrm{i}\phi_1 T_o) + D_O(2) \exp(-\mathrm{i}\phi_2 T_o) \Big] ,$$
$$D_G(T_o) = \exp[-2\pi \mathrm{i} (\boldsymbol{k}_O + \boldsymbol{g}_G) \cdot \boldsymbol{r}] \Big[B_1 D_O(1) \exp(-\mathrm{i}\phi_1 T_o) + B_2 D_O(2) \exp(-\mathrm{i}\phi_2 T_o) \Big] , \tag{7.98}$$

where T_o is the crystal thickness, and

$$\phi_1 = -2\pi k \zeta(1) , \qquad \phi_2 = -2\pi k \zeta(2) . \tag{7.99}$$

The accommodation ζ takes the form

$$\left. \begin{matrix} \zeta(1) \\ \zeta(2) \end{matrix} \right\} = \frac{1}{2\gamma_O} \left(z \mp \sqrt{z^2 + q} + \frac{1}{2} C_{\sigma O} - \chi_{O\text{-}O} \right) \tag{7.100}$$

according to (7.77) and (7.91).

The intensity ratios of the diffracted waves with respect to the incident wave are then

$$\frac{I_O}{I_O^e} = \frac{|D_O(T_o)|^2}{|D_O^e|^2} = \left| \frac{B_1 B_2}{B_2 - B_1} (p_1 - p_2) \right|^2 ,$$
$$\frac{I_G}{I_O^e} = \frac{|D_G(T_o)|^2}{|D_O^e|^2} = \left| \frac{B_2 p_1 - B_1 p_2}{B_2 - B_1} \right|^2 , \tag{7.101}$$

where

$$p_1 = \exp(-\mathrm{i}\phi_1 T_o) , \qquad p_2 = \exp(-\mathrm{i}\phi_2 T_o) . \tag{7.102}$$

For convenience, let us define

$$z^2 + q = m + \mathrm{i}n , \qquad \sqrt{z^2 + q} = v + \mathrm{i}w , \tag{7.103}$$

where

$$v = \frac{1}{\sqrt{2}} \sqrt{\sqrt{m^2 + n^2} + m} , \qquad w = \frac{1}{\sqrt{2}} \sqrt{\sqrt{m^2 + n^2} - m} ,$$
$$m = z_r^2 - z_i^2 + q_r , \qquad n = 2 z_r z_i + q_i . \tag{7.104}$$

The quantities z_r, z_i, and q_r, q_i are the real and imaginary parts of z and q, respectively. Combining the equations from (7.95) to (7.104), we can write the intensity ratios as

$$\frac{I_G}{I_O^e} = \left(\frac{\gamma_O}{\gamma_G}\right)^2 |\chi_{GO,HO}|^2 \exp(-\mu T) \left(\frac{\sin^2(av) + \sinh^2(aw)}{v^2 + w^2}\right) , \quad (7.105)$$

where the following relationships and definitions

$$\begin{aligned}
B_1 B_2 &= -\frac{(\chi_{GO,HO})^2}{q} , \\
B_2 - B_1 &= -\frac{2(v + iw)}{\left(\frac{\gamma_O}{\gamma_G}\right) q} , \\
\cosh\left(\frac{\phi^+}{2}\right) &+ \sinh\left(\frac{\phi^+}{2}\right) = \exp(-\mu T) , \\
\sin^2\left(\frac{\phi^-}{2}\right) &+ \sinh^2\left(\frac{\phi^+}{2}\right) = \sin^2(av) + \sinh^2(aw) , \\
\phi^+ &= (\phi_1 + \phi_2) T_o , \qquad \phi^- = (\phi_1 - \phi_2) T_o , \\
T &= \frac{1}{2}\left(\frac{1}{\gamma_O} + \frac{1}{\gamma_G}\right) T_o , \\
\mu &= \mu_0 + \sigma_o , \qquad \mu_0 = -\frac{2\pi k}{\gamma_O} \chi_{O,i} , \\
\sigma_o &= \frac{\pi k}{\gamma_O^2} \frac{1}{\left(\frac{1}{\gamma_O} + \frac{1}{\gamma_G}\right)} \left(\frac{\gamma_O}{\gamma_G} C_{\sigma G,i} - C_{\sigma O,i}\right) , \\
a &= \frac{\pi k}{\gamma_O} T_o , \qquad (7.106)
\end{aligned}$$

have been used. $C_{\sigma O,i}$ and $C_{\sigma G,i}$ are the imaginary parts of $C_{\sigma O}$ and $C_{\sigma G}$, respectively. The relation between the normal linear absorption coefficient μ_0 and the correction σ_o indicates that modification in absorption is mainly due to the three-wave interaction and the presence of the third reflection L in the diffraction process. This modification often leads to anomalous transmission in multiple-wave diffraction. Most importantly, (7.105), the well-known expression for the intensity of two-wave transmission [48], is adopted here for the first- and second-order Bethe approximations for a three-wave diffraction.

(ii) Three-Wave Bragg Diffraction Case

There are also four permitted modes for two polarized wavefields in a three-wave Bragg diffraction under a two-wave approximation. Suppose that at least one Bragg reflection, say, reflection G, is involved in a three-wave case and the corresponding direction cosine γ_G is negative. In the two-wave approximation, the boundary conditions at the entrance surface for a given polarization are

$$D_O(1) + D_O(2) = D_O^e \,, \qquad D_G(1) + D_G(2) = D_G^e \,, \tag{7.107}$$

where D_G^e is the amplitude of the reflected wavefield in vacuum. At the exit surface, the boundary conditions give

$$D_G(1)p_1 + D_G(2)p_2 = 0 \,. \tag{7.108}$$

By combining (7.102), (7.107), and (7.108), the wavefield amplitudes of the direct waves are obtained as

$$D_O(1) = \frac{p_2 B_2 D_O^e}{p_2 B_2 - p_1 B_1} \,, \qquad D_O(2) = -\frac{p_1 B_1 D_O^e}{p_2 B_2 - p_1 B_1} \,. \tag{7.109}$$

The reflected intensity at the entrance surface of the crystal can be expressed, according to (7.107) and (7.109), as

$$\frac{I_G}{I_O^e} = \left|\frac{D_G^e}{D_O^e}\right|^2 = \left|\frac{B_1 B_2 (p_1 - p_2)}{p_2 B_2 - p_1 B_1}\right|^2 \,. \tag{7.110}$$

The transmitted intensity at the exit surface is

$$\frac{I_O(T_o)}{I_O^e} = \left|\frac{p_1 p_2 (B_2 - B_1)}{p_2 B_2 - p_1 B_1}\right|^2 \,. \tag{7.111}$$

Equation (7.110) may be written explicitly in terms of the modified polarizability as

$$\frac{I_G}{I_O^e} = \frac{1}{R}\left(\frac{\gamma_O}{\gamma_G}\right)^2 |\chi_{GO,HO}|^2 \left[\sin^2(av) + \sinh^2(aw)\right] \,, \tag{7.112}$$

where

$$\begin{aligned} R = {} & \left|q + z^2\right| + R_1 \sinh^2(aw) - R_2 \sin^2(av) + \frac{1}{2}\sqrt{|R_1^2 - |q^2||} \\ & \times \sinh|2aw| + \frac{1}{2}\sqrt{|R_2^2 - |q^2||}\sin|2av| \,, \end{aligned} \tag{7.113}$$

and

$$R_1 = \left|q + z^2\right| + |z|^2 \,, \qquad R_2 = \left|q + z^2\right| - |z|^2 \,. \tag{7.114}$$

This expression is again the well-known two-wave intensity equation [48]. For the second-order approximation, the modified polarizability, q, and z need to be replaced by the corresponding $\overline{\chi}_{GO,LO}$, $\bar{q}$, and $\bar{z}$.

(iii) Integrated Reflection for Non-Absorbing Crystals

Integrated intensity is always measured in the X-ray diffraction experiments, which can be accounted for by integrating the intensity expressions, (7.105) and (7.112), over an appropriate angular range. Under some circumstances, integrated intensity or integrated reflection may have a much simpler form than that of (7.105) or (7.112). Since the expressions for the intensities of the Laue and the Bragg reflection are exactly the same as for the two-wave cases, we shall only present the final expressions for the integrated reflection for various specific situations without detailed derivation. Interested readers are suggested to refer to those textbooks on the dynamical theory of X-ray diffraction [48,52,55].

The integrated reflection R_G^θ for a given reflection G is defined as the integration of the power ratio P_G/P_O^e between the reflected and the incident waves over an incident angle θ:

$$R_G^\theta = \int \frac{P_G}{P_O^e} \mathrm{d}\theta \, . \tag{7.115}$$

The power ratio takes the form

$$\frac{P_G}{P_O^e} = \frac{I_G}{I_O^e} \frac{\Lambda_G}{\Lambda_O} = \frac{I_G}{I_O^e} \frac{\gamma_G}{\gamma_O} = \frac{I_G}{|b| I_O^e} \tag{7.116}$$

because the diffraction power is the product of the reflected intensity and the cross section Λ of the reflected wave. The ratio of the direction cosines is denoted as the asymmetry parameter $1/b$.

The electric susceptibility, namely, the X-ray polarizability, of a non-absorbing crystal is a real number, so that $\chi^*_{GO,LO} = \chi_{OG,OL}$ and $|\chi_{GO}\chi_{OG}| = |\chi_{OG}|^2$. The power ratios of non-absorbing crystals, which are different for different situations, are outlined below:

a) Non-absorbing Laue Case

The parameter b is positive for a Laue transmission case. The q and z are real because the polarizability is real. Namely, the imaginary part of $\sqrt{z^2+q}$ is zero, i.e., $w = 0$ and $z = z_r$. Similarly, the linear absorption coefficient μ is also zero. The power ratio then takes the form

$$\frac{P_G}{P_O^e} = \frac{\sin^2\left(A\sqrt{1+\varrho^2}\right)}{1+\varrho^2} \, , \tag{7.117}$$

where

$$\begin{aligned} \varrho &= \frac{z_r}{M} \\ &= \frac{1}{\sqrt{b}\,|\chi_{GO,LO}|}\left[-b(\Delta\theta)\sin 2\theta_G + \frac{(1-b)}{2}\chi_{O,r} + \frac{b}{2}C_{\sigma G,r} - \frac{1}{2}C_{\sigma O,r}\right] , \\ v &= \sqrt{z^2+q} = M\sqrt{1+\varrho^2} \, , \end{aligned} \tag{7.118}$$

with M and A defined as

$$\begin{aligned} M &= \sqrt{|b|}\,|\chi_{GO,HO}| \, , \\ A = aM &= \frac{\pi k T_o \,|\chi_{GO,HO}|}{\sqrt{|\gamma_O\gamma_G|}} \, . \end{aligned} \tag{7.119}$$

The subscripts r's in $C_{\sigma O,r}$ and $C_{\sigma G,r}$ indicate the real parts of $C_{\sigma O}$ and $C_{\sigma G}$. The power ratio is therefore a function of the asymmetry parameter and the angular deviation from the exact Bragg angle θ_G of the G reflection.

b) Non-absorbing Bragg Case

The parameter b is negative in a Bragg case, because $\gamma_G < 0$. Hence $q = -|b||\chi_{GO,LO}|^2 < 0$, and $\sqrt{z^2+q}$ can be real or imaginary depending on whether $z^2 + q$ is greater or smaller than zero.

For $z^2 + q > 0$, the real part of $\sqrt{z^2+q}$ is $v = A\sqrt{\varrho^2-1}$ and the imaginary part $w = 0$. The power ratio becomes

$$\begin{aligned}\frac{P_G}{P_O^e} &= \frac{\sin^2\left(A\sqrt{\varrho^2-1}\right)}{\varrho^2-1+\sinh^2\left(A\sqrt{\varrho^2-1}\right)}\\ &= \frac{1}{\varrho^2+(\varrho^2-1)\cot^2\left(A\sqrt{\varrho^2-1}\right)}\ .\end{aligned} \tag{7.120}$$

For $z^2 + q < 0$, $v = 0$ and $\sqrt{z^2+q} = \mathrm{i}A\sqrt{1-\varrho^2}$. The power ratio is

$$\begin{aligned}\frac{P_G}{P_O^e} &= \frac{\sinh^2\left(A\sqrt{1-\varrho^2}\right)}{1-\varrho^2+\sinh^2\left(A\sqrt{1-\varrho^2}\right)}\\ &= \frac{1}{\varrho^2+(1-\varrho^2)\coth^2\left(A\sqrt{1-\varrho^2}\right)}\ .\end{aligned} \tag{7.121}$$

The power ratios given above as functions of ϱ can be considered as the diffraction patterns for the modified two-wave Laue and Bragg reflections. They are symmetric about $\varrho = 0$. The centers of the diffraction patterns are at $\varrho = 0$, which deviate from the exact two-wave Bragg angle θ_G by

$$\begin{aligned}\Delta\theta_{3-\text{wave}} &= \frac{(1-b)\chi_{O,r}}{2b\sin 2\theta_G} + \frac{C_{\sigma G,r}}{2\sin 2\theta_G} - \frac{C_{\sigma O,r}}{2b\sin 2\theta_G}\\ &= \Delta\theta_{2-\text{wave}} + \theta_3\ ,\end{aligned} \tag{7.122}$$

according to (7.118), where $\Delta\theta_{2-\text{wave}}$ is the angular shift from θ_G for two-wave Bragg cases, defined as

$$\Delta\theta_{2-\text{wave}} = \frac{(1-b)\chi_{O,r}}{2b\sin 2\theta_G}\ . \tag{7.123}$$

There is no angular shift of the center of the diffraction pattern for a two-wave symmetric Laue diffraction, because $b = 1$. The further shift, θ_3, of the center results from the presence of the additional reflection L and the coupling G–L between the primary reflection G and the secondary reflection L.

The formulation for the power ratios just derived is exactly the same as for the two-wave Laue and Bragg diffractions, except that the polarizability and the parameter A are modified for the three-wave situation. The parameter A is usually employed as a reduced parameter to classify the crystal to be 'thick' or 'thin' in a sense of diffraction. According to Zachariasen [48], a crystal being 'thick' or 'thin' depends on the value of A being greater or smaller than unity, respectively. A crystal is regarded as intermediately thick if $A \simeq 1$. According to (7.119), the diffraction from a thick crystal might involve a large thick crystal or a strong reflection with a large polarizability, whereas the diffraction from a thin crystal might involve crystals of small thicknesses or weak reflections. For intermediately thick crystals, the crystals may be considered thin for a strong reflection or thick for a weak reflection. We shall discuss these three categories separately as follows.

c) Laue Diffraction in Thick Crystals
The power ratio of (7.117) does not reach any finite values for increasing A, unless A is very large. When the variation of A is much greater than $\pi/2$, the square of the sine function may be replaced by its average value, 1/2. Under this situation, the average power ratio is

$$\left\langle \frac{P_G}{P_O^e} \right\rangle = \frac{1}{2\,(1+\varrho^2)} , \tag{7.124}$$

which has the maximum of 1/2 at $\varrho = 0$. The full width at half maximum of the diffraction pattern is then

$$\Delta W = \Delta\theta(\varrho = 1) - \Delta\theta(\varrho = -1) = \frac{2\,|\chi_{GO,LO}|}{\sqrt{|b|}\sin 2\theta_G} . \tag{7.125}$$

The integrated reflection over θ is

$$R_G^\theta = \frac{\mathrm{d}\theta}{\mathrm{d}\varrho} \int \left\langle \frac{P_G}{P_O^e} \right\rangle \mathrm{d}\varrho = \frac{\pi\,|\chi_{GO,LO}|}{2\sqrt{|b|}\sin 2\theta_G} , \tag{7.126}$$

where

$$\int_{-\infty}^{\infty} \frac{\mathrm{d}\varrho}{1+\varrho^2} = \pi , \qquad \frac{\mathrm{d}\theta}{\mathrm{d}\varrho} = \frac{|\chi_{GO,LO}|}{\sqrt{|b|}\sin 2\theta_G} . \tag{7.127}$$

For an unpolarized incident wave, both the σ and π components of wavefields are included in the integrated reflection as

$$R_G^\theta = \frac{1}{2}\left(R_{G\sigma}^\theta + R_{G\pi}^\theta\right) , \tag{7.128}$$

where $R_{G\sigma}^\theta$ and $R_{G\pi}^\theta$ are the integrated reflections of the σ and π waves related to $|\chi_{GO,LO}^\sigma|$ and $|\chi_{GO,LO}^\pi|$, respectively.

d) Bragg Reflection in Thick Crystals
The angular range of a Bragg reflection is usually divided into three regions in terms of ϱ. Region II is the range for $|\varrho| < 1$ and regions III and I with $|\varrho| > 1$ are the angular ranges below and above region II. For $|\varrho| < 1$, (7.121) holds. The power ratio reaches its limit as A approaches infinity:

$$\frac{P_G}{P_O^e} = \lim_{A\to\infty} \frac{\sinh^2\left(A\sqrt{1-\varrho^2}\right)}{1-\varrho^2+\sinh^2\left(A\sqrt{1-\varrho^2}\right)} = 1 . \tag{7.129}$$

This means that the reflected power is equal to the incident power and there is no diffracted power for the transmitted wave, i.e., $P_O/P_O^e = 0$. Region II is therefore called the range of total reflection.

For $|\varrho| > 1$, the power ratio has no definite limit as A increases, as already mentioned in the Laue diffraction case. Following the same argument that the variation of A is much larger than $\pi/2$, we can express the power ratio as its average value:

$$\frac{P_G}{P_O^e} = 1 - \sqrt{1 - \varrho^{-2}} \,. \tag{7.130}$$

Since

$$\int_1^\infty \left(1 - \sqrt{1 - \varrho^{-2}}\right) \mathrm{d}\varrho = \frac{\pi}{2} - 1 \,, \tag{7.131}$$

the integrated reflection then takes the form

$$R_G^\theta = \frac{\pi \left|\chi_{GO,LO}\right|}{\sqrt{|b|}\sin 2\theta_G} \,. \tag{7.132}$$

e) Diffraction in Thin Crystals

The power ratios for both the Laue and Bragg cases are

$$\frac{P_G}{P_O^e} \simeq \frac{\sin^2(A\varrho)}{\varrho^2} \tag{7.133}$$

for very small A. The integrated reflection is

$$R_G^\theta = \frac{\pi A \left|\chi_{GO,LO}\right|}{\sqrt{|b|}\sin 2\theta_G} \,, \tag{7.134}$$

where the relation

$$\int_0^\infty \frac{\sin^2(A\varrho)}{\varrho^2} \mathrm{d}\varrho = \frac{\pi}{2} A \tag{7.135}$$

has been used. Substituting A of (7.119) into (7.134), we obtain the integrated reflection

$$R_G^\theta = \frac{Q_{GO} T_o}{\gamma_G} \,, \tag{7.136}$$

where the reflectivity takes the following usual form:

$$Q_{GO} = \frac{\pi^2}{\lambda} \frac{\left|\chi_{GO,LO}\right|^2}{\sin 2\theta_G} \,. \tag{7.137}$$

That the reflectivity depends on $|\chi_{GO,LO}|^2$ implies that the diffraction in thin crystals is kinematical rather than dynamical.

f) Laue Diffraction in Crystals of Intermediate Thickness

The approximate solutions of the power ratios for the extreme situations, $A \gg 1$ and $A \ll 1$, are no longer valid for the intermediate situation where

$A \simeq 1$. In the intermediate situation, no analytical expression for the power ratio can be found. Numerical calculation is usually used to describe the diffraction patterns for a two-wave diffraction [48]. The calculated patterns depend very much on whether A is in the vicinity of $\pi/2$. If, however, the integrated reflection is concerned, the integrated reflection

$$R_G^{\varrho} = \int_{-\infty}^{\infty} \frac{\sin^2\left(A\sqrt{1+\varrho^2}\right)}{1+\varrho^2} \mathrm{d}\varrho \tag{7.138}$$

may be replaced [506] by

$$R_G^{\varrho} = \int_0^{\pi/2} \frac{\sin\left(2A\sin\zeta_0\right)}{\sin\zeta_0} \mathrm{d}\zeta_0 \ . \tag{7.139}$$

The sine function $\sin(2A\sin\zeta_0)$ in (7.139) is related to the zeroth-order Bessel function $J_0(\xi_1)$ of $\xi_1 = 2A$ as

$$J_0\left(\xi_1\right) = \frac{2}{\pi}\int_0^{\pi/2} \cos\left(\xi_1 \sin\zeta_0\right) \mathrm{d}\zeta_0 \tag{7.140}$$

and

$$\frac{\mathrm{d}R_G^{\varrho}}{\mathrm{d}\xi_1} = \frac{\pi}{2} J_0\left(\xi_1\right) \ . \tag{7.141}$$

The integrated reflection is then expressed as

$$R_G^{\varrho} = \int_0^{2A} \frac{\mathrm{d}R_G^{\varrho}}{\mathrm{d}\xi_1} \mathrm{d}\xi_1 = \frac{\pi}{2}\int_0^{2A} J_0\left(\xi_1\right) \mathrm{d}\xi_1 = \pi W_0 \ , \tag{7.142}$$

where

$$W_0 = \sum_{n=0}^{n=\infty} J_{2n+1}(2A) = \begin{cases} A & \text{for} \quad A \ll 1 \ , \\ \frac{1}{2} & \text{for} \quad A \gg 1 \ . \end{cases} \tag{7.143}$$

The integrated reflection over θ takes the form

$$R_G^{\theta} = \frac{\pi \left|\chi_{GO,LO}\right| W_0}{\sqrt{|b|}\sin 2\theta_G} \ , \tag{7.144}$$

which approaches the expressions for the diffraction in thin and thick crystals at the limits $A \ll 1$ and $A \gg 1$, respectively.

g) Bragg Reflection in Crystals of Intermediate Thickness
The integrated reflection in this case can be written as

$$R_G^{\theta} = \frac{\left|\chi_{GO,LO}\right|}{\sqrt{|b|}\sin 2\theta_G} \pi \tanh A \ , \tag{7.145}$$

since, according to Darwin [70],

$$\int_{-\infty}^{\infty} \frac{\mathrm{d}\varrho}{\varrho^2 + (1-\varrho^2)\coth^2\left(A\sqrt{1-\varrho^2}\right)} = \pi \tanh A \ . \tag{7.146}$$

If $A \gg 1$, then $\tanh A \simeq 1$ and if $A \ll 1$, then $\tanh A = A$. To these two extreme conditions, (7.145) approaches (7.132) and (7.134), respectively.

(iv) Diffraction in Absorbing Crystals

The strength of a diffraction depends on either the structure factor or the polarizability, which is usually a complex. At an absorption edge or at an resonance photon energy, both the real and imaginary parts of the polarizability are still complex due to the phase shifts at resonance. Therefore, the polarizability may be defined without losing generality, as

$$\chi_{GO} = \chi_{GO,r} + \mathrm{i}\chi_{GO,i} , \tag{7.147}$$

where

$$\begin{aligned} \chi_{GO,r} &= \chi'_{GO,r} + \mathrm{i}\chi''_{GO,r} , \\ \chi_{GO,i} &= \chi'_{GO,i} + \mathrm{i}\chi''_{GO,i} . \end{aligned} \tag{7.148}$$

The quantities $|\chi_{GO}|^2$, $|\chi_{\overline{GO}}|^2$, and $|\chi_{GO}\chi_{\overline{GO}}|$, which are all identical in non-absorbing crystals, take the following different forms

$$\begin{aligned} |\chi_{GO}|^2 &= |\chi_{GO,r}|^2 + |\chi_{GO,i}|^2 + 2\left(\chi''_{GO,r}\chi'_{GO,i} - \chi'_{GO,r}\chi''_{GO,i}\right) \\ &= |\chi_{GO,r}|^2\left(1 + \kappa^2 + 2s\right) , \\ \left|\chi_{\overline{GO}}\right|^2 &= |\chi_{GO,r}|^2\left(1 + \kappa^2 - 2s\right) , \\ \left|\chi_{GO}\chi_{\overline{GO}}\right| &= |\chi_{GO,r}|^2\left(1 - \kappa^2 - 2\mathrm{i}p_0\right) , \end{aligned} \tag{7.149}$$

where

$$\begin{aligned} \kappa &= \frac{|\chi_{GO,i}|}{|\chi_{GO,r}|} , \\ s &= \frac{\chi''_{GO,r}\chi'_{GO,i} - \chi'_{GO,r}\chi''_{GO,i}}{|\chi_{GO,r}|^2} , \\ p_0 &= \frac{\chi'_{GO,r}\chi'_{GO,i} + \chi''_{GO,r}\chi''_{GO,i}}{|\chi_{GO,r}|^2} . \end{aligned} \tag{7.150}$$

Similarly, the polarizabilities of the Friedel's pair, G and $-G$, are

$$\begin{aligned} \chi_{GO} &= |\chi_{GO}|\,\mathrm{e}^{\mathrm{i}\delta_{G\text{–}O}} , \\ \chi_{\overline{GO}} &= \left|\chi_{\overline{GO}}\right|\mathrm{e}^{\mathrm{i}\delta_{O\text{–}G}} , \end{aligned} \tag{7.151}$$

where the phases are defined as

$$\begin{aligned} \delta_{G\text{–}O} &= \arctan\frac{\chi''_{GO,r} + \chi'_{GO,i}}{\chi'_{GO,r} - \chi''_{GO,i}} , \\ \delta_{O\text{–}G} &= \arctan\frac{\chi''_{GO,r} - \chi'_{GO,i}}{\chi'_{GO,r} + \chi''_{GO,i}} . \end{aligned} \tag{7.152}$$

For the photon energies far from absorption edges, $\chi''_{GO,r}$ and $\chi''_{GO,i}$ tend to be zero and the phases are then given by

$$\delta_{G\text{–}O} = \arctan \frac{\chi'_{GO,i}}{\chi'_{GO,r}} ,$$

$$\delta_{O\text{–}G} = -\arctan \frac{\chi'_{GO,i}}{\chi'_{GO,r}} , \tag{7.153}$$

which are related by

$$\delta_{O\text{–}G} = -\delta_{G\text{–}O} . \tag{7.154}$$

For the photon energies near an absorption edge, the phases, as given in (7.152), do not follow (7.154). The differences in $\delta_{G\text{–}O}$ and $\delta_{O\text{–}G}$ between (7.152) and (7.153) are the phase shifts $\Delta_{G\text{–}O}$ and $\Delta_{O\text{–}G}$ due to resonance.

For a centrosymmetric structure possessing a center of inversion, $s = 0$, $\kappa = p_0$, and $|\chi_{GO}\chi_{\overline{GO}}| = |\chi_{GO}|^2$. For noncentrosymmetric crystals, the expressions given in (7.149) and (7.150) need to be used.

Assume that the three-wave diffraction is not very weak and the X-ray energies used are far from absorption edges so that $|\chi_{GO,LO,r}| \gg |\chi_{GO,LO,i}|$. The quantities M, A, ϱ, m_0, and g_0 are redefined, in terms of the real and imaginary parts, as

$$M = \sqrt{|b|}\sqrt{\mathrm{Re}\{\chi_{GO,LO}\chi_{OG,OL}\}} ,$$

$$m_0 = \frac{\mathrm{Im}\{\chi_{GO,LO}\chi_{OG,OL}\}}{\mathrm{Re}\{\chi_{GO,LO}\chi_{OG,OL}\}} ,$$

$$A = \frac{\pi k T_o}{\sqrt{|\gamma_O\gamma_G|}}\sqrt{\mathrm{Re}\{\chi_{GO,LO}\chi_{OG,OL}\}} ,$$

$$\varrho = \frac{z_r}{M} , \qquad g_0 = \frac{z_i}{M} . \tag{7.155}$$

The variables v and w of (7.104) for Laue cases are

$$\left.\begin{matrix} v \\ w \end{matrix}\right\} = \frac{M}{\sqrt{2}}\sqrt{\Xi_1 \pm (\varrho^2 - g_0^2 + 1)} , \tag{7.156}$$

where

$$\Xi_1 = \sqrt{(\varrho^2 - g_0^2 + 1)^2 + (2\varrho g_0 + m_0)^2} . \tag{7.157}$$

For Bragg cases,

$$\left.\begin{matrix} v \\ w \end{matrix}\right\} = \frac{M}{\sqrt{2}}\sqrt{\Xi_2 \pm (\varrho^2 - g_0^2 - 1)} , \tag{7.158}$$

with

$$\Xi_2 = \sqrt{(\varrho^2 - g_0^2 - 1)^2 + (2\varrho g_0 - m_0)^2} . \tag{7.159}$$

In the following, we shall discuss some special diffraction situations:

a) Laue diffraction with $g_0 \ll m_0 < 1$ and $b = 1$
The variables v and w of (7.156) reduce to

$$v = M\sqrt{1+\varrho^2}\,, \qquad w = \frac{Mm_0}{\sqrt{1+\varrho^2}}\,, \tag{7.160}$$

and

$$v^2 + w^2 \simeq M^2\left(1+\varrho^2\right) . \tag{7.161}$$

Substituting (7.160) into (7.105), we obtain the power ratio

$$\frac{P_G}{P_O^e} = \frac{\left|\chi_{GO,LO}\right|^2}{\mathrm{Re}\left\{\chi_{GO,LO}\chi_{OG,OL}\right\}}\frac{\exp(-\mu T)}{(1+\varrho^2)}\left[\sin^2\left(A\sqrt{1+\varrho^2}\right)\right.$$
$$\left. + \sinh^2\left(\frac{m_0 A}{\sqrt{1+\varrho^2}}\right)\right] . \tag{7.162}$$

If A and the variation of A are very large, then the square of the sine function can be replaced by its average value, i.e., 1/2. Thus the reflection power ratio for a thick crystal takes the following simple form

$$\frac{P_G}{P_O^e} \simeq \frac{\left|\chi_{GO,LO}\right|^2}{2\left|\mathrm{Re}\left\{\chi_{GO,LO}\chi_{OG,OL}\right\}\right|}\frac{\exp(-\mu T)}{(1+\varrho^2)}$$
$$\times\left[1 + 2\sinh^2\left(\frac{m_0 A}{\sqrt{1+\varrho^2}}\right)\right] . \tag{7.163}$$

If the imaginary part of $\chi_{GO,LO}\chi_{OG,OL}$ is much smaller than its real part $\mathrm{Re}\{\chi_{GO,LO}\chi_{OG,OL}\}$, so that $|m_0|A < 0.4$, then (7.162) reduces further to

$$\frac{P_G}{P_O^e} = \frac{\left|\chi_{GO,LO,r}\right|^2}{2\left|\mathrm{Re}\left\{\chi_{GO,LO}\chi_{OG,OL}\right\}\right|}\frac{\exp\left(-\mu T\right)}{(1+\varrho^2)}\left[1 + \frac{2\left(m_0 A\right)^2}{1+\varrho^2}\right] . \tag{7.164}$$

The corresponding integrated reflection is

$$R_G^\theta = \frac{\pi}{2\sin 2\theta_G}\frac{\left|\chi_{GO,LO,r}\right|}{\sqrt{\mathrm{Re}\left\{\chi_{GO,LO}\chi_{OG,OL}\right\}}}$$
$$\times\left[1 + \left(m_0 A\right)^2\right]\exp\left(-\mu T\right) . \tag{7.165}$$

b) Bragg Reflection in Thick Crystals

When the crystal thickness is so large that the hyperbolic sine functions have very large values, the intensity ratio given in (7.112) can be greatly simplified as

$$\frac{I_G}{I_O^e} = \frac{b^2\left|\chi_{GO,LO}\right|^2}{|q+z^2| + |z|^2 + \sqrt{\left(|q+z^2| + |z|^2\right)^2 - |q|^2}} . \tag{7.166}$$

The same equation as (7.166) can be easily derived by considering only the permitted mode associated with the two-wave diffraction in a very thick crystal and by employing the appropriate boundary conditions.

An alternative form of the intensity ratio, according to Miller [507], is

$$\frac{I_G}{I_O^e} = \frac{|b|\,|\chi_{GO,LO}|^2}{|\chi_{GO,LO}\chi_{OG,OL}|}\frac{1}{|q|}\left[|q+z^2| + |z|^2 - \sqrt{(|q+z^2|+|z|^2)^2 - |q|^2}\right] . \tag{7.167}$$

By defining

$$Na = \frac{|\chi_{GO,LO}|^2}{|\chi_{GO,LO}\chi_{OG,OL}|} \tag{7.168}$$

and employing (7.158), the intensity ratio can be simplified as

$$\frac{I_G}{I_O^e} = |b| Na \left(\mathcal{L} - \sqrt{\mathcal{L}^2 - 1}\right) , \tag{7.169}$$

where

$$\mathcal{L} = \frac{\varrho^2 + g_0^2 + \sqrt{(\varrho^2 - g_0^2 - 1)^2 + (2\varrho g_0 - m_0)^2}}{\sqrt{1+m_0^2}} . \tag{7.170}$$

The integrated reflection is simply the integration of the intensity ratio over ϱ:

$$R_G^\theta = \frac{Na\sqrt{\mathrm{Re}\{\chi_{GO,LO}\chi_{OG,OL}\}}}{\sqrt{|b|}\sin 2\theta_G}\int_{-\infty}^{\infty}\left(\mathcal{L} - \sqrt{\mathcal{L}^2-1}\right)\mathrm{d}\varrho . \tag{7.171}$$

For a strong three-wave diffraction, $m_0 \ll 1$ and $g_0 < 1$, namely, $z_i < \sqrt{|b|}\sqrt{\mathrm{Re}\{\chi_{GO,LO}\chi_{OG,OL}\}}$. The integral in (7.171) then is

$$\int_{-\infty}^{\infty}\left(\mathcal{L} - \sqrt{\mathcal{L}^2-1}\right)\mathrm{d}\varrho = \frac{8}{3} . \tag{7.172}$$

This leads the expression for the integrated reflection to

$$R_G^\theta = \frac{8Na}{3\sqrt{|b|}\sin 2\theta_G}\,\frac{\sqrt{\mathrm{Re}\{\chi_{GO,LO}\chi_{OG,OL}\}}}{\sqrt{1+m_0^2}} , \tag{7.173}$$

which is proportional to $\sqrt{\mathrm{Re}\{\chi_{GO,LO}\chi_{OG,OL}\}}$.

For a weak three-wave diffraction, $g_0 \gg 1$ and $g_0 > m_0$, namely, $z_i > \sqrt{|b|}\sqrt{\mathrm{Re}\{\chi_{GO,LO}\chi_{OG,OL}\}}$. The function L defined in (7.170) has a value much greater than unity. The integral then becomes

$$\int_{-\infty}^{\infty}\left(\mathcal{L} - \sqrt{\mathcal{L}^2-1}\right)\mathrm{d}\varrho = \frac{1+m_0^2}{4}\int_{-\infty}^{\infty}\frac{1}{\varrho^2+g_0^2}\mathrm{d}\varrho = \frac{\pi\left(1+m_0^2\right)}{4g_0} . \tag{7.174}$$

The integrated reflection is thus written as

$$R_G^\theta = \frac{\pi Na\,(1+m_0^2)}{4\sqrt{|b|}|g_0|\sin 2\theta_G}\left(\sqrt{\mathrm{Re}\{\chi_{GO,LO}\chi_{OG,OL}\}}\right)^2 , \tag{7.175}$$

which is proportion al to $(\sqrt{\mathrm{Re}\{\chi_{GO,LO}\chi_{OG,OL}\}})^2$.

Equations (7.173) and (7.175) indicate respectively the dynamical and the kinematical characteristics of a three-wave diffraction under a Bethe approximation. For a strong three-wave case, the integrated reflection is proportional to the first power of $\sqrt{\mathrm{Re}\{\chi_{GO,LO}\chi_{OG,OL}\}}$, while for a weak diffraction, it depends on the second power of $\sqrt{\mathrm{Re}\{\chi_{GO,LO}\chi_{OG,OL}\}}$, provided that $|\mathrm{Re}\{\chi_{GO,LO}\chi_{OG,OL}\}| \gg |\mathrm{Im}\{\chi_{GO,LO}\chi_{OG,OL}\}|$. Here, the polarizability could be of a first- or second-order, depending on the approximation adopted.

The imaginary part of the polarizability may be comparable with or even greater than its real part, when the X-ray photon energy is very close to an absorption edge or when the three-wave case involves very weak reflections. Under this circumstance, the integrated reflection is proportional to the imaginary part of the polarizability in the following situations:

i) For $|\varrho' - g_0'| < 1$ and $|m_0'| \ll 1$, or for $|\varrho' + g_0'| < 1$ and $|m_0'| \ll 1$,

$$R_G^\theta = \frac{8Na}{3\sqrt{|b|}\sin 2\theta_G}\,\frac{\sqrt{|\mathrm{Im}\{\chi_{GO,LO}\chi_{OG,OL}\}|}}{\sqrt{1+m_0'^2}} , \tag{7.176}$$

where

$$\varrho' = \frac{z_r}{M'} , \qquad g_0' = \frac{z_i}{M'} , \qquad m_0' = \frac{1}{m} , \tag{7.177}$$

and

$$M' = \sqrt{|b|}\sqrt{\sqrt{|\mathrm{Im}\{\chi_{GO,LO}\chi_{OG,OL}\}|}} . \tag{7.178}$$

ii) For $|\varrho' - g_0'| \gg 1$ and $|\varrho' - g_0'| > |m_0'|$, or for $|\varrho' + g_0'| \gg 1$ and $|\varrho' + g_0'| > |m_0'|$,

$$R_G^\theta = \frac{\pi Na\left(1+m_0'^2\right)}{4\sqrt{|b|}\sin 2\theta_G}\,\frac{\sqrt{|\mathrm{Im}\{\chi_{GO,LO}\chi_{OG,OL}\}|}}{|\varrho' \mp g_0'|} . \tag{7.179}$$

When the imaginary part of $\chi_{GO,LO}$ is comparable with its real part, the phase shift is different from 0 or $\pi/2$, and the integrated reflection cannot be expressed as being solely proportional to either the real or the imaginary part. In this case, numerical calculations may be necessary to exhibit the behavior of the integrated reflection. One of the examples is the multiple-wave diffraction at resonance, where additional phase shifts become important. We will discuss this resonance diffraction situation in detail later in Chap. 9.

7.2.2 Born Approximation

An alternative of treating the perturbation to a two-wave diffraction due to the presence of a third reflection in a three-wave diffraction is the adoption of the Born approximation of quantum mechanics, proposed by Shen [285], where Green's function is used. Since the Born approximation deals with single scattering of particles, the approach presented below is of the kinematical nature.

The dielectric constant of a crystal is assumed to be

$$\varepsilon(\boldsymbol{r}) = \varepsilon_o + \delta\varepsilon(\boldsymbol{r}) \;, \tag{7.180}$$

where

$$\varepsilon_o = 1 \tag{7.181}$$

and

$$\delta\varepsilon(\boldsymbol{r}) = \begin{cases} 0 & \text{for empty space }, \\ \chi(\boldsymbol{r}) & \text{for crystals }. \end{cases} \tag{7.182}$$

The quantity $\chi(\boldsymbol{r})/4\pi$ is the usual electric susceptibility, whose Fourier components χ_G of all the reflections are defined in (6.11). For simplicity, the incident wave is again assumed to be a plane wave, i.e.,

$$\boldsymbol{D}_O(\boldsymbol{r}) = \boldsymbol{D}_O \exp\left(-2\pi \mathrm{i} \boldsymbol{k}_O \cdot \boldsymbol{r}\right) \;, \tag{7.183}$$

where $\boldsymbol{k}_O$ is the wavevector. The fundamental equation of wavefield remains the same as (6.14), whose solution can be found by following the Green's function method [508] as

$$\begin{aligned} \boldsymbol{D}(\boldsymbol{r}) = \boldsymbol{D}^{(0)}(\boldsymbol{r}) + \frac{1}{4\pi} \int \mathrm{d}\boldsymbol{r}' \frac{\exp\left(-2\pi \mathrm{i} k_O |\boldsymbol{r}-\boldsymbol{r}'|\right)}{|\boldsymbol{r}-\boldsymbol{r}'|} \\ \times \boldsymbol{\nabla}' \times \boldsymbol{\nabla}' \times [\chi(\boldsymbol{r}')\boldsymbol{D}(\boldsymbol{r}')] \;. \end{aligned} \tag{7.184}$$

$\boldsymbol{D}^{(0)}(\boldsymbol{r})$, the special solution to the differential equation (6.14), is the unperturbed incident wavefield, i.e.,

$$\boldsymbol{D}^{(0)}(\boldsymbol{r}) = \boldsymbol{D}_O \exp\left(-2\pi \mathrm{i} \boldsymbol{k}_O \cdot \boldsymbol{r}\right) \;. \tag{7.185}$$

$\boldsymbol{r}$ is the position vector of the observation point, and $\boldsymbol{r}'$ is the position vector of the scatterer, i.e., the atoms. The expression for the wavefield given in (7.184) is similar to the wavefunction used in the Born approximation in quantum mechanics. With successive approximations up to higher-order terms, multiple-wave diffraction can be treated in a systematic way: In the first-order approximation, we substitute $\boldsymbol{D}^{(0)}(\boldsymbol{r})$ for $\boldsymbol{D}(\boldsymbol{r}')$ in the integrand and obtain the first-order solution. This solution can, in turn, be treated as the $\boldsymbol{D}(\boldsymbol{r}')$ in the integrand of (7.184), which gives the second-order solution. As the iterative substitution goes on, the resultant solution for $\boldsymbol{D}(\boldsymbol{r})$ is

$$\boldsymbol{D} = \boldsymbol{D}^{(0)} + \boldsymbol{D}^{(1)} + \boldsymbol{D}^{(2)} + \cdots \;, \tag{7.186}$$

where $\boldsymbol{D}^{(j)}$ for $j = 1, 2, \ldots$ is the j-th order iteration correction to the unperturbed wavefield $\boldsymbol{D}^{(0)}(\boldsymbol{r})$. It will be clearly shown in the following that the first-order approximation leads to the two-wave kinematical formulation, and the second-order solution yields the three-wave expressions for the wavefields and diffraction intensities.

First-Order Approximation. The first-order correction $\boldsymbol{D}^{(1)}$ takes the form

$$\begin{aligned}\boldsymbol{D}^{(1)}(\boldsymbol{r}) = \frac{1}{4\pi}\int \mathrm{d}\boldsymbol{r}'\frac{\exp\left(-2\pi k_O|\boldsymbol{r}-\boldsymbol{r}'|\right)}{|\boldsymbol{r}-\boldsymbol{r}'|} \\ \times\boldsymbol{\nabla}'\times\boldsymbol{\nabla}'\times\left[\chi(\boldsymbol{r}')\boldsymbol{D}_O\exp\left(-2\pi\mathrm{i}\boldsymbol{k}_O\cdot\boldsymbol{r}'\right)\right] ,\end{aligned} \tag{7.187}$$

where $\boldsymbol{D}(\boldsymbol{r}')$ in the integrand of (7.184) is approximated by $\boldsymbol{D}^{(0)}(\boldsymbol{r})$. In the far-field region, the modulus of $\boldsymbol{r}$ of the observation point is much larger than that of $\boldsymbol{r}'$ of the scatterer. Therefore, $|\boldsymbol{r}| \gg |\boldsymbol{r}'|$ and

$$|\boldsymbol{r}-\boldsymbol{r}'| \simeq r - \hat{\boldsymbol{r}}\cdot\boldsymbol{r}' , \tag{7.188}$$

where $\hat{\boldsymbol{r}}$ is the unit vector of $\boldsymbol{r}$, which can be considered approximately as the vector in the direction of the scattered wave. With this consideration, (7.187) becomes

$$\begin{aligned}\boldsymbol{D}^{(1)}(\boldsymbol{r}) = \frac{\exp\left(-2\pi\mathrm{i}k_O r\right)}{4\pi r}\int_{V_o}\mathrm{d}\boldsymbol{r}'\exp\left(2\pi\mathrm{i}k_O\hat{\boldsymbol{r}}\cdot\boldsymbol{r}'\right) \\ \times\boldsymbol{\nabla}'\times\boldsymbol{\nabla}'\times\left[\chi(\boldsymbol{r}')\boldsymbol{D}_O\exp\left(-2\pi\mathrm{i}\boldsymbol{k}_O\cdot\boldsymbol{r}'\right)\right] .\end{aligned} \tag{7.189}$$

The integration is taken over the entire volume V_o of the crystal. By integrating by parts twice and ignoring the surface integral due to the contribution of surface scattering, the above equation assumes the form

$$\begin{aligned}\boldsymbol{D}^{(1)}(\boldsymbol{r}) = \frac{\exp\left(-2\pi\mathrm{i}k_O r\right)}{r}k_O^2\hat{\boldsymbol{r}}\times(\hat{\boldsymbol{r}}\times\boldsymbol{D}_O) \\ \times\sum_G\chi_{G\text{–}O}\int_{V_o}\mathrm{d}\boldsymbol{r}'\exp\left\{2\pi\mathrm{i}\left[k_O\hat{\boldsymbol{r}}-(\boldsymbol{k}_O+\boldsymbol{g}_G)\right]\cdot\boldsymbol{r}'\right\} ,\end{aligned} \tag{7.190}$$

where $\boldsymbol{g}_G$ is the reciprocal lattice vector of the G reflection. The integral does not vanish if and only if

$$k_O\hat{\boldsymbol{r}} = \boldsymbol{k}_O + \boldsymbol{g}_G = \boldsymbol{k}_G , \tag{7.191}$$

which is exactly the Bragg diffraction condition. Now let's assume that only the G reflection fulfills the diffraction condition so that

$$\begin{aligned}\boldsymbol{D}^{(1)}(\boldsymbol{r}) &= \frac{k_O^2\exp\left(-2\pi\mathrm{i}k_O r\right)}{r}\hat{\boldsymbol{r}}\times(\hat{\boldsymbol{r}}\times\boldsymbol{D}_O)V_o\chi_{G\text{–}O} \\ &= Nr_eF_G\hat{\boldsymbol{r}}\times(\hat{\boldsymbol{r}}\times\boldsymbol{D}_O)\frac{\exp\left(-2\pi\mathrm{i}k_O r\right)}{r} ,\end{aligned} \tag{7.192}$$

where F_G is the structure factor of the G reflection, $N = V_o/V_c$ is the number of unit cells in the crystal, and V_c is the volume of the unit cell. The wavefield given in (7.192) is exactly the one for the reflected wave G of the two-wave (O, G) diffraction. $\boldsymbol{r}$ is now along the wavevector $\boldsymbol{k}_G$ of the reflected wave G. Following the definition of polarization unit vectors in two-wave cases, for a σ-polarized incident wave $\boldsymbol{D}_O = D_O\hat{\boldsymbol{\sigma}}_O$, the polarization unit vector of the G reflected wave is $\hat{\boldsymbol{r}} \times (\hat{\boldsymbol{r}} \times \hat{\boldsymbol{\sigma}}_O) = -\hat{\boldsymbol{\sigma}}_G$, where $\boldsymbol{r}$ lies in and $\hat{\boldsymbol{\sigma}}_O$ is perpendicular to the plane of incidence. For a π-polarized incident wave $\boldsymbol{D}_O = D_O\hat{\boldsymbol{\pi}}_O$ the polarization unit vector is $\hat{\boldsymbol{r}} \times (\hat{\boldsymbol{r}} \times \hat{\boldsymbol{\pi}}_O) = -\cos 2\theta_G \hat{\boldsymbol{\pi}}_G$, where $\hat{\boldsymbol{\pi}}_O$ lies in the plane of incidence and is perpendicular to $\hat{\boldsymbol{\sigma}}_O$. θ_G is the Bragg angle of the G reflection. Therefore, for an unpolarized incident wave, the diffraction intensity takes the familiar kinematical form as (See Sect. 2.4)

$$\begin{aligned} I_{G(2-\mathrm{wave})} &= \frac{\boldsymbol{D}^{(1)} \cdot \boldsymbol{D}^{(1)*}}{|D_O|^2} \\ &= \frac{N^2 r_e^2 |F_G|^2}{r^2} \frac{\left(1 + \cos^2 2\theta_G\right)}{2} . \end{aligned} \tag{7.193}$$

Second-Order Approximation. The second-order correction $\boldsymbol{D}^{(2)}$ can be expressed by substituting $\boldsymbol{D}^{(1)}$ for $\boldsymbol{D}(\boldsymbol{r}')$ in the integrand of (7.184) as

$$\begin{aligned} \boldsymbol{D}^{(2)}(\boldsymbol{r}) = \frac{1}{4\pi} \int_{V_o} \mathrm{d}\boldsymbol{r}' \frac{\exp\left(-2\pi \mathrm{i} k_O |\boldsymbol{r} - \boldsymbol{r}'|\right)}{|\boldsymbol{r} - \boldsymbol{r}'|} \times \boldsymbol{\nabla}' \times \boldsymbol{\nabla}' \\ \times \left[\chi(\boldsymbol{r}')\boldsymbol{D}^{(1)}(\boldsymbol{r}')\right] . \end{aligned} \tag{7.194}$$

After integrating by parts twice, the above integral becomes

$$\boldsymbol{D}^{(2)}(\boldsymbol{r}) = -\frac{k_O^2 \pi \exp\left(-2\pi \mathrm{i} k_O r\right)}{r} \hat{\boldsymbol{r}} \times (\hat{\boldsymbol{r}} \times \boldsymbol{D}_V) , \tag{7.195}$$

where

$$\begin{aligned} \boldsymbol{D}_V = \int_{V_o} & \mathrm{d}\boldsymbol{r}' \exp\left(2\pi \mathrm{i} k_O \hat{\boldsymbol{u}} \cdot \boldsymbol{r}'\right) \chi(\boldsymbol{r}') \\ & \times \frac{1}{4\pi} \int \mathrm{d}\boldsymbol{r}'' \left\{ \frac{\exp\left(-2\pi k_O |\boldsymbol{r}' - \boldsymbol{r}''|\right)}{|\boldsymbol{r}' - \boldsymbol{r}''|} \times \boldsymbol{\nabla}' \times \boldsymbol{\nabla}' \right. \\ & \left. \times \left[\chi(\boldsymbol{r}'') D_O \exp\left(-2\pi \mathrm{i} \boldsymbol{k}_O \cdot \boldsymbol{r}''\right)\right] \right\} . \end{aligned} \tag{7.196}$$

In terms of the Fourier integral

$$\frac{\exp\left(-2\pi \mathrm{i} k_O r\right)}{4\pi r} = -\frac{1}{4\pi^2} \int \mathrm{d}\boldsymbol{k} \frac{\exp(-2\pi \mathrm{i} \boldsymbol{k} \cdot \boldsymbol{r})}{k_O^2 - k^2} \tag{7.197}$$

and the Fourier series expansion for $\chi(\boldsymbol{r})$, the integral for $\boldsymbol{D}_V$ can be expressed as

$$\boldsymbol{D}_V = \sum_M \sum_L \chi_{M-O} \chi_{L-O} \boldsymbol{k}_L \times (\boldsymbol{k}_L \times \boldsymbol{D}_O)$$

$$\times \int_{V_o} \mathrm{d}\boldsymbol{r}' \exp\left[2\pi\mathrm{i}\left(k_O\hat{\boldsymbol{r}} - \boldsymbol{g}_M\right)\cdot\boldsymbol{r}'\right] \int_{V_o} \mathrm{d}\boldsymbol{r}'' \exp\left(-2\pi\mathrm{i}\boldsymbol{k}_L\cdot\boldsymbol{r}''\right)$$
$$\times \int \mathrm{d}\boldsymbol{k} \frac{\exp[-2\pi\mathrm{i}\boldsymbol{k}\cdot(\boldsymbol{r}' - \boldsymbol{r}'')]}{k_O^2 - k^2} , \tag{7.198}$$

where $\boldsymbol{k}_L = \boldsymbol{k}_O + \boldsymbol{g}_L$ and L is an additional reciprocal lattice point approaching the Ewald sphere. By applying the relation

$$\int_{V_o} \mathrm{d}\boldsymbol{r}'' \exp\left[2\pi\mathrm{i}\left(\boldsymbol{k} - \boldsymbol{k}_L\right)\cdot\boldsymbol{r}''\right] = \delta\left(\boldsymbol{k} - \boldsymbol{k}_L\right) ,$$
$$\int_{V_o} \mathrm{d}\boldsymbol{r}'' \exp\left[2\pi\mathrm{i}\left(k_O\hat{\boldsymbol{r}} - \boldsymbol{k}_L - \boldsymbol{g}_M\right)\cdot\boldsymbol{r}'\right] = V\delta_{M,G\text{–}L} , \tag{7.199}$$

where $k_O\hat{\boldsymbol{r}} - \boldsymbol{k}_L = k_O\hat{\boldsymbol{r}} - \boldsymbol{k}_O - \boldsymbol{g}_L = \boldsymbol{g}_G - \boldsymbol{g}_L$, the integral for $\boldsymbol{D}_V$ reduces to

$$\boldsymbol{D}_V = V_o \sum_L \chi_{G\text{–}L}\chi_{L\text{–}O} \frac{\boldsymbol{k}_L \times (\boldsymbol{k}_L \times \boldsymbol{D}_O)}{k_O^2 - k_L^2} , \tag{7.200}$$

where $L \neq O, G$. The reciprocal lattice point L cannot be exactly on the surface of the Ewald sphere, because we are dealing with a small perturbation to the two-wave situation owing to this additional reflection. By substituting (7.200) into (7.195), the second-order correction becomes

$$\boldsymbol{D}^{(2)}(\boldsymbol{r}) = -\frac{NV_c\pi k_O^2 \exp\left(-2\pi\mathrm{i}k_O r\right)}{r}\hat{\boldsymbol{r}}$$
$$\times \left[\hat{\boldsymbol{r}} \times \sum_L \chi_{G\text{–}L}\chi_{L\text{–}O} \frac{\boldsymbol{k}_L \times (\boldsymbol{k}_L \times \boldsymbol{D}_O)}{k_O^2 - k_L^2}\right] . \tag{7.201}$$

The resultant diffracted wavefield of the G reflection then takes the form

$$\boldsymbol{D}(\boldsymbol{r}) = \frac{NV_c\pi k_O^2 \exp\left(-2\pi\mathrm{i}k_O r\right)}{r}\hat{\boldsymbol{r}} \times \left\{\hat{\boldsymbol{r}} \times \left[\chi_{G\text{–}O}\boldsymbol{D}_O\right.\right.$$
$$\left.\left. - \sum_L \chi_{G\text{–}L}\chi_{L\text{–}O} \frac{\boldsymbol{k}_L \times (\boldsymbol{k}_L \times \boldsymbol{D}_O)}{k_O^2 - k_L^2}\right]\right\} , \tag{7.202}$$

which consists of the first-order term, χ, and the second-order term χ^2. The terms in the square brackets can be considered as the product of the modified polarizability $\chi'_{GO,LO}$ and $\boldsymbol{D}_O$, i.e.,

$$\chi'_{GO,LO}\boldsymbol{D}_O = \chi_{G\text{–}O}\boldsymbol{D}_O - \sum_L \chi_{G\text{–}L}\chi_{L\text{–}O} \frac{\boldsymbol{k}_L \times (\boldsymbol{k}_L \times \boldsymbol{D}_O)}{k_O^2 - k_L^2} . \tag{7.203}$$

Accordingly, the intensity of the diffracted G wave can be calculated as

$$I_G = \frac{\boldsymbol{D}^*(\boldsymbol{r})\cdot\boldsymbol{D}(\boldsymbol{r})}{|D_O|^2} . \tag{7.204}$$

This procedure can be extended to higher-order approximations, where the interaction among all the L reflections and the involvement of more reflections in the diffraction processes are considered.

In comparison with the Bethe approximation, the first-order Born approximation gives the expression for the wavefield of a two-wave diffraction, whereas the second-order Born approximation is equivalent to the first-order Bethe approximation, which shows that the diffraction intensity is proportional to the modified polarizability. The validity of the second-order Born approximation is similar to that for the first-order Bethe approximation. The difference between the two approximations lies in that the Bethe approximation derives the wavefield from the equation of dispersion and the boundary conditions, while the Born approximation deals directly with the wavefields without invoking the boundary conditions and the equation of dispersion.

7.2.3 Quantum-Mechanical Approach

The quantum mechanical approach for X-ray diffraction in crystals, as mentioned in Chap. 1, has been developed by Kohler [71], Molière [72–74], and many others to provide a more strictly microscopic view of the interaction between the radiation field and the electron distribution around the atomic cores. The absorption of X-rays in crystals and X-ray diffraction at resonance could be described by this approach as well. In 1964, Ohtsuki [75,76] adopted Yoshioka's dynamical theory [509] of inelastic scattering of electrons and derived a quantum mechanical theory to describe the elementary processes of inelastic scattering of X-rays by phonons in crystals. A similar theory to account for the interaction of γ-rays and neutrons with crystals has been developed independently by Afanas've and Kagan [83]. Quantum field theory has also been used for the same purpose by Ashkin and Kuriyama [78], Ohtsuki and Yanagawa [84], and others [85–92]. In this section, we shall first briefly describe the quantum mechanical approach of Molière [72–74] and Ohtsuki [75, 76] and then present the approximation based on the quantum field theory of Ohtsuki and Yanagawa [84].

Fundamental Equation of Wavefields. The interaction of X-rays with crystals can be described, based on Maxwell's equations (6.1a)–(6.1d), by the relationship between the radiation field represented by the vector potential A given in (2.4) and the current density $\boldsymbol{J}$ as

$$\nabla^2 \boldsymbol{A} - \frac{1}{c^2}\frac{\partial^2 \boldsymbol{A}}{\partial t^2} = -\frac{4\pi}{c}\boldsymbol{J}\,. \tag{7.205}$$

There is no interaction between the radiation and the crystals if $\boldsymbol{J}$ is null. For $J \neq 0$, by following (2.6), the interaction can be described by the time-dependent Schrödinger equation

$$\begin{aligned} \mathrm{i}\hbar\frac{\partial \Psi}{\partial t} = \sum_j \Big\{ & -\frac{\hbar^2}{2m}\nabla_j^2 + \frac{\mathrm{i}e\hbar}{mc}\boldsymbol{A}(\boldsymbol{r}_j)\cdot\boldsymbol{\nabla}_j + \frac{\mathrm{i}e\hbar}{2mc}\left[\boldsymbol{\nabla}\cdot\boldsymbol{A}(\boldsymbol{r}_j)\right] \\ & + \frac{e^2}{2mc^2}A^2(\boldsymbol{r}_j) + e\phi(\boldsymbol{r}_j) + V(\boldsymbol{r}_j)\Big\}\Psi\,, \end{aligned} \tag{7.206}$$

where Ψ is the wavefunction of the whole system interested. $\boldsymbol{r}_j$ is the position vector of the electron j in the crystal and $\boldsymbol{\nabla}_j$ is the gradient with respect to $\boldsymbol{r}_j$. The electromagnetic fields $\boldsymbol{A}$ and ϕ are very weak for X-rays, so that $\boldsymbol{\nabla} \cdot \boldsymbol{A}$, ϕ, and A^2 are neglected. Equation (7.206) then becomes

$$\mathrm{i}\hbar \frac{\partial \Psi}{\partial t} = (H_0 + H')\,\Psi \,, \tag{7.207}$$

where

$$\begin{aligned} H_0 &= -\frac{\hbar^2}{2m} \sum_j \nabla_j^2 + \sum_j V(\boldsymbol{r}_j) \,, \\ H' &= \frac{\mathrm{i}e\hbar}{mc} \sum_j \boldsymbol{A}(\boldsymbol{r}_j) \cdot \boldsymbol{\nabla}_j \,. \end{aligned} \tag{7.208}$$

The solution to this simplified equation can be found according to the time-dependent perturbation theory [510]. The wavefunction Ψ can then be expressed as the superposition of the stationary eigenfunctions $u_l(\boldsymbol{r}) = u_l(\boldsymbol{r}_1, \boldsymbol{r}_2, \ldots, \boldsymbol{r}_n)$ of the unperturbed Hamiltonian H_0 with time-dependent coefficients $a_l(t)$, i.e.,

$$\Psi(\boldsymbol{r}, t) = \sum_l a_l(t) u_l(\boldsymbol{r}) \exp\left(\frac{\mathrm{i}E_l^0 t}{\hbar}\right) \,, \tag{7.209}$$

where

$$\begin{aligned} a_l(t) &= a_l^{(0)} + a_l^{(1)}(t) \,, \\ a_l^{(0)} &= \langle l|0\rangle = \delta_{l0} \,, \\ a_l^{(1)}(t) &= \frac{1}{\mathrm{i}\hbar} \int_{-\infty}^{t} \langle l|H'(t')|0\rangle \exp(\mathrm{i}\omega_{l0} t')\,\mathrm{d}t' \,. \end{aligned} \tag{7.210}$$

The 0 and l stand for the initial and final states, respectively. The 'ket' and 'bra' have their usual meanings in quantum mechanics. δ_{l0} is the Kronecker delta, and the Bohr angular frequency $\omega_{\ell 0}$ is defined as

$$\omega_{l0} = \frac{E_l^0 - E_0^0}{\hbar} \,, \tag{7.211}$$

where E^0 is the energy. By substituting the second equation of (7.208) into (7.210), the time-dependent coefficient takes the form

$$a_l^{(1)}(t) = -\frac{\langle l|H_1|0\rangle \exp[\mathrm{i}(\omega_{l0} - \omega)t]}{\hbar(\omega_{l0} - \omega - \mathrm{i}\epsilon)} - \frac{\langle l|H_2|0\rangle \exp[\mathrm{i}(\omega_{l0} + \omega)t]}{\hbar(\omega_{l0} + \omega - \mathrm{i}\epsilon)} \,, \tag{7.212}$$

where

$$\begin{aligned} \langle l|H_1|0\rangle &= \frac{\mathrm{i}e\hbar}{mc} \sum_j \langle l| [\exp(2\pi\mathrm{i}\boldsymbol{k} \cdot \boldsymbol{r}_j)]\, \boldsymbol{A}_0^*(\boldsymbol{r}_j) \cdot \boldsymbol{\nabla}_j |0\rangle \,, \\ \langle l|H_2|0\rangle &= \frac{\mathrm{i}e\hbar}{mc} \sum_j \langle l| [\exp(-2\pi\mathrm{i}\boldsymbol{k} \cdot \boldsymbol{r}_j)]\, \boldsymbol{A}_0(\boldsymbol{r}_j) \cdot \boldsymbol{\nabla}_j |0\rangle \,, \end{aligned} \tag{7.213}$$

and ϵ is an infinitesimal positive quantity, which is inversely proportional to the transition time from the initial state to the final state.

The current density of the perturbed system due to the radiation field has the following well-known form

$$\begin{aligned} \boldsymbol{J}(\boldsymbol{r}) = & \frac{e}{2m} \sum_j \left[\boldsymbol{p}_j \delta\left(\boldsymbol{r} - \boldsymbol{r}_j\right) + \delta\left(\boldsymbol{r} - \boldsymbol{r}_j\right) \boldsymbol{p}_j \right] \\ & - \frac{e^2}{mc} \sum_j \boldsymbol{A}\left(\boldsymbol{r}_j, t\right) \delta\left(\boldsymbol{r} - \boldsymbol{r}_j\right) , \end{aligned} \tag{7.214}$$

where the sum is over all the electrons. The averaged current density, or the expectation value of $\boldsymbol{J}$, is then

$$\langle \boldsymbol{J} \rangle = \sum_l \left[\boldsymbol{j}_{0l}(\boldsymbol{r}) a_l^*(t) + \boldsymbol{j}_{0l}^*(\boldsymbol{r}) a_l(t) \right] - \frac{e}{mc} \varrho(\boldsymbol{r}) \boldsymbol{A}(\boldsymbol{r}, t) , \tag{7.215}$$

where

$$\begin{aligned} \boldsymbol{j}_{l0}(\boldsymbol{r}) = \boldsymbol{j}_{0l}^*(\boldsymbol{r}) &= \frac{e\hbar}{2m\mathrm{i}} \sum_j \int \left(u_0 \boldsymbol{\nabla}_j u_l^* - u_0^* \boldsymbol{\nabla}_j u_l \right) \delta\left(\boldsymbol{r} - \boldsymbol{r}_j\right) \mathrm{d}\boldsymbol{r}_j , \\ \varrho(\boldsymbol{r}) = eN(\boldsymbol{r}) &= \sum_j e \int \Psi^* \Psi \delta\left(\boldsymbol{r} - \boldsymbol{r}_j\right) \mathrm{d}\boldsymbol{r}_j . \end{aligned} \tag{7.216}$$

Alternatively, (7.215) can be written in terms of the dyadic $[\boldsymbol{A}, \boldsymbol{j}]$, a second-rank tensor, as

$$\begin{aligned} \langle \boldsymbol{J} \rangle = & -\frac{e\varrho(\boldsymbol{r})}{mc} \boldsymbol{A}(\boldsymbol{r}, t) + \frac{1}{\hbar c} \sum_l \left\{ \frac{\boldsymbol{j}_{0l}(\boldsymbol{r}) \int \left[\boldsymbol{A}(\boldsymbol{r}', t), \boldsymbol{j}_{0l}^*(\boldsymbol{r}') \right] \mathrm{d}\boldsymbol{r}'}{\omega_{l0} - \omega - \mathrm{i}\epsilon} \right. \\ & \left. + \frac{\boldsymbol{j}_{0l}^*(\boldsymbol{r}) \int \left[\boldsymbol{A}(\boldsymbol{r}', t), \boldsymbol{j}_{0l}(\boldsymbol{r}') \right] \mathrm{d}\boldsymbol{r}'}{\omega_{l0} + \omega - \mathrm{i}\epsilon} \right\} , \end{aligned} \tag{7.217}$$

where the coefficient $a_l^{(1)}(t)$ has been transformed into j_{0l} via (7.212), (7.213), and (7.216). By using this expression for $\langle \boldsymbol{J} \rangle$ and setting $\phi = 0$ in (7.201), Maxwell's equations (6.1a)–(6.1d) related to to $\boldsymbol{A}$ and $\boldsymbol{J}$ are given by

$$\begin{aligned} \boldsymbol{E} &= -\frac{\mathrm{i}\omega}{c} \boldsymbol{A} , \\ \boldsymbol{\nabla} \times \boldsymbol{H} &= \frac{\mathrm{i}\omega}{c} \int \tilde{\varepsilon}(\boldsymbol{r}, \boldsymbol{r}', \omega) \boldsymbol{E}(\boldsymbol{r}') \mathrm{d}\boldsymbol{r}' , \end{aligned} \tag{7.218}$$

where the dielectric tensor is defined as

$$\tilde{\varepsilon}(\boldsymbol{r}, \boldsymbol{r}', \omega) \simeq \delta\left(\boldsymbol{r} - \boldsymbol{r}'\right) \left(1 - \frac{4\pi e \varrho(\boldsymbol{r}')}{m\omega^2} \right) + \frac{4\pi}{\hbar\omega^2} \sum_l \boldsymbol{A}_l(\boldsymbol{r}, \boldsymbol{r}', \omega) \tag{7.219}$$

with

$$\boldsymbol{A}_\ell(\boldsymbol{r}, \boldsymbol{r}', \omega) = \frac{\left[\boldsymbol{j}_{0l}(\boldsymbol{r}), \boldsymbol{j}_{0l}^*(\boldsymbol{r}') \right]}{\omega_{l0} - \omega - \mathrm{i}\epsilon} + \frac{\left[\boldsymbol{j}_{0l}^*(\boldsymbol{r}), \boldsymbol{j}_{0l}(\boldsymbol{r}') \right]}{\omega_{l0} + \omega - \mathrm{i}\epsilon} . \tag{7.220}$$

By substituting the dielectric tensor into (6.6), the fundamental equation (6.14) becomes

$$\frac{K_G^2 - k^2}{K_G^2}\boldsymbol{D}_G + \sum_L \chi_{G-L}\boldsymbol{D}_{L[G]} + \sum_L \tilde{C}_{GL}\boldsymbol{D}_{L[G]} = 0\,, \tag{7.221}$$

where χ_{G-L} is defined in (6.11) for the unperturbed electron density distribution, and the tensor $\tilde{C}_{GL}$ is due to the perturbed current density, defined as

$$\begin{aligned}\tilde{C}_{GL}(\omega) = {}& \frac{4\pi}{\hbar\omega^2}\sum_l \iint \exp\left(2\pi \mathrm{i}\boldsymbol{K}_G\cdot\boldsymbol{r}\right)\\ & \times \boldsymbol{A}_l(\boldsymbol{r},\boldsymbol{r}',\omega)\exp\left(-2\pi\mathrm{i}\boldsymbol{K}_L\cdot\boldsymbol{r}'\right)\mathrm{d}\boldsymbol{r}\mathrm{d}\boldsymbol{r}'\,.\end{aligned} \tag{7.222}$$

By considering the σ and π polarized components of wavefields, the fundamental equation can be written as

$$\frac{K_G^2 - k^2}{K_G^2}D_G + \sum_L \left(\Phi_{G-L} + C_{GL}\right)D_L = 0\,, \tag{7.223}$$

where

$$\begin{aligned}\Phi_{G-L} &= (\hat{\boldsymbol{\sigma}}_G\cdot\hat{\boldsymbol{\sigma}}_L)\,\chi_{G-L}\,,\\ C_{GL} &= \hat{\boldsymbol{\sigma}}_G\cdot\tilde{C}_{GL}\cdot\hat{\boldsymbol{\sigma}}_L\,.\end{aligned} \tag{7.224}$$

The polarization unit vectors $\hat{\boldsymbol{\sigma}}_G$ and $\hat{\boldsymbol{\pi}}_G$ represent all possible σ's and π's unit vectors, and D_G and D_L are the corresponding $D_{\sigma G}$, $D_{\pi G}$ and $D_{\sigma L}$, $D_{\pi L}$, respectively.

Equation (7.223) reduces to its classical form, (6.21), when C_{GL} is negligibly small. This situation occurs either when the angular frequency ω is far from the frequency ω_{ln} or when the current density is normal to the polarization unit vectors. If the photon energies of the X-rays used are close to a critical absorption edge or a transition energy, the quantum mechanical correction will become appreciable compared with the electric susceptibility $\chi/(4\pi)$. Under this circumstance, the diffraction may be anisotropic if the involved reflections have a nonzero $\tilde{C}_{GL}$. This type of diffraction is the so-called resonance scattering or diffraction, such as diffraction anomalous fine structure (DAFS) and anisotropic X-ray scattering (AXS), which becomes increasingly important when the information of both the electronic and atomic structures of a substance is needed. For these situations the quantum mechanical approach or the like should be employed.

Quantum Field Theoretical Approach. The interaction of multiply diffracted waves inside a crystal can be described in a more fundamental fashion by the quantum field theory. In the following, we shall follow the theoretical approach of Ohtsuki and Yanagowa [84] for a general situation and then extend the formulation to multiple-wave diffraction [275]. The Green functions usually used in many-body problems are employed.

The Hamiltonian (2.7) of the interaction, without the contribution of electron spin, can be written as

$$H = H_0 + H_{\mathrm{cf}} + H_{\mathrm{ff}} + H_{\mathrm{rad}} \;, \tag{7.225}$$

where H_0 and H_{rad} are the Hamiltonians of the crystal and the radiation field, respectively; H_{cf} and H_{ff} are the Hamiltonians related to the interactions of the crystal-field and of the field-field, respectively. H_{rad} has been given in the last term of (2.8). The rest terms are defined as

$$H_0 = \sum_{j=1}^{N} \left[\frac{p_j^2}{2m} + V\left(\boldsymbol{r}_j\right) \right] \;,$$

$$H_{\mathrm{cf}} = \sum_{j=1}^{N} \frac{e}{mc} \left[\boldsymbol{A}\left(\boldsymbol{r}_j, t\right) \cdot \boldsymbol{p}_j \right] \;,$$

$$H_{\mathrm{ff}} = \sum_{j=1}^{N} \frac{e^2}{2mc^2} A^2\left(\boldsymbol{r}_j, t\right) \;, \tag{7.226}$$

where the summation is taken over the N electrons in the crystal. The eigenfunction of the unperturbed H_0 of the crystal satisfies the condition

$$H_0|n\rangle = E_n^0|n\rangle \;, \tag{7.227}$$

where $|n\rangle$ and E_n are the n-th excited state and its energy of the crystal. The ionization states of the atoms in the crystal are not considered.

The vector potential $\boldsymbol{A}(\boldsymbol{r})$ can be expressed as a Fourier series,

$$\boldsymbol{A}(\boldsymbol{r}) = c\sqrt{4\pi} \sum_{K,G} q_{K,G} \exp\left(2\pi \mathrm{i} \boldsymbol{K}_G \cdot \boldsymbol{r}\right) \hat{\boldsymbol{\sigma}}_{K,G} \;, \tag{7.228}$$

where $\hat{\boldsymbol{\sigma}}$ is the polarization unit vector such that $\hat{\boldsymbol{\sigma}}_{K,G} = \hat{\boldsymbol{\sigma}}_{-K,-G}$. $\hat{\boldsymbol{\sigma}}$ can be associated with either σ or π polarized wavefields. The wavevector $\boldsymbol{K}_G$ satisfies Bragg's law, i.e., $\boldsymbol{K}_G = \boldsymbol{K}_O + \boldsymbol{g}$. The second-order quantized field variable $q_{K,G}$ related to coordinates and its conjugate $p_{K,G}$ are defined as

$$q_{K,G} = \sqrt{\frac{\hbar}{2\omega_{K,G}}} \left(b_{K,G} + b^{\dagger}_{-K,-G} \right) \;,$$

$$p_{K,G} = \mathrm{i}\sqrt{\frac{\hbar\omega_{K,G}}{2}} \left(b^{\dagger}_{-K,-G} - b_{K,G} \right) \;, \tag{7.229}$$

where

$$\omega_{K,G} = 2\pi c K_G \;. \tag{7.230}$$

The interaction H_{cf} between the electrons of the crystal and the radiation field can then be written accordingly as

$$H_{\mathrm{cf}} = \sqrt{4\pi} \left(\frac{e}{m} \right) \sum_{K,G} \sum_{j=1}^{N} \left(\boldsymbol{p}_j \cdot \hat{\boldsymbol{\sigma}}_{K,G} \right) \exp\left(2\pi \mathrm{i} \boldsymbol{K}_G \cdot \boldsymbol{r}_j\right) q_{K,G} \;. \tag{7.231}$$

Similarly, H_{ff} and H_{rad} take the following forms:

$$\begin{aligned} H_{\mathrm{ff}} &= \frac{2\pi e^2}{m} \sum_{K,G} \sum_{K',G'} (\hat{\boldsymbol{\sigma}}_{K',G'} \cdot \hat{\boldsymbol{\sigma}}_{K,G}) \times \sum_j \exp\left[2\pi \mathrm{i} \left(\boldsymbol{K}'_{G'} + \boldsymbol{K}_G\right) \cdot \boldsymbol{r}_j\right] \\ &\quad \times q_{K',G'} q_{K,G} \; , \\ H_{\mathrm{rad}} &= \frac{1}{2} \sum_{K,G} p_{K,G} p_{-K,-G} + \frac{1}{2} \sum_{K,G} c^2 K_G^2 q_{K,G} q_{-K,-G} \; . \end{aligned} \tag{7.232}$$

For convenience, the sum of the annihilation and creation operators, $b_{K,G}$ and $b^{\dagger}_{K,G}$, is denoted by

$$B_{K,G} = b_{K,G} + b^{\dagger}_{-K,-G} \; . \tag{7.233}$$

In terms of this operator the Hamiltonians can be rewritten as

$$\begin{aligned} H_{\mathrm{cf}} &= \sum_{K,G} \sum_{n,l} \frac{e}{m} \sqrt{\frac{2\pi\hbar}{\omega_{K,G}}} \langle n \Big| \sum_j (\boldsymbol{p}_j \cdot \hat{\boldsymbol{\sigma}}_{K,G}) \exp\left(2\pi \mathrm{i} \boldsymbol{K}_G \cdot \boldsymbol{r}_j\right) \Big| l \rangle B_{K,G} \; , \\ H_{\mathrm{ff}} &= \frac{2\pi e^2 \hbar}{m} \sum_{K,G,K',G',n,l} \frac{(\hat{\boldsymbol{\sigma}}_{K',G'} \cdot \hat{\boldsymbol{\sigma}}_{K,G})}{\sqrt{\omega_{K',G'} \omega_{K,G}}} \\ &\quad \times \langle n \Big| \sum_j \exp\left[2\pi \mathrm{i} \left(\boldsymbol{K}'_{G'} + \boldsymbol{K}_G\right) \cdot \boldsymbol{r}_j\right] \Big| l \rangle B_{K',G'} B_{K,G} \; , \\ H_{\mathrm{rad}} &= \sum_{K,G} \hbar \omega_{K,G} B_{-K,-G} B_{K,G} \; . \end{aligned} \tag{7.234}$$

If only the elastic scattering of X-rays by electrons is considered, the scattering can be treated as a transition from the ground state $|0\rangle$ with a photon of $\boldsymbol{K}_G$ to $|0\rangle$ with a $\boldsymbol{K}_L$ photon. The wavefunction $|0\rangle$ satisfies the condition of (7.227). An effective Hamiltonian $\mathcal{H}$ is the sum $\mathcal{H} = \mathcal{H}_0 + \mathcal{H}_1$, where $\mathcal{H}_0$ can be deduced from treating H_{ff} to the first-order and $\mathcal{H}_1$ from H_{cf} to the second-order perturbation theory in the interaction picture [84, 392, 510], by using the photon Green function as follows:

(i) From H_{ff} given in (7.234), the effective Hamiltonian denoted by $\mathcal{H}_0$ is

$$\begin{aligned} \mathcal{H}_0 &= \frac{2\pi e^2 \hbar}{m} \frac{\hat{\boldsymbol{\sigma}}_{K,G} \cdot \hat{\boldsymbol{\sigma}}_{K,L}}{\sqrt{\omega_{K,G} \omega_{K,L}}} \langle 0 \Big| \sum_j \exp\left[-2\pi \mathrm{i} (\boldsymbol{g} - \boldsymbol{l}) \cdot \boldsymbol{r}_j\right] \Big| 0 \rangle \\ &\quad \times B_{-K,-G} B_{K,L} \; , \end{aligned} \tag{7.235}$$

which is related to the electric susceptibility $\chi_{G\text{–}L}$ because

$$\chi_{G\text{–}L} = -\frac{4\pi e^2}{m\omega^2} \int \varrho(\boldsymbol{r}) \exp[-2\pi \mathrm{i} (\boldsymbol{g} - \boldsymbol{l}) \cdot \boldsymbol{r}] \mathrm{d}\boldsymbol{r} \; , \tag{7.236}$$

where

$$\varrho(\boldsymbol{r}) = \langle 0 \Big| \sum_j \delta\left(\boldsymbol{r} - \boldsymbol{r}_j\right) \Big| 0 \rangle \; . \tag{7.237}$$

(ii) The effective Hamiltonian $\mathcal{H}_1$, due to H_{cf}, can be derived from the Green function defined as

$$G\left(\boldsymbol{K}_G, t; \boldsymbol{K}_L, t'\right) \equiv \mathrm{i}\langle 0\left|T\mathcal{B}_{K,G}(t)\mathcal{B}_{-K,-L}(t')\right| 0\rangle \ , \tag{7.238}$$

where T is a chronological operator and $\mathcal{B}_{k,G}(t')$ is a Heisenberg operator given by

$$\mathcal{B}_{K,G}(t) = \exp\left[\left(\frac{\mathrm{i}}{\hbar}\right) Ht\right] B_{K,G} \exp\left[-\left(\frac{\mathrm{i}}{\hbar}\right) Ht\right] . \tag{7.239}$$

$\langle\rangle$ in (7.238) means averaging over the ground state of the Hamiltonian H. For convenience, we define $H = H_0' + H'$, where $H_0' = H_0 + H_{\mathrm{rad}}$ and $H' = H_{\mathrm{cf}} + H_{\mathrm{ff}}$. When the interaction part H' is zero, only the incident X-ray wave exists. The Green function is then the free Green function, i.e.,

$$\begin{aligned} G_o\left(\boldsymbol{K}_G, t; \boldsymbol{K}_L, t'\right) &= -\mathrm{i}\langle 0\left|T\mathcal{B}_{K,G}(t)\mathcal{B}_{-K,-L}(t')\right| 0\rangle \\ &= G_o\left(\boldsymbol{K}_G; t-t'\right)\delta_{K_G,K_L} \ , \end{aligned} \tag{7.240}$$

where

$$G_o\left(\boldsymbol{K}_G; t-t'\right) = \begin{cases} -\mathrm{i}\exp\left(-\mathrm{i}\omega_{K,G}|t-t'|\right) & \text{for} \quad t-t' > 0 \ , \\ 0 & \text{for} \quad t-t' < 0 \ . \end{cases} \tag{7.241}$$

Its Fourier transform is

$$G_o\left(\boldsymbol{K}_G, \omega\right) = \int_{-\infty}^{\infty} G_o\left(K_G\tau\right)\exp(\mathrm{i}\omega\tau - \epsilon|\tau|)\mathrm{d}\tau \ , \tag{7.242}$$

where $\tau = t - t'$, and a damping factor, $\exp(-\epsilon|\tau|)$, with a small positive ϵ is introduced to converge the integral. The integral leads to

$$G_o\left(\boldsymbol{K}_G, \omega\right) = \frac{2\omega_{K,G}}{\omega^2 - \omega_{K,G}^2} \ . \tag{7.243}$$

When the interaction turns on, i.e., $H' \neq 0$, the Green function $G(K_H, t; K_L, t')$ is usually calculated as [511]

$$\begin{aligned} & G\left(K_G, t; K_L, t'\right) \\ & = \sum_{n=0}^{\infty} \frac{1}{n!}\left(\frac{1}{\mathrm{i}\hbar}\right)^n \int_{-\infty}^{\infty}\int_{-\infty}^{\infty} \cdots \int_{-\infty}^{\infty} \mathrm{d}t_1\mathrm{d}t_2\cdots\mathrm{d}t_n \\ & \quad \times\left\langle 0\left|T\left\{\mathcal{B}_{K,G}(t)\mathcal{B}_{-K,-G}(t')\mathcal{H}'\left(t_1\right)\mathcal{H}'\left(t_2\right)\cdots\mathcal{H}'\left(t_n\right)\right\}\right| 0\right\rangle , \end{aligned} \tag{7.244}$$

where

$$\mathcal{H}'(t) = \exp\left[\left(\frac{\mathrm{i}}{\hbar}\right) H_0' t\right] H' \exp\left[-\left(\frac{\mathrm{i}}{\hbar}\right) H_0' t\right] \tag{7.245}$$

and $t > t_n > t_{n-1} > \cdots > t_1 > t'$. The perturbed terms for each order of approximation in the expansion of (7.244) can be represented in diagrams.

The scattering of the photon with the wavevector $\boldsymbol{K}_{\boldsymbol{L}}$ from the state $|0\rangle$ to the state $|n\rangle$ with the wavevector $\boldsymbol{K}_{\boldsymbol{G}}$ can be represented by the Green function $G^{(n,0)}(K_G,\omega_n;K_L,\omega_0)$, where the energies $E_n = \hbar\omega_n$ and $E_0 = \hbar\omega_0$. According to the schematic representation given in Fig. 7.3, the Green function is written up to a second order as

$$\begin{aligned} &G^{(n,0)}\left(K_G,\omega_n;K_L,\omega_0\right) \\ &= \delta_{n,0}\delta_{K_G,K_L}G_o^{(0)}\left(K_L,\omega_0\right)+\frac{1}{\hbar}G_o^{(n)}\left(K_G,\omega_n\right)G_o^{(0)}\left(K_L,\omega_0\right) \\ &\times\left\{H'_{no}\left(K_G,K_L\right)+\sum_{\substack{K',G'\\ m\neq 0}}\frac{H'_{nm}\left(K_G,K_M\right)H'_{m0}\left(K_M,K_L\right)}{\hbar\omega_{m0}-\omega_{K,M}-\mathrm{i}\epsilon}\right\}, \end{aligned} \tag{7.246}$$

where

$$H'_{nm}\left(K_G,K_L\right)=\int_{V_o}\langle m|H'|n\rangle\exp\left[-2\pi\mathrm{i}\left(\boldsymbol{K}_G-\boldsymbol{K}_L\right)\cdot\boldsymbol{r}\right]\mathrm{d}\boldsymbol{r}\,. \tag{7.247}$$

$\omega_{K,M}$ is the angular frequency associated with the photon of $\boldsymbol{K}_M$ and $\omega_{n0} = (E_n - E_0)/\hbar$. The quantity in the braces of (7.246) is the self-energy part, denoted by S, of the Green function, to which the effective Hamiltonian is related.

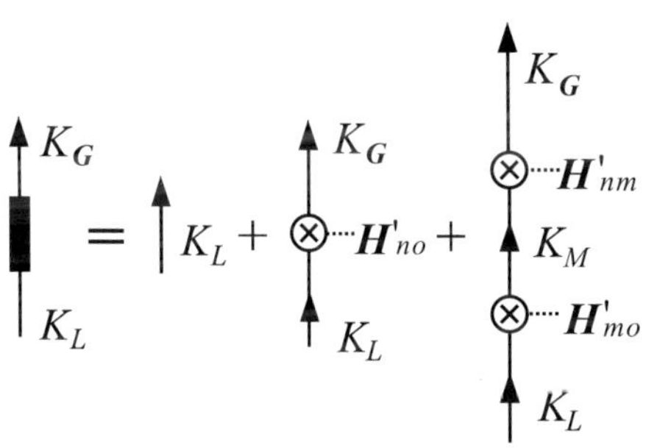

Fig. 7.3. Schematic representation of the Green function defined in (7.246) [84]

When H' only includes H_{cf}, the first term in the braces, $H'_{n,0}$, is zero and the second term leads to

$$\begin{aligned} \mathcal{H}_1 = &-\frac{2\pi e^2}{m^2\sqrt{\omega_{K,G}\omega_{K,L}}}\sum_{n=0}\left\{\frac{\left\langle 0\left|\sum_j\left(\boldsymbol{p}_j\cdot\hat{\boldsymbol{\sigma}}_{K,G}\right)\exp\left(-2\pi\mathrm{i}\boldsymbol{K}_G\cdot\boldsymbol{r}_j\right)\right|n\right\rangle}{\omega_{n0}-\omega_{K,L}-\mathrm{i}\epsilon}\right. \\ &\times\left\langle n\left|\sum_j\left(\boldsymbol{p}_j\cdot\hat{\boldsymbol{\sigma}}_{K,L}\right)\exp\left(2\pi\mathrm{i}\boldsymbol{K}_L\cdot\boldsymbol{r}_j\right)\right|0\right\rangle \\ &+\frac{\left\langle 0\left|\sum_j\left(\boldsymbol{p}_j\cdot\hat{\boldsymbol{\sigma}}_{K,L}\right)\exp\left(2\pi\mathrm{i}\boldsymbol{K}_L\cdot\boldsymbol{r}_j\right)\right|n\right\rangle}{\omega_{n0}+\omega_{K,G}-\mathrm{i}\epsilon} \end{aligned}$$

$$\times\left\langle n\left|\sum_j (\boldsymbol{p}_j\cdot\hat{\boldsymbol{\sigma}}_{K,G})\exp\left(-2\pi\mathrm{i}\boldsymbol{K}_G\cdot\boldsymbol{r}_j\right)\right|0\right\rangle\Bigg\}$$
$$\times B_{-K,-G}B_{K,L}\;. \tag{7.248}$$

When the wavevectors $\boldsymbol{K_L}$ and $\boldsymbol{K_G}$ satisfy diffraction conditions, only the elastic scattering needs to be considered. That is,

$$\omega_{K,G}\cong\omega_{K,L}=\omega=2\pi ck\;. \tag{7.249}$$

Taking into account H_{rad}, $\mathcal{H}_0$, and $\mathcal{H}_1$, we obtain the effective Hamiltonian as

$$\mathcal{H}=\sum_G\hbar\left(\omega_{K,G}-\frac{1}{2}\omega\varphi_{GG}\right)B_{-K,-G}B_{K,G}$$
$$-\sum_{\substack{G,L\\ G\neq L\neq O}}\frac{1}{2}\hbar\omega\varphi_{GL}B_{-K,-G}B_{K,L}\;, \tag{7.250}$$

where

$$\varphi_{GL}=\Phi_{GL}+C_{GL} \tag{7.251}$$

with

$$\Phi_{GL}=p_{GL}\chi_{G\text{–}L}\;,$$

$$C_{GL}=\frac{4\pi e^2}{\hbar m^2\omega^2}\sum_{n=0}\left\{\frac{\langle 0\left|\sum_j(\boldsymbol{p}_j\cdot\hat{\boldsymbol{\sigma}}_{K,G})\exp\left(-2\pi\mathrm{i}\boldsymbol{K}_G\cdot\boldsymbol{r}_j\right)\right|n\rangle}{\omega_{n0}-\omega_{K,L}-\mathrm{i}\epsilon}\right.$$
$$\times\langle n\left|\sum_j(\boldsymbol{p}_j\cdot\hat{\boldsymbol{\sigma}}_{K,L})\exp\left(2\pi\mathrm{i}\boldsymbol{K}_L\cdot\boldsymbol{r}_j\right)\right|0\rangle$$
$$+\frac{\langle 0\left|\sum_j(\boldsymbol{p}_j\cdot\hat{\boldsymbol{\sigma}}_{K,L})\exp\left(2\pi\mathrm{i}\boldsymbol{K}_L\cdot\boldsymbol{r}_j\right)\right|n\rangle}{\omega_{n0}+\omega_{K,G}-\mathrm{i}\epsilon}$$
$$\left.\times\langle n\left|\sum_j(\boldsymbol{p}_j\cdot\hat{\boldsymbol{\sigma}}_{K,G})\exp\left(-2\pi\mathrm{i}\boldsymbol{K}_G\cdot\boldsymbol{r}_j\right)\right|0\rangle\right\}\;,$$
$$p_{GL}=\hat{\boldsymbol{\sigma}}_{K,G}\cdot\hat{\boldsymbol{\sigma}}_{K,L}\;. \tag{7.252}$$

$\varphi_{GG}=\varphi_{OO}$ and φ_{OO} is proportional to χ_O.

The equation of dispersion can be determined from the pole of the corresponding Green function. For a given Hamiltonian $\mathcal{H}$ as in (7.250), the Green function for K_G by following the relation between the time-derivative of an operator A and the commutation bracket

$$\mathrm{i}\frac{\partial}{\partial t}A(t)=[A(t),\mathcal{H}] \tag{7.253}$$

is governed by

$$
\begin{aligned}
&\mathrm{i}\frac{\partial}{\partial t}G\left(K_G,t;K_L,t'\right) \\
&= \delta_{G,L}\delta(t-t') + \hbar\left(\omega_{K,G} - \frac{1}{2}\omega\varphi_{GG}\right)G\left(K_G,t;K_G,t'\right) \\
&\quad - \sum_L \frac{1}{2}\hbar\omega\varphi_{GL}G\left(K_G,t;K_L,t'\right) . \qquad (7.254)
\end{aligned}
$$

The Green function reduces to the free Green function $G_0(K_G;t-t')\delta_{G,L}$ for all $\varphi_{GL}=0$, so that

$$
\mathrm{i}\frac{\partial}{\partial t}G_o\left(K_G;t-t'\right) = \delta(t-t') + \hbar\left(\omega_{K,G} - \frac{1}{2}\omega\varphi_{GG}\right)G_o\left(K_G;t-t'\right) . \qquad (7.255)
$$

The solution to (7.255) is

$$
\begin{aligned}
G\left(K_G,t;K_L,t'\right) = \delta_{G,L}G_o\left(K_G;t-t'\right) - \sum_L \frac{1}{\hbar}\int \mathrm{d}t''G_o\left(K_G;t-t''\right) \\
\times\left[\frac{1}{2}\hbar\omega\varphi_{GL}\right]G\left(K_G,t'';K_L,t'\right) . \qquad (7.256)
\end{aligned}
$$

In the following, we shall consider one-, two-, three-, four-, and N-wave cases separately.

(1) One-wave diffraction
For a one-wave O diffraction, (7.256) leads to

$$
\begin{aligned}
G\left(K_O,t;K_O,t'\right) = \delta(t-t')G_o\left(K_O;t-t'\right) - \frac{1}{\hbar}\int \mathrm{d}t''G_o\left(K_O;t-t''\right) \\
\times\left[\frac{1}{2}\hbar\omega\varphi_{OO}\right]G\left(K_O,t'';K_O,t'\right) , \qquad (7.257)
\end{aligned}
$$

whose Fourier transform in the frequency domain is

$$
G\left(K_O;\omega\right) = G_o\left(K_O;\omega\right) + G_o\left(K_O;\omega\right)S\left(K_O,K_O;\omega\right)G\left(K_O;\omega\right) . \qquad (7.258)
$$

The last term of the above equation is related to the so-called self-energy part denoted by

$$
S\left(K_O,K_O;\omega\right) = -\frac{1}{\hbar}\left(\frac{\hbar\omega\varphi_{OO}}{2}\right) . \qquad (7.259)
$$

The Green function $G(K_O;\omega)$ for the one-wave case then takes the form

$$
G\left(K_O;\omega\right) = \frac{G_o\left(K_O;\omega\right)}{1 - G_o\left(K_O;\omega\right)S\left(K_O;\omega\right)} , \qquad (7.260)
$$

or

$$
G\left(K_O;\omega\right) = \frac{1}{G_o^{-1}\left(K_O;\omega\right) - S\left(K_O;\omega\right)} . \qquad (7.261)
$$

$G_0^{-1}(K_O;\omega)$ denotes the reciprocal of $G_0(K_O;\omega)$ given in (7.243). The dispersion equation is determined from the pole of the Green function, i.e.,

$$G_o^{-1}(K_O;\omega) - S(K_O;\omega) = 0\,, \tag{7.262}$$

which leads to

$$\omega^2 - \omega_{K,O}^2 + \omega_{K,O}\omega\varphi_{OO} \simeq \omega^2 - \omega_{K,O}^2 + \omega^2\varphi_{OO} = 0\,. \tag{7.263}$$

Referring to (7.249), we can find that the above relation is the usual dispersion equation for a one-wave diffraction

$$K_O^2 - k^2(1+\varphi_{OO}) = 0\,. \tag{7.264}$$

Absorption coefficients can also be determined from the dispersion equation, as in the dynamical theory. In other words, the imaginary parts of the solution to the dispersion equation yield the linear absorption coefficients. Thus, the Green function can be expressed as

$$G(K_O;\omega) = \frac{2\omega_{K,O} - \varphi_{OO}\omega}{\omega^2 - \left[\omega_{K,O} - \frac{\omega}{2}\varphi_{OO}\right]^2} \approx \frac{2\omega_{K,O} - \varphi_{OO}\omega}{\omega^2 - \omega_{K,O}^2 + \omega^2\varphi_{OO}^2}\,. \tag{7.265}$$

(2) Two-wave diffraction

For a two-wave (O,G) diffraction, (7.256) leads to

$$\begin{aligned} G(K_O,t;K_O,t') = \delta(t-t')G_o(K_O;t-t') - \frac{1}{\hbar}\int \mathrm{d}t'' G_o(K_O;t-t'') \\ \times\frac{1}{2}\hbar\omega\left\{\varphi_{OO}G(K_O,t'';K_O,t')\right. \\ \left. + \varphi_{OG}G(K_G,t'';K_O,t')\right\} \end{aligned} \tag{7.266}$$

for $G = L = O$, and

$$\begin{aligned} G(K_G,t;K_O,t') = -\frac{1}{\hbar}\int \mathrm{d}t'' G_o(K_G;t-t'')\frac{\hbar\omega}{2}\Big\{\varphi_{GO}G(K_O,t'';K_O,t') \\ + \varphi_{GG}G(K_G,t'';K_O,t')\Big\} \end{aligned} \tag{7.267}$$

for $G = G$, and $L = O$. The frequency-components of the Fourier transform of the above two equations give

$$\begin{aligned} G(K_O,K_O;\omega) &= G_o(K_O;\omega) - \frac{\omega}{2}G_o(K_O;\omega)\left\{\varphi_{OO}G(K_O;\omega)\right. \\ &\quad \left. + \varphi_{OG}G(K_G,K_O;\omega)\right\}\,, \\ G(K_G,K_O;\omega) &= -\frac{\omega}{2}G_o(K_G;\omega)\left\{\varphi_{GO}G(K_O;\omega)\right. \\ &\quad \left. + \varphi_{GG}G(K_G,K_O;\omega)\right\}\,. \end{aligned} \tag{7.268}$$

$G(K_O,K_O;\omega)$ and $G(K_G,K_O;\omega)$ can be solved by the matrix equation

$$\begin{bmatrix} 1+\frac{\omega}{2}\varphi_{OO}G_o(K_O;\omega) & \frac{\omega}{2}\varphi_{OG}G_o(K_O;\omega) \\ \frac{\omega}{2}\varphi_{GO}G_o(K_G;\omega) & 1+\frac{\omega}{2}\varphi_{GG}G_o(K_G;\omega) \end{bmatrix} \begin{bmatrix} G(K_O,K_O;\omega) \\ G(K_G,K_O;\omega) \end{bmatrix} = \begin{bmatrix} G_o(K_O;\omega) \\ 0 \end{bmatrix} . \tag{7.269}$$

The determinant of the 2×2 matrix gives the following dispersion equation

$$\left(\omega^2 - \omega_{K,O}^2 + \omega^2\varphi_{OO}\right)\left(\omega^2 - \omega_{K,G}^2 + \omega^2\varphi_{GG}\right) = \omega^4\varphi_{OG}\varphi_{GO} , \tag{7.270}$$

or

$$\left[K_O^2 - k^2(1+\varphi_{OO})\right]\left[K_G^2 - k^2(1+\varphi_{GG})\right] = k^4\varphi_{OG}\varphi_{GO} . \tag{7.271}$$

The frequency-components of the Green functions, $G(K_O, K_O;\omega)$ and $G(K_G, K_O;\omega)$ can be obtained to a second order with ease from the schematic representation shown in Fig. 7 4 as

$$\begin{aligned} G(K_O,K_O;\omega) &= G_o(K_O;\omega) + G_o(K_O;\omega)\frac{\omega^2}{4}\varphi_{GO}\varphi_{OG}G_o(K_G;\omega) \\ &\quad \times G(K_O,K_O;\omega) , \\ G(K_G,K_O;\omega) &= G_o(K_O;\omega)\frac{\omega}{2}\varphi_{GO}G_o(K_G;\omega) + G_o(K_O;\omega)\frac{\omega^2}{4} \\ &\quad \times\varphi_{GO}\varphi_{OG}G_o(K_G;\omega)G(K_G,K_O;\omega) . \end{aligned} \tag{7.272}$$

The arrows with the bold line and solid line in Fig. 7.4(a) represent $G(K_O, K_O;\omega)$ and $G_0(K_O;\omega)$, respectively. The vertices indicate the polarizabilities φ_{OG} and φ_{OG}. In Fig. 7.4(b), $G(K_G, K_O;\omega)$ is represented as an arrow with a bold solid bar in the middle.

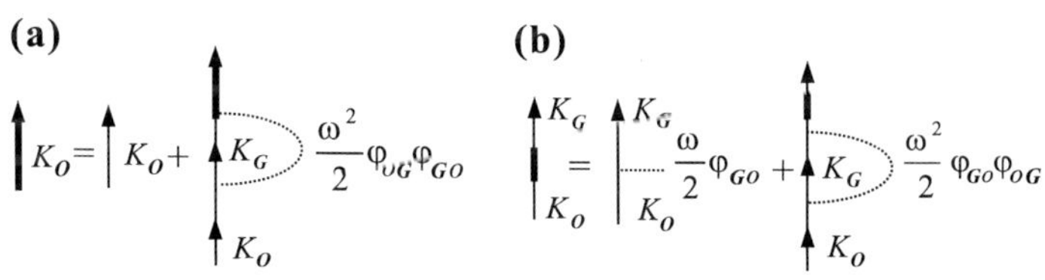

Fig. 7.4. Schematic representation of the Green functions for two-wave diffraction: (**a**) $G(K_O, K_O;\omega)$ and (**b**) $G(K_G, K_O;\omega)$ [84, 275]

The first equation of (7.272) can be written in a similar form to (7.258) as

$$\begin{aligned} G(K_O,K_O;\omega) &= G_o(K_O;\omega) + G_o(K_O;\omega)S(K_G,K_O;\omega) \\ &\quad \times G(K_O,K_O;\omega) , \end{aligned} \tag{7.273}$$

where

$$S(K_G,K_O;\omega) = \frac{1}{\hbar^2}\left(\frac{\hbar\omega}{2}\right)^2\varphi_{GO}\varphi_{OG}G_o(K_G;\omega) . \tag{7.274}$$

The Green function $G(K_O, K_O; \omega)$ then takes the same mathematical form as (7.261) for the one-wave case, except that $G_0(K_O; \omega)$ needs to be replaced by the one-wave Green function $G(K_O; \omega)$ given in (7.265) and $S(K_O; \omega)$ is substituted by $S(K_G, K_O; \omega)$. The pole of $G(K_G, K_O; \omega)$ give the same dispersion equation as (7.271), which has exactly the same form as that derived from the dynamical theory.

The second equation of (7.272) gives

$$G(K_G, K_O; \omega) = \frac{\frac{\omega}{2}\varphi_{GO} G_o(K_O; \omega)}{G_o^{-1}(K_O; \omega) - \frac{\omega^2}{4}\varphi_{GO}\varphi_{OG} G_o(K_G; \omega)} . \quad (7.275)$$

The wavefield-amplitude ratios among the diffracted waves, according to (7.269), are determined as the ratios among the corresponding Green functions [512], i.e.,

$$\frac{D_G(j)}{D_O(j)} = \left. \frac{G(K_G, K_O; \omega)}{G(K_O, K_O; \omega)} \right|_j \quad (7.276)$$

for a given mode j. From the appropriate boundary conditions given in the dynamical theory, the intensities of the diffracted waves can be calculated straightforwardly.

The above approach, following closely the procedures adopted by the dynamical theory, is therefore dynamical. If the derivation of the dispersion equation is skipped, then the diffracted intensities can be calculated kinematically. This kinematical approach is described below.

The first-order iterations of the Green functions of (7.272) give the relationship between $G(K_O, K_O; \omega)$ and $G_0(K_O; \omega)$:

$$G(K_G, K_O; \omega) = \frac{\omega}{2}\varphi_{GO} G_o(K_G; \omega) G(K_O, K_O; \omega) . \quad (7.277)$$

Accordingly, the intensity I_G of the diffracted wave G is

$$I_G = \left| \frac{G(K_G, K_O; \omega)}{G(K_O, K_O; \omega)} \right|^2 = \left| \frac{\omega}{2}\varphi_{GO} G_o(K_G; \Omega) \right|^2 . \quad (7.278)$$

With the aid of (7.265), the intensity I_G becomes

$$I_G = \frac{1}{4} \left| \frac{k^2 \varphi_{GO}}{K_G^2 - K'^2} \right|^2 , \quad (7.279)$$

where $k = \omega/(2\pi c)$, and $K'^2 = k^2(1 + \varphi_{OO})$ is the square of the modulus of the one-wave wavevector. This expression of the two-wave peak intensity is exactly identical to that derived from the kinematical theory.

(3) Three-wave diffraction

The Green functions, $G(K_O, K_O; \omega)$ and $G(K_G, K_O; \omega)$, of a three-wave (O, G, L) diffraction are schematically shown in Fig. 7.5 and Fig. 7.6, respectively, where the bold arrow representing a Green function is also shown in Fig. 7.5(b). In comparison with the two-wave diagram, the interactions $K_O \to K_L \to K_O$, $K_O \to K_G \to K_L \to K_O$, and $K_O \to K_L \to K_G \to K_O$ in Fig. 7.5(a), $K_O \to K_L \to K_G$ and the interactions involving $\varphi_{OL}\varphi_{LO}$ and $\varphi_{OL}\varphi_{LG}\varphi_{GO}$ in Fig. 7.6 are extra terms. The dispersion equation derived from Fig. 7.5(a) is

$$\begin{aligned}&\left(K_O^2 - K'^2\right)\left(K_G^2 - K'^2\right)\left(K_L^2 - K'^2\right) - k^4\left(K_O^2 - K'^2\right)\varphi_{GL}\varphi_{LG}\\&- k^4\left(K_G^2 - K'^2\right)\varphi_{OL}\varphi_{LO} - k^4\left(K_L^2 - K'^2\right)\varphi_{OG}\varphi_{GO}\\&- k^6\left(\varphi_{OG}\varphi_{GL}\varphi_{LO} + \varphi_{OL}\varphi_{LG}\varphi_{GO}\right) = 0\,.\end{aligned} \tag{7.280}$$

This dispersion equation is the same as that derived from the dynamical theory at the exact three-wave diffraction position, except that the usual Fourier component χ_{GL} is replaced by φ_{GL}.

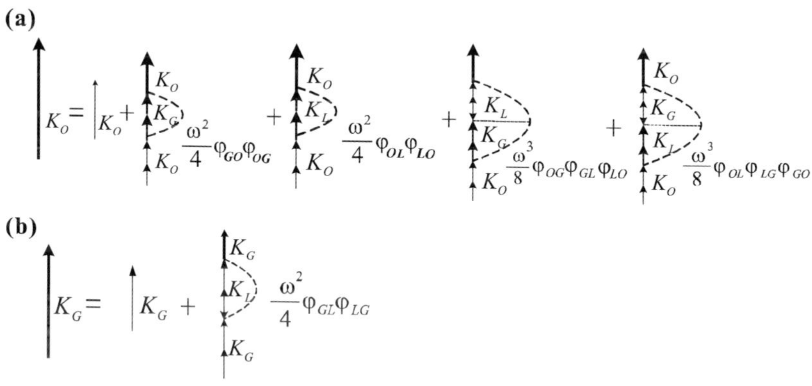

Fig. 7.5. Schematic representation of the Green function for three wave diffraction: (**a**) $G(K_O, K_O; \omega)$ and (**b**) $G(K_G; \omega)$ [84, 275]

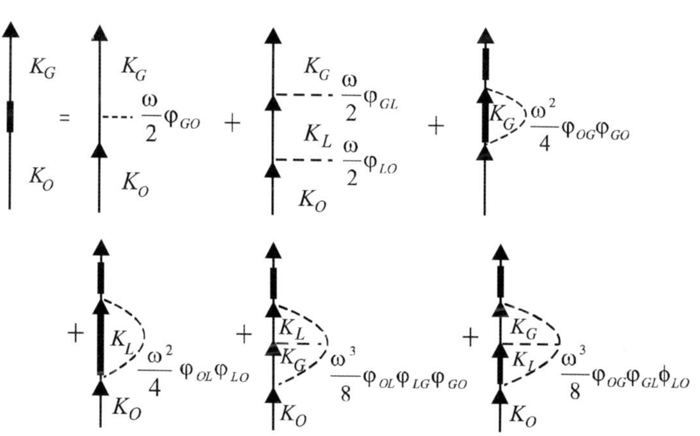

Fig. 7.6. Schematic representation of the Green function $G(K_G, K_O; \omega)$ for three-wave diffraction [275]

The Green function $G(K_G, K_O; \omega)$ in terms of $G(K_O, K_O; \omega)$ to a first-order approximation is

$$G(K_G, K_O; \omega) \cong G(K_O, K_O; \omega) \left[\frac{\omega}{2} \varphi_{GO} G_o(K_G; \omega) + \frac{\omega^2}{4} \varphi_{GL} \varphi_{LO} G_o(K_G; \omega) G_o(K_L; \omega) \right] . \quad (7.281)$$

The kinematical expression for the diffracted intensity I_G is then written as

$$\begin{aligned} I_G &= \left| \frac{k^2 \varphi_{GO}}{2(K_G^2 - K'^2)} + \frac{k^4 \varphi_{GL} \varphi_{LO}}{4(K_G^2 - K'^2)(K_L^2 - K'^2)} \right|^2 \\ &= \frac{1}{4} \left| \frac{k^2 \varphi_{GO}}{K_G^2 - K'^2} \right|^2 + \frac{1}{16} \left| \frac{k^4 \varphi_{GL} \varphi_{LO}}{(K_G^2 - K'^2)(K_L^2 - K'^2)} \right|^2 \\ &\quad + \frac{1}{4} \frac{k^6 \left| \varphi_{OG} \varphi_{GL} \varphi_{LO} \right|}{(K_G^2 - K'^2)^2 (K_L^2 - K'^2)} \cos \delta_3 , \end{aligned} \quad (7.282)$$

where δ_3 is the triplet phase, i.e., $\delta_3 = \delta_{O\text{–}G} + \delta_{G\text{–}L} + \delta_{L\text{–}O}$.

(4) Four-wave diffraction
An additional wavevector K_M is involved in a four-wave (O, G, L, M) diffraction. There are two equally possible diffraction channels involving the interactions $K_O \to K_L \to K_M \to K_G$ and $K_O \to K_M \to K_L \to K_G$, which are depicted in the schematic of the Green function $G(K_G, K_O; \omega)$, shown in Fig. 7.7. The last term of the Green function shown in this figure stands for the higher-order contribution to the Green function. Similarly, the Green function $G(K_O, K_O; \omega)$ can be deduced from Fig. 7.5(a) by adding more terms related to K_M. The dispersion equation can be obtained as easily as in the three-wave case. We shall skip this equation and go directly to write the expression for I_G at the exact four-wave diffraction point as

$$\begin{aligned} I_G = \Big| & \frac{\omega}{2} \varphi_{GO} G_o(K_G; \omega) + \frac{\omega^2}{4} \varphi_{GL} \varphi_{LO} G_o(K_L; \omega) G_o(K_G; \omega) \\ & + \frac{\omega^2}{4} \varphi_{GM} \varphi_{MO} G_o(K_M; \omega) G_o(K_G; \omega) \\ & + \frac{\omega^3}{8} \varphi_{GM} \varphi_{ML} \varphi_{LO} G_o(K_L; \omega) G_o(K_M; \omega) G_o(K_G; \omega) \\ & + \frac{\omega^3}{8} \varphi_{GL} \varphi_{LM} \varphi_{MO} G_o(K_M; \omega) G_o(K_L; \omega) G_o(K_G; \omega) \Big|^2 . \end{aligned} \quad (7.283)$$

The expansion of this equation is very lengthy and cumbersome. We shall discuss the expanded terms in details in Chap. 8.

(5) N-wave diffraction
There are N waves, $K_O, K_G, K_1, K_2, \cdots, K_{N-2}$, involved in a general N-wave diffraction. Let's define $P_{ijk}(N-2)$ as the permutation of (ijk) with i,

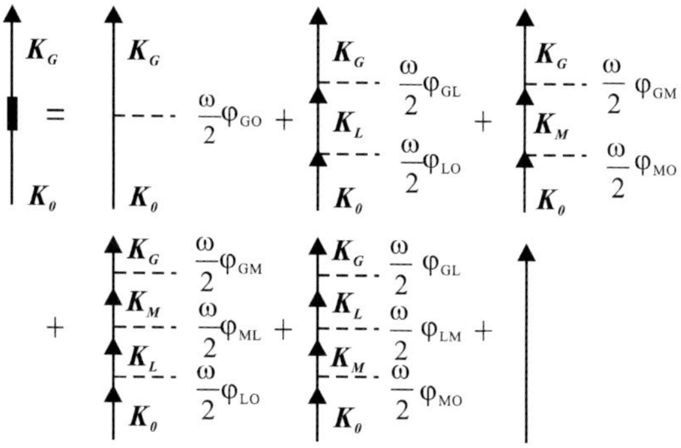

Fig. 7.7. Schematic representation of the Green function $G(K_G, K_O; \omega)$ for four-wave diffraction [275]

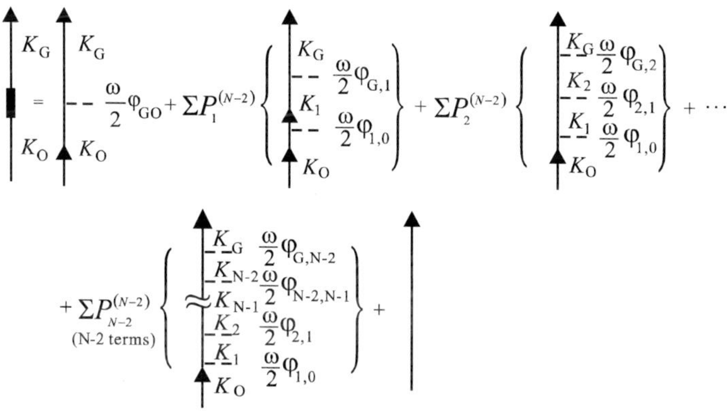

Fig. 7.8. Schematic representation of the Green function $G(K_G, K_O; \omega)$ for N-wave diffraction [275]

j, and k taken from the $N-2$ wavevectors $\boldsymbol{k}_1, \boldsymbol{k}_2, \cdots, \boldsymbol{k}_{N-2}$. For simplicity, the following notations are used in the schematic representation of the Green function $G(K_G, K_O; \omega)$ shown in Fig. 7.8: $P_i(N-2) = P_1(N-2, 1) = P_1^{N-2}$, $P_{ij}(N-2) = P(N-2, 2) = P_2^{N-2}$, $P_{ijk}(N-2) = P(N-2, 3) = P_3^{N-2}$, etc., where $P_n^{N-2} = (N-2)!/(N-n-2)!$. n is the number of the subscripts of $P_{ijk\ldots z}(N-2)$.

The Green function $G(K_G, K_O; \omega)$ takes the form

$$\begin{aligned} G_N = {} & G\left(K_O, K_O; \omega\right) \left\{ \frac{\omega}{2} \varphi_{GO} G_o\left(K_G; \omega\right) + \sum_{n=1}^{N-2} P_n^{(N-2)} \left[\left(\frac{\omega}{2}\right)^{n+1} \varphi_{GS_1} \right.\right. \\ & \times \varphi_{S_1 S_2} \cdots \varphi_{S_{n-1} S_n} G_o\left(K_G; \omega\right) G_o\left(K_{S_1}; \omega\right) G_o\left(K_{S_2}; \omega\right) \cdots \\ & \left.\left. \times G_o\left(K_{S_n}; \omega\right) \right] \right\} \\ = {} & G\left(K_O, K_O; \omega\right) A_G(N) \,. \end{aligned} \tag{7.284}$$

The intensity I_G of the N-wave diffraction can then be written as

$$I_G(N) = |A_G(N)|^2 \ . \tag{7.285}$$

7.2.4 Quantitative Approach for Phasing – An Iterative Approximation

The intensity I_G of the diffracted wave G in a three-wave (O, G, L) diffraction, as described in (7.282), depends explicitly on the cosine function of the triplet phase δ_3. To solve the phase problem, it is also necessary to obtain the information about the sine function of δ_3. For this purpose, Hümmer and Billy [279] in 1986 considered the phase shift Δ due to the presence of the third reflection L and obtained

$$R_L = \frac{k^2}{K_L^2 - K'^2} = |R_L| \exp(\mathrm{i}\triangle) \ , \tag{7.286}$$

where R_L is involved in (7.282). The phase shift Δ results from the consideration of introducing a small positive imaginary part of K_L due to absorption and scattering, which is similar to the $-i\epsilon$ in (7.248). By substituting (7.286) into the first equation of (7.282), the $\cos\delta_3$ in the second equation for I_G of (7.282) is then replaced by $\cos(\delta_3 + \Delta)$, where Δ varies from 0° to 180° as the $K_L^2 - K'^2$ changes from negative values to positive ones. This approach has led to phase determination on a semi-quantitative basis, where the values of δ_3 can be estimated from three-wave diffraction profiles by inspection [288, 289, 293, 305, 328]. The details about phase determination will be described in Chap. 9.

A theoretical approach for quantitative analysis of a reflection phase was developed in 1988 by Chang and Tang [296, 454], based on an iterative Born approximation. A brief description of this theoretical development is given below:

The fundamental equation of wavefield for a three-wave case can be expressed, by following (6.37a) and (6.37b), as

$$\begin{aligned} 2\varepsilon_O \boldsymbol{D}_O &= \chi_O \boldsymbol{D}_{O[O]} + \chi_{O\text{-}G} \boldsymbol{D}_{G[O]} + \chi_{O\text{-}L} \boldsymbol{D}_{L[O]} \ , \\ 2\varepsilon_G \boldsymbol{D}_G &= \chi_{G\text{-}O} \boldsymbol{D}_{O[G]} + \chi_O \boldsymbol{D}_{G[G]} + \chi_{G\text{-}L} \boldsymbol{D}_{L[G]} \ , \\ 2\varepsilon_L \boldsymbol{D}_L &= \chi_{L\text{-}O} \boldsymbol{D}_{O[L]} + \chi_{L\text{-}G} \boldsymbol{D}_{G[L]} + \chi_O \boldsymbol{D}_{L[L]} \ , \end{aligned} \tag{7.287}$$

where 2ε's have been defined in (6.38). By introducing the resonance term, similar to the R_L in (7.282),

$$A_M = \frac{1}{2\varepsilon_M - \chi_O} \cong \frac{k^2}{K_M^2 - K'^2} \ , \tag{7.288}$$

the fundamental equation becomes

$$\begin{aligned} \boldsymbol{D}_O &= A_O \chi_{O\text{-}G} \boldsymbol{D}_{G[O]} + A_O \chi_{O\text{-}L} \boldsymbol{D}_{L[O]} \ , \\ \boldsymbol{D}_G &= A_G \chi_{G\text{-}O} \boldsymbol{D}_{O[G]} + A_G \chi_{G\text{-}L} \boldsymbol{D}_{L[G]} \ , \\ \boldsymbol{D}_L &= A_L \chi_{L\text{-}O} \boldsymbol{D}_{O[L]} + A_L \chi_{L\text{-}G} \boldsymbol{D}_{G[L]} \ . \end{aligned} \tag{7.289}$$

The three equations of (7.289) can be combined into the following two equations by substituting the third equation into the first two equations:

$$\begin{aligned}\boldsymbol{D}_O &= A_O\chi_{O\text{–}G}\boldsymbol{D}_{G[O]} + A_O A_L\chi_{L\text{–}O}\chi_{O\text{–}L}\boldsymbol{D}_{O[L][O]} \\ &\quad + A_O A_L\chi_{L\text{–}G}\chi_{O\text{–}L}\boldsymbol{D}_{G[L][O]}\ , \end{aligned} \tag{7.290a}$$

$$\begin{aligned}\boldsymbol{D}_G &= A_G\chi_{G\text{–}O}\boldsymbol{D}_{O[G]} + A_G A_L\chi_{L\text{–}O}\chi_{G\text{–}L}\boldsymbol{D}_{O[L][G]} \\ &\quad + A_G A_L\chi_{L\text{–}G}\chi_{G\text{–}L}\boldsymbol{D}_{G[L][G]}\ , \end{aligned} \tag{7.290b}$$

where

$$\boldsymbol{D}_{O[L][G]} = -\left[\hat{\boldsymbol{K}}_G \times \hat{\boldsymbol{K}}_G \times \boldsymbol{D}_{O[L]}\right]\ . \tag{7.291}$$

We now consider the three-wave diffraction as a perturbed two-wave (O, G) case. The first term on the right-hand side of the equality in (7.290a) can be approximated by the two-wave wavefield $\boldsymbol{D}_O^{(2)}$ of the O reflection, and the second term $\boldsymbol{D}_{O[L][O]}$ can be replaced by $\boldsymbol{D}^{(2)}_{O[L][O]}$. The $\boldsymbol{D}_O$ then takes the approximate form

$$\boldsymbol{D}_O \cong \boldsymbol{D}_O^{(2)} + A_O A_L\chi_{L\text{–}O}\chi_{O\text{–}L}\boldsymbol{D}^{(2)}_{O[L][O]} + A_O A_L\chi_{L\text{–}G}\chi_{O\text{–}L}\boldsymbol{D}_{G[L][O]}\ . \tag{7.292}$$

With the same token, by substituting (7.292) and the two-wave wavefield

$$\boldsymbol{D}_G^{(2)} = A_G\chi_{G\text{–}O}\boldsymbol{D}^{(2)}_{O[G]} \tag{7.293}$$

of the G reflection into (7.290b), the approximate expression for $\boldsymbol{D}_G$ up to a second-order iteration is obtained as

$$\begin{aligned}\boldsymbol{D}_G^{(3)} &\cong A_G\chi_{G\text{–}O}\boldsymbol{D}^{(2)}_{O[G]} + A_G A_L\chi_{L\text{–}O}\chi_{G\text{–}L}\boldsymbol{D}^{(2)}_{O[L][G]} - A_G A_O A_L \\ &\qquad \times\chi_{G\text{–}O}\chi_{L\text{–}O}\chi_{O\text{–}L}\boldsymbol{D}^{(2)}_{O[L][O][G]} - A_G^2 A_L\chi_{G\text{–}O}\chi_{L\text{–}G}\chi_{G\text{–}L}\boldsymbol{D}^{(2)}_{O[G][L][G]} \\ &\equiv \boldsymbol{D}_2 + \boldsymbol{D}_{\mathrm{Um}} + \boldsymbol{D}_{\mathrm{Au1}} + \boldsymbol{D}_{\mathrm{Au2}}\ , \end{aligned} \tag{7.294}$$

where

$$\boldsymbol{D}_{O[G][L][M]} = -\left(\hat{\boldsymbol{K}}_M \times \hat{\boldsymbol{K}}_M \times \boldsymbol{D}_{O[G][L]}\right) \tag{7.295}$$

for $M = O, G$, and L. The first two terms on the right of (7.294) are due to the first-order iteration, which is equivalent to the second-order Born approximation. Hence this second-order iteration is equivalent to a third-order Born approximation.

Equation (7.294) can be rewritten as

$$\begin{aligned}\boldsymbol{D}_G^{(3)} &\cong A_G\chi_{G\text{–}O}\Big(\boldsymbol{D}^{(2)}_{O[G]} + A_L\left|\chi_{L\text{–}O}\right|\left|\chi_{G\text{–}L}\right|\left|\chi_{G\text{–}O}\right|^{-1}\exp\left(\mathrm{i}\delta_3\right)\boldsymbol{D}^{(2)}_{O[L][G]} \\ &\qquad - A_O A_L\left|\chi_{L\text{–}O}\right|\left|\chi_{O\text{–}L}\right|\boldsymbol{D}^{(2)}_{O[L][O][G]} - A_G A_L\left|\chi_{L\text{–}G}\right|\left|\chi_{G\text{–}L}\right| \\ &\qquad \times\boldsymbol{D}^{(2)}_{O[G][L][G]}\Big) \\ &\equiv \boldsymbol{D}_2 + \boldsymbol{D}_{\mathrm{Um}} + \boldsymbol{D}_{\mathrm{Au1}} + \boldsymbol{D}_{\mathrm{Au2}}\ , \end{aligned} \tag{7.296}$$

where the triplet phase invariant δ_3 is defined as

$$\delta_3 = -\delta_{G\text{–}O} + \delta_{L\text{–}O} + \delta_{G\text{–}L} \,. \tag{7.297}$$

The first term on the right of (7.296) is the two-wave contribution, the second term is due to the 'Umweganregung' process [4], and the third and fourth terms are attributed to the 'Aufhellung' process [1,3]. They are abbreviated as $\boldsymbol{D}_2$, $\boldsymbol{D}_{\mathrm{Um}}$, $\boldsymbol{D}_{\mathrm{Au1}}$, and $\boldsymbol{D}_{\mathrm{Au2}}$, respectively.

In the following, we shall concentrate on the Bragg-type three-wave diffraction, where the primary reflection G is a symmetric Bragg reflection. The condition $A_O = A_G$ can be used for the modes of propagation that make a major contribution to the diffracted intensities [52]. According to (7.62) and (7.77), the resonance terms A_G and A_L may take the forms

$$\begin{aligned} A_G &= \frac{1}{-z_G + \sqrt{z_G^2 - p_G^2 \chi_{O\text{–}G}\chi_{G\text{–}O}}} \,, \\ A_L &= \frac{1}{-z_L + \sqrt{z_L^2 + p_L^2 \chi_{O\text{–}L}\chi_{L\text{–}O}\left(\frac{\gamma_O}{\gamma_L}\right)}} \,, \end{aligned} \tag{7.298}$$

where

$$\begin{aligned} z_G &= (\Delta\theta_G)\sin 2\theta_G + \chi_O \,, \\ z_L &= \frac{\chi_O}{2}\left(1 - \frac{\gamma_L}{\gamma_O}\right) + W_\psi(\Delta\psi) + W_\theta(\Delta\theta) \,. \end{aligned} \tag{7.299}$$

p_G is the usual two-wave polarization factor, and

$$\begin{aligned} p_L^2 &= 1 - \left(\hat{\boldsymbol{\sigma}}_O \cdot \hat{\boldsymbol{K}}_L\right)^2 \,, \\ W_\psi &= \left(\hat{\boldsymbol{\sigma}}_O \cdot \hat{\boldsymbol{K}}_L\right)\cos\theta_G \,, \\ W_\theta &= \sqrt{(p_L^2 - \gamma_L^2)}\sin\theta_G + \frac{1}{2}\sin 2\theta_G\left(\frac{\gamma_L}{\gamma_G}\right) \,. \end{aligned} \tag{7.300}$$

The resonance terms in (7.298) are too complicated to be useful in simplifying the expression of I_G for the three-wave diffraction. An alternative is to assume

$$A_G = \frac{L_F(\theta_G)}{2\Delta\theta - \mathrm{i}\eta_\theta} \,, \qquad A_L = \frac{L_F(\psi)}{2\Delta\psi - \mathrm{i}\eta_\psi} \,, \tag{7.301}$$

where the Lorenz factors L_F and intrinsic widths η of the two-wave and three-wave diffractions are given by

$$\begin{aligned} L_F(\Delta\theta) &= \frac{1}{\sin 2\theta_G} \,, \qquad & \eta_\theta &= \frac{|\chi_{G\text{–}O}|}{\sin 2\theta_G} \,, \\ L_F(\Delta\psi) &= \frac{1}{W_\psi} \,, \qquad & \eta_\psi &= |\chi_O|\, L_F(\psi) \,. \end{aligned} \tag{7.302}$$

$L_F(\psi)$ has already been defined in (3.10). The position at $\Delta\theta = 0$ and $\Delta\psi = 0$ is the exact three-wave diffraction point.

This assumptions for A_G and A_L resemble the relation given in (7.286). Moreover, $A_G^* A_G$ and $A_L^* A_L$ are Lorentzian, which reflect the characteristics of the diffraction. Introducing the reduced angular parameters

$$\psi_u = \frac{2\Delta\psi}{\eta_\psi}\,, \qquad \theta_u = \frac{2\Delta\theta}{\eta_\theta}\,, \tag{7.303}$$

and using (7.301), we obtain the three-wave intensity as

$$\begin{aligned} I_G^{(3)}(\theta_u, \psi_u) = {} & \frac{d_2^2}{\theta_u^2 + 1} + \frac{2d_{2\mathrm{Um}}\left(\psi_u \cos\delta_3 - \sin\delta_3\right) + d_{\mathrm{Um}}^2}{\left(\psi_u^2 + 1\right)\left(\theta_u^2 + 1\right)} \\ & + \frac{2d_{2\mathrm{Au}}\left(\psi_u\theta_u - 1\right) + 2d_{\mathrm{UmAu}}\left(\theta_u \cos\delta_3 + \sin\delta_3\right) + d_{\mathrm{Au}}^2}{\left(\psi_u^2 + 1\right)\left(\theta_u^2 + 1\right)^2}\,, \end{aligned} \tag{7.304}$$

where

$$\begin{aligned} d_2^2 &= \left|\boldsymbol{D}_{O[G]}^{(2)}\right|^2, \\ d_{2\mathrm{Um}} &= f\left(\boldsymbol{D}_{O[G]}^{(2)} \cdot \boldsymbol{D}_{O[L][G]}^{(2)}\right), \\ d_{2\mathrm{Au}} &= f_L\left(\boldsymbol{D}_{O[G]}^{(2)} \cdot \boldsymbol{D}_{O[L][O][G]}^{(2)}\right) + f_{G\text{–}L}\left(\boldsymbol{D}_{O[G]}^{(2)} \cdot \boldsymbol{D}_{O[G][L][G]}^{(2)}\right), \\ d_{\mathrm{Um}}^2 &= f^2\left|\boldsymbol{D}_{O[L][G]}^{(2)}\right|^2, \\ d_{\mathrm{UmAu}} &= f f_L\left(\boldsymbol{D}_{O[L][G]}^{(2)} \cdot \boldsymbol{D}_{O[L][O][G]}^{(2)}\right) + f f_{G\text{–}L}\left(\boldsymbol{D}_{O[L][G]}^{(2)} \cdot \boldsymbol{D}_{O[G][L][G]}^{(2)}\right), \\ d_{\mathrm{Au}}^2 &= \left|f_L \boldsymbol{D}_{O[L][O][G]}^{(2)} + f_{G\text{–}L}\boldsymbol{D}_{O[G][L][G]}^{(2)}\right|^2, \\ f &= \frac{|\chi_{L\text{–}O}|\,|\chi_{G\text{–}L}|}{|\chi_{G\text{–}O}|\,|\chi_O|} \equiv \frac{|F_{L\text{–}O}|\,|F_{G\text{–}L}|}{|F_{G\text{–}O}|\,|F_O|}\,, \\ f_L &= \frac{|\chi_{L\text{–}O}|\,|\chi_{-L}|}{|\chi_{G\text{–}O}|\,|\chi_O|} \equiv \frac{|F_{L\text{–}O}|\,|F_{-L}|}{|F_{G\text{–}O}|\,|F_O|}\,, \\ f_{G\text{–}L} &= \frac{|\chi_{L\text{–}G}|\,|\chi_{G\text{–}L}|}{|\chi_{G\text{–}O}|\,|\chi_O|} \equiv \frac{|F_{L\text{–}G}|\,|F_{G\text{–}L}|}{|F_{G\text{–}O}|\,|F_O|}\,. \end{aligned} \tag{7.305}$$

The semi-integral over the beam-divergence in θ_u,

$$I_G^{(3)}(\psi_u) = \int_{-\infty}^{\infty} I_G^{(3)}(\theta_u, \psi_u)\,\mathrm{d}\theta_u\,, \tag{7.306}$$

leads to

$$\begin{aligned} I_G^{(3)}(\psi_u) &= I_G^{(2)} + I_D(\delta_3, \psi_u) + I_K(\psi_u) \\ &= \pi d_2^2 + \frac{\pi\left(A\psi_u \cos\delta_3 - B\sin\delta_3\right)}{\psi_u^2 + 1} + \frac{\pi C}{\psi_u^2 + 1}\,, \end{aligned} \tag{7.307}$$

where

$$A = 2d_{2\mathrm{Um}} , \qquad B = 2d_{2\mathrm{Um}} - d_{\mathrm{UmAu}} ,$$
$$C = d_{\mathrm{Um}}^2 + \frac{d_{\mathrm{Au}}^2}{2} - d_{2\mathrm{Au}} . \tag{7.308}$$

$I_G^{(2)}$ is the two-wave intensity and I_D and I_K are the phase-dependent and phase-independent intensities, respectively. This expression facilitates the analysis of the phase of a reflection on a quantitative basis. Moreover, the polarization also affects the the three-wave diffraction intensity. The explicit forms of the polarization factors involved in the scalar products of (7.305) are given in (8.113) of Sect. 8.3.

7.2.5 Approach Using Takagi-Taupin Equation

The expression for the diffracted intensity I_G obtained from the various approximation approaches based on the plane-wave dynamical theory is usually not valid at the exact N-wave diffraction position. The calculated intensity diverges at this exact point when the phase shift Δ is not introduced into the formulation. This is, however, not the case when the Takagi-Taupin equation [128–131] is employed for a finite crystal. As already described in (6.222), the imaginary parts of wavefields are taken into account explicitly in the Takagi-Taupin equation. Thus, the sine function of δ_3 should, in principle, automatically emerge from the Takagi-Taupin equation.

Thorkildsen and coworkers adopted the formulation of the Takagi-Taupin equation and applied it to deal with the three-wave diffraction in finite [327] and semi-infinite perfect crystals [513–516]. This theoretical approach using the boundary-value Green function is described below.

The Takagi-Taupin equation (6.229) can be rewritten as

$$\frac{\partial D_{\sigma G}}{\partial x_G} = 2\pi \mathrm{i} k \beta_G D_{\sigma G} + \mathrm{i} \sum_{\pi M \neq \sigma G} \kappa_{GM}^{\sigma,\pi} D_{\pi M} , \tag{7.309}$$

where σG and πM represent all possible polarized components and the summation is taken over all the σ and π components. The quantity $\kappa_{GM}^{\sigma,\pi}$ is defined as

$$\kappa_{GM}^{\sigma,\pi} = -\pi k \chi_{G-M} \left(\hat{\boldsymbol{\sigma}}_G \cdot \hat{\boldsymbol{\pi}}_M \right) . \tag{7.310}$$

The deviation parameter $k\beta_G$ can be expressed [517] as

$$k\beta_G \cong K_G - K' = k\alpha_G - k\zeta\gamma_G - \frac{1}{2}k\chi_O , \tag{7.311}$$

where $k\alpha_G = k_G - k$ is the excitation error of the G reflection. β_G and α_G are defined in (6.225) and (6.72), respectively. The square of K' is equal to $k^2(1+\chi_O)$ defined in (6.49). ζ is the Anpassung, and γ_G is the direction cosine of $\boldsymbol{K}_G$ with respect to the inward surface normal of the entrance surface.

By performing a unitary transformation

$$D_{\sigma G} = \bar{D}_{\sigma G} \exp\left(2\pi \mathrm{i} \sum_M k\beta_M x_M\right) , \tag{7.312}$$

the Takagi-Taupin equation is reduced to

$$\frac{\partial \bar{D}_{\sigma G}}{\partial x_G} = \mathrm{i} \sum_{\pi M \neq \sigma G} \kappa_{GM}^{\sigma,\pi} \bar{D}_{\pi M} , \tag{7.313}$$

where x_G is a positional coordinate along $\hat{\boldsymbol{K}}_G$.

Since the unitary transformation treats the Lorentz point as the excitation point for the wavefields, what we are dealing with is the wavefields at the exact three-wave diffraction position. For simplicity, the polarization coupling among the wavefields is neglected.

The Takagi-Taupin equations for a three-wave (O, G, L) diffraction are

$$\begin{aligned}
\frac{\partial \bar{D}_O}{\partial x_O} &= \mathrm{i}\kappa_{OG}\bar{D}_G + \mathrm{i}\kappa_{OL}\bar{D}_L , \\
\frac{\partial \bar{D}_G}{\partial x_G} &= \mathrm{i}\kappa_{GO}\bar{D}_O + \mathrm{i}\kappa_{GL}\bar{D}_L , \\
\frac{\partial \bar{D}_L}{\partial x_L} &= \mathrm{i}\kappa_{LO}\bar{D}_O + \mathrm{i}\kappa_{LG}\bar{D}_G ,
\end{aligned} \tag{7.314}$$

where the subscript and superscript indicating the polarization have been dropped out. The solutions to these partial differential equations can be obtained by integration. For convenience, let's define a linear operator $\mathcal{L}_{GL}$ of scattering as

$$\mathcal{L}_{GL}\bar{D}_L(x_O, x_G, x_L) = \mathrm{i}\kappa_{GL} \int_{x_G^b}^{x_G} \mathrm{d}x'_G \bar{D}_L(x_O, x'_G, x_L) , \tag{7.315}$$

where the superscript b indicates a boundary point. Equation (7.314) thus becomes

$$\begin{aligned}
\bar{D}_O &= \bar{D}_O^{(b)} + \mathcal{L}_{OG}\bar{D}_G + \mathcal{L}_{OL}\bar{D}_L , \\
\bar{D}_G &= \bar{D}_G^{(b)} + \mathcal{L}_{GO}\bar{D}_O + \mathcal{L}_{GL}\bar{D}_L , \\
\bar{D}_L &= \bar{D}_L^{(b)} + \mathcal{L}_{LO}\bar{D}_O + \mathcal{L}_{LG}\bar{D}_G .
\end{aligned} \tag{7.316}$$

The boundary values for a given point source S with the wavefield-amplitude $D_O^{(e)}$ along $\boldsymbol{k}_O$ at the entrance surface are governed by the boundary conditions

$$\begin{aligned}
\bar{D}_O^{(b)}(S) &= \bar{D}_O\left[x_O^b(x_G, x_L), x_G, x_L\right] \\
&= D_O^{(e)}\delta\left[x_G - x_G(S)\right]\delta\left[x_L - x_L(S)\right] , \\
\bar{D}_G^{(b)}(S) &= \bar{D}_G\left[x_O, x_G^b(x_O, x_L), x_L\right] = 0 , \\
\bar{D}_L^{(b)}(S) &= \bar{D}_L\left[x_O, x_G, x_L^b(x_O, x_G)\right] = 0 ,
\end{aligned} \tag{7.317}$$

where $\delta(x)$ is a Dirac delta function.

The solutions to this boundary-value problem can be found by employing Laplace transformation or series expansion. Following Thorkildsen's approach [513], the usually employed series expansions [141, 518] are adopted as the solutions to (7.316), i.e.,

$$\bar{D}_M\left(x_O, x_G, x_L\right) = D_O^{(e)} \sum_{n=0}^{\infty} \varepsilon^n d_M^{(n)}\left(x_O, x_G, x_L\right) \tag{7.318}$$

for $M = O, G, L$, where ε is a counting variable reflecting the number of scattering events. The quantity d_M is related to the wavefield-amplitude. The final expression for $\bar{D}$'s is for $\varepsilon \to 1$.

By substituting (7.318) into (7.316), the coefficients of equal power of ε give the following recurrent relations for $n \geq 1$:

$$\begin{aligned} d_O^{(n)} &= \mathcal{L}_{OG} d_G^{(n-1)} + \mathcal{L}_{OL} d_L^{(n-1)} , \\ d_G^{(n)} &= \mathcal{L}_{GO} d_O^{(n-1)} + \mathcal{L}_{GL} d_L^{(n-1)} , \\ d_L^{(n)} &= \mathcal{L}_{LO} d_O^{(n-1)} + \mathcal{L}_{LG} d_G^{(n-1)} . \end{aligned} \tag{7.319}$$

Together with the boundary conditions for d's:

$$d_O^{(0)} = \delta\left(x_G\right) \delta\left(x_L\right) , \qquad d_G^{(0)} = d_L^{(0)} = 0 , \tag{7.320}$$

the amplitude $d_G^{(n)}$ to any order n can be calculated, reflecting the routes of scattering taken by multiple reflection occurring for the G diffraction. This is equivalent to the route consideration in Kato's statistical theory [141, 142].

In the following, three-wave diffraction with and without absorption are considered separately.

(i) Non-absorbing crystals

The solution $\bar{D}_G(x_O, x_G, x_L)$ is the amplitude of the G diffracted wave at (x_O, x_G, x_L) due to the source at the origin, i.e., $S = (0, 0, 0)$. For a non-absorbing crystal, the amplitude $\bar{D}_G$ at a point $P = (x_O(P), x_G(P), x_L(P))$ due to the same source at S is given by

$$\begin{aligned} D_G(P \leftarrow S) = {} & J_c \bar{D}_G\left(\Pi_O, \Pi_G, \Pi_L\right) \exp\left(2\pi \mathrm{i} k \alpha_G \Pi_G\right) \\ & \times \exp\left(2\pi \mathrm{i} k \alpha_L \Pi_L\right) \Theta\left(\Pi_O\right) \Theta\left(\Pi_G\right) \Theta\left(\Pi_L\right) , \end{aligned} \tag{7.321}$$

where $k\beta_G$ has been replaced by $k\alpha_G$ without considering the absorption effects in (7.311). The Jacobian J_c is a geometrical factor resulting from the transformation of the source Dirac field density function from one coordinate system to another system. The parameter Π_M for $M = O, G$, and L is defined as

$$\Pi_M = x_M(P) - x_M(S) . \tag{7.322}$$

The Heaviside unit step function $\Theta(\Pi_M)$ adopted for the wavefields is zero outside the pyramid formed by unit vectors, $\hat{\boldsymbol{K}}_O$, $\hat{\boldsymbol{K}}_G$, and $\hat{\boldsymbol{K}}_L$ with a top at the source point S.

The resultant wavefield of the G diffracted wave at an exit point P due to the incoming plane-wave is a superposition of the amplitudes from all the source points, S, on the entrance surface [327, 518–520]:

$$D_G(P) = \int \mathrm{d}\boldsymbol{S} \cdot \hat{\boldsymbol{K}}_O D_G(P \leftarrow S) , \tag{7.323}$$

where $\boldsymbol{S}$ is the surface vector representing the entrance surface. The diffracted intensity then takes the form

$$I_G(P) = \frac{c}{4\pi} |D_G(P)|^2 . \tag{7.324}$$

Consequently, the power of the diffracted wave G due to the contribution from all the points on the exit surface, P, is

$$P_G = \int \mathrm{d}\boldsymbol{P} \cdot \hat{\boldsymbol{K}}_G I_G . \tag{7.325}$$

The integrated power over the effective beam divergence, i.e., the horizontal divergence ϵ_1 [267], can be calculated accordingly as

$$\mathcal{P}_G = \int_{-\infty}^{\infty} \mathrm{d}\epsilon_1 P_G \left[\alpha_G\left(\epsilon_1\right)\right]\Big|_{\alpha_L} , \tag{7.326}$$

where the excitation error α_G depends on the Lorentz factor $1/\sin 2\theta_G$,

$$\alpha_G = \left(\frac{\sin 2\theta_G}{\lambda}\right) \epsilon_1 , \tag{7.327}$$

while α_L of the secondary reflection L is considered as almost constant.

Two crystal geometries for the Laue-Laue and the Bragg-Laue diffraction are considered for mathematical simplicity. The crystals are all confined by the unit vectors $\hat{\boldsymbol{K}}_O$, $\hat{\boldsymbol{K}}_G$, and $\hat{\boldsymbol{K}}_L$ as shown in Figs. 7.9(a) and (b). The sizes of the crystal are $\ell_O \times \ell_G \times \ell_L$. The geometric factor J_c for both cases is

$$J_c = \frac{1}{V_K} , \tag{7.328}$$

where the volume V_K spanned by the unit vectors $\hat{\boldsymbol{K}}_O$, $\hat{\boldsymbol{K}}_G$, and $\hat{\boldsymbol{K}}_L$ is

$$\begin{aligned} V_K = \{1 &- \cos^2 2\theta_G - \cos^2 2\theta_L - \cos^2 2\theta_{GL} \\ &+ 2\cos 2\theta_G \cos 2\theta_L \cos 2\theta_{GL}\}^{\frac{1}{2}} , \end{aligned} \tag{7.329}$$

$2\theta_{GL}$ being the angle between $\hat{\boldsymbol{K}}_G$ and $\hat{\boldsymbol{K}}_L$.

For the Laue-Laue case, where all the diffractions are Laue transmission, the entrance surface S is $OACB$ and the exit surface P is $BCVU$ (Fig. 7.9(a)). The parameters Π's are given, according to Fig. 7.9(a), by

$$\begin{aligned} \Pi_O &= x_O(P) , \\ \Pi_G &= \ell_G - x_G(S) , \\ \Pi_L &= x_L(P) - x_L(S) . \end{aligned} \tag{7.330}$$

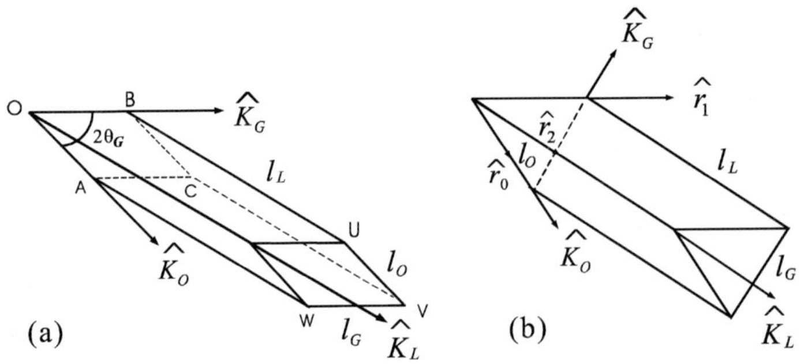

Fig. 7.9. Crystal geometries: (**a**) Laue-Laue case and (**b**) Bragg-Laue case [513]

The wavefield at P due to the incoming wave and the diffracted power at the exit surface for the G reflection are, respectively,

$$\begin{aligned} D_G(P) &= V_K \int_0^{\ell_G} \mathrm{d}\Pi_G \int_0^{x_L(P)} \mathrm{d}\Pi_L D_G(P \leftarrow S) \,, \\ P_G &= V_K \int_0^{\ell_O} \mathrm{d}x_O(P) \int_0^{\ell_L} \mathrm{d}x_L(P) I_G(P) \,. \end{aligned} \tag{7.331}$$

In the Bragg-Laue case, G is a Bragg reflection and L a Laue transmission. For convenience, the coordinate system of $\hat{\boldsymbol{r}}_0$, $\hat{\boldsymbol{r}}_1$, and $\hat{\boldsymbol{r}}_2$ is used, where $\hat{\boldsymbol{r}}_0$ and $\hat{\boldsymbol{r}}_2$ coincide with $\hat{\boldsymbol{K}}_O$ and $\hat{\boldsymbol{K}}_L$, respectively. The unit vector $\hat{\boldsymbol{r}}_1$ bisects the angle $2\theta_G$ between $\hat{\boldsymbol{K}}_O$ and $\hat{\boldsymbol{K}}_G$. The geometrical relation among the $\hat{\boldsymbol{r}}$'s and $\hat{\boldsymbol{K}}$'s is shown in Fig. 7.9(b). The transformation from $\hat{\boldsymbol{K}}$ to $\hat{\boldsymbol{r}}$ given by

$$\begin{aligned} x_O &= r_O + \frac{r_1}{2\cos\theta_G} \,, \\ x_G &= \frac{r_1}{2\cos\theta_G} \,, \\ x_L &= r_2 \,, \end{aligned} \tag{7.332}$$

leads to

$$\begin{aligned} \Pi_O = \Pi_G &= \frac{1}{2\cos\theta_G} \{r_1(P) - r_1(S)\} \equiv \frac{\Delta_1}{2\cos\theta_G} \,, \\ \Pi_L &= r_2(P) - r_2(S) \equiv \Delta_2 \,. \end{aligned} \tag{7.333}$$

The wavefield-amplitude at P for an incident plane wave and the diffracted power at the exit surface are then formulated as

$$\begin{aligned} D_G(P) &= \left(\frac{V_K}{2\cos\theta_G}\right) \int_0^{r_1(P)} \mathrm{d}\Delta_1 \int_0^{r_2(P)} \mathrm{d}\Delta_2 D_G(P \leftarrow S) \,, \\ P_G &= \left(\frac{V_K}{2\cos\theta_G}\right) \int_0^{2\ell_O \cos\theta_G} \mathrm{d}r_1(P) \int_0^{\ell_L} \mathrm{d}r_2(P) I_G(P) \,. \end{aligned} \tag{7.334}$$

The following summarizes for the Laue-Laue and Bragg-Laue cases the expressions of the functions $d^{(n)}$ up to a sixth-order perturbation, the diffracted power P_G, the integrated power $\mathcal{P}_G$, and the change in the integrated power relative to the two-wave diffraction power:

Laue-Laue case:
The $d^{(n)}$ functions are

$$
\begin{aligned}
d_G^{(1)} &= \mathrm{i}\kappa_{GO}\delta(x_L)\ , \\
d_G^{(2)} &= \mathrm{i}\kappa_{GO}\Gamma_0\left(\mathrm{i}\cos\delta_3 - \sin\delta_3\right)\ , \\
d_G^{(3)} &= -\mathrm{i}\kappa_{GO}\left[\Pi_O\Pi_G u\delta\left(x_L\right) + \Pi_O v + \Pi_G w\right]\ , \\
d_G^{(4)} &= \mathrm{i}\kappa_{GO}\Gamma_0\left\{\left[\Pi_G\Pi_O u + \Pi_L\Pi_O v + \Pi_L\Pi_G w\right]\sin\delta_3\right. \\
&\quad \left. -\,\mathrm{i}\left[3\Pi_G\Pi_O u + \Pi_L\Pi_O v + \Pi_L\Pi_G w\right]\cos\delta_3\right\}\ , \\
d_G^{(5)} &= \mathrm{i}\kappa_{GO}\left\{\frac{1}{4}\Pi_G^2\Pi_O^2u^2\delta\left(x_L\right) + \Pi_G\Pi_O^2uv + \frac{1}{2}\Pi_L\Pi_O^2v^2\right. \\
&\quad + \Pi_G^2\Pi_O uw + 3\Pi_L\Pi_G\Pi_O vw + \frac{1}{2}\Pi_L\Pi_G^2w^2 \\
&\quad \left. + \,\Pi_L\Pi_G\Pi_O vw\cos 2\delta_3 + \mathrm{i}\Pi_L\Pi_G\Pi_O vw\sin 2\delta_3\right\}\ , \\
d_G^{(6)} &= -\mathrm{i}\kappa_{GO}\Gamma_0\left\{\left[\frac{1}{4}\Pi_G^2\Pi_O^2u^2 + \Pi_L\Pi_G\Pi_O^2uv + \frac{1}{4}\Pi_L^2\Pi_O^2v^2\right.\right. \\
&\quad \left. + \,\Pi_L\Pi_G^2\Pi_O uw + \Pi_L^2\Pi_G\Pi_O vw + \frac{1}{4}\Pi_L^2\Pi_G^2w^2\right]\sin\delta_3 \\
&\quad - \mathrm{i}\left[\frac{5}{4}\Pi_G^2\Pi_O^2u^2 + 3\Pi_L\Pi_G\Pi_O^2uv + \frac{1}{4}\Pi_L^2\Pi_O^2v^2\right. \\
&\quad \left.\left. + \,3\Pi_L\Pi_G^2\Pi_O uw + \Pi_L^2\Pi_G\Pi_O vw + \frac{1}{4}\Pi_L^2\Pi_G^2w^2\right]\cos\delta_3\right\}.
\end{aligned}
\tag{7.335}
$$

The diffraction power is

$$
\begin{aligned}
P_G = I_O\left|\kappa_{GO}\right|^2 V_L\ell_G &\left\{2f_1\left(\xi_G'\right) - \left|\eta_{OG}\right|^2 f_1\left(\xi_G'\right) - 4\left(\frac{\left|\eta_{GL}\right|\left|\eta_{LO}\right|}{\eta_{GO}}\right)\right. \\
&\times\left[f_2\left(\xi_G'\right)\cos\delta_3 + f_1\left(\xi_G'\right)\sin\delta_3\right] f_1\left(\xi_G'\right) - 2\left|\eta_{LO}\right|^2 f_1\left(\xi_L'\right) f_1\left(\xi_G'\right) \\
&- 2\left|\eta_{GL}\right|^2\left[f_1\left(\xi_L'\right) f_1\left(\xi_G'\right) + f_2\left(\xi_L'\right) f_4\left(\xi_G'\right)\right] \\
&\left. + 4\left(\frac{\left|\eta_{GL}\right|^2\left|\eta_{LO}\right|^2}{\left|\eta_{GO}\right|^2}\right) f_3\left(\xi_L'\right) f_1\left(\xi_G'\right)\right\}.
\end{aligned}
\tag{7.336}
$$

The relative change in integrated diffraction power is

$$
\begin{aligned}
\frac{\Delta\mathcal{P}_G\left(\xi_L'\right)}{\mathcal{P}_G^{(2)}} &= \frac{\mathcal{P}_G\left(\xi_L'\right) - \mathcal{P}_G^{(2)}}{\mathcal{P}_G^{(2)}} \\
&= -2\left(\frac{\left|\eta_{GL}\right|\left|\eta_{LO}\right|}{\left|\eta_{GO}\right|}\right)\left[f_2\left(\xi_L'\right)\cos\delta_3 + f_1\left(\xi_L'\right)\sin\delta_3\right] - \left(\left|\eta_{LO}\right|^2\right. \\
&\quad \left. + \left|\eta_{GL}\right|^2\right) f_1\left(\xi_L'\right) + 2\left(\frac{\left|\eta_{GL}\right|^2\left|\eta_{LO}\right|^2}{\left|\eta_{GO}\right|^2}\right) f_3\left(\xi_L'\right)\ .
\end{aligned}
\tag{7.337}
$$

The notations used are defined as

$$u = \kappa_{OG}\kappa_{GO} , \qquad v = \kappa_{OL}\kappa_{LO} , \qquad w = \kappa_{GL}\kappa_{LG} ,$$

$$\Gamma_0 = \frac{|\kappa_{GL}|\,|\kappa_{LO}|}{|\kappa_{GO}|} ,$$

$$\delta_3 = \delta_{G\text{–}L} + \delta_{L\text{–}O} - \delta_{G\text{–}O} ,$$

$$I_O = \frac{c}{4\pi}\left|D_O^{(e)}\right|^2 ,$$

$$|\eta_{GL}|^2 = |\kappa_{GL}|^2 \ell_G \ell_L ,$$

$$\xi'_G = 2\pi\alpha_G\ell_G , \tag{7.338}$$

where I_O is the intensity of the incident wave and V_L is the volume of the crystal.

The functions f's employed are

$$f_1(u) = \left(\frac{1}{u^2}\right)(1-\cos u) ,$$

$$f_2(u) = \left(\frac{1}{u}\right)\left[1-\left(\frac{1}{u}\right)\sin u\right] ,$$

$$f_3(u) = \left(\frac{1}{u^2}\right)\left[1-\left(\frac{1}{u}\right)\sin u\right] ,$$

$$f_4(u) = \left(\frac{1}{u^2}\right)\left[\sin u-\left(\frac{2}{u}\right)(1-\cos u)\right] ,$$

$$f_5(u) = \left(\frac{1}{3u^5}\right)\left(-12u+u^3-12u\cos u+24\sin u-3u^2\sin u\right) ,$$

$$f_6(u) = \left(\frac{1}{2u^4}\right)\left(2+u^2-2\cos u-2u\sin u\right) ,$$

$$f_7(u) = \left(\frac{1}{u^4}\right)(-2u-u\cos u+3\sin u) . \tag{7.339}$$

The integrated power of the G reflection in the two-wave diffraction is given by

$$\mathcal{P}_G^{(2)} = I_O v\,|\kappa_{GO}|^2\left(\frac{\lambda}{\sin 2\theta_G}\right) . \tag{7.340}$$

Bragg-Laue case:
The $d^{(n)}$ functions are

$$d_G^{(1)} = \mathrm{i}\kappa_{GO}\delta(x_L) ,$$

$$d_G^{(2)} = \mathrm{i}\kappa_{GO}\Gamma_0\,(\mathrm{i}\cos\delta_3 - \sin\delta_3) ,$$

$$
\begin{aligned}
d_G^{(3)} &= -\mathrm{i}\kappa_{GO}\left[\frac{1}{2}\Pi_G^2 u\delta\left(x_L\right)+\Pi_G v+\Pi_G w\right] , \\
d_G^{(4)} &= \mathrm{i}\kappa_{GO}\Gamma_0\left\{\left[\frac{1}{2}\Pi_G^2 u+\Pi_G\Pi_L v+\Pi_G\Pi_L w\right]\sin\delta_3\right. \\
&\qquad \left. -\mathrm{i}\left[\frac{3}{2}\Pi_G^2 u+\Pi_G\Pi_L v+\Pi_G\Pi_L w\right]\cos\delta_3\right\} , \\
d_G^{(5)} &= \mathrm{i}\kappa_{GO}\left\{\frac{1}{12}\Pi_G^4 u^2\delta\left(x_L\right)+\frac{1}{2}\Pi_G^3 uv+\frac{1}{2}\Pi_G^2\Pi_L v^2\right. \\
&\qquad +\frac{1}{2}\Pi_G^3 uw+2\Pi_G^2 vw+\frac{1}{2}\Pi_G^2\Pi_L w^2 \\
&\qquad \left. +\frac{1}{2}\Pi_G^2\Pi_L vw\cos 2\delta_3+\frac{\mathrm{i}}{2}\Pi_G^2\Pi_L vw\sin 2\delta_3\right\} , \\
d_G^{(6)} &= -\mathrm{i}\kappa_{GO}\Gamma_0\left\{\left[\frac{1}{12}\Pi_G^4 u^2+\frac{1}{2}\Pi_G^3\Pi_L uv+\frac{1}{4}\Pi_G^2\Pi_L^2 v^2\right.\right. \\
&\qquad \left. +\frac{1}{2}\Pi_G^3\Pi_L uw+\frac{3}{4}\Pi_G^2\Pi_L^2 vw+\frac{1}{4}\Pi_G^2\Pi_L^2 w^2\right]\sin\delta_3 \\
&\qquad -\mathrm{i}\left[\frac{5}{12}\Pi_G^4 u^2+\frac{3}{2}\Pi_G^3\Pi_L uv+\frac{1}{4}\Pi_G^2\Pi_L^2 v^2\right. \\
&\qquad \left.\left. +\frac{3}{2}\Pi_G^3\Pi_L uw+\frac{3}{4}\Pi_G^2\Pi_L^2 vw+\frac{1}{4}\Pi_G^2\Pi_L^2 w^2\right]\cos\delta_3\right\} .
\end{aligned}
\tag{7.341}
$$

The diffraction power is

$$
\begin{aligned}
P_G = 2I_O\left|K_{GO}\right|^2 V_B\ell_O &\left\{2f_3\left(\xi_G'\right)-\left|\eta_{OG}\right|^2 f_5\left(\xi_G'\right)\right. \\
&-4\left(\frac{\left|\eta_{GL}\right|\left|\eta_{LO}\right|}{\left|\eta_{GO}\right|}\right)\left[f_2\left(\xi_L'\right)\cos\delta_3+f_1\left(\xi_L'\right)\sin\delta_3\right]f_3\left(\xi_G'\right) \\
&-2\left(\left|\eta_{LO}\right|^2+\left|\eta_{GL}\right|^2\right)\left[f_1\left(\xi_L'\right)f_6\left(\xi_G'\right)-f_2\left(\xi_L'\right)f_7\left(\xi_G'\right)\right] \\
&\left. +2\left(\frac{\left|\eta_{GL}\right|^2\left|\eta_{LO}\right|^2}{\left|\eta_{GO}\right|^2}\right)f_3\left(\xi_L'\right)f_3\left(\xi_G'\right)\right\} .
\end{aligned}
\tag{7.342}
$$

The relative change in the integrated diffraction power is

$$
\begin{aligned}
\frac{\Delta\mathcal{P}_G\left(\xi_L'\right)}{\mathcal{P}_G^{(2)}} &= -2\left(\frac{\left|\eta_{GL}\right|\left|\eta_{LO}\right|}{\left|\eta_{GO}\right|}\right)\left[f_2\left(\xi_L'\right)\cos\delta_3+f_1\left(\xi_L'\right)\sin\delta_3\right] \\
&\quad -\frac{2}{3}\left(\left|\eta_{LO}\right|^2+\left|\eta_{GL}\right|^2\right)f_1\left(\xi_L'\right)+\left(\frac{\left|\eta_{GL}\right|^2\left|\eta_{LO}\right|^2}{\left|\eta_{GO}\right|^2}\right)f_3\left(\xi_L'\right) .
\end{aligned}
\tag{7.343}
$$

The crystal volume is denoted by V_B.

All the relative changes in integrated diffraction power, given in (7.337) and (7.343), for the Laue-Laue and Bragg-Laue cases consist of phase-dependent and phase-independent parts. The Aufhellung contribution (the second term) and the Umweganregung contribution (the third term) are the two components of the phase-independent part. This feature is similar to that given in (7.296) and (7.307). Moreover, that both the Aufhellung and Umweg are represented by symmetric functions with respect to $u = 0$ is consistent with the symmetric function ψ_u for the phase-independent intensity profile of (7.307) in the quantitative approach.

(ii) Absorbing crystals
The $k\beta_G$ given in (7.311) without approximation is used as the magnitude of the wavevector of the G reflection inside a given absorbing crystal. By considering the absorption of the incident wave O, the first equation of the boundary conditions (7.317) is replaced by

$$\begin{aligned} \bar{D}_O^{(b)} &= D_O^{(e)} \exp\left[-2\pi i k \alpha_G x_G(S)\right] \exp\left[-2\pi i k \alpha_L x_L(S)\right] \\ &\quad \times \exp\left\{i\pi k \chi_O \left[x_O(S) + x_G(S) + x_L(S)\right]\right\} . \end{aligned} \tag{7.344}$$

Moreover, anomalous dispersion corrections in the real and the imaginary atomic scattering factors, f' and f'', are included in the electric susceptibilities, $\chi_O/4\pi$ and $\chi_{G\text{–}L}/4\pi$, which give rise to an additional phase shift Δ to $\delta_{G\text{–}L}$ in the boundary-value Green functions d'_G for all the involved G and L reflections.

The solution for the wavefield at a point P in the crystal is

$$\begin{aligned} D_M(P) = J_c D_O^{(e)} \int \mathrm{d}S d'_M\left(\Pi_O, \Pi_G, \Pi_L\right) \exp\left[2\pi i k\left(\alpha'_O \Pi_O + \alpha'_G \Pi_G\right.\right. \\ \left.\left. + \, \alpha'_L \Pi_L\right)\right] \exp\left[-\left(\frac{\mu}{2}\right)\left(\Pi_O + \Pi_G + \Pi_L\right)\right] , \end{aligned} \tag{7.345}$$

where the effective excitation error α'_M is defined as

$$\alpha'_M = \alpha_M - \frac{k\chi_{O,r}}{2} , \tag{7.346}$$

and the linear absorption coefficient μ takes its usual form:

$$\mu = -2\pi k \chi_{O,i} . \tag{7.347}$$

Now the Friedel's law is no longer valid due to absorption, namely,

$$|\kappa_{GL}| \neq |\kappa_{LG}| , \qquad \delta_{G\text{–}L} \neq \delta_{L\text{–}G} . \tag{7.348}$$

Following the same procedure of derivation, the diffraction powers are given, up to a third order, for the Laue-Laue case as

$$\begin{aligned} P_G = P_L^0 \Bigg\{ & g_1\left(\xi'_{O,i}\right) g_2\left(\xi'_{G,r}, \xi'_{G,i}\right) \left\{ 1 - 2\left(\frac{|\eta_{GL}|\,|\eta_{LO}|}{|\eta_{GO}|}\right) \right. \\ & \left. \times \left[g_4\left(\xi'_{L,r}, \xi'_{L,i}\right) \cos\left(\delta_3 - \Delta_G\right)\right] \right\} + \left(\frac{|\eta_{GL}|^2\,|\eta_{LO}|^2}{|\eta_{GO}|^2}\right) \\ & \times g_1\left(\xi'_{O,i}\right) g_5\left(\xi'_{L,r}, \xi'_{L,i}\right) g_2\left(\xi'_{G,r}, \xi'_{G,i}\right) + 2\,|\eta_{LO}|\,|\eta_{OL}| \end{aligned}$$

$$\times g_6\left(\xi'_{O,i}\right) g_2\left(\xi'_{G,r},\xi'_{G,i}\right)\left[-g_3\left(\xi'_{L,r},\xi'_{L,i}\right)\cos\Delta_L\right.$$
$$\left.+\,g_4\left(\xi'_{L,r},\xi'_{L,i}\right)\sin\Delta_L\right]+2\left|\eta_{LG}\right|\left|\eta_{GL}\right| g_1\left(\xi'_{O,i}\right)$$
$$\times\Big\{\left[g_4\left(\xi'_{L,r},\xi'_{L,i}\right) g_8\left(\xi'_{G,r},\xi'_{G,i}\right)-g_3\left(\xi'_{L,r},\xi'_{L,i}\right) g_7\left(\xi'_{G,r},\xi'_{G,i}\right)\right]$$
$$\times\cos\Delta_{G\text{–}L}+\Big[g_4\left(\xi'_{L,r},\xi'_{L,i}\right) g_7\left(\xi'_{G,r},\xi'_{G,i}\right)$$
$$+\,g_3\left(\xi'_{L,r},\xi'_{L,i}\right) g_8\left(\xi'_{G,r},\xi'_{G,i}\right)\Big]\sin\Delta_{G\text{–}L}\Big\}\Bigg\}\,, \tag{7.349}$$

and for the Bragg-Laue case as

$$P_G=P_B^0\Bigg\{g_9\left(\xi'_{G,r},\xi'_{G,i}\right)\Big\{1-2\left(\frac{\left|\eta_{GL}\right|\left|\eta_{LO}\right|}{\left|\eta_{GO}\right|}\right)\Big[g_4\left(\xi'_{L,r},\xi'_{L,i}\right)$$
$$\times\cos\left(\delta_3-\Delta_G\right)+g_3\left(\xi'_{L,r},\xi'_{L,i}\right)\sin\left(\delta_3-\Delta_G\right)\Big]\Big\}$$
$$+\left(\frac{\left|\eta_{GL}\right|^2\left|\eta_{LO}\right|^2}{\left|\eta_{GO}\right|^2}\right) g_5\left(\xi'_{L,r},\xi'_{L,i}\right) g_9\left(\xi'_{G,r},\xi'_{G,i}\right)+2\left|\eta_{LO}\right|\left|\eta_{OL}\right|$$
$$\times\left\{\left[g_4\left(\xi'_{L,r},\xi'_{L,i}\right) g_{10}\left(\xi'_{G,r},\xi'_{G,i}\right)-g_3\left(\xi'_{L,r},\xi'_{L,i}\right) g_{11}\left(\xi'_{G,r},\xi'_{G,i}\right)\right]\right.$$
$$\times\cos\Delta_L+\left[g_3\left(\xi'_{L,r},\xi'_{L,i}\right) g_{10}\left(\xi'_{G,r},\xi'_{G,i}\right)\right.$$
$$\left.\left.+\,g_4\left(\xi'_{L,r},\xi'_{L,i}\right) g_{11}\left(\xi'_{G,r},\xi'_{G,i}\right)\right]\sin\Delta_L\right\}$$
$$+2\left|\eta_{LG}\right|\left|\eta_{GL}\right|\left\{\left[g_4\left(\xi'_{L,r},\xi'_{L,i}\right) g_{10}\left(\xi'_{G,r},\xi'_{G,i}\right)\right.\right.$$
$$\left.-\,g_3\left(\xi'_{L,r},\xi'_{L,i}\right) g_{11}\left(\xi'_{G,r},\xi'_{G,i}\right)\right]\cos\Delta_{G\text{–}L}$$
$$+\left[g_3\left(\xi'_{L,r},\xi'_{L,i}\right) g_{10}\left(\xi'_{G,r},\xi'_{G,i}\right)+g_4\left(\xi'_{L,r},\xi'_{L,i}\right) g_{11}\left(\xi'_{G,r},\xi'_{G,i}\right)\right]$$
$$\left.\times\sin\Delta_{G\text{–}L}\right\}\Bigg\}\,. \tag{7.350}$$

The terms involving $\left|\eta_{GO}\right|^2$ have been left out in (7.349) and (7.350), that amounts to neglect the contribution to the primary extinction. The quantities P_L^0 and P_B^0 are the kinematical diffraction powers of the two-wave Laue and Bragg reflections, respectively. The parameters $\xi'_{G,r}$ and $\xi'_{G,i}$ are the real and imaginary parts of ξ'_G, defined as

$$\xi'_{G,r}=2\pi\alpha'_G\ell_G\,,\qquad \xi'_{G,i}=\mu\ell_G\,. \tag{7.351}$$

The additional phase shifts due to anomalous dispersion are given by

$$\begin{aligned}\Delta_G&=\delta_{G\text{–}O}+\delta_{O\text{–}G}\,,\\ \Delta_L&=\delta_{L\text{–}O}+\delta_{O\text{–}L}\,,\\ \Delta_{G\text{–}L}&=\delta_{G\text{–}L}+\delta_{L\text{–}G}\,,\end{aligned} \tag{7.352}$$

which have values nearly equal to zero for photon energies far from the absorption edges. But they have appreciable values at the absorption edges. The functions g's are

$$
\begin{aligned}
g_1(u) &= \frac{1-\exp(-2u)}{2u} \, , \\
g_2(u,v) &= \frac{1+\exp(-2v)-2\exp(-v)\cos u}{u^2+v^2} \, , \\
g_3(u,v) &= \frac{1}{\left(u^2+v^2\right)^2}\left[-v^2+v^3+u^2+u^2v-\left(u^2-v^2\right)\exp(-v)\cos u\right. \\
&\quad \left. -\,2uv\exp(-v)\sin u\right] \, , \\
g_4(u,v) &= \frac{1}{\left(u^2+v^2\right)^2}\left[u\left(-2v+u^2+v^2\right)+2uv\exp(-v)\cos u\right. \\
&\quad \left. +\left(-u^2+v^2\right)\exp(-v)\sin u\right] \, , \\
g_5(u,v) &= \frac{1}{2v\left(u^2+v^2\right)^2}\left[-3v^2+2v^3+u^2+2u^2v-\left(u^2+v^2\right)\right. \\
&\quad \left. \times\exp(-2v)+4v^2\exp(-v)\cos u-4uv\exp(-v)\sin u\right] \, , \\
g_6(u) &= \frac{1-(1+2u)\exp(-2u)}{4u^2} \, , \\
g_7(u,v) &= \frac{v+\left(v+u^2+v^2\right)\exp(-2v)-\left(2v+u^2+v^2\right)\exp(-v)\cos u}{\left(u^2+v^2\right)^2} \, , \\
g_8(u,v) &= \frac{1}{\left(u^2+v^2\right)^2}\left[u+u\exp(-2v)\right. \\
&\quad \left. -\,2u\exp(-v)\cos u-\left(u^2+v^2\right)\exp(-v)\sin u\right] \, , \\
g_9(u,v) &= \frac{1}{4v\left(u^2+4v^2\right)^2}\left[-12v^2+16v^3+u^2+4u^2v-\left(u^2+4v^2\right)\right. \\
&\quad \left. \times\exp(-4v)+16v^2\exp(-2v)\cos u-8uv\exp(-2v)\sin u\right] \, , \\
g_{10}(u,v) &= \frac{1}{4v\left(u^2+4v^2\right)^3}\left[-u\left(u^2+4v^2\right)\exp(-4v)+u\left(-28v^2+16v^3\right.\right. \\
&\quad \left. +\,4u^2v+u^2\right)+4uv\left(8v+u^2+4v^2\right)\exp(-2v)\cos u+4v \\
&\quad \left. \times\left(4v^2+8v^3-3u^2+2u^2v\right)\exp(-2v)\sin u\right] \, , \\
g_{11}(u,v) &= \frac{1}{16v^2\left(u^2+4v^2\right)^3}\left[-144v^4+128v^5+32u^2v^2+32u^2v^3+u^4\right. \\
&\quad -\left(u^2+4v^2\right)\left(12v^2+16v^3+u^2+4u^2v\right)\exp(-4v) \\
&\quad -\,16v^2\left(-12v^2-8v^3+u^2-2u^2v\right)\exp(-2v)\cos u \\
&\quad \left. -\,16uv^2\left(8v+4v^2+u^2\right)\exp(-2v)\sin u\right] \, . \qquad (7.353)
\end{aligned}
$$

If the polarization coupling is considered, then the Takagi-Taupin equations for a multiple-wave case are written as

$$
\frac{\partial \bar{D}_{\sigma G}}{\partial x_G} = \mathrm{i}\sum_{M\neq G}\kappa_{GM}^{\sigma,\sigma}\bar{D}_{\sigma M}+\mathrm{i}\sum_{M\neq G}\kappa_{GM}^{\sigma,\pi}\bar{D}_{\pi M} \, ,
$$

$$\frac{\partial \bar{D}_{\pi G}}{\partial x_G} = \mathrm{i} \sum_{M \neq G} \kappa_{GM}^{\pi,\sigma} \bar{D}_{\sigma M} + \mathrm{i} \sum_{M \neq G} \kappa_{GM}^{\pi,\pi} \bar{D}_{\pi M} \; , \tag{7.354}$$

for the σ- and π- components of all the involved G reflections. The summation is taken over the rest of reflections.

If the absorption effects are excluded, the relative change of the integrated diffraction power of the G reflection can be expressed as

$$\begin{aligned}\frac{\Delta \mathcal{P}_G(\xi_L)}{\mathcal{P}_G^{(2)}} =& -2\left(\frac{|\eta_{GL}|\,|\eta_{LO}|}{|\eta_{GO}|}\right) p_1 \left[f_2(\xi_L)\cos\delta_3 + f_1(\xi_L)\sin\delta_3\right] \\ &- \left(|\eta_{GL}|^2 p_2 + |\eta_{LO}|^2 p_3\right) f_1(\xi_L) \\ &+ 2\left(\frac{|\eta_{GL}|^2\,|\eta_{LO}|^2}{|\eta_{GO}|^2}\right) p_4 f_3(\xi_L)\end{aligned} \tag{7.355}$$

with appropriate polarization factors p's according to the diffraction geometry and experimental conditions.

7.2.6 Expanded Distorted-Wave Born Approximation

The distorted-wave Born approximation (DWBA) for scattering is a well known method to treat nuclear collisions and related phenomena quantum mechanically [510,521,522]. This approximation has been adopted to describe the grazing incidence X-ray diffraction and scattering(GIXD) from surfaces and interfaces, where the electric susceptibility of the scattering medium, considered initially as an average homogeneous substance, is represented by its zeroth-order Fourier component of the incident wave O. The scattering from this homogeneous medium with a distorted component gives rise to a distorted wave, which is governed by the usual Fresnel optical theory. This distorted wave is then re-scattered by surface features, such as roughness, reconstructed structure, and atomic over-layer. The Born approximation is employed to handle the re-scattered wave.

For a three-wave (O, G, L) diffraction, the distorted component can be expanded to include the Fourier components of the Bragg reflection G, which is, as usual, a sinusoidal function. In other words, the two-wave (O, G) diffraction generates a distorted wave which is then re-scattered by the atomic planes of the L reflection via a multiple-wave process. The behavior of this re-scattered wave can, in principle, be described in the same way as for the grazing incidence scattering. This idea was proposed by Shen to account for the reference-beam technique for the phase determination of structure-factor triplets [314]. The details of this approach is given below:

Let's rewrite the fundamental equation (7.184) in its integral form as

$$\begin{aligned}\boldsymbol{D}(\boldsymbol{r}) = \boldsymbol{D}^{(0)}(\boldsymbol{r}) + \frac{1}{4\pi}\int \mathrm{d}\boldsymbol{r}' \frac{\exp\left(-2\pi \mathrm{i} k_O \left|\boldsymbol{r}-\boldsymbol{r}'\right|\right)}{\left|\boldsymbol{r}-\boldsymbol{r}'\right|} \\ \times \nabla' \times \nabla' \times \left[\chi(\boldsymbol{r}')\boldsymbol{D}(\boldsymbol{r}')\right] \; ,\end{aligned} \tag{7.356}$$

where the electric susceptibility χ, following the distorted-wave approximation for GIXD [166], is composed of a distorting component χ_1 and the remaining part χ_2:

$$\chi(\boldsymbol{r}) = \chi_1(\boldsymbol{r}) + \chi_2(\boldsymbol{r}) \,. \tag{7.357}$$

In contrast to the GIXD case where $\chi_1 = \chi_O$ of the O wave, the distorting component χ_1 for the three-wave diffraction is given by

$$\chi_1(\boldsymbol{r}) = \chi_O + \chi_{G\text{–}O} \exp(-2\pi \mathrm{i} \boldsymbol{g} \cdot \boldsymbol{r}) + \chi_{O\text{–}G} \exp(2\pi \mathrm{i} \boldsymbol{g} \cdot \boldsymbol{r}) \tag{7.358}$$

to include the components due to the two-wave diffraction. The remaining χ_2, taking care of the contribution of the additional L reflection, is

$$\chi_2(\boldsymbol{r}) = - \sum_{L \neq O, \pm G} \chi_{L\text{–}O}\, \mathrm{e}^{-2\pi \mathrm{i} \boldsymbol{\ell} \cdot \boldsymbol{r}} \,. \tag{7.359}$$

The solution to (7.356) can be written by following the conventional distorted-wave Born approximation as

$$\begin{aligned} \boldsymbol{D}(\boldsymbol{r}) = \boldsymbol{D}_1(\boldsymbol{r}) + \frac{1}{4\pi} \int \mathrm{d}\boldsymbol{r}' \frac{\exp\left(-2\pi \mathrm{i} k_O \left|\boldsymbol{r} - \boldsymbol{r}'\right|\right)}{|\boldsymbol{r} - \boldsymbol{r}'|} \\ \times \nabla' \times \nabla' \times \left[\chi_2(\boldsymbol{r}') \boldsymbol{D}_1(\boldsymbol{r}')\right] \,, \end{aligned} \tag{7.360}$$

where the first term $\boldsymbol{D}_1(\boldsymbol{r})$ on the right is the distorted wave due only to χ_1, and the second term is related to the scattering of $\boldsymbol{D}_1(\boldsymbol{r})$ due to χ_2.

$\boldsymbol{D}_1(\boldsymbol{r})$ can be determined by solving

$$\left(\nabla^2 + k_O^2\right) \boldsymbol{D}_1 = -\nabla \times \nabla \times \left(\chi_1 \boldsymbol{D}_1\right) \,. \tag{7.361}$$

The solution is the usual two-wave wavefield

$$\boldsymbol{D}_1(\boldsymbol{r}) = \boldsymbol{D}_O \exp\left(-2\pi \mathrm{i} \boldsymbol{K}_O \cdot \boldsymbol{r}\right) + \boldsymbol{D}_G \exp\left(-2\pi \mathrm{i} \boldsymbol{K}_G \cdot \boldsymbol{r}\right) \,, \tag{7.362}$$

where $\boldsymbol{D}_G$, derived from (7.64), (7.78), and (7.79) by setting $C_{\sigma O} = C_{\sigma G} = 0$, $\chi_{OG,OL} = \chi_{O\text{–}G}$ and $\chi_{GO,LO} = \chi_{G\text{–}O}$, is given in

$$\frac{D_G}{D_O} = \frac{-b}{\sqrt{|b|}\,|p_G|} \sqrt{\left|\frac{\chi_{G\text{–}O}}{\chi_{O\text{–}G}}\right|} \exp^{\mathrm{i}\delta_{G\text{–}O}} \left[z'_G \mp \sqrt{z'^2_G + \frac{b}{|b|}}\,\right] \,, \tag{7.363}$$

with

$$z'_G = \frac{z}{\sqrt{|b| \chi_{G\text{–}O} \chi_{O\text{–}G}}} = \frac{b \Delta\theta \sin 2\theta_G + \frac{\chi_O (1-b)}{2}}{\sqrt{|b| \chi_{G\text{–}O} \chi_{O\text{–}G}}\,|p_G|} \,, \tag{7.364}$$

b being γ_O/γ_G. The smaller absolute ratio of (7.363) is chosen for $\boldsymbol{D}_G$. $|p_G|$ is the polarization factor of the G-reflection.

The expanded distorted-wave solution, resulting from the scattering of $\boldsymbol{D}_1(\boldsymbol{r})$ of (7.362) due to χ_2, can be derived from the second term on the right side of (7.360). Following the Born approximation given in the previous section, the total scattered wave of the L reflection, calculated from (7.360), takes the form

$$\boldsymbol{D}_L(\boldsymbol{r}) = k_O^2 V_o \hat{\boldsymbol{r}} \times [\hat{\boldsymbol{r}} \times (\chi_{L\text{-}O}\boldsymbol{D}_O + \chi_{L\text{-}G}\boldsymbol{D}_G)] \frac{\exp(-2\pi \mathrm{i} k_O r)}{r} , \quad (7.365)$$

where the amplitudes $\boldsymbol{D}_O$ and $\boldsymbol{D}_G$ of (7.362), which depend on the extinction length, have been factored out of the volume integral, because the amplitudes are slowly varying functions in space compared with the phase factors $\boldsymbol{K}_O \cdot \boldsymbol{r}$ and $\boldsymbol{K}_G \cdot \boldsymbol{r}$.

Equation (7.365) consists of a singly scattered wave ($O \to L$), i.e., the diffracted wave of the L reflection, and a doubly scattered wave ($O \to G \to L$), i.e., the detoured diffracted wave via the primary (reference) reflection G and the coupling $L - G$ reflection. Normalizing (7.365) to the two-wavefield of the L reflection yields the first-order Born wavefield

$$\boldsymbol{D}_L^{(1)} = k_O^2 \chi_{G\text{-}O} V_o \hat{\boldsymbol{r}} \times (\hat{\boldsymbol{r}} \times \boldsymbol{D}_O) \frac{\exp(-2\pi \mathrm{i} k_O r)}{r} . \quad (7.366)$$

Equation (7.365) then reduces to

$$\boldsymbol{D}_L(\boldsymbol{r}) = D_L^{(1)} [D_O + \tau D_G \exp(\mathrm{i}\delta_3)] , \quad (7.367)$$

where τ and the triplet phase δ_3 are

$$\tau = \left| \frac{\chi_{L\text{-}G}}{\chi_{L\text{-}O}} \right| , \qquad \delta_3 = \delta_{G\text{-}O} + \delta_{L\text{-}G} - \delta_{L\text{-}O} . \quad (7.368)$$

The integrated intensity over the crystal thickness T for a crystal plate, normalized to the two-wave kinematical intensity $|D_L^{(1)}|^2$, is

$$I_L = \frac{1}{T} \int_0^T |D_O(z) + \tau D_G(z) \exp(\mathrm{i}\delta_3)|^2 \, \mathrm{d}z , \quad (7.369)$$

where z is the depth from the crystal entrance surface.

If the three-wave diffraction is a Laue-Laue case with a symmetric Laue diffraction G, then the boundary conditions (7.95) hold. The total distorted waves with the wavevectors equal to the averaged vectors of the involved two modes for a single polarizatin, i.e.,

$$\boldsymbol{K}_O = \frac{\boldsymbol{K}_O(1) + \boldsymbol{K}_O(2)}{2} ,$$
$$\boldsymbol{K}_G = \frac{\boldsymbol{K}_G(1) + \boldsymbol{K}_G(2)}{2} , \quad (7.370)$$

are

$$D_O(z) = D_O^{(e)} \cos \frac{(\Delta k) z}{2} - \mathrm{i} \frac{z'_G}{\sqrt{z'^2_G + 1}} \sin \frac{(\Delta k) z}{2} ,$$
$$D_G(z) = D_O^{(e)} \frac{\mathrm{i}}{\sqrt{b}} \frac{1}{\sqrt{z'^2_G + 1}} \sin \frac{(\Delta k) z}{2} , \quad (7.371)$$

where

$$\Delta k = k \, |\chi_{G\text{-}O}| \frac{\sqrt{z'^2_G + 1}}{\sqrt{\gamma_O \gamma_G}} . \quad (7.372)$$

The integrated intensity (7.369) is

$$\frac{I_L}{I_O^{(e)}} = \frac{1}{2}\left[1+\frac{\sin(T\Delta k)}{T\Delta k}\right] - \frac{\tau\sin\delta_3}{2\sqrt{b}\sqrt{z_G'^2+1}}\frac{[1-\cos(T\Delta k)]}{T\Delta k}$$
$$+\frac{\tau^2+4bz_G'^2-4\sqrt{b}z_G'^2\tau\cos\delta_3}{8b\left(z_G'^2+1\right)}\left[1-\frac{\sin(T\Delta k)}{T\Delta k}\right] . \tag{7.373}$$

If the primary reflection G is a symmetric Bragg reflection and the crystal is much thicker than the extinction length, then the boundary conditions (7.107) and (7.108) hold for a plane-parallel crystal plate with a thickness T. The wavefield amplitudes of the distorted waves are

$$D_O(z) = D_O^{(e)}\left\{\frac{z_G'\sin\left[\frac{\Delta k(T-z)}{2}\right]+\mathrm{i}\sqrt{z_G'^2-1}\cos\left[\frac{\Delta k(T-z)}{2}\right]}{z_G'\sin\left(\frac{\Delta kT}{2}\right)+\mathrm{i}\sqrt{z_G'^2-1}\cos\left(\frac{\Delta kT}{2}\right)}\right\} ,$$

$$D_G(z) = D_O^{(e)}\left\{\frac{-\sqrt{|b|}\sin\left[\frac{\Delta k(T-z)}{2}\right]}{z_G'\sin\left(\frac{\Delta kT}{2}\right)+\mathrm{i}\sqrt{z_G'^2-1}\cos\left(\frac{\Delta kT}{2}\right)}\right\} , \tag{7.374}$$

where

$$\Delta k = k\left|\chi_{G-O}\right|\frac{\sqrt{z_G'^2-1}}{\sqrt{\gamma_O\left|\gamma_G\right|}} . \tag{7.375}$$

The angular range for z_G' in $\sqrt{z_G'^2-1}$ could be either $|z_G'|>1$ or $|z_G'|<1$. Following a similar procedure to that for (7.112), we obtain the integrated intensities with respect to the incident intensity $I_O^{(e)}$ for these two ranges as

$$\frac{I_L}{I_O^{(e)}} = \frac{2}{|\zeta_o|^2|b|}\left(z_G'^2|b|+\tau^2-2z_G'\tau\sqrt{|b|}\cos\delta_3\right)\left[1-\frac{\sin(\Delta kT)}{\Delta kT}\right]$$
$$+\frac{2\left(z_G'^2-1\right)}{|\zeta_o|^2}\times\left[1+\frac{\sin(\Delta kT)}{\Delta kT}\right]$$
$$+\frac{4\tau}{|\zeta_o|^2\sqrt{|b|}}\sqrt{z_G'^2-1}\sin\delta_3\frac{\cos(\Delta kT)-1}{\Delta kT} \tag{7.376}$$

for $|z_G'|>1$, and

$$\frac{I_L}{I_O^{(e)}} = \frac{\exp(2\Delta k'T)-1-2\Delta k'T\exp(\Delta k'T)}{\Delta k'T|\zeta_o'|^2|b|}\left(\tau^2+z_G'^2|b|\right.$$
$$\left.-2z_G'\tau\sqrt{|b|}\cos\delta_3\right)+\frac{\exp(2\Delta k'T)-1-2\Delta k'T\exp(\Delta k'T)}{\Delta k'T|\zeta_o'|^2}$$
$$\times\left(1-z_G'^2\right)-\frac{2\left(\exp(\Delta k'T)-1\right)^2}{\Delta k'T|\zeta_o'|^2\sqrt{|b|}}\tau\sqrt{1-z_G'^2}\sin\delta_3 \tag{7.377}$$

for $|z'_G| < 1$, with

$$|\zeta_o|^2 = 4z'^2_G - 2[1 + \cos(T\Delta k)] ,$$
$$|\zeta'_o|^2 = \exp(2\Delta k'T) + \left(2 - 4z'^2_G\right)\exp(\Delta k'T) + 1 ,$$
$$\Delta k' = k\,|\chi_{G-O}|\,\frac{\sqrt{1 - z'^2_G}}{\sqrt{\gamma_O\,|\gamma_G|}} . \tag{7.378}$$

This expanded DWBA approach has recently been shown to be equivalent to the approach of using Takagi-Taupin equation [523].

7.2.7 Resonance Perturbation Bethe Approximation

Three-wave multiple diffraction at the absorption edges for a singly polarized incident wave has been described in the previous section by using the Takagi-Taupin equation without considering the polarization coupling. Since polarization is one of the important factors affecting the diffracted intensities of a multiple-wave diffraction, a formulation for resonance three-wave diffraction with the consideration of polarization is needed to interpret the intensity change of multiply diffracted waves. The so-called resonance perturbation Bethe approach (RPB) has recently been developed by Stetsko et al. to account for the measured phase shift due to resonance [323].

The resonance perturbation Bethe approach starts with employing (7.290a) and (7.290b) for a three-wave (O, G, L) diffraction. From (7.290b), the wavefield $\boldsymbol{D}^{(3)}_G$ of the diffracted wave G in the three-wave situation, is given by

$$\boldsymbol{D}^{(3)}_G = \frac{\chi_{G-O}\boldsymbol{D}_{O[G]} + A_L\chi_{G-L}\chi_{L-O}\boldsymbol{D}_{O[L][G]}}{\mathcal{A}_G}$$
$$= \frac{\left(\chi_{G\ O}\boldsymbol{p}_{G(2)} + A_L\chi_{G-L}\chi_{L-O}\boldsymbol{p}_{G(d)}\right)D_O}{\mathcal{A}_G} , \tag{7.379}$$

where $\boldsymbol{D}_O = D_O\boldsymbol{p}$ is the incident wave with the vector $\boldsymbol{p}$ in the direction of either $\hat{\boldsymbol{\sigma}}_O$ or $\hat{\boldsymbol{\pi}}_O$. The perturbed term $\mathcal{A}_G$ related to the reciprocal of the resonance terms for a symmetric Bragg reflection is

$$\mathcal{A}_G = A_G^{-1} - A_L\chi_{G-L}\chi_{L-G}p^2_{GL} , \tag{7.380}$$

where A_G and A_L are the ordinary resonance terms given in (7.288), and p^2_{GL} is defined as

$$p^2_{GL} = 1 - \left(\hat{\boldsymbol{K}}_L \cdot \hat{\boldsymbol{\sigma}}\right)^2 \tag{7.381}$$

and

$$p^2_{GL} = 1 - \left(\hat{\boldsymbol{K}}_L \cdot \hat{\boldsymbol{\pi}}_G\right)^2 \tag{7.382}$$

for the σ and π polarized incident wave, respectively. The polarization vectors $\boldsymbol{p}_{G(2)}$ and $\boldsymbol{p}_{G(d)}$ are

$$\boldsymbol{p}_{G(2)} = -\hat{\boldsymbol{K}}_G \times \left(\hat{\boldsymbol{K}}_G \times \boldsymbol{p}\right) ,$$
$$\boldsymbol{p}_{G(d)} = \hat{\boldsymbol{K}}_G \times \left[\hat{\boldsymbol{K}}_G \times \left(\hat{\boldsymbol{K}}_G \times \boldsymbol{p}\right)\right] . \tag{7.383}$$

Thus, the wavefield $\boldsymbol{D}_G^{(3)}$ of the primary G reflection in a three-wave (O, G, L) diffraction situation is the sum of the two-wave $\boldsymbol{D}_G^{(2)}$ and the detoured wavefield $\boldsymbol{D}_G^{(d)}$ via the doubly reflections L and $G - L$:

$$\boldsymbol{D}_G^{(3)} = \boldsymbol{D}_G^{(2)} + \boldsymbol{D}_G^{(d)} , \tag{7.384}$$

where

$$\boldsymbol{D}_G^{(2)} = \frac{\chi_{G\text{–}O}\boldsymbol{p}_{G(2)} D_O}{\mathcal{A}_G} ,$$
$$\boldsymbol{D}_G^{(d)} = \frac{A_L \chi_{G\text{–}L}\chi_{L\text{–}O}\boldsymbol{p}_{G(d)} D_O}{\mathcal{A}_G} . \tag{7.385}$$

The perturbed term $\mathcal{A}_G$ can be determined according to the Bethe approximation from the perturbed dispersion equation derived from (7.385). For a general three-wave diffraction with an asymmetric Bragg reflection G as the primary reflection, $\mathcal{A}_G$ is given by

$$\mathcal{A}_G = -z_G'' + \sqrt{z_G''^2 + \left(\frac{\gamma_G}{\gamma_O}\right) \chi_{G\text{–}O}\chi_{O\text{–}G} p_+^p p_-^p} , \tag{7.386}$$

where

$$p_+^p = \boldsymbol{p}_+ \cdot \boldsymbol{p} , \qquad p_-^p = \boldsymbol{p}_- \cdot \boldsymbol{p} ,$$
$$\boldsymbol{p}_\pm = \boldsymbol{p}_{G(2)} + A_L \chi_\pm \boldsymbol{p}_{G(d)} . \tag{7.387}$$

The vector $\boldsymbol{p}$ is the polarization unit vector of the incident wave and $\chi_\pm$ are the Fourier components of the triplets defined as

$$\chi_+ = \frac{\chi_{G\text{–}L}\chi_{L\text{–}O}}{\chi_{G\text{–}O}} , \qquad \chi_- = \frac{\chi_{L\text{–}G}\chi_{O\text{–}L}}{\chi_{O\text{–}G}} . \tag{7.388}$$

The function z_G'' of the angular deviation $\Delta\theta_G$ from the Bragg angle θ_G, defined in (7.78), is rewritten as

$$z_G'' = (\Delta\theta_G)\sin 2\theta_G + \frac{\chi_O\left(1 - \frac{\gamma_G}{\gamma_O}\right)}{2} + \frac{A_L\left(-\frac{\gamma_G}{\gamma_O}\chi_{L\text{–}O}\chi_{O\text{–}L}p_L^2 + \chi_{G\text{–}L}\chi_{L\text{–}G}p_{GL}^2\right)}{2} , \tag{7.389}$$

where

$$p_L^2 = 1 - \left(\hat{\boldsymbol{K}}_L \cdot \boldsymbol{p}\right)^2 . \tag{7.390}$$

The resonance term A_L can be defined approximately by the fundamental equation of the two-wave (O, L) diffraction, i.e.,

$$\boldsymbol{D}_O = A_O \chi_{O\text{–}L} \boldsymbol{D}_{L[O]} , \qquad \boldsymbol{D}_L = A_L \chi_{L\text{–}O} \boldsymbol{D}_{O[L]} , \tag{7.391}$$

where the L is generally an asymmetric reflection with respect to the surface of the crystal in a three-wave diffraction. Under such a circumstance, the polarization vector $\boldsymbol{p}$ of the incident wave O is the vector sum of its components projected onto the polarization unit vectors $\hat{\boldsymbol{\sigma}}_L$ and $\hat{\boldsymbol{\pi}}_L$ of the L reflection,

$$\boldsymbol{p} = \alpha \hat{\boldsymbol{\sigma}}_{O,L} + \beta \hat{\boldsymbol{\pi}}_{O,L} , \tag{7.392}$$

where

$$\hat{\boldsymbol{\sigma}}_{O,L} = \hat{\boldsymbol{K}}_O \times \hat{\boldsymbol{K}}_L , \quad \hat{\boldsymbol{\pi}}_{O,L} = \hat{\boldsymbol{K}}_O \times \hat{\boldsymbol{\sigma}}_L , \quad \alpha^2 + \beta^2 = 1 . \tag{7.393}$$

The wavefield $\boldsymbol{D}_L$ of the L reflection is then

$$\boldsymbol{D}_L = \chi_{G\text{–}O} D_O \left(\alpha p_{L,\sigma} \hat{\boldsymbol{\sigma}}_L A_{L,\sigma} + \beta p_{L,\pi} \hat{\boldsymbol{\pi}}_L A_{L,\pi} \right) , \tag{7.394}$$

where the unit vectors are

$$\hat{\boldsymbol{\sigma}}_L = \hat{\boldsymbol{\sigma}}_{O,L} , \qquad \hat{\boldsymbol{\pi}}_L = \hat{\boldsymbol{K}}_L \times \hat{\boldsymbol{\sigma}}_L , \tag{7.395}$$

and the polarization factors are

$$\begin{aligned} p_{L,\sigma} &= \left| \hat{\boldsymbol{K}}_L \times \left(\hat{\boldsymbol{K}}_L \times \hat{\boldsymbol{\sigma}}_{O,L} \right) \right| = 1 , \\ p_{L,\pi} &= \left| \hat{\boldsymbol{K}}_L \times \left(\hat{\boldsymbol{K}}_L \times \hat{\boldsymbol{\pi}}_{O,L} \right) \right| = \cos 2\theta_L . \end{aligned} \tag{7.396}$$

The resonance term $A_{L,p}$ of (7.394) is defined as

$$A_{L,p} = \frac{1}{-z_L + \sqrt{z_L^2 + \left(\frac{\gamma_L}{\gamma_O} \right) \chi_{L\text{–}O} \chi_{O\text{–}L} p_{L,p}^2}} \tag{7.397}$$

for $p = \sigma$ and π, where z_L has been given in (7.299). With the kinematical approximation for the polarization factors and unit vectors

$$\begin{aligned} & p_{L,\sigma} \cong p_{L,\pi} \cong p_L = \left| \hat{\boldsymbol{K}}_L \times \left(\hat{\boldsymbol{K}}_L \times \boldsymbol{p} \right) \right| , \\ & A_{L,\sigma} \cong A_{L,\pi} , \\ & \alpha p_{L,\sigma} \hat{\boldsymbol{\sigma}}_L + \beta p_{L,\pi} \hat{\boldsymbol{\pi}}_L \cong -\hat{\boldsymbol{K}}_L \times \left(\hat{\boldsymbol{K}}_L \times \boldsymbol{p} \right) = \boldsymbol{p}_{L(2)} , \end{aligned} \tag{7.398}$$

the wavefield $\boldsymbol{D}_L$ reduces to

$$\boldsymbol{D}_L = A_L \chi_{G\text{–}O} D_O \boldsymbol{p}_{L(2)} , \tag{7.399}$$

where

$$A_L = \frac{1}{-z_L + \sqrt{z_L^2 + \left(\frac{\gamma_L}{\gamma_O} \right) \chi_{L\text{–}O} \chi_{O\text{–}L} p_L^2}} . \tag{7.400}$$

In both (7.386) and (7.400), the square roots with positive imaginary parts are chosen.

The integrated intensity of the G reflection at a given azimuth angle $\Delta\psi_G$ deviated from the exact three-wave point in the three-wave diffraction condition is then proportional to

$$I_G^{(3)}(\Delta\psi_G) \propto \int \left| \boldsymbol{D}_G^{(3)*} \cdot \boldsymbol{D}_G^{(3)} \right| \mathrm{d}\,(\Delta\theta_G) \ , \tag{7.401}$$

where the integration is over the angular deviation $\Delta\theta_G$ from the Bragg angle θ_G. This intensity, related to χ_+ and χ_-, depends therefore on the triplet phases

$$\begin{aligned} \delta_{3+} &= -\delta_{G\text{–}O} + \delta_{L\text{–}O} + \delta_{G\text{–}L} \ , \\ \delta_{3-} &= -\delta_{O\text{–}G} + \delta_{O\text{–}L} + \delta_{L\text{–}G} \end{aligned} \tag{7.402}$$

as well as the resonance phase shifts defined as

$$\begin{aligned} \Delta_G &= \delta_{G\text{–}O} + \delta_{O\text{–}G} \ , \qquad \Delta_L = \delta_{L\text{–}O} + \delta_{O\text{–}L} \ , \\ \Delta_{G\text{–}L} &= \delta_{G\text{–}L} + \delta_{L\text{–}G} \ , \qquad \Delta_3 = \delta_{3+} + \delta_{3-} \ . \end{aligned} \tag{7.403}$$

These resonance phases usually have appreciable values for X-ray photon energies near the absorption edges of the constituent atoms of a crystal.

8. Dynamical Diffraction Properties and Behaviors

The characteristics and behaviors of the dynamical aspects of multiple-wave diffraction will be described in the light of theoretical considerations and experimental observations. A brief summary of two-wave diffraction in symmetric and asymmetric geometries, including grazing incidence and back scattering, will be given first. The diffractions of three-, four-, and higher-order multiple-wave situations follow. The polarization effect on the diffracted intensity profile under various experimental conditions will be specifically addressed. The so-called 'crystal truncation rod diffraction' from surfaces and interfaces will be discussed in this chapter as well.

8.1 Two-Wave Diffraction

Two-wave diffraction is a special case of multiple-wave diffraction. Its fundamental equation of wavefield with two reciprocal lattice points involved follows (6.37a) and (6.37b) and the dispersion relation is given in (6.56). There are, in general, $4N$ dispersion sheets with $N = 2$ for the two-wave diffraction. The number of dispersion sheets, or tie points, reduces from 8 to 4 for diffractions with wide angles of incidence and emergence. Boundary conditions and expressions for diffracted intensities of this kind of diffraction can be found in the section of Bethe approximation (Sect. 7.2.1), except that the quantities involving the L reflection need to be set to zero. Many excellent books, including the recently published one by Authier [55], have given a thorough description of two-wave Bragg and two-wave Laue diffractions under various geometric conditions. Here we shall only mention the essential dynamical properties and behaviors briefly, and thus only the diffraction from perfect crystals will be considered. For extremely asymmetric diffractions where the contribution of grazing incident or emergent waves to the wavefields is appreciable, a detailed description of the equation of dispersion and boundary conditions will be given.

8.1.1 Wide-Angle Two-Wave Diffraction

Consider a two-wave (O, G) diffraction, in which both the angle of incidence and the angle of emergence are much larger than the critical angle of total external reflection. Both σ_O- and σ_G-polarized components are perpendicular to the π_O and π_G components. According to (7.73), the dispersion equations for both σ- and π-polarized wavefields take the same form as

$$4\xi_{O\pi}\xi_{G\pi} - P_\pi^2 \chi_{O\text{-}G}\chi_{G\text{-}O} = 0 \,, \tag{8.1}$$

where $P_\pi = \cos 2\theta_G$ for the π polarization. For the σ polarization, the subscripts π change to σ and $P_\sigma = 1$. The surface of the Ewald sphere in the vicinity of the Lorentz point is assumed to be flat. This assumption is justified in the wide angle diffraction situation. Since all the σ-polarized wavefields are perpendicular to all the π-polarized ones, there is no coupling between the σ- and π-polarized wavefields. By following the same derivation for (7.76), (8.1) becomes

$$\xi_{O\pi}^2 + \xi_{O\pi} b\left[(\Delta\theta)\sin 2\theta_G + \frac{\chi_O}{2}(1-b)\right] - \frac{1}{4} b P_\pi^2 \chi_{O\text{-}G}\chi_{G\text{-}O} = 0 \,, \tag{8.2}$$

where

$$\left.\begin{array}{l}\xi_{O\pi}(1)\\ \xi_{O\pi}(2)\end{array}\right\} = \frac{1}{2}\left(z_\pi \mp \sqrt{z_\pi^2 + q_\pi}\right) \,, \tag{8.3}$$

with

$$\begin{aligned} z_\pi &= -b\left[(\Delta\theta)\sin 2\theta_G + \frac{\chi_O}{2}(1-b)\right] \,, \\ q_\pi &= b P_\pi^2 \chi_{O\text{-}G}\chi_{G\text{-}O} \,. \end{aligned} \tag{8.4}$$

The wavefield-amplitude ratios of the two π modes are

$$\begin{aligned} \frac{D_{\pi G}(1)}{D_{\pi O}(1)} &= -\frac{b P_\pi \chi_{G\text{-}O}}{z_\pi + \sqrt{z_\pi^2 + q_\pi}} \,, \\ \frac{D_{\pi G}(2)}{D_{\pi O}(2)} &= -\frac{b P_\pi \chi_{G\text{-}O}}{z_\pi - \sqrt{z_\pi^2 + q_\pi}} \,. \end{aligned} \tag{8.5}$$

The expressions given above hold for a general Laue diffraction and a Bragg reflection in a thin crystal. The difference between a Laue and a Bragg case is mainly in the sign of the asymmetry factor $b = \gamma_O/\gamma_G$. That is, $b > 0$ for Laue cases and $b < 0$ for Bragg cases. The corresponding scattering matrices are therefore hermitian and non-hermitian, respectively. If a two-wave Bragg reflection from an infinitely thick crystal is considered, there is only one permitted mode for each polarization. The wavefield-amplitude ratio is the one with the smallest absolute value.

The expressions, from (7.95) to (7.179), of diffracted intensities and integrated powers for different crystal thicknesses and absorption conditions given in Sect. 7.2.1 are valid for two-wave Bragg and Laue situations, except that all the quantities C_{σ_O}, C_{π_O}, C_{σ_G}, and C_{π_G} and the electric susceptibilities related to the L reflection and its coupling $G - L$ are set to zero.

As an illustration, the dispersion surface, linear absorption coefficient, mode excitation, wavefield intensity and diffracted intensities of the symmetric two-wave (000)(111) diffraction of germanium with CuKα_1 radiation

($\lambda = 1.540562\,\text{Å}$) are calculated dynamically for both Laue and Bragg cases, where (000) and (111) are the O and G reflections. The calculated results are given below:

(i) Symmetric Laue diffraction ($b = 1$) for a plane-parallel crystal of thickness T_0: The calculated dispersion surface, linear absorption coefficient, mode excitation, and effective linear absorption coefficient as a function of $\Delta\theta_G$ are shown in Fig. 8.1. $\Delta\theta$ is the angular deviation from the Bragg angle θ_G. There are four permitted modes for the two-wave case. Because there is no correlation between the σ- and π-polarized wavefields, the corresponding dispersion curves are separated from each other. The dispersion curves 1 and 3 are the σ branches, and curves 2 and 4 the π branches (Fig. 8.1(a)). The Laue point La is located at $k\zeta_r = 0$ and $\Delta\theta = 0$, and the Lorentz point Lo is at the midpoint of the pairs of dispersion hyperbolae along $\Delta\theta = 0$. The minimum value of the linear absorption coefficient among the four modes is $106\,\text{cm}^{-1}$, which follows (6.166) and (6.167) for the Debye parameter $B = 0.56\,\text{Å}^2$ at room temperature. The average linear absorption coefficient over the four modes is $352\,\text{cm}^{-1}$ for all the $\Delta\theta$, which is the one-wave value of the (000) reflection (Fig. 8.1(b)). The excitations of modes 1 and 2 are nearly the same, which have a maximum of 0.50% on the lower angle side and a minimum of 0% on the higher angle side (Fig. 8.1(c)). The situation is reversed for modes 3 and 4. At the exact two-wave position, $\Delta\theta = 0$, all the four modes are equally

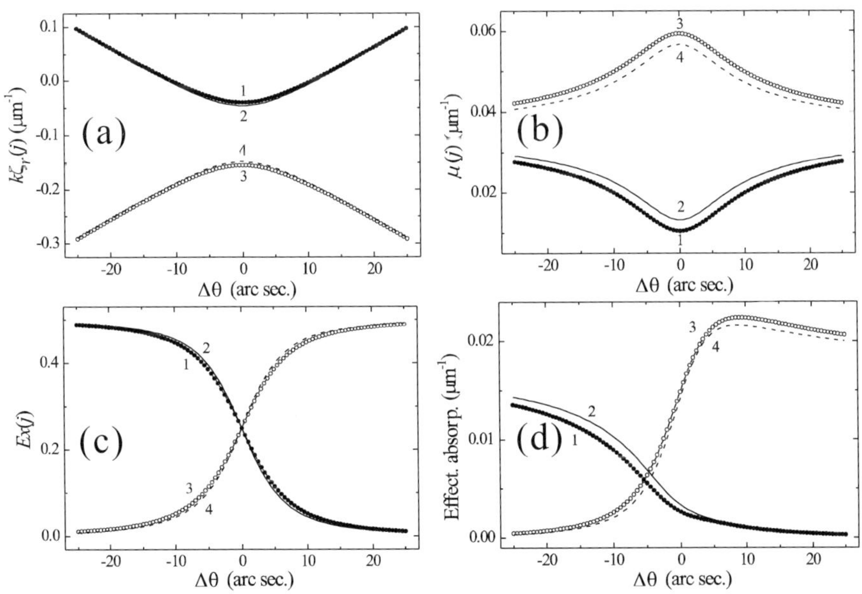

Fig. 8.1. Dynamical calculation for a symmetric two-wave (000)(111) transmission of germanium and CuKα_1: (**a**) dispersion surface, (**b**) linear absorption coefficients, (**c**) excitation of mode, and (**d**) effective absorption

excited. The effective linear absorption, the product of the linear absorption coefficient and the excitation of mode, shown in Fig. 8.1(d), indicates that the effective absorptions of modes 1 and 2 are higher on the lower angles and are lower on the higher angles. The reversed situation is seen for that of modes 3 and 4. The resultant absorption averaged over the four modes is lower for the lower angles than for the higher angles.

The reflection coefficients, the ratios between the $G(111)$ diffracted intensities and that of the incident $O(000)$ wave, for a σ-polarized incident radiation and various crystal thicknesses T_0 are shown in Fig. 8.2. For a very thin crystal of $2\,\mu$m, as expected from the kinematical theory, the intensity of the (000) wave decreases and that of the (111) increases at the exact two-wave position (Fig. 8.2(a)). The intensities also oscillate with respect to $\Delta\theta$ for thin crystals. The oscillation period decreases as the crystal thickness increases. The nearly symmetric intensity distributions against $\Delta\theta = 0$, calculated for thin crystals, become asymmetric as the crystal plate gets thicker (Figs. 8.2(b) and (c)). When $\mu T_0 \gg 1$, the intensity oscillation diminishes and anomalous transmission results (Fig. 8.2(c)).

(ii) Symmetric Bragg reflection ($b = -1$) : The wavefield-amplitude ratios, (8.5), reduce to (7.363) with $b = -1$, i.e.,

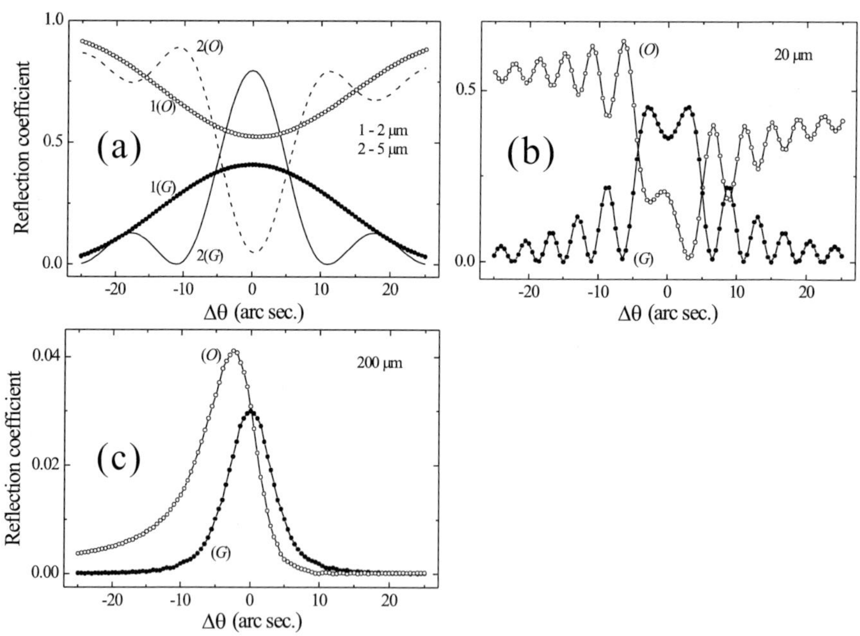

Fig. 8.2. Reflection coefficients of a symmetric two-wave (000)(111) transmission of germanium and CuKα_1 for various crystal thicknesses T_0: (**a**) $T_0 = 1$ and $2\,\mu$m, (**b**) $T_0 = 20\,\mu$m, and (**c**) $T_0 = 200\,\mu$m. (O = (000) and G = (111))

$$\frac{D_G(j)}{D_O(j)} = \sqrt{|b|}\sqrt{\left|\frac{\chi_{G-O}}{\chi_{O-G}}\right|}\exp(i\delta_{G-O})\left[z'_G \mp \sqrt{{z'_G}^2 - 1}\right] , \tag{8.6}$$

where z'_G has been given in (7.364). The corresponding reflected intensity, the reflectivity, is given by

$$\frac{I_G}{I^e_O} = \frac{1}{|b|}\left|\frac{D_G(j)}{D_O(j)}\right|^2 = \left|\frac{\chi_{G-O}}{\chi_{O-G}}\right|\left[z'_G \mp \sqrt{{z'_G}^2 - 1}\right]^2 . \tag{8.7}$$

The sign in (8.6) is chosen so that the absolute ratio has the smallest value. By comparing the value of z'_G with unity, the angular range in a Bragg reflection is divided into three regions, i.e., the total reflection region for $|z'_G| < 1$, and the two regions with angles lower ($z'_G > 1$) and higher ($z'_G < -1$) than the total reflection region. These three regions are usually called range III, II, and I, in the order of the angular setting.

The intensity of wavefiled of mode j, $I_F(j)$, in a two-wave symmetric Bragg reflection for a σ polarization, derived from (6.211), takes the following simple form:

$$I_F(j) = \frac{|D_{\sigma O}|^2}{|D^e_{\sigma O}|^2}\left\{1 + |Y_j|^2 + 2\mathrm{Re}\left[Y_j \exp(2\pi \boldsymbol{g}\cdot\boldsymbol{r})\right]\right\}$$
$$\times \exp\left[-4\pi k\zeta_i(j)\hat{\boldsymbol{n}}_e\cdot\boldsymbol{r}\right] , \tag{8.8}$$

where

$$Y_j = \frac{D_{\sigma G}(j)}{D_{\sigma O}(j)} . \tag{8.9}$$

For an absorbing crystal, there is only one permitted mode, say, $j = 1$, to be considered. In this case, Y_1 can be expressed as

$$\begin{aligned} Y_1 &= -X\exp(-v) && \text{for range I} , \\ Y_1 &= X\exp(-iv) && \text{for range II} , \\ Y_1 &= X\exp(-v) && \text{for range III} , \end{aligned} \tag{8.10}$$

where

$$X = (-1)^\gamma\left(\frac{\gamma_O}{|\gamma_G|}\right)^{\frac{1}{2}}\exp\left(i\delta_{G-O}\right) , \tag{8.11}$$

v is given by $z'_G = -\cosh v$ for $z'_G < -1$ (range I), $z'_G = \cos v$ for $|z'_G| < 1$ (range II), and $z'_G = \cosh v$ for $z'_G > 1$ (range III). $\gamma = 0$ for the phase $\delta_{G-O} = 0°$ and $\gamma = 1$ for $\delta_{G-O} = 180°$. Therefore, I_F depends on the phase term $\exp(i\delta_{G-O})$ and the angular region. For a positive real part of $\exp(i\delta_{G-O})$, I_F for ranges I, II, and III are

$$\begin{aligned} I_F &= [1 + |X|\exp(-v)]^2 , \\ I_F &= [1 - |X|\exp(-v)]^2 , \\ I_F &= [1 + |X|^2 - 2|X|\cos v] , \end{aligned} \tag{8.12}$$

respectively, where $D_{\sigma O}(1) = D^{e}_{\sigma O}$ for $j = 1$ has be used in deriving this equation. For a negative real part of $\exp(i\delta_{G\text{-}O})$, the sign in front of $|X|$ in (8.12) needs to be changed to the opposite sign for the I_F.

The linear absorption coefficients are usually lower than the ordinary value μ_0 in range III and have maximum values at the center of range II, i.e., $z'_G = 0$. The absorption drops monotonically in range I. The minimum linear absorption coefficient in range III and the maximum value in range II are given by

$$\mu^{III}_{\min} = \frac{\mu_0}{\gamma_O}\sqrt{1 - \Omega \cos^2 \delta_{G\text{-}O}}\ ,$$
$$\mu^{II}_{\max} = \frac{2\pi P_G\, |\chi_{G\text{-}O}|}{\lambda \sin\theta_G}\ . \tag{8.13}$$

The calculated dispersion surface, linear absorption coefficients, excitation of mode for zero and infinitely thick crystals, and the effective linear absorption coefficients of the Ge (000)(111) symmetric Bragg reflection for CuKα_1 radiation are shown in Fig. 8.3. In Fig. 8.3(a) the Laue point La and Lorentz point Lo are located at $\Delta\theta = 0$ and $13''$, respectively, along $k\zeta_r = 0$. The two pairs of dispersion hyperbolae of the two-wave transmission case (Fig. 8.1(a)) lie on both sides of the Lorentz point. The region between $\Delta\theta = 5''$ and $20''$ is the range of the total reflection, range II. The angular region below and above are ranges III and II, respectively. There are four possible modes of propagation. In the total reflection range, there are no intersection between the crystal surface normals along the $k\zeta_r$ direction and the dispersion surface. However, the equation of dispersion does give solutions for the eigenvalue ζ_r [40, 524], leading to the so-called 'solution dispersion surface' [414]. Therefore, the dispersion surface is modified according to the presence of the 'solution' surface as four curves crossing with each other near the boundaries between ranges III and II and between II and I. Detailed calculations show that the crossing occurs in the region between $\Delta\theta = 7.5''$ and $8.75''$ and at $\Delta\theta = 16.0835''$ [414]. These crossings also indicate the degeneracy of eigenvalues for the Bragg-type reflection. The degeneracy results from the nonhermitian scattering matrix involved.

The linear absorption coefficients of modes 1 and 2 are positive, and that of modes 3 and 4 are negative (Fig. 8.3(b)). For crystals with zero thickness, the four modes are effective, while for an infinitely thick crystal, only modes 1 and 2 need to be considered for the sake of conservation of energy. Since the number of modes involved in the diffraction depends on the crystal thickness, the excitations of mode are calculated for crystals of zero and infinite thickness (Fig. 8.3(c) and (d)). The drop of the excitation in the central angular region is due to the total reflection, and the negative excitation is associated with the Poynting vector of the (111) reflected wave, where the direction cosine γ_G is negative (Fig. 8.3(c)). The profile of the excitation and of the effective absorption for crystals of infinite thickness are shown in Figs. 8.3(d) and (e), respectively.

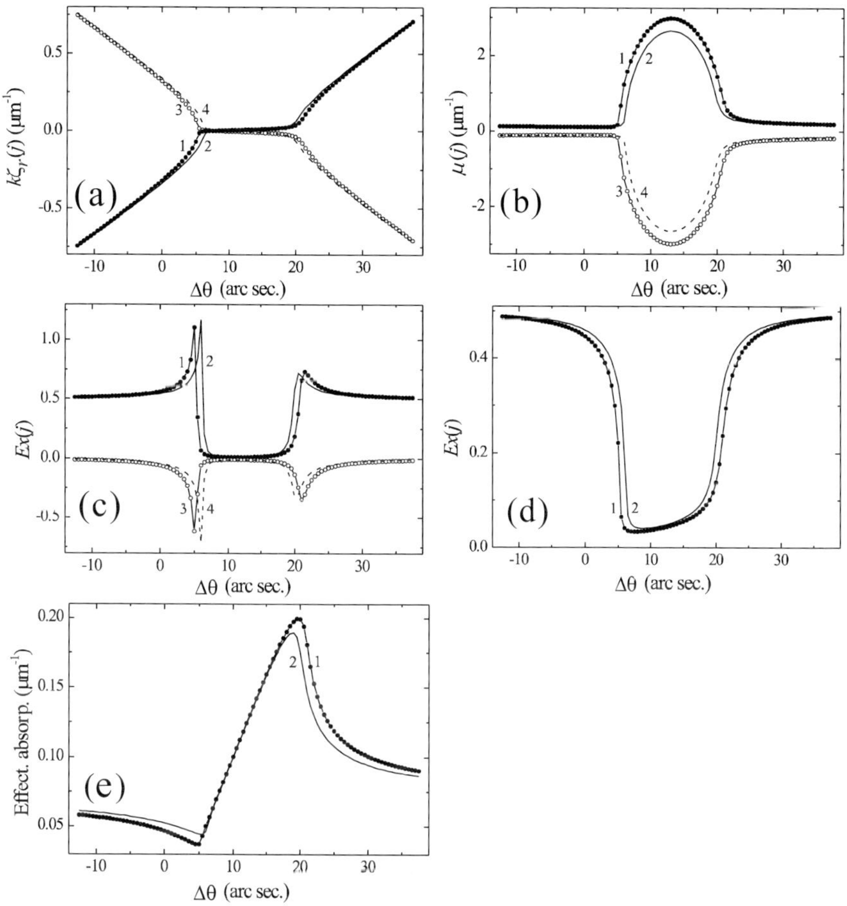

Fig. 8.3. Dynamical calculation for a symmetric two-wave (000)(111) Bragg reflection of germanium and CuKα_1: (**a**) dispersion surface, (**b**) linear absorption coefficients, (**c**) excitation of mode for zero thickness, (**d**) excitation of mode for infinite thickness, (**e**) effective absorption for infinitely thick crystals

Figure 8.4(a) depicts the phase of the (111) reflected wave versus $\Delta\theta$ for each mode. A phase change of 180° for modes 1 and 2 is anticipated for the diffraction from a crystal moving from the lower angles to the higher angles, i.e., for the reciprocal lattice point G moving from the outside to the inside of the Ewald sphere. The phase change of modes 3 and 4 is opposite to that of modes 1 and 2, mainly because the waves of the former propagate backward from the lower surface to the upper surface of the crystal. The intensities of wavefield for an infinitely thick crystal are also shown for the crystallographic phase $\delta_{G\text{–}O} = 180°$ and 0° (Figs. 8.4(b) and (c)). A reversal of intensity asymmetry due to phase change is clearly seen. Because the

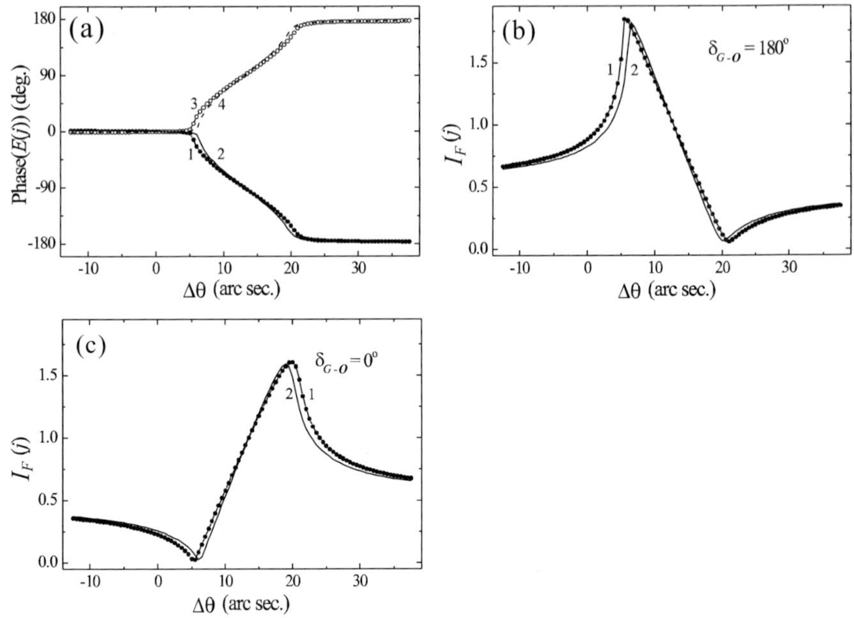

Fig. 8.4. (**a**) Calculated phases of the (111) reflected wave, (**b**) wavefield intensities with $\delta_{G\text{–}O} = 180^\circ$ and (**c**) $\delta_{G\text{–}O} = 0^\circ$ for a symmetric two-wave (000)(111) Bragg reflection of germanium and CuKα_1

wavefield intensities are closely related to the effective absorption, Fig. 8.4(c) resembles the effective linear absorption coefficients shown in Fig. 8.3(e).

The reflection coefficients of the σ and π components for an infinitely thick crystal are plotted in Fig. 8.5(a), which resemble the profiles of excitation shown in Fig. 8.3(d), except that the latter is upside-down. To illustrate the thickness effect on the reflection coefficient, the calculated intensities versus $\Delta\theta$ for $T_0 = 0.5, 1, 2\,\mu\text{m}$, and semi-infinite with a σ-polarized incident radiation are shown in Fig. 8.5(b). The intensity oscillation against $\Delta\theta$ are clearly observed for thin crystals.

8.1.2 Extremely Asymmetric Two-Wave Diffraction

Extremely asymmetric diffraction occurs when either the angle of incidence or the angle of emergence is close to the critical angle of the total reflection. Many studies have been devoted to extremely asymmetric Bragg and Laue diffractions, either grazing incidence or grazing emergence [525–548]. In this kind of diffraction, the curvature of the surface of the Ewald sphere and the tie points associated with the grazing incident or emergent waves need to be considered.

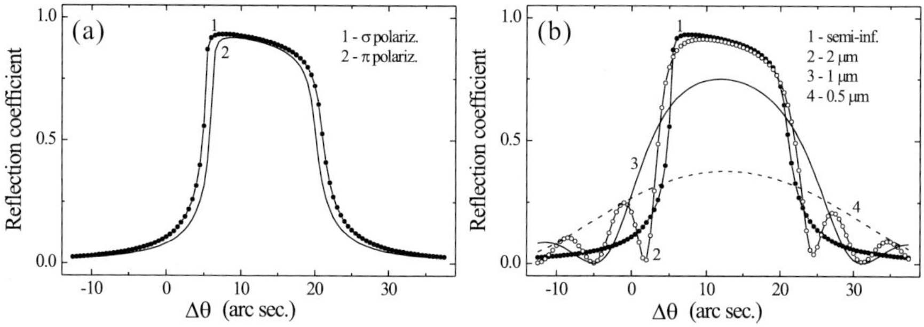

Fig. 8.5. The effect of crystal thickness on the reflection coefficient for a symmetric two-wave (000)(111) Bragg reflection of germanium and CuKα_1: (**a**) infinite thickness and (**b**) $T_0 = 0.5$, 1, 2 μm, and semi-infinite

Let us rewrite the dispersion equation for a singly polarized wavefield as

$$(\chi_O - 2\varepsilon_O)(\chi_O - 2\varepsilon_G) = p_G^2 \chi_{O\text{-}G}\chi_{G\text{-}O} \,, \tag{8.14}$$

where

$$2\varepsilon_M \simeq \frac{K_M^2 - k^2}{k^2} \tag{8.15}$$

for $M = O, G$; $p_G = 1$ and $\cos 2\theta$ for the σ and π polarizations, respectively. The linear approximation for the difference between K_M and k is no longer valid because the difference is not a small quantity for an extremely asymmetric diffraction. Taking into account this fact and considering a generic expression for the dispersion equation, we introduce a characteristic angle ϕ_C, which is related to the generalized refractive index n_O by

$$n_O = \frac{K_O}{k} = \cos\phi_C \,, \qquad n_G = \frac{K_G}{k} \,. \tag{8.16}$$

In the following, a Bragg reflection with a surface specular reflection is described.

Referring to the general situation of an asymmetric diffraction shown in Fig. 8.6, we can write the accommodation ζ as

$$k\zeta = -k\left(\sin\phi_O - n_O \sin u_O\right) \,. \tag{8.17}$$

The dispersion equation then takes the form

$$\left(\chi_O + \sin^2\phi_C\right)\left(\chi_O + \sin^2\phi_C - \alpha'_G\right) - p_G^2 \chi_{O\text{-}G}\chi_{G\text{-}O} = 0 \,, \tag{8.18}$$

where

$$\begin{aligned}
\sin^2\phi_C &= -2\varepsilon_O = -2\zeta\sin\phi_O - \zeta^2 \,, \\
\alpha'_G &= \frac{K_G^2 - K_O^2}{k^2} = 4\sin\theta_G\left[\sin\theta_G - n_O\sin\left(u_O + \beta\right)\right] \,, \\
\alpha_G &= \frac{(\boldsymbol{k}+\boldsymbol{g})^2 - k^2}{k^2} = 4\sin\theta_G\left[\sin\theta_G - \sin\left(\phi_O + \beta\right)\right] \,.
\end{aligned} \tag{8.19}$$

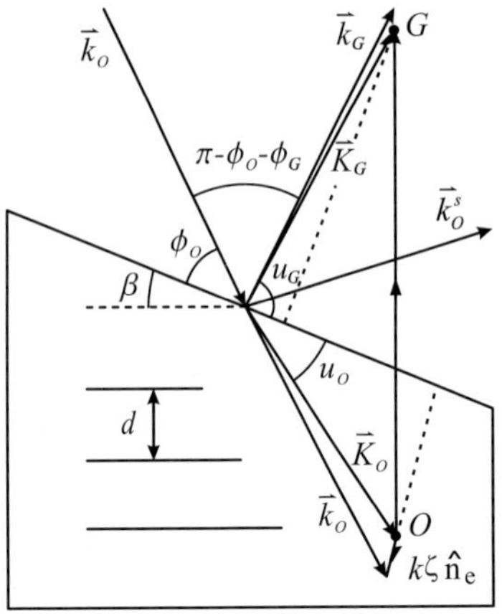

Fig. 8.6. Geometry of an asymmetric Bragg reflection [546]

Because the dispersion surface reflects the intrinsic property of a dynamical diffraction, it is independent of the crystal shape and boundaries. To simplify the situation, we assume the angle β between the crystal surface and the atomic plane G to be zero. This implies a symmetric Bragg reflection situation. By using the relations given in (8.19) with $\beta = 0$, the dispersion equation can be written as a 4th-order polynomial equation of ζ, which can be simplified as

$$y^4 - 2\left[\sin^2\theta_G + (\sin^2\phi_O + \chi_O)\right] y^2 + \Big[(\sin^2\phi_O + \chi_O - \sin^2\theta_G)^2 - p_G^2 \chi_{O-G}\chi_{G-O} \Big] = 0 \,, \tag{8.20}$$

where

$$y = \frac{\alpha'_G}{4\sin\theta_G} = \sin\theta_G - n_O \sin u_O \,,$$
$$n_O \sin u_O = \sqrt{\sin^2\phi_O - \sin^2\phi_C} \,. \tag{8.21}$$

The variable y in the first equation of (8.21) is related to both refraction and diffraction. Moreover, equation (8.20) can be further decomposed into two hyperbola equations:

$$Y^2 - Z^2 = 1 \,,$$
$$\left(Y + \frac{2\sin\theta_G}{U}\right)^2 - Z^2 = 1 \,, \tag{8.22}$$

where

$$Z = \frac{y}{U} \,,$$
$$Y = \frac{\sqrt{\sin^2\phi_O + \chi_O + U^2} - \sin\theta_G}{U} \,,$$

$$U = \frac{\sqrt{p_G^2 \chi_{O-G} \chi_{G-O}}}{\sin\theta_G} \,. \tag{8.23}$$

The solutions to (8.22) derived from Z are

$$n_O \sin u_O = \sin\theta_G \mp U\sqrt{Y^2 - 1} \,, \tag{8.24}$$

or

$$n_O \sin u_O = \sin\theta_G \mp \sqrt{\left(YU + 2\sin\theta_G\right)^2 - U^2} \,. \tag{8.25}$$

In terms of ky, (8.24) and (8.25) can be expressed as

$$ky = k\left(\sin\theta_G - n_O \sin u\right) = \begin{cases} \pm k\sqrt{(YU)^2 - U^2} \,, \\ \pm k\sqrt{\left(YU + 2\sin\theta_G\right)^2 - U^2} \,. \end{cases} \tag{8.26}$$

For a given incident angle ϕ_O, namely, for a given Y, the real part of ky determines the coordinates of tie points on the dispersion surface, and the imaginary part yields the absorption. To illustrate the geometry of the dispersion surface, we consider an asymptotic surface which is composed of two spheres S_O^{eff} and S_G^{eff} centered at the reciprocal lattice points O and G of radius r_{eff} equal to the magnitude of the effective wavevector,

$$r_{\text{eff}} = k\sqrt{1 + \chi_O + U^2} \,. \tag{8.27}$$

This implies that the critical angle ϕ_C^{eff} of the effective wavevector can be defined as

$$\sin\phi_C^{\text{eff}} = \sqrt{-\chi_O - U^2} \tag{8.28}$$

so that the total in the square root of (8.23) is zero. Similarly, the usual critical angle for refraction, ϕ_C^{opt}, is

$$\sin\phi_C^{\text{opt}} = \sqrt{-\chi_O} \,, \tag{8.29}$$

and the radius of the refraction spheres S_O^{opt} and S_G^{opt} (the one-wave dispersion surface) is

$$r = k\sqrt{1 + \chi_O} \,, \tag{8.30}$$

resulting from (8.27) with $U = 0$.

Figure 8.7 shows the spheres S_O, S_G, S_O^{opt}, S_G^{opt}, S_O^{eff}, and S_G^{eff} and the critical angles for a singly polarized wavefield, where the radius of S_O and S_G is k. Points La and Lo are the Laue and Lorentz points. The intersection of S_O^{eff} and S_G^{eff} is denoted by L_R. The actual dispersion surface approaches asymptotically the surface of the spheres S_O^{eff} and S_G^{eff}. The Bragg angle θ_G is $\angle LaOW$.

Figure 8.8 depicts the dispersion curve (in dashed lines) in the plane of incidence for an infinitely thick crystal. The scale near the intersection region has been exaggerated to show the details. In Fig. 8.7, for an entrance point E^0 on the sphere S_O with the incident angle ϕ_O, the surface normal

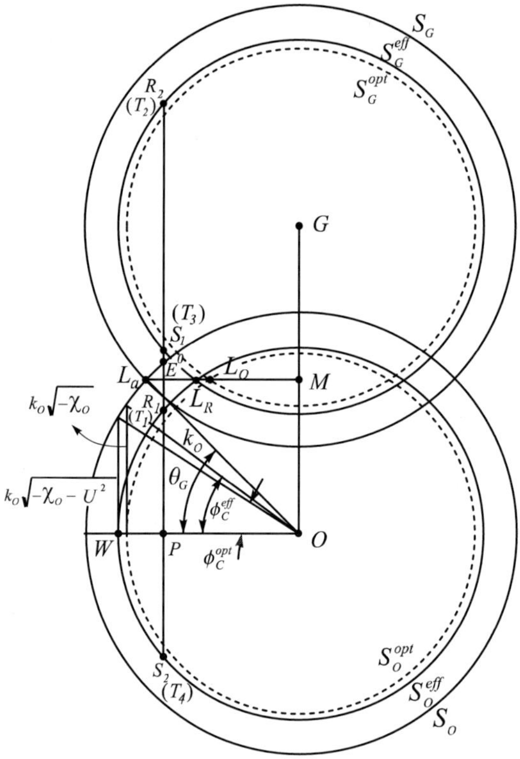

Fig. 8.7. Schematic representation of the dispersion surface for a symmetric Bragg reflection

intersects S_O^{eff} and S_G^{eff} at R_1, S_2 and R_2, S_1, respectively. The corresponding tie points, T_1, T_4, and T_2, T_3, also along the same normal, are the solutions of (8.26), where ky and kYU are represented by QT_i and QR_i, point Q being the intersection point of LaM and E^0P. The tie points are symmetric about the line LaM. For an infinitely thick crystal, only T_1 and T_2 with Poynting vectors towards the interior of the crystal fulfill the requirement of energy conservation. As ϕ_O increases, the excited T_i with $i = 1$ to 4 moves along the dashed curve following the dispersion relation

$$\overline{T_iR_i} \cdot \overline{T_iS_i} = \frac{k^2 p_G^2 \chi_{O\text{–}G}\chi_{G\text{–}O}}{(2\sin\theta_G)^2} \,. \tag{8.31}$$

This relation is valid for all the angular ranges except for the total-reflection region between A_1 and A_2. This is different from the relation

$$\overline{T_1B_O} \cdot \overline{T_1B_G} \simeq \frac{k^2 p_G^2 \chi_{O\text{–}G}\chi_{G\text{–}O}}{4} \tag{8.32}$$

derived from the conventional dynamical theory, where T_1B_O and T_1B_G are the normal distances from T_1 to S_O^{opt} and S_G^{opt}, respectively.

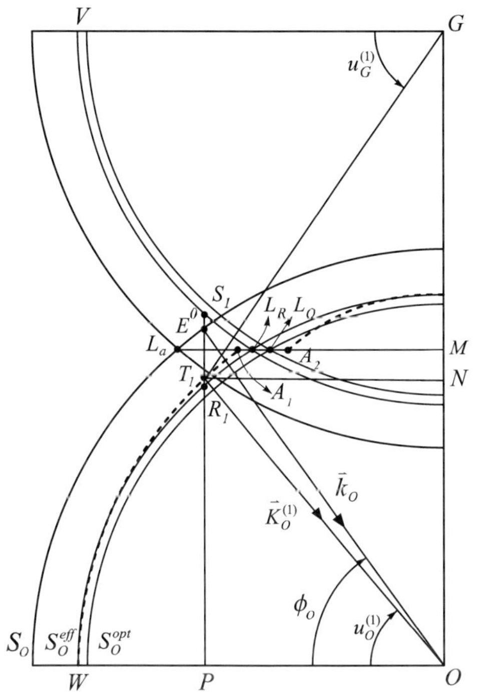

Fig. 8.8. Details of Fig. 8.7 [546]

If the incident angle is denoted as ϕ_R for $Y = 0$, then the excitation point on the dispersion surface at L_R by the incident wave, corresponds to the center of the total-reflection region. Under this circumstance, the following relation holds:

$$\sin^2 \phi_R = \sin^2 \theta_G + \sin^2 \phi_C^{\text{eff}} \tag{8.33}$$

for a symmetric Bragg case, and

$$\sin^2 \phi_R = \sin^2 \phi_B + \cos^2 \beta \sin^2 \phi_C^{\text{eff}} + \sin 2\beta \sin \theta_G (\cos \theta_G - \cos \theta_R) \tag{8.34}$$

for an asymmetric reflection situation, where $\phi_B = \theta_G - \beta$ and θ_R is equal to ϕ_R when $\beta = 0$. For the symmetric Bragg case, the angular deviations $\Delta\theta_{RB}$ can be calculated directly from (8.33) as

$$\Delta\theta_{RB} = \phi_R - \theta_G \, , \tag{8.35}$$

and for the asymmetric case $\Delta\phi_{RB}$ are given by [531, 549]

$$\Delta\phi_{RB} = \phi_R - \phi_B \cong \frac{-\sin \phi_B + \sqrt{\sin \phi_B + \frac{\cos \beta}{\cos \theta_G} \sin^2 \phi_C^{\text{eff}} \cos \phi_B}}{\cos \phi_B} \, . \tag{8.36}$$

The small difference between ϕ_R and ϕ_B has been considered in deriving (8.36). It should be noted that in the conventional dynamical theory, $\Delta\theta_{RB} \approx -\chi_O / \sin 2\theta_G$.

The wavefield-amplitude ratios of the two modes T_1 and T_2 derived from the fundamental equation are

$$c(j) = \frac{E_G(j)}{E_O(j)} = (Z \pm \sqrt{Z^2+1})\sqrt{\frac{\chi_{G\text{–}O}}{\chi_{O\text{–}G}}} , \tag{8.37}$$

for $j = 1, 2$. Here we use the electric fields $\boldsymbol{E}$'s instead of the electric displacements $\boldsymbol{D}$'s, because the normal component of $\boldsymbol{E}$ and that of $\boldsymbol{D}$ can differ considerably for grazing incidence geometry [536,538]. The boundary conditions for the continuities of the normal components of the electric displacements and the tangential components of the magnetic fields at the crystal surface lead to

$$\begin{aligned} &\sum_{j=1}^{2} E_M(j) = E(M) + E^s_G , \\ &\sum_{j=1}^{2} n_M(j) \sin u_M(j) E_M(j) = \sin\phi_M \left[E(M) - E^s_G\right] , \end{aligned} \tag{8.38}$$

for $M = O$ and G, where $E(O) = E^e_O$ and $E(G) = 0$. E^e_O is the wavefield amplitude of the incident wave and E^s_M of the surface-reflected one. From (8.37) and (8.38), the wavefield amplitudes are

$$\begin{aligned} E_O(j) &= G(j)\left[\frac{2\sin\phi_O}{\sin\phi_O + n_O(j)\sin u_O(j)}\right] E^e_O , \\ E_G(j) &= G(j)\left[\frac{2\sin\phi_G}{\sin\phi_G + n_G(j)\sin u_G(j)}\right] E^e_O \end{aligned} \tag{8.39}$$

for $j = 1, 2$, where

$$\begin{aligned} G(1) &= \frac{1}{1+\tau} , \\ G(2) &= \frac{\tau}{1+\tau} , \end{aligned} \tag{8.40}$$

with

$$\begin{aligned} &\tau = -\frac{c(1)\omega_O(2)\omega_G(1)}{c(2)\omega_O(1)\omega_G(2)} , \\ &\omega_M(j) = \sin\phi_M + n_M(j) \qquad \text{for} \quad j = 1, 2 \quad \text{and} \quad M = O, G . \end{aligned} \tag{8.41}$$

Clearly, the sum of $G(1)$ and $G(2)$ is unity. The amplitudes of the specularly reflected waves from the crystal surface are

$$\begin{aligned} E^s_O &= \sum_{j=1}^{2} G(j)\left[\frac{\sin\phi_O - n_O(j)\sin u_O(j)}{\sin\phi_O + n_O(j)\sin u_O(j)}\right] E^e_O , \\ E^s_G &= \sum_{j=1}^{2} c(j) E_O(j) , \end{aligned} \tag{8.42}$$

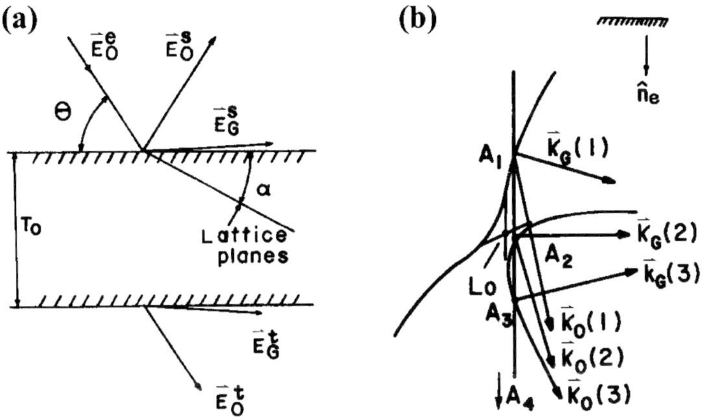

Fig. 8.9. Two-wave asymmetric Laue diffraction: (**a**) real-space geometry, (**b**) the dispersion surface [528]

and the corresponding reflection powers are then

$$P_G = \mathrm{Re}\left\{\frac{\sin\phi_G}{\sin\phi_O}\right\}\left|\frac{E_G^s}{E_O^e}\right|^2 ,$$

$$P_O^s = \left|\frac{E_O^s}{E_O^e}\right|^2 . \tag{8.43}$$

The terms in the square brackets of (8.39) and (8.42) are Fresnel's transmission and reflection coefficients, and the mode of excitation is merely the product of the geometric factor and Fresnel's coefficients. Similar results for the π-polarized waves can be derived in the same way as for the σ-waves.

For a thin crystal plate, the dispersion curve is symmetric about the line LaM. The dispersion surface above the VG and below the WO coincide with the surface of the spheres S_O^{eff} and S_G^{eff}. This is because for an angle of incidence smaller than the critical angle ϕ_C^{eff}, there are no real tie points, i.e., the surface normal and the dispersion surface have no intersections. For an asymmetric diffraction with $\beta \neq 0$, the surface normal is not parallel to the reciprocal lattice vector $\boldsymbol{OG}$. Points V and W will move upward or downward along the surface of S_G^{eff} and S_O^{eff}, respectively, depending on the inclination of the crystal surface relative to the atomic planes of the G reflection.

The above derivation can be equally applied to cases of asymmetric Laue diffraction geometry. A two-wave transmission diffraction with specularly diffracted and reflected waves is schematically shown in Fig. 8.9(a) and the dispersion surface in Fig. 8.9(b). E_O^s and E_G^s are the specularly reflected waves of the incident wave E_O^e and the transmitted E_G^t, respectively. E_O^t is the transmitted wave of the O reflection penetrating the crystal. In the following, we shall consider a more general situation involving grazing incidence X-ray diffraction (GIXD) instead of the details of an asymmetric transmission.

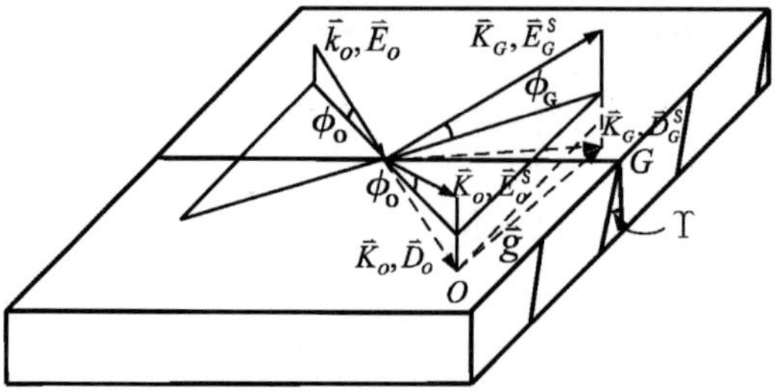

Fig. 8.10. Schematic representation of GIXD

8.1.3 Grazing Incidence Two-Wave Diffraction

Grazing incidence X-ray scattering or diffraction, GIXS or GIXD, first demonstrated in 1979 by Marra, Eisenberger, and Cho [165], has been frequently employed as a tool for probing the structures of crystal surfaces and interfaces. Theoretically, Vineyard described GIXS with a distorted-wave Born approximation in a kinematical approach [166]. In terms of the conventional dynamical theory, Afanas'ev and Melkonyan [167] worked out a formalism for the dynamical GIXD and many investigations then followed [255, 312, 544, 545, 547, 550–566].

Figure 8.10 is a schematic representation of GIXD of the two-wave (O, G) case. For simplicity, only the σ-polarized incident wave $\boldsymbol{E}_O$ is considered for an infinitely thick perfect crystal. Under the grazing incidence condition, i.e., the incident angle $\phi_O \sim |\chi_O|^{1/2}$ being much smaller than unity, the internal Bloch waves D_O and D_G generated by the diffraction G are specularly reflected by the crystal surface, thus giving rise to the specularly reflected and diffracted wavefields $\boldsymbol{E}_O^s$ and $\boldsymbol{E}_G^s$. $\boldsymbol{k}_O$, $\boldsymbol{K}_O$, $\boldsymbol{k}_G$, $\boldsymbol{K}_G$, $\boldsymbol{k}_O^s$, and $\boldsymbol{k}_G^s$ are the wavevectors of $\boldsymbol{E}_O$, $\boldsymbol{D}_O$, $\boldsymbol{E}_G$, $\boldsymbol{D}_G$, $\boldsymbol{E}_O^s$, and $\boldsymbol{E}_G^s$, respectively. $\boldsymbol{g} = \boldsymbol{OG}$ is the reciprocal lattice vector of reflection G. The emergence angles of the specularly reflected $\boldsymbol{k}_O^s$ and $\boldsymbol{k}_G^s$ with respect to the crystal surface are ϕ_O and ϕ_G. The angle between the diffracting atomic planes and the surface normal is denoted as Υ.

The dispersion equation, similar to (8.18), can be easily derived as [55,550]

$$\left(\chi_O + \sin^2\phi_O - \sin^2 u_O\right)\left[\chi_O + \sin^2\phi_G - (\sin u_O + 2\sin\Upsilon\sin\theta_G)^2\right]$$
$$= p_G^2\chi_{O-G}\chi_{G-O}\ , \tag{8.44}$$

which is reduced to

$$\left(\chi_O + \phi_O^2 - u_O^2\right)\left[\chi_O + \phi_G^2 - (u_O + 2\Upsilon\sin\theta_G)^2\right] = p_G^2\chi_{O-G}\chi_{G-O} \tag{8.45}$$

for small ϕ_O and Υ, where

$$\sin u_O \cong u_O = \frac{K_{Oz}}{k}\ , \tag{8.46}$$

K_{Oz} is the component of $\boldsymbol{K}_O$ normal to the surface. The relation between ϕ_O and ϕ_G is

$$\phi_G = (\phi_O + 2\Upsilon \sin\theta_G)^2 - \alpha_G\ , \tag{8.47}$$

where α_G is the angular deviation from the exact Bragg angle θ_G, defined in (8.19).

The wavevectors involved in a GIXD are usually distributed in a wide angular range. Hence the three-dimensional aspects of the dispersion surface need to be considered to clearly show the geometric relationship among the refracted and diffracted waves for various experimental conditions, for example, during an azimuthal scan or a Bragg-angle ω scan. Figures 8.11(a) and (b) show, respectively, the dispersion surface cut by the crystal surface and by the plane perpendicular to the crystal surface, which bisects the vector $\boldsymbol{g}$. In Fig. 8.11(a), La and Lo are the Laue and the Lorentz point, and A_1 and A_2 are the tie points. Point M lies in the middle of $\boldsymbol{g}$. The relationship between the incident angle ϕ_O and the azimuthal angle ψ during an azimuthal rotation around $\boldsymbol{g}$ is shown in Fig. 8.11(b), where the concentric circles are the loci of La, Lo, A_1, and A_2 during the azimuthal rotation. The ϕ_{C1}, ϕ_{C2}, and ϕ_C, at which the tangent lines to the loci of A_1, A_2, and Lo intersect with the locus of La, are the critical angles for ϕ_O.

The relation between ϕ_O and ψ is given by

$$\sin\phi_O = \cos\theta_G \sin\psi\ . \tag{8.48}$$

Assume that $\psi = \psi_{C1}$ and ψ_{C2} for $\phi_O = \phi_{C1}$ and ϕ_{C2} at which the tie points A_1 and A_2 are excited. The details of the dispersion surface near Lo and its excitation are shown in Fig. 8.12. In Fig. 8.12(a), the distance PQ between the incident wavefronts Σ_O in vacuum and Σ'_O in the crystal is $k\chi_O/2$. The relations among A_1, A_2, Q, La, and Lo are $A_1Q = k\chi_{G\text{-}O}/2$, $A_1La = k(\chi_O - \chi_{G\text{-}O})/(2\cos\theta_G)$, $A_1A_2 = k\chi_{G\text{-}O}\cos\theta_G$, and $LoLa = k\chi_O/(2\cos\theta_G)$. The projection of Fig. 8.12(a) normal to the crystal surface is shown in Fig. 8.12(b). The entrance point E on the incident wavefront represented by the arc NLa excites the dispersion curve N_1A_1 at B. The azimuthal angle for this incident wave is ψ. Similarly, the incident waves from the entrance points E_1 and E_2, whose azimuthal angles are ψ_{C1} and ψ_{C2} in Fig. 8.12(b), excite the dispersion curves at A_1 and A_2. The angles in terms of the electric susceptibilities are

$$\sin\phi_{C1} \cong \sqrt{\left|\chi_O - \sqrt{\chi_{G\text{-}O}\chi_{O\text{-}G}}\right|}\ , \qquad \sin\phi_{C2} \cong \sqrt{\left|\chi_O + \sqrt{\chi_{G\text{-}O}\chi_{O\text{-}G}}\right|}\ ,$$
$$\cos\psi_{C1} = 1 - \frac{\chi_O - \sqrt{\chi_{G\text{-}O}\chi_{O\text{-}G}}}{2\cos^2\theta_G}\ , \qquad \cos\psi_{C2} = 1 - \frac{\chi_O + \sqrt{\chi_{G\text{-}O}\chi_{O\text{-}G}}}{2\cos^2\theta_G}\ . \tag{8.49}$$

From Fig. 8.12(b), an incident wave with the wavevector outside the crystal starting from the entrance point E impinges on the crystal surface at the

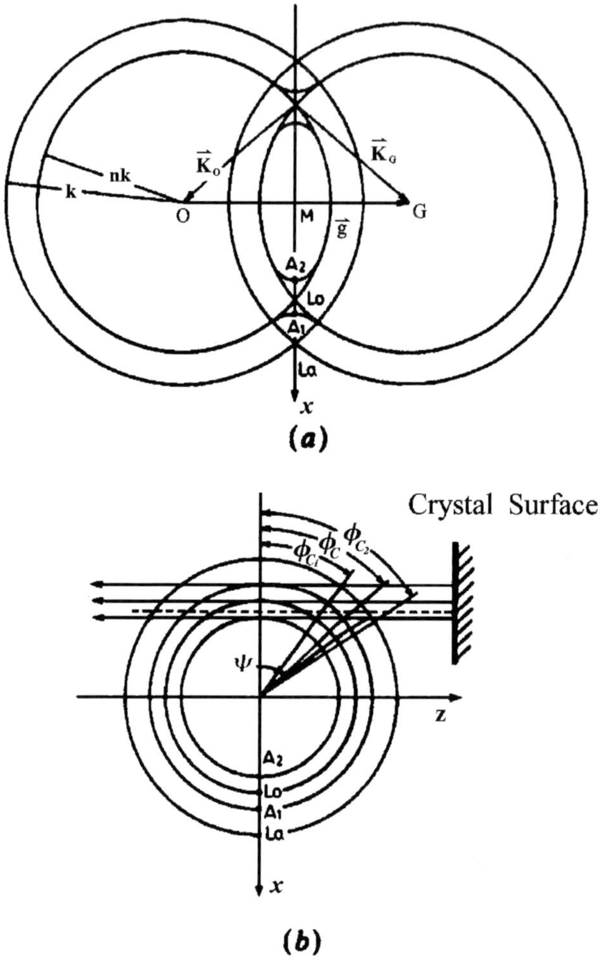

Fig. 8.11. (**a**) Section of the dispersion surface in the crystal-surface plane, (**b**) The dispersion curves in the plane perpendicular to the surface and bisecting OG for azimuthal scans [544]

incident angle ϕ_O, where $\phi_{C1} < \phi_O < \phi_{C2}$. The corresponding accommodation is $\overline{EB} = k\zeta_1$ and the normal component of the incident wavevector inside the crystal is $\overline{BA} = k \sin u_O(1) \simeq k u_O$. Since AE does not intersect the arc A_2N_2, $u_O(2) = 0$ and $k\zeta_2 = k \sin \phi_O \simeq k\phi_O$. For $\phi_O < \phi_{C1}$, there is no intersection between the surface normal and the dispersion curves, so that $u_O(1) = u_O(2) = 0$ and $k\zeta_1 = k\zeta_2 = k\phi_O$. If $\phi_O > \phi_{C2}$, AE always intersects the arcs A_1N_1 and A_2N_2 at B_1 and B_2, i.e., $k\zeta_1 = B_1E$ and $k\zeta_2 = B_2E$. Clearly, the modes that do not intersect the surface normal are evanescence waves, because their amplitudes decay exponentially with $k\zeta$ which is linearly

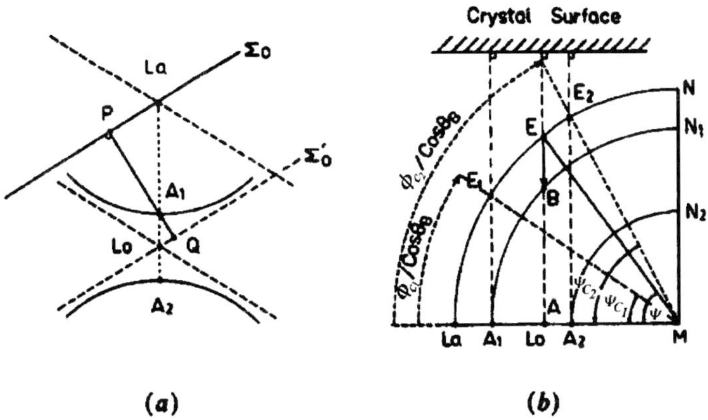

Fig. 8.12. (**a**) Section of the dispersion surface near the Lorentz point in the crystal surface. (**b**) The dispersion curves of (a) in the plane perpendicular to the surface and bisecting OG for azimuthal scans [544]

proportional to ϕ_O. Waves generated from other tie points are transmission waves.

The accommodation ζ can be obtained analytically from the first equation of (8.19), i.e.,

$$\zeta_j = \sin\phi_O - \sin u_O(j) , \tag{8.50}$$

where

$$\sin u_O(j) = \sqrt{\sin^2\phi_O - \sin^2\phi_j} \sim \sqrt{\phi_O^2 - \phi_j^2} . \tag{8.51}$$

The imaginary part of $u_O(j)$ gives the penetration depth t_j for mode j,

$$t_j = \frac{1}{2\pi k\,\mathrm{Im}\,[u_O(j)]} . \tag{8.52}$$

The wavefield-amplitude ratios c_j for a given ζ_j can be calculated as usual from the fundamental equation, and the absolute amplitudes can be determined from the boundary conditions given in (7.36) with the polarization direction along the x-axis for a two-wave Laue diffraction in a semi-infinite crystal as

$$E_O^e + E_O^s = \sum_{j=1}^{2} E_O(j) ,$$

$$E_G^s = \sum_{j=1}^{2} c_j E_O(j) ,$$

$$k_{Oz}^s E_O^e - k_{Oz}^s E_O^s = \sum_{j=1}^{2} K_{Oz}(j) E_O(j) ,$$

$$-k_{Gz}^s E_G^s = \sum_{j=1}^{2} c_j K_{Oz}(j) E_O(j) \, . \tag{8.53}$$

The related wavefield amplitudes are

$$\begin{aligned}
E_O(1) &= -E_G(1) = \left(\frac{\sin\phi_O}{C_1}\right) E_O^e \, , \\
E_O(2) &= E_G(2) = \left(\frac{\sin\phi_O}{C_2}\right) E_O^e \, , \\
E_O^s &= -E_O^e + E_O(1) + E_O(2) \\
&= \left\{\frac{\sin^2\phi_O - \sqrt{\sin^2\phi_O - \sin^2\phi_{C1}}\sqrt{\sin^2\phi_O - \sin^2\phi_{C2}}}{C_1 C_2}\right\} E_O^e \, , \\
E_G^s &= E_G(1) + E_G(2) = -\left[\frac{(C_2 - C_1)\sin\phi_O}{C_1 C_2}\right] E_O^e \, ,
\end{aligned} \tag{8.54}$$

where

$$C_j = \sin\phi_O + \sqrt{\sin^2\phi_O - \sin^2\phi_{Cj}} \, . \tag{8.55}$$

The corresponding reflection powers can be calculated according to (8.43) as

$$\begin{aligned}
P_G^s(\phi_O) = \Bigg| &\frac{1}{2}\left\{\frac{2\sin\phi_O}{\sin\phi_O + \sqrt{\sin^2\phi_O + \chi_O - |p_G \chi_{G\text{-}O}|}}\right\} \\
&- \frac{1}{2}\left\{\frac{2\sin\phi_O}{\sin\phi_O + \sqrt{\sin^2\phi_O + \chi_O + |p_G \chi_{G\text{-}O}|}}\right\}\Bigg|^2 \, .
\end{aligned} \tag{8.56}$$

For an inclined GIXD from a semi-infinite crystal with $\Upsilon \neq 0$, the boundary conditions (8.53) are modified as

$$\begin{aligned}
E_O^e + E_O^s &= \sum_{j=1}^{2} E_O(j) \, , \\
E_G^s &= \sum_{j=1}^{2} c_j E_O(j) \, , \\
\sin\phi_O \left(E_O^e - E_O^s\right) &= \sum_{j=1}^{2} \sin u_O(j) E_O(j) \, , \\
-\sin\phi_G E_G^s &= \left(\sin 2\Upsilon + \sin u_O(1)\right) c_1 E_O(1) \\
&\quad + \left(2\Upsilon \sin\theta_G + \sin u_O(2)\right) c_j E_O(2) \, .
\end{aligned} \tag{8.57}$$

The corresponding wavefield amplitudes and reflection powers can be found in the same way as for the symmetric case.

The symmetric two-wave (000)(220) GIXD of germanium for $\lambda = 3.463683$ Å without absorption is used as an example (Fig. 8.13) to illustrate the specularly reflected powers, dispersion surface, mode excitation, and penetration depth. The mode excitation Ex(j) and the average penetration depth $\bar{t}$ are defined as

$$\mathrm{Ex}(j) = \sum_{M} \left| \frac{E_M(j)}{E_O(j)} \right|^2 , \qquad \bar{t} = \sum_{j=1}^{2} t_j \mathrm{Ex}(j) . \tag{8.58}$$

In Fig. 8.13, the critical angles ϕ_{C1} and ϕ_{C2} are 6.542 and 15.962 mrad, respectively, according to (8.49). At these two positions, the dispersion curves of modes 1 and 2 begin to have non-zero values (Fig. 8.13(b)). The normalized intensity P^s_{000} starts with unity at the zero position, then gradually drops to 0.1680 at ϕ_{C1}, recovers afterwards slightly to the value of 0.2735 at ϕ_{C2}, and then decreases monotonically to zero for large ϕ_O. P^s_{220} behaves like a Darwin curve with a maximum of unity at ϕ_{C1} for non-absorbing cases and reaches a plateau of total reflection for ϕ_O in between the two critical angles ϕ_{C1} and ϕ_{C2}. Note that the sum of P^s_{000} and P^s_{220}, denoted by ΣP^s, is equal to unity for $\phi_O \le \phi_{C1}$. The Fresnel reflectivity P_{op}, the one-wave intensity, defined as

$$P_{\mathrm{op}} = \left| \frac{\sin\phi_O - \sqrt{n^2 - \cos^2\phi_O}}{\sin\phi_O + \sqrt{n^2 - \cos^2\phi_O}} \right|^2 , \tag{8.59}$$

is also shown in Fig. 8.13(a), $n = \sqrt{1 + \chi_O}$ being the refractive index. The dispersion curves (the solid lines) and the wavefronts (the dashed lines) of the incident (000) and the reflected (220) reflections are depicted in Fig. 8.13(b),

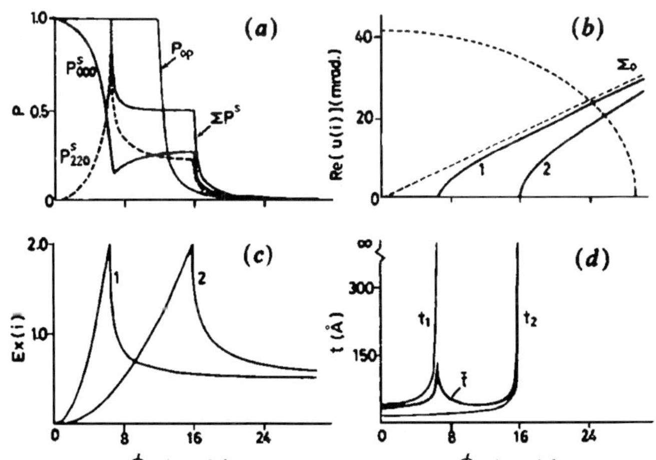

Fig. 8.13. (**a**) Calculated intensities, (**b**) dispersion surface, (**c**) mode excitation, and (**d**) penetration depth for Ge (000)(220) GIXD and 3.463683 Å [544]

which are consistent with the explanation for Fig. 8.12(b). The excitations of modes 1 and 2 have their maximum values at ϕ_{C1} and ϕ_{C2}, respectively, and approach 0.5 asymptotically due to the contribution of the incident wave (Fig. 8.13(c)). The penetration depth t_j of each mode j and the average value $\bar{t}$, shown in Fig. 8.13(d), indicate that the penetration depth starts with a few tens of Angstroms below ϕ_{C1} and increases drastically to thousands of Angstroms at the two critical angles, while the average penetration depth $\bar{t}$ peaks at ϕ_{C1} and reaches maximum at ϕ_{C2}.

This three-dimensional feature of the dispersion surface and the calculated reflectivity and excitation of mode of a two-wave GIXD will be useful in understanding a multiple-wave GIXD. Very recently, a universal computation algorithm has been developed for a general two-wave diffraction for grazing-angle and wide-angle incidence and emergence [567]. This algorithm can also give the coordinates of the dispersion surface and reflectivities for the two-wave GIXD.

8.1.4 Two-Wave Back Diffraction

Two-wave back diffraction, similar to the extremely asymmetric diffraction, is a special diffraction situation where the Bragg angle is close or equal to $\pi/2$. Theoretical and experimental investigations have been very active recently [55, 568–593], mainly for the possibility of using this type of diffraction for making high-resolution monochromators and X-ray resonators of the Fabry-Perot type. Owing to the special geometry, the dispersion equation of a back diffraction, derived from the general formulation (8.14), is

$$\left(K_O^2 - n^2k^2\right)\left(K_G^2 - n^2k^2\right) = k^4 p_G^2 \chi_{G\text{–}O}\chi_{O\text{–}G}\,, \tag{8.60}$$

where the curvature of the Ewald sphere must be taken into account. By following the coordinates system of Kohra and Matsushita [568] and Authier [55], the plane of incidence is in the XZ-plane and the position of a tie point is given by $(X, 0, Z)$, with the inward surface normal being along the Z-axis (see Fig. 8.14). The reciprocal lattice points O and G lie on the Z-axis and $\overline{OC} = \overline{CG} = g/2$. For a given entrance point E with the incident angle $\theta = \pi/2 - \epsilon$, tie points T_1, T_2, T_3, and T_4 are excited. The moduli of the wavevectors $\boldsymbol{K}_O = \boldsymbol{TO}$ and $\boldsymbol{K}_G = \boldsymbol{TG}$ are given by

$$\begin{aligned} K_O^2 &= X^2 + \left(Z - \frac{g}{2}\right)^2 , \\ K_G^2 &= X^2 + \left(Z + \frac{g}{2}\right)^2 . \end{aligned} \tag{8.61}$$

For a small ϵ, $\epsilon \simeq 2X/g$ and Bragg's law becomes

$$k = \frac{g}{2}\left(1 + \frac{\epsilon_G^2}{2}\right) , \tag{8.62}$$

where $\epsilon_G = \pi/2 - \theta_G$. The equation of dispersion can be written as [55]

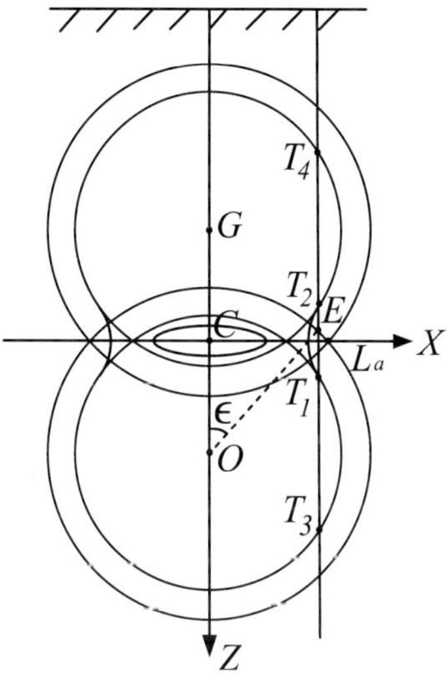

Fig. 8.14. Section of the dispersion surface in the plane of incidence for a back diffraction

$$\left[X^2 + Z^2 - \frac{g^2}{4}\left(\epsilon_G^2 + \chi_O\right)\right]^2 - g^2 Z^2 = k^4 p_G^2 \chi_{G\text{-}O}\chi_{O\text{-}G} \,. \tag{8.63}$$

For θ_G close to $\pi/2$, the Darwin widths and the ranges of total reflection depend only on the X coordinate of tie points. Hence, by taking $Z = 0$, X becomes

$$X^2 = \frac{g^2}{4}\left(\epsilon_G^2 + \chi_O \pm |p_G|\sqrt{\chi_{G\text{-}O}\chi_{O\text{-}G}}\right) \,. \tag{8.64}$$

For $\epsilon_G < \sqrt{-\chi_O - |p_G|\sqrt{\chi_{G\text{-}O}\chi_{O\text{-}G}}}$, the two spheres of radius nk do not intersect each other. No reflection occurs. For

$$\sqrt{-\chi_O - |p_G|\sqrt{\chi_{G\text{-}O}\chi_{O\text{-}G}}} < \epsilon_G < \sqrt{-\chi_O + |p_G|\sqrt{\chi_{G\text{-}O}\chi_{O\text{-}G}}} \,,$$

the two spheres of radius nk nearly touch each other. There are two roots together with one total-reflection region centered at point C. The angular width in 2θ of this region is [55]

$$\Delta(2\theta) = 2\sqrt{\epsilon_G^2 + \chi_O + |p_G|\sqrt{\chi_{G\text{-}O}\chi_{O\text{-}G}}} \,. \tag{8.65}$$

For $\epsilon_G > \sqrt{-\chi_O + |p_G|\sqrt{\chi_{G\text{-}O}\chi_{O\text{-}G}}}$, there are four roots and two ranges of total reflection. The width of the total-reflection ranges is [55]

$$\Delta(2\theta) = \sqrt{\epsilon_G^2 + \chi_O + |p_G|\sqrt{\chi_{G\text{-}O}\chi_{O\text{-}G}}} - \sqrt{\epsilon_G^2 + \chi_O - |p_G|\sqrt{\chi_{G\text{-}O}\chi_{O\text{-}G}}} \,. \tag{8.66}$$

The angular deviation of the center of the total reflection region from the Bragg angle θ_G is given by [55]

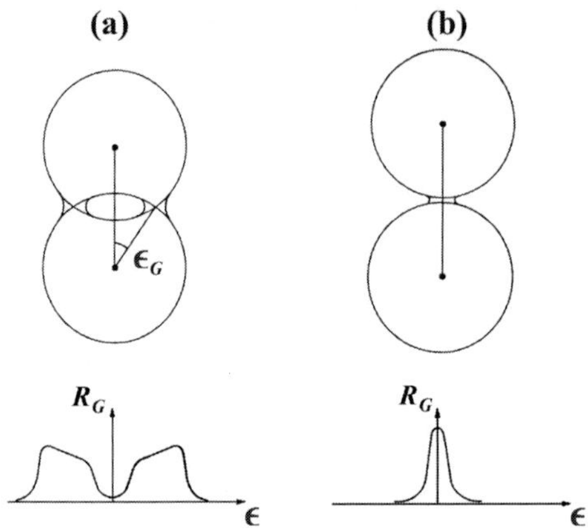

Fig. 8.15. The dispersion surface and reflectivity for back diffraction: (**a**) when $\epsilon_G > \sqrt{-\chi_O + |p_G|\sqrt{\chi_{G\text{-}O}\chi_{O\text{-}G}}}$ and (**b**) when $\epsilon_G < \sqrt{-\chi_O - |p_G|\sqrt{\chi_{G\text{-}O}\chi_{O\text{-}G}}}$

$$\Delta\theta_{RB} = \epsilon_G - \sqrt{\epsilon_G^2 + \chi_O}\ . \tag{8.67}$$

The solutions to the dispersion equation (8.63) can be easily found. Since tie points T_3 and T_4 on the circles of radius nk are far from the entrance point E, their contributions to the wavefields are negligible. Hence, by neglecting higher-order terms, the solutions for T_1 and T_2 are

$$Z_j = \pm\frac{g}{4}\sqrt{\left(\epsilon_G^2 - \epsilon^2 + \chi_O\right)^2 - p_G^2\chi_{G\text{-}O}\chi_{O\text{-}G}}\ , \tag{8.68}$$

for $j = 1, 2$. The corresponding amplitude ratio then takes the form [55]

$$c_j = \frac{E_G}{E_O} = \frac{4\left[Z_j^2 + gZ_j + \frac{g^2}{4}\left(\epsilon^2 - \epsilon_G^2 + \chi_O\right)\right]}{g^2 p_G \chi_{O\text{-}G}}\ . \tag{8.69}$$

The reflected and transmitted intensities can be calculated as in the ordinary two-wave Bragg diffraction by employing the same boundary conditions.

Figures 8.15(a) and (b) show schematically the dispersion surface and the reflectivity versus the incident angular deviation ϵ for the two extreme situations mentioned, where $\epsilon_G > \sqrt{-\chi_O + |p_G|\sqrt{\chi_{G\text{-}O}\chi_{O\text{-}G}}}$ and $\epsilon_G < \sqrt{-\chi_O - |p_G|\sqrt{\chi_{G\text{-}O}\chi_{O\text{-}G}}}$, respectively. The symmetric Bragg geometry is assumed that the reflectivity is symmetric with respect to $\epsilon = 0$.

8.2 Three-Wave Diffraction

Three-wave diffraction can be categorized as Laue-Laue, Bragg-Laue, Bragg-surface, and grazing incidence, according to the geometric relations between a given diffracted wave and the crystal boundaries. Whether a case is a Laue or a Bragg depends on the diffracted wave being transmitted- or reflected. The 'surface' case involves a diffracted wave propagating along the crystal surface or in the direction slightly deviated from the crystal surface, i.e., the angle deviated from the surface is smaller than the critical angle of total reflection. If the angle of incidence or emergence is of the order of the critical angle, the diffraction is of grazing incidence. The two-wave grazing incidence diffraction has already been discussed in the previous section. Various types of three-wave diffraction will be described below.

8.2.1 Three-Wave Laue-Laue Diffraction

Three-wave Laue-Laue diffraction, also called 'three-wave Borrmann diffraction' [27, 28], involves all diffracted waves transmitting through the crystal, including the direct wave. Two three-wave cases, (a) $(000)(111)(11\bar{1})$ and (b) $(000)(111)(1\bar{1}\bar{1})$ of germanium and $\mathrm{CuK}\alpha_1$ radiation are illustrated here as examples. Case (a) is a classic three-wave diffraction, which was used to demonstrate the anomalous transmission of X-rays through a crystal at the three-wave diffraction position by Borrmann and Hartwig in 1965 [27]. Although both cases involve the reflections of the same $\{111\}$ family, the coupling reflection for case (a) is a forbidden (002) reflection and for case (b) a strong (220). The difference in three-wave interaction between the two cases can be clearly seen when we discuss the excitations of mode and diffraction intensities of both cases. Since some aspects of dynamical diffractions have been described in Chap. 6, and since analytical expressions are not possible to obtain for a general three-wave diffraction, we shall only present the results of the numerical calculation based on the procedures described in Chap. 7.

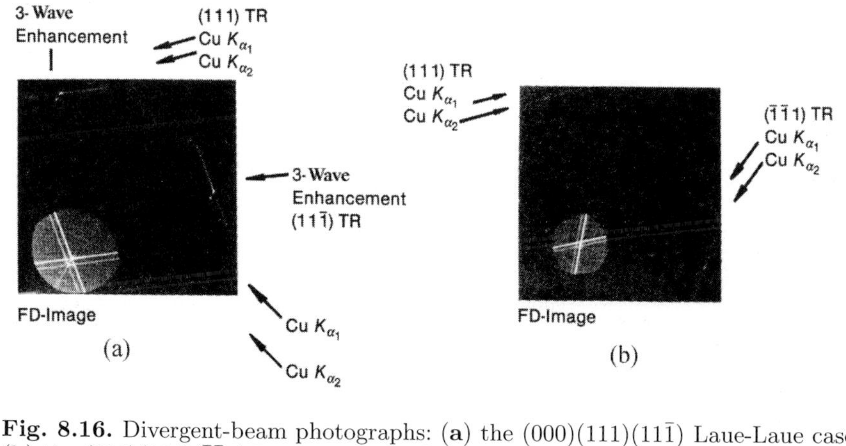

Fig. 8.16. Divergent-beam photographs: (**a**) the $(000)(111)(11\bar{1})$ Laue-Laue case, (**b**) the $(000)(111)(\bar{1}\bar{1}1)$ Laue-Laue case [226]

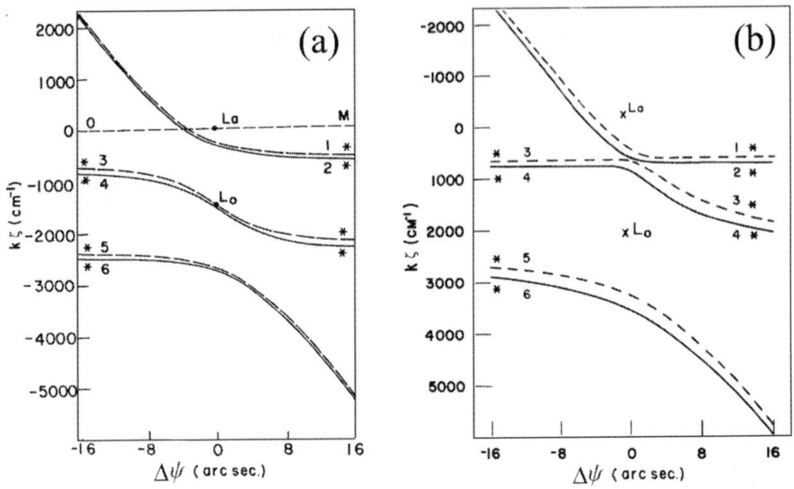

Fig. 8.17. Section of the dispersion surface: (**a**) the (000)(111)(11$\bar{1}$) Laue-Laue case, (**b**) the (000)(111)($\bar{1}\bar{1}$1) Laue-Laue case [53]

Figures 8.16(a) and (b) show the intensity distributions of the forward diffracted (FD) (000) wave and the transmitted reflected (TR) (111) wave for cases (a) and (b), respectively. The crystal is a 0.5 mm thick plane-parallel plate with its large face parallel to the involved reciprocal lattice vectors. The images are obtained by taking the divergent-beam photographs of CuKα radiation, described in Chap. 3. The distances between the crystal and the X-ray films and between the X-ray source and the crystal are about 50 mm and 1000 mm, respectively. The refraction and reflection lines of CuKα_1 and CuKα_2 appear in both FD and TR images. The brighter lines are due to CuKα_1. The separation between the CuKα_1 and CuKα_2 lines is proportional to the difference in $2\theta_{111}$. Compared with the intensity in case (b), appreciable intensity enhancement in case (a) is observed at the exact three-wave diffraction position. This difference in anomalous transmission for three-wave diffraction can be understood from the following dynamical calculations.

Figures 8.17(a) and (b) depict the sections of the dispersion surface cut by the plane of incidence of the (111) reflection along the (111) reflection line for both case (a) and case (b). There are six dispersion sheets, which are grouped into pairs due to the σ and π polarizations. In Fig. 8.17(a), La and Lo are the three-wave Laue point and Lorentz point, respectively. The horizontal axis, $\Delta\psi$, represents the azimuth angle of rotation around the reciprocal lattice vector [111]. The horizontal line passing through point La is therefore the locus of the two-wave Laue point ($\Delta\theta = 0$). The point at $\Delta\psi = 0$ and $\Delta\theta = 0$ is the exact three-wave diffraction position. The ordinate is the accommodation $k\zeta$, which is normal to the entrance surface of the crystal. The distance from point La to the reciprocal lattice point of the (111) reflection is the reciprocal of the wavelength λ, i.e., about $6.5 \times 10^7\,\mathrm{cm}^{-1}$. Far from $\Delta\psi = 0$, only the dispersion sheets of modes 3, 4, 5, and 6 on the lower angle side and of modes 1, 2, 3, and 4 on the higher angle side are effectively excited (see the relative excitation of mode shown in Fig. 8.18(a)). This is the two-wave characteristics of having four dispersion sheets. The separation between the two effective pairs of dispersion sheets therefore approaches the asymptotic value $r_e\lambda p_G|F_G|/[\pi V|\cos\theta_G|]$.

If we draw a line (not shown) from the upper left to the lower right through point Lo, then it is clear that the dispersion sheets of modes 1 and 2 for negative $\Delta\psi$ and that of modes 5 and 6 for positive $\Delta\psi$ approach asymptotically this line as $|\Delta\psi|$ increases. This asymptotic line is the dispersion curve of the one-wave $(11\bar{1})$ reflection. Hence, near points La and Lo, the two-wave dispersion surface of (111) and one wave dispersion surface of $(11\bar{1})$ intersect each other. The modified dispersion surface shown in the vicinity of $\Delta\psi = 0$ indicates the three-wave characteristics, which are described below.

The dispersion equation for a singly polarized three-wave diffraction at the exact three-wave point, i.e., $\Delta\theta = 0$ and $\Delta\psi = 0$, can be written, according to (6.63), as

$$
\begin{aligned}
8y^3 &- 2yk^2\left[C_1\chi_{O-G_1}\chi_{G_1-O} + C_2\chi_{O-G_2}\chi_{G_2-O} + C_3\chi_{G_1-G_2}\chi_{G_2-G_1}\right] \\
&+ k^3C_1C_2C_3\left[\chi_{O-G_1}\chi_{G_1-G_2}\chi_{G_2-O}+\chi_{G_1-O}\chi_{O-G_2}\chi_{G_2-G_1}\right] = 0\,,
\end{aligned}
\tag{8.70}
$$

where $y = y_O = y_{G_1} = y_{G_2}$ and the origin of y is the Lorentz point. For a centrosymmetric crystal like germanium and for the X-ray wavelength far from an absorption edge,

$$
\begin{aligned}
&\chi_{O-G_1}\chi_{G_1-G_2}\chi_{G_2-O} + \chi_{G_1-O}\chi_{O-G_2}\chi_{G_2-G_1} \\
&\quad = 2\left|\chi_{O-G_1}\chi_{G_1-G_2}\chi_{G_2-O}\right|\cos\delta_3\,,
\end{aligned}
\tag{8.71}
$$

where δ_3 is the phase of the triplet $\chi_{O-G_1}\chi_{G_1-G_2}\chi_{G_2-O}$, or $F_{O-G_1}F_{G_1-G_2}F_{G_2-O}$. For case (a), since the coupling reflection $G_2 - G_1 = 11\bar{1} - 111 = 00\bar{2}$ is a forbidden reflection, i.e., the structure factor $F(002) = 0$, one of the solutions to (8.70) is $y = 0$. This implies that a dispersion sheet goes through point Lo, which agrees with the calculated dispersion sheets of modes 3 and 4 shown in Fig. 8.17(a). For case (b), the coupling reflection (220) is a strong reflection and the triplet phase $\delta_3 = 0$. According to Ewald [594], the dispersion sheets of modes 3 and 4 lie in between the La and Lo points for a positive $\cos\delta_3$ and lie below the Lo point for a negative $\cos\delta_3$. Hence the dispersion sheets of modes 3 and 4 in case (b) is close to that of modes 1 and 2. We shall discuss this phase effect on multiple-wave diffraction in detail in Chap. 9.

Figures 8.18(a) and (b) are the calculated relative excitation of mode $\mathrm{Ex}_G(j)$ with $G = (111)$ along the (111) reflection line for both cases. Again, the three-wave diffraction takes place at $\Delta\psi = 0$, and the regions far from $\Delta\psi = 0$ are the two-wave regions. Assume that the total excitation of the

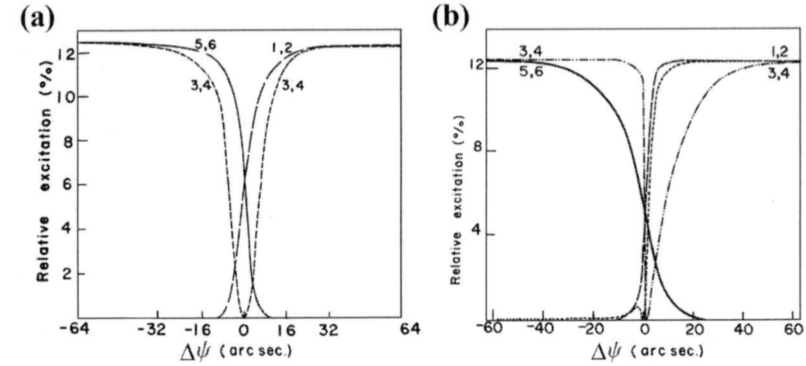

Fig. 8.18. Relative excitations of modes along the (111) line: (**a**) the (000)(111)(11$\bar{1}$) Laue-Laue case, (**b**) the (000)(111)($\bar{1}\bar{1}$1) Laue-Laue case [53]

dispersion surface is 100%. In the two-wave region, there are two reflections, (000) and (111), and four modes to share this excitation. Thus, on the average the maximum excitation for each reflection involved is 12.5%. The excitations of modes 1, 2, 3, and 4 for large positive $\Delta\psi$ and that of modes 3, 4, 5, and 6 for large negative $\Delta\psi$ remain their two-wave characteristics as shown in Figure 8.18(a). At the exact three-wave position, both modes 1, 2 and 5, 6 have the same excitation of 6.125%, while that of modes 3 and 4 are zero. Similar behavior is found for the excitation in case (b) (Fig. 8.18(b)), except that the excitation of mode 1 is zero at the exact three-wave point.

The linear absorption coefficients along the (111) reflection line are shown in Figure 8.19 for both cases. The absorption is known to have lower values

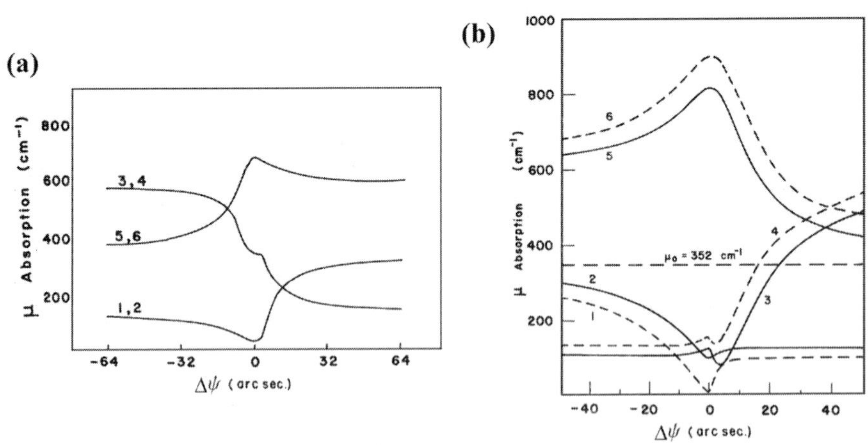

Fig. 8.19. Linear absorption coefficients along the (111) line: (**a**) the (000)(111)(11$\bar{1}$) Laue-Laue case, (**b**) the (000)(111)($\bar{1}\bar{1}$1) Laue-Laue case [53]

for the tie points closer to the Laue point [594]. Therefore, at the three-wave point of case (a), modes 1 and 2 have the lowest absorption, i.e., 17 and 28 cm^{-1}, respectively, compared to 356, 357, 672, and 683 cm^{-1} for modes 3, 4, 5, and 6. Far from the three-wave point, the linear absorption coefficients of modes 1 and 2 for negative $\Delta\psi$ and of modes 3 and 4 for positive $\Delta\psi$ approach asymptotically the minimum two-wave value, 106 cm^{-1}, of (111), in agreement with (6.167). The absorption approaches the maximum two-wave value, 600 cm^{-1}, of (111) for modes 3 and 4 in lower $\Delta\psi$ region and modes 5 and 6 in higher $\Delta\psi$ region. As indicated in (6.165), the average value of the linear absorption coefficient over the six modes at any angular position is equal to the ordinary value μ_0, 352 cm^{-1}.

The absorption of case (b) (Fig. 8.19(b)) is different from that of case (a). Because the dispersion curves of modes 1, 2, 3, and 4 are close to the Laue point near the three-wave diffraction position (Fig. 8.17(b)), the corresponding linear absorption coefficients are relatively low. The lowest value, 4.2 cm^{-1}, of mode 1 at $\Delta\psi = 0$ is the minimum two-wave absorption of (220), the coupling reflection between (111) and $(\bar{1}\bar{1}1)$. The asymptotic behavior of the modes in case (b) for large $|\Delta\psi|$ is also different from that in case (a), though the asymptotic values are the same, i.e, 106, 352, and 600 cm^{-1}.

The transmitted intensity of (111) for various crystal thicknesses is shown in Fig. 8.20 for cases (a) and (b). For a thin crystal plate, all the modes of wave propagation have appreciable transmitted intensities. The interaction

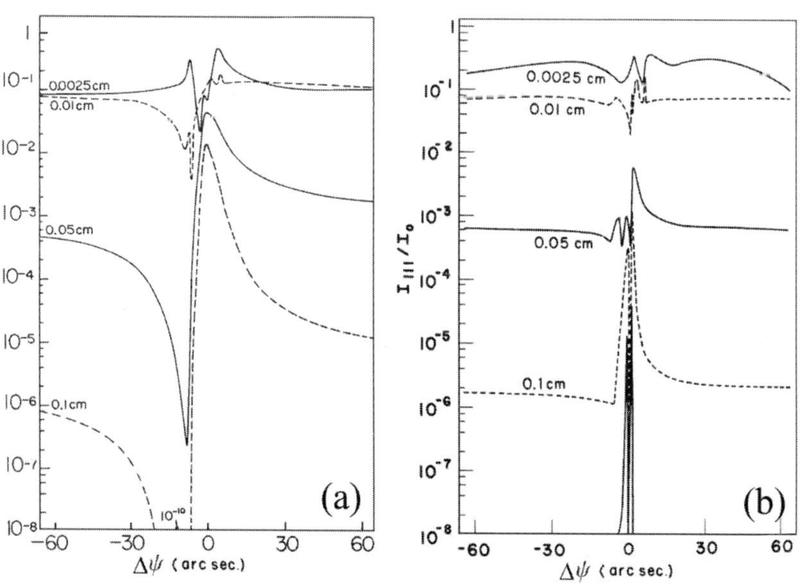

Fig. 8.20. Transmitted intensity of (111) along the (111) line: (**a**) the (000)(111)$(11\bar{1})$ Laue-Laue case, (**b**) the (000)(111)$(\bar{1}\bar{1}1)$ Laue-Laue case [53]

among these modes within the crystal gives rise to intensity oscillation for the transmitted (111) reflection. As the crystal thickness increases, only those modes with lower absorption survive, thus producing resonance-like intensity profiles. The asymmetry of the intensity profile usually provides the phase information of the involved structure-factor triplets [257, 595]. The feature of having a decrease in intensity for lower $\Delta\psi$ and an increase for higher $\Delta\psi$ becomes much more pronounced for even thicker crystals, as shown in Fig. 8.20(a). The calculated intensity profile for a 0.5 mm thick crystal is in good agreement with the experimental result shown in Fig. 8.16(a), where an intensity enhancement on one side of the three-wave point is observable. This result is anticipated from the viewpoint of the dynamical interaction of X-rays in thick crystals. For a thick crystal with $\mu T > 10$, only the modes of the lowest absorption, i.e., modes 1 and 2, contribute to the intensity transmission. As shown in Figs. 8.18(a) and 8.19(a), the asymmetric behaviors of absorption and excitation lead to the asymmetry of intensity distribution, the decrease and enhancement of the intensity are shown in Fig. 8.20(a). The intensity enhancement is usually called 'the anomalous Borrmann effect', or 'super Borrmann effect'.

The pronounced intensity asymmetry is not seen in the (111) transmitted intensity of case (b) (see Fig. 8.20(b)). In fact, intensity oscillation appears for thin crystals of thickness up to 0.5 mm. For thicker crystals, two distinct peaks, one or two seconds apart, show up. This is attributed to the contribution of mode 1, because the excitation of this mode resembles the intensity distribution near the three-wave point. The experimental observation of this three-wave diffraction image shows a similar intensity distribution to the calculated one.

The three-wave anomalous transmission of case (a) can be understood from either the intensity of wavefield, I_F, given in (6.209) for the modes of propagation or the linear absorption coefficients with respect to the atomic sites of the crystal. According to (6.158) and (6.209), photoelectric absorption is directly proportional to I_F, provided that thermal vibration is negligibly small, i.e., $\Omega \simeq 0$ in (6.168). Anomalous absorption occurs when the position of maximal I_F coincides with the atomic sites where all electrons are concentrated. On the contrary, absorption is minimal or nearly null when the minima of I_F appear at the atomic sites, thus anomalous transmission taking place. Figure 8.21(a) shows the calculated linear absorption coefficients without thermal vibration along the [110] direction versus the atomic sites of germanium for the one-wave case, the two-wave (111) reflection, and the three-wave case (a). The respective atomic positions projected onto the $(1\bar{1}0)$ plane is shown in Fig. 8.21(b). The absorption coefficient indicated by the horizontal line in Fig. 8.21(a) is 352 cm^{-1} for the one-wave case, which is independent of the atomic positions. This absorption coefficient is then modified to behave as that indicated by the dashed curves for the two-wave situation and the solid curves for the three-wave diffraction. The minimum absorption

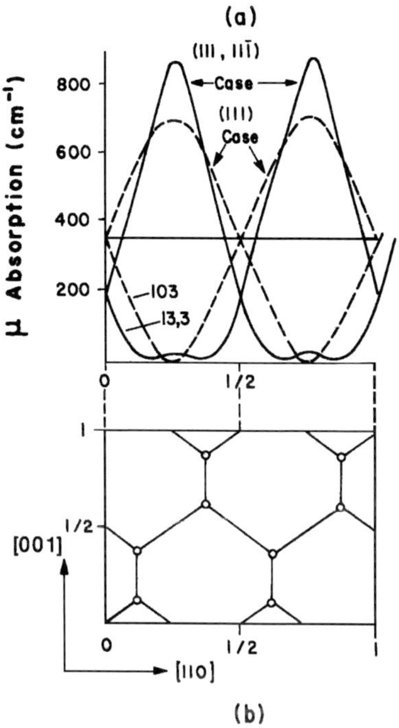

Fig. 8.21. (**a**) Calculated linear absorption coefficients along the [110] direction versus the atomic sites of the germanium (000)(111)(11$\bar{1}$) Laue-Laue case. (**b**) Projected atomic positions in the (1$\bar{1}$0) plane [53]

coefficient at the germanium site is 13.3 cm^{-1} for the three-wave case (a) and 103 cm^{-1} for the two-wave situation. This explains why the transmitted intensity of the (111) reflection is further enhanced in the three-wave diffraction situation. The detailed calculation can be found in the reference [53].

The minimum linear absorption coefficient of this three-wave diffraction for CuKα_1 at room temperature at the exact three-wave point has been extensively investigated experimentally and theoretically. The reported values are 15.5 [27], 17.0 [184], 18.7 [180, 187], 19.0 [195], 18.4 [206], and 17.9 cm^{-1} [205].

8.2.2 Three-Wave Bragg-Laue Diffraction

The dynamical interaction of three-wave diffractions involving a Bragg reflection is more complicated than that of a three-wave Laue case, because of the presence of total reflection and the nonhermitian scattering matrix. Here, following the reference [53], we will illustrate the three-wave (000)(111)($\bar{1}\bar{1}$1) Bragg-Laue diffraction of germanium for CuKα_1 radiation, where (111) is a symmetric Bragg reflection and ($\bar{1}\bar{1}$1) a Laue transmission. The coupling reflection is (220).

To discuss the dispersion surface of this Bragg-Laue case, three sections corresponding to the three ranges defined by the total reflection need to be considered. The range of total reflection for (111), as shown in Fig. 8.3, is the region between $\Delta\theta = 5''$ and $\Delta\theta = 20''$. Three sections of the dispersion surface at the Laue point ($\Delta\theta = 0''$), Lorentz point ($\Delta\theta = 13''$), and $\Delta\theta = 25''$ along the (111) reflection are shown in Fig. 8.22. There are six dispersion sheets for the six possible modes of propagation. For a semi-infinitely thick crystal plate, according to (6.152), the number of permitted modes is 4. Those modes not permitted are labelled with a '×'. The dispersion curves running parallel to the horizontal $\Delta\psi$ axis are associated with the (111) reflection, while the inclined lines or curves from the upper left to lower right are due to the $(\bar{1}\bar{1}1)$ reflection. At $\Delta\theta = 0$, the dispersion curves approach $\pm 3275\,\mathrm{cm}^{-1}$,

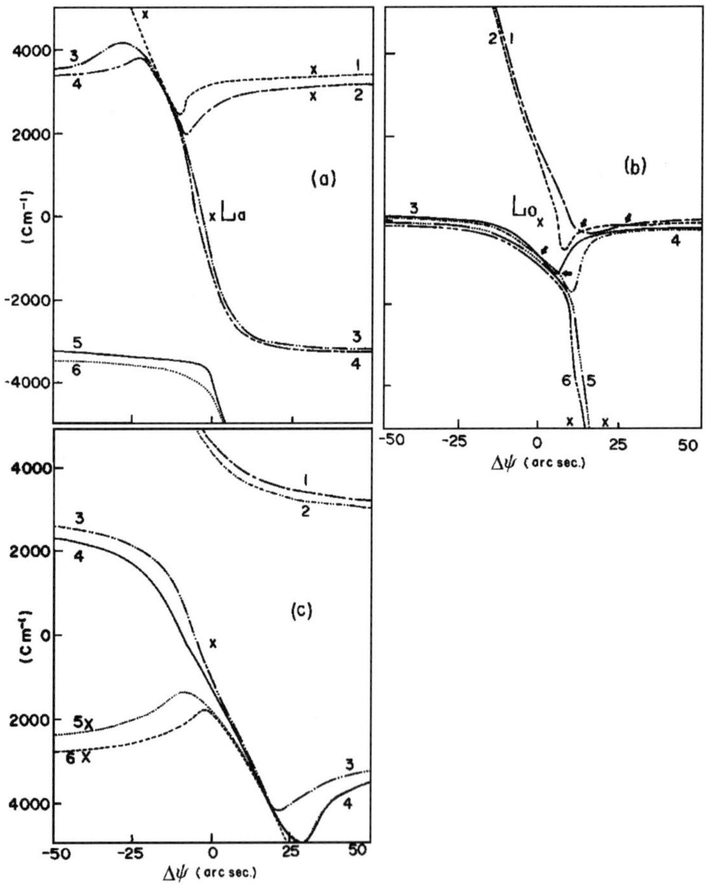

Fig. 8.22. Sections of the dispersion surface at (**a**) $\Delta\theta = 0$, (**b**) $\Delta\theta = 13''$, and (**c**) $\Delta\theta = 25''$ along the (111) line for the germanium (000)(111)$(\bar{1}\bar{1}1)$ Bragg-Laue case with CuKα_1 [53]

the two-wave values of the accommodation $k\zeta$, for large $|\Delta\psi|$ (Fig. 8.22(a)). In comparison with the two-wave dispersion surface of (111) (Fig. 8.3), the separation between the two pairs of dispersion sheets of modes 3 and 4 and modes 5 and 6 for large negative $\Delta\psi$ and that between modes 1 and 2 and modes 3 and 4 for large positive $\Delta\psi$ are nearly equal to the two-wave separation, 6550 cm^{-1} in this case. This separation is almost zero at the Lorentz point Lo, i.e., $\Delta\theta = 13''$, because the 'solution' dispersion sheets of the (111) reflection and the dispersion sheets of $(\bar{1}\bar{1}1)$ intersect at Lo (Fig. 8.22(b)). At $\Delta\theta = 25''$, the separation between the two pairs of dispersion sheets approach again the two-wave value.

As discussed in the beginning of this chapter, the so-called 'solution dispersion surface' accompanies the total reflection. Since the Lorentz point is the center of the total reflection range, the intersection of the solution dispersion sheets usually occurs near this point. This intersection also appears in the three-wave Bragg-Laue diffraction. In Fig. 8.22(b), the intersections of the dispersion sheets of modes 1 and 2 and modes 3 and 4 are marked with arrows. However, modes 1 and 3 are σ-polarized and modes 2 and 4 are π-polarized. The intersections indicate the polarization-dependent degeneracy of the eigenvalues of (6.46). This degeneracy sometimes causes ambiguity in the mode assignment in the dynamical calculation.

The dispersion curves of the permitted modes for $\Delta\theta = 0$ are all below the Laue point La at $\Delta\psi = 0$, so the corresponding absorption coefficients are positive. This is consistent with the definition of permitted mode given in Sect. 6.7. Figures 8.23(a) and (c) show the linear absorption coefficients for the four permitted modes at $\Delta\theta = 0$ and $25''$. For large $|\Delta\psi|$, the asymptotic two-wave value is 257 cm^{-1} for $\Delta\theta = 0$ and 695 cm^{-1} for $\Delta\theta = 25''$. The maximum of absorption occurs at $\Delta\psi = -15''$ for $\Delta\theta = 0$. This is due to the interaction of modes 1 and 2 and modes 3 and 4. That is, at this position, the two pairs of dispersion sheets nearly touch each other (see Fig. 8.22(a)). The minimum of absorption at $\Delta\psi = -2''$ for $\Delta\theta = 0$ is due to the anomalous transmission of the $(\bar{1}\bar{1}1)$ Laue reflection, because the dispersion sheets are the closest to the Laue point at this angular position. The absorption in the total reflection region is usually very high as expected (Fig. 8.23(b)). The interplay of absorptions among the modes takes place in the vicinity of the Lorentz point at $\Delta\theta = 13''$ and $\Delta\psi = 0$, as clearly shown in Fig. 8.23(b).

The excitation of mode is best illustrated for $\Delta\theta = 13''$, because the reflected intensity is maximal at the Lorentz point. Figure 8.24 shows the excitation for the four allowed modes. In the range far from the center of the total reflection range ($\Delta\psi = 13''$), modes 1 and 2 for positive $\Delta\psi$ and modes 3 and 4 for negative $\Delta\psi$ are excited with the excitation of 50%. The four modes are equally excited, 25% each, at $\Delta\psi = 0$ and maximal excitations occur in the range from $\Delta\psi = 10''$ to $\Delta\psi = 15''$. These increases in excitation are due to the interaction between the (111) and the $(\bar{1}\bar{1}1)$ reflections, as can be seen clearly from the dispersion surface shown in Fig. 8.22(b).

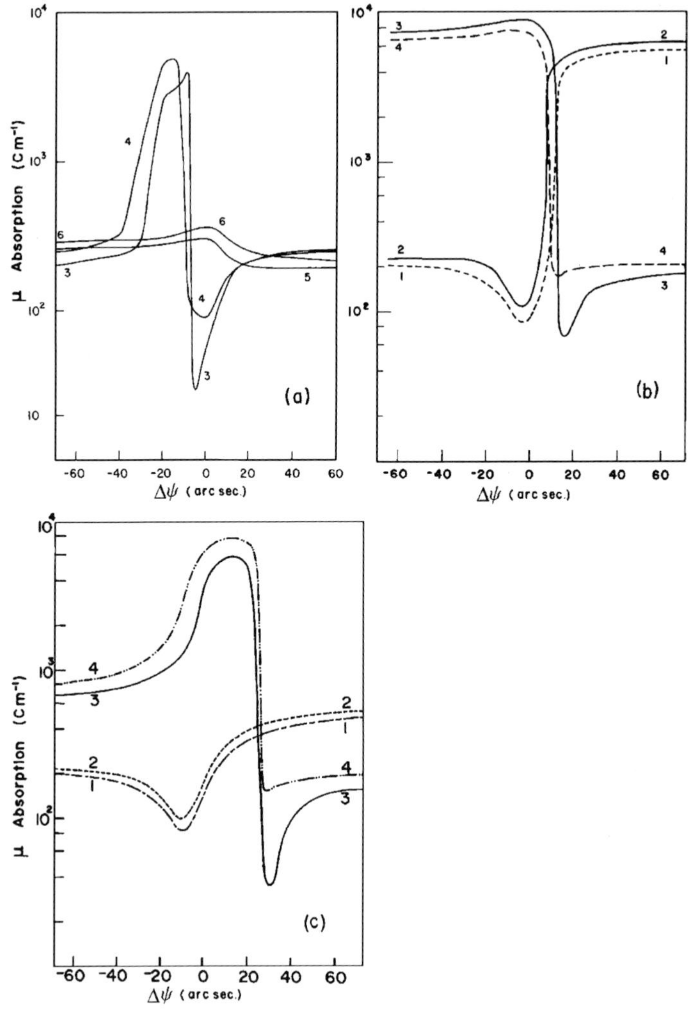

Fig. 8.23. Linear absorption coefficients for Fig. 8.22 [53]

The calculated intensity of the (111) reflection, I_{111}/I_O, in terms of $\Delta\theta$ and $\Delta\psi$ is shown in Fig. 8.25. The direction of the reflection lines of (111) and $(\bar{1}\bar{1}1)$ are also indicated. The intensity modification of the (111) reflection due to the presence of the $(\bar{1}\bar{1}1)$ is clearly seen. The integrated intensity over one arc minute in $\Delta\theta$ versus $\Delta\psi$ is plotted in Fig. 8.26(a). Additional integration over $\Delta\psi$ is necessary to compare with the measured intensity profiles. Nevertheless, the single integral is sufficient to describe the intensity profile qualitatively. The divergent-beam image of the (111) reflection at the three-wave Bragg-Laue diffraction position, which shows the intensity distribution, is similar to that for Ge (000)(004)$(\bar{1}\bar{1}1)$$(\bar{1}\bar{1}3)$ four-wave case shown

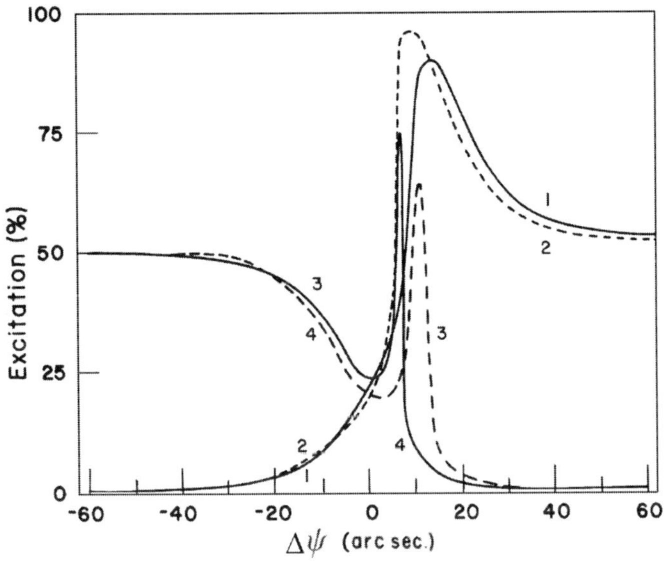

Fig. 8.24. Mode excitation for Fig. 8.22(b) [53]

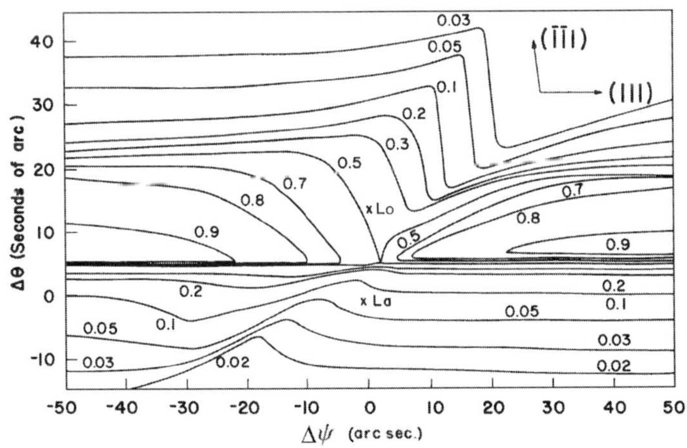

Fig. 8.25. Calculated intensity distribution for the germanium (000)(111)($\bar{1}\bar{1}1$) Bragg-Laue case and CuKα_1 [53]

in Fig. 4.7(a). The experimental integrated intensity profile (Fig 8.26(b)) obtained with the collimated-beam Renninger scan resembles the calculated one (Fig. 8.26(a)). The asymmetry of the profile, as will be discussed in Chap. 9, results from the phase of the involved structure-factor triplets.

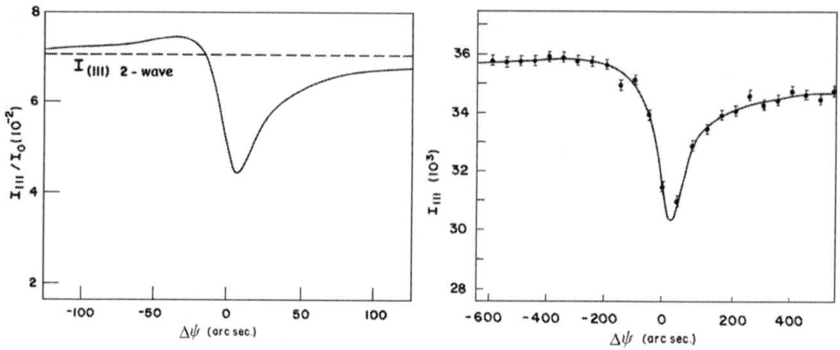

Fig. 8.26. Integrated intensity of Fig. 8.25: (**a**) calculated and (**b**) measured profiles [53]

8.2.3 Three-Wave Bragg-Bragg Diffraction

Three-wave Bragg-Bragg diffraction involves two Bragg reflections, aside from the direct (000) reflection. If an inversion operator is applied to the Bragg-Laue case just described, then a Bragg-Bragg diffraction emerges. Figure 8.27 shows this inversion operation. In other words, the Bragg-Bragg (000)(111)(220) case is an inversion case of the Bragg-Laue (000)(111)($\bar{1}\bar{1}1$) diffraction. Hence, the reciprocity theorem of optics holds.

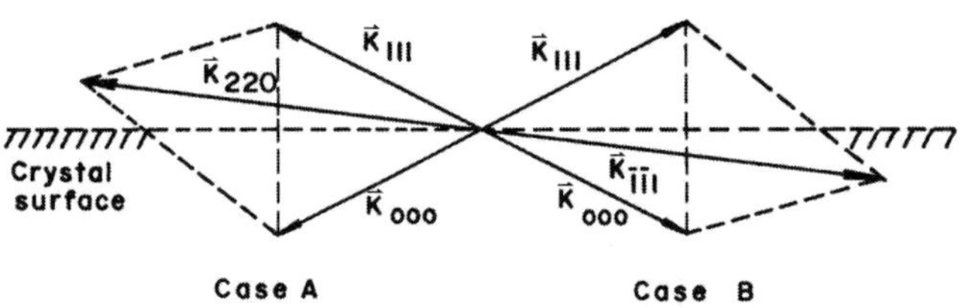

Fig. 8.27. Inversion relation between the (000)(111)($\bar{1}\bar{1}1$) Bragg-Laue case and the (000)(111)(220) Bragg-Bragg case [53]

Similarly, the dispersion surface also follows this reciprocity theorem. The dispersion surface at $\Delta\theta = 13''$ shown in Fig. 8.28 is the inversion of that shown in Fig. 8.22(b) for the Bragg-Laue case. The number of permitted modes for this Bragg-Bragg case is 2 for semi-infinitely thick crystals. Only modes 1 and 2 are allowed. The corresponding linear absorption coefficient and mode excitation are shown in Figs. 8.29(a) and (b). The resultant effects in absorption and excitation are similar to that for the Bragg-Laue case, except for the number of permitted modes. Moreover, the reflected intensity distribution and the integrated intensity of the (111) reflection are the same

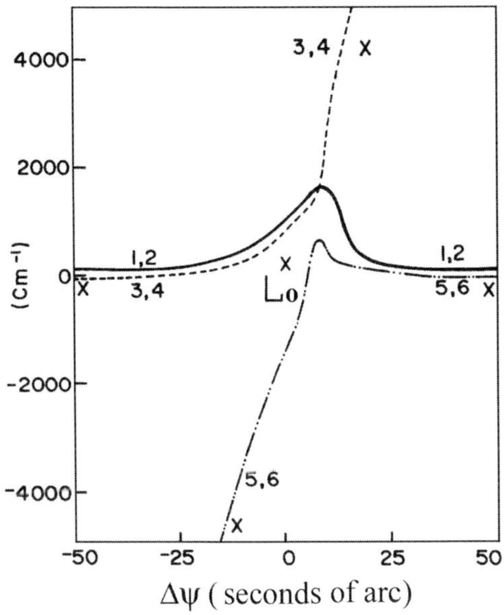

Fig. 8.28. Section of the dispersion surface at $\Delta\theta = 13''$ of the Ge (000)(111)(220) Bragg-Bragg case for CuKα_1 [53]

as that of its counter part, the Bragg-Laue diffraction, shown in Figs. 8.25 and 8.26. This indicates once again the two cases are linked by the reciprocity theorem.

The dispersion surface, linear absorption, and diffracted intensities of a three-wave Bragg-type diffraction with a forbidden reflection as the primary or secondary reflection can also be calculated in the same way [338] as described in this section.

8.2.4 Three-Wave Bragg-Surface Diffraction

The primary reflection G and the secondary reflection L of a three-wave (O, G, L) Bragg-surface diffraction are a symmetric Bragg and a surface reflection, respectively. The diffracted wave of the latter is along the crystal surface. Its reciprocal lattice point lies on the equatorial plane bisecting the reciprocal lattice vector of the primary reflection, i.e., $\ell_p = g/2$, where ℓ_p is the component of the reciprocal lattice vector $\boldsymbol{\ell}$ of the secondary reflection L parallel to $\boldsymbol{g}$. The direction cosine $\gamma_L = 0$ of the surface diffraction invalidates the dynamical calculation under a wide-angle approximation, where only the first-order terms of the accommodation ζ are considered. Attempts to solve this dynamical problem by treating a Bragg-surface case as the asymptote of either a Bragg-Bragg or a Bragg-Laue case may not give correct intensity

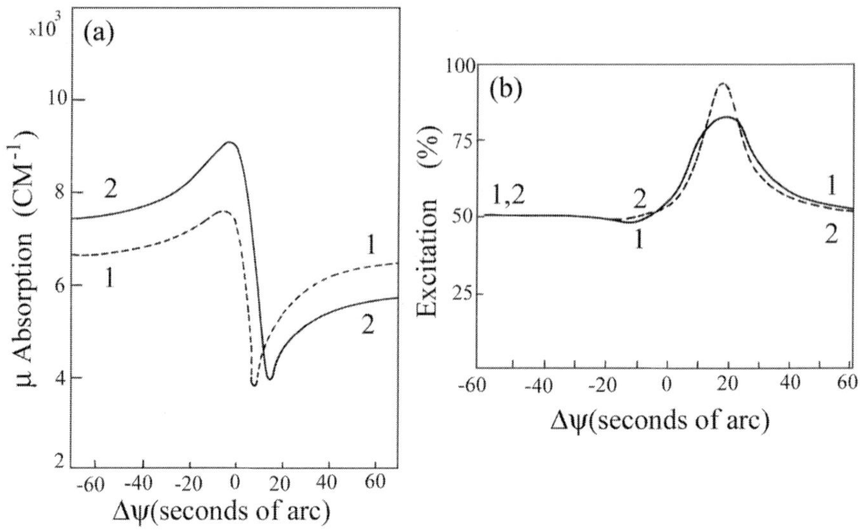

Fig. 8.29. (**a**) Linear absorption coefficients and (**b**) mode excitation for Fig. 8.28 [53]

distributions and dispersion surface [596]. Therefore, the dynamical calculation in the Cartesian coordinates representation given in Sect. 7.1.3 needs to be employed.

Figure 8.30 shows the intensity distributions of the primary G (222) reflection and the secondary L $(31\bar{1})$ reflection for the three-wave $(000)(222)(31\bar{1})$ diffraction of Ge and $\lambda = 1.540562$ Å with a π-polarized incident radiation. The horizontal line with $\Delta\theta = 0$ represents the (222) reflection line and the inclined direction is parallel to the surface $(31\bar{1})$ reflection line. The intensity

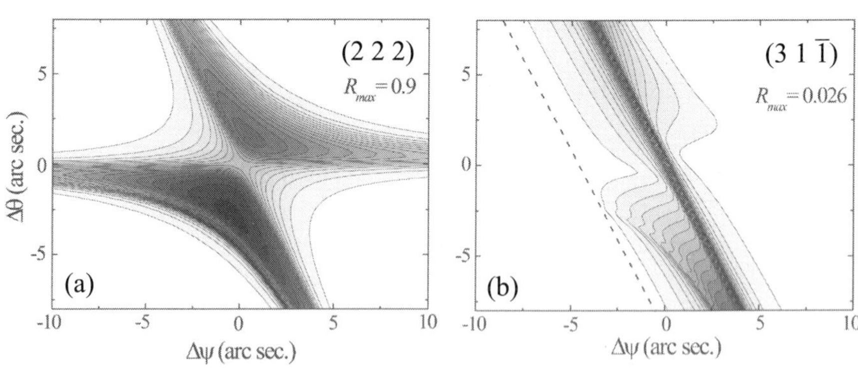

Fig. 8.30. Calculated intensity distributions of (**a**) G (222) and (**b**) L $(3\ 1\ \bar{1})$ for the three-wave Ge $(000)(222)(31\bar{1})$ diffraction and $\lambda = 1.540562$ Å with a π-polarized incident radiation

of (222) is redistributed in the vicinity of the exact three-wave point, $\Delta\theta = 0$ and $\Delta\psi = 0$, due to the presence of the $(31\bar{1})$ reflection (Fig. 8.30(a)). Similar situation happens to the $(31\bar{1})$ reflection (Fig. 8.30(b)). The maximal reflectivities plotted are 0.9 for the (222) and 0.026 for the $(31\bar{1})$. The dashed line shown in Fig. 8.30(b) indicates the exact angular positions at which the wavevector of the $(31\bar{1})$ lies on the crystal surface. Since there is no intensity along this line, the exact surface-diffracted wave has no intensity. This implies that the surface $(31\bar{1})$ reflection is more 'Bragg' (reflection) than 'Laue' (transmission) in the sense of dynamical diffraction.

The calculated dispersion surface and linear absorption coefficients for each mode along the (222) reflection line are shown in Figs. 8.31(a) and (b), and the details around the three-wave Lorentz point, $\Delta\theta = 0$ and $\Delta\psi = 0$, are plotted in Figs. 8.32(a) and (b), respectively.

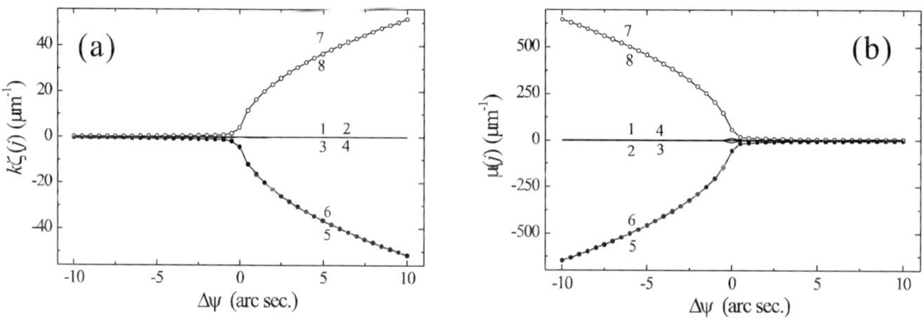

Fig. 8.31. (**a**) Calculated dispersion surface and (**b**) calculated linear absorption coefficients for the three-wave Ge $(000)(222)(31\bar{1})$ diffraction and $\lambda = 1.540562$ Å with a π-polarized incident radiation

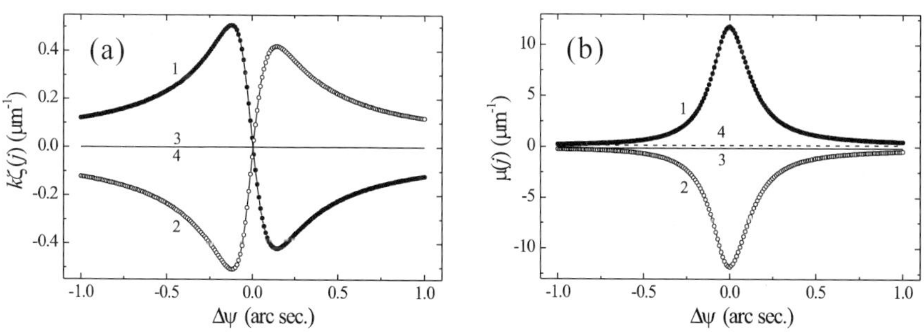

Fig. 8.32. The details of Fig. 8.31

There are 12 possible modes of propagation, among which modes 1 to 4 are associated with the symmetric (222) Bragg reflection and modes 5 to 8 with the $(31\bar{1})$. The tie points for modes 9 to 12 associated with the (000) reflection are very far from the Lorentz point, which are not shown in Figs. 8.31(a) and (b). Since the cross section of the dispersion surface is at the Lorentz point along the (222), the accommodations $k\zeta$ of modes 1 to 4, which define the solution surface at the center of the total reflection region, are, as expected, very close to zero (Figs. 8.31(a) and 8.32(a)). Half of the modes have negative linear absorption coefficients. They are modes 1, 4, 7, and 8 (see Figs. 8.31(b) and 8.32(b)). These modes are not allowed in the diffraction process for an infinitely thick crystal. The peak in the absorption shown in Fig. 8.32(b) is attributed to the total reflection at the center of the total reflection region of the (222) reflection. Similar calculations for the three-wave Bragg-surface diffraction have been reported recently [360].

8.2.5 Three-Wave Grazing Incidence Diffraction

The geometry of a three-wave (O, G, L) grazing incidence X-ray diffraction (GIXD) in the reciprocal space is shown in Fig. 8.33, where the three reciprocal lattice points O, G, and L lie in the crystal surface OGL. Point O, the origin of the reciprocal lattice, represents the direct reflection. Vectors $\boldsymbol{g} = \boldsymbol{OG}$ and $\boldsymbol{\ell} = \boldsymbol{OL}$ are the reciprocal lattice vectors of the G and L reflections. Point C is the center of the triangle OGL and M is the middle point of OG. The semi-circle is the locus of the Laue point of the two-wave G reflection around $\boldsymbol{g}$. For a given entrance point E on the semi-circle, $EO = EG = k = 1/\lambda$. $\angle EOP$ and $\angle EMP$ are the incident angle ϕ_O and the azimuthal angle ψ around $\boldsymbol{g}$, respectively.

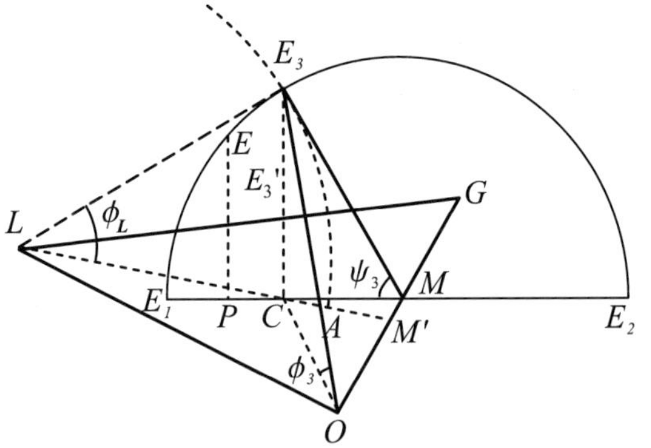

Fig. 8.33. Geometry of three-wave GIXD in reciprocal space [566]

The three-wave GIXD takes place at point E_3. The wavevectors $\boldsymbol{k}_O$, $\boldsymbol{k}_G$, and $\boldsymbol{k}_L$ outside the crystal are $\boldsymbol{E_3O}$, $\boldsymbol{E_3G}$, and $\boldsymbol{E_3L}$. Lines EP and E_3C are normal to the crystal surface, and E_3' is the tie point on the dispersion surface with respect to the entrance point E_3. The angles $\angle E_3OC$ and $\angle E_3MC$ are the incident angle ϕ_3 and the azimuthal angle ψ_3 at the exact three-wave diffraction position, respectively. The corresponding reflection angles of the G and L reflections are denoted as ϕ_G and ϕ_L. If $E_3C = 0$, then E_3 coincides with C, i.e., $\phi_3 = 0$ and $\psi_3 = 0$, thus the diffracted intensities are null. Only the three-wave GIXD's with $\phi_3 \neq 0$ and $\psi_3 \neq 0$ have appreciable intensities. In the following, we shall deal with this kind of GIXD.

Point C is the projection of E_3 onto the crystal surface and lies inside the triangle OGL, as shown in Fig. 8.33. If the ψ rotation is carried out from E_1 to E_3, then the three-wave GIXD occurring at E_3 is in the 'OUT' situation, because point L is moving away from E_3. If the point C is brought by the ψ rotation to a position outside the triangle OGL, then the three-wave case is in its 'IN' position, because point L is moving towards E_3 during the ψ rotation.

The diffraction condition for a general three-wave GIXD is

$$\lambda = 2d_G \sin \angle OLG \cos \phi_3 = 2d_G \sin \theta_G \ , \tag{8.72}$$

where θ_G is the Bragg angle and d_G the interatomic spacing of the G reflection. When $\phi_L = 0$, the corresponding azimuthal angle ψ_Δ is

$$\cos \psi_\Delta \cong \frac{\left[\cos \phi_3 \cos \angle OLG - \frac{\sin^2 \phi_3}{2 \cos \phi_3 \cos \nu}\right]}{\cos \theta_G} \ , \tag{8.73}$$

where $\nu = \angle MCM'$, and the corresponding incident angle ϕ_Δ is defined as

$$\sin \phi_\Delta = \cos \theta_G \sin \psi_\Delta \ , \tag{8.74}$$

where ψ_3 is assumed to be very small. According to (8.72), the wavelength λ varies as the incident angle ϕ_3 changes. For a given λ, the angle ϕ_O varies during the ψ scan. The surface specularly reflected intensities P_O^s, P_G^s, and P_L^s versus ψ are the diffraction profiles of interest.

The fundamental equations of wavefield take a similar form to (7.54) as

$$\begin{aligned}
&2\xi_O E_{\sigma O} + P_\sigma \chi_{O\text{-}G} E_{\sigma G} + d_1 \chi_{O\text{-}L} E_{\sigma L} = 0 \ , \\
&2\xi_O E_{\pi O} + P_\pi \chi_{O\text{-}G} E_{\pi G} + d_2 \chi_{O\text{-}L} E_{\sigma L} + d_3 \chi_{O\text{-}L} E_{\pi L} = 0 \ , \\
&P_\sigma \chi_{G\text{-}O} E_{\sigma O} + 2\xi_G E_{\sigma G} + d_4 \chi_{G\text{-}L} E_{\sigma L} = 0 \ , \\
&P_\pi \chi_{G\text{-}O} E_{\pi O} + 2\xi_G E_{\pi G} + d_5 \chi_{G\text{-}L} E_{\sigma L} + d_6 \chi_{G\text{-}L} E_{\pi L} = 0 \ , \\
&d_1 \chi_{L\text{-}O} E_{\sigma O} + d_2 \chi_{L\text{-}O} E_{\pi O} + d_4 \chi_{L\text{-}G} E_{\sigma G} + d_5 \chi_{L\text{-}G} E_{\pi G} + 2\xi_L E_{\sigma L} = 0 \ , \\
&d_3 \chi_{L\text{-}O} E_{\pi O} + d_6 \chi_{L\text{-}G} E_{\pi G} + 2\xi_L E_{\pi L} = 0 \ ,
\end{aligned} \tag{8.75}$$

where the polarization factors are defined as

$$\begin{aligned}
&P_\sigma = \hat{\boldsymbol{\sigma}}_O \cdot \hat{\boldsymbol{\sigma}}_G = 1 \ , \\
&P_\pi = \hat{\boldsymbol{\pi}}_O \cdot \hat{\boldsymbol{\pi}}_G \approx \cos 2\theta_G \ , \\
&d_1 = \hat{\boldsymbol{\sigma}}_O \cdot \hat{\boldsymbol{\sigma}}_L \ , \quad d_2 = \hat{\boldsymbol{\pi}}_O \cdot \hat{\boldsymbol{\sigma}}_L \ , \quad d_3 = \hat{\boldsymbol{\pi}}_O \cdot \hat{\boldsymbol{\pi}}_L \ , \\
&d_4 = \hat{\boldsymbol{\sigma}}_G \cdot \hat{\boldsymbol{\sigma}}_L \ , \quad d_5 = \hat{\boldsymbol{\pi}}_G \cdot \hat{\boldsymbol{\sigma}}_L \ , \quad d_6 = \hat{\boldsymbol{\pi}}_G \cdot \hat{\boldsymbol{\pi}}_L \ .
\end{aligned} \tag{8.76}$$

The unit vectors $\hat{\boldsymbol{\sigma}}_O$ and $\hat{\boldsymbol{\sigma}}_G$ are chosen to be perpendicular to the E_3OG plane and $\hat{\boldsymbol{\pi}}_O$ and $\hat{\boldsymbol{\pi}}_G$ lie in that plane. The unit vector $\hat{\boldsymbol{\pi}}_L$ is chosen to be normal to the $E_3'ML$ plane, so that it is perpendicular to both $\hat{\boldsymbol{\sigma}}_O$ and $\hat{\boldsymbol{\sigma}}_G$. Assume that E_3' is a tie point on a branch of the dispersion surface. Then $\boldsymbol{E_3E_3'} = k\zeta\hat{\boldsymbol{n}}_{\boldsymbol{e}}$, where the unit vector $\hat{\boldsymbol{n}}_{\boldsymbol{e}}$ is normal to the surface and pointing towards the interior of the crystal. $\boldsymbol{K}_O = \boldsymbol{E_3'O}$, $\boldsymbol{K}_G = \boldsymbol{E_3'G}$, and $\boldsymbol{K}_L = \boldsymbol{E_3'L}$ are the wavevectors inside the crystal. The $\hat{\boldsymbol{\sigma}}$'s, $\hat{\boldsymbol{\pi}}$'s, and $\hat{\boldsymbol{K}}$'s are related, as usual, by

$$\hat{\boldsymbol{\sigma}}_L = \hat{\boldsymbol{K}}_L \times \hat{\boldsymbol{\pi}}_L \, . \tag{8.77}$$

The angles ψ_3 and ϕ_3 are very small for the grazing incidence situation, so that the wavevectors $\boldsymbol{k}$'s are nearly parallel to the crystal surface. Under this circumstance, the polarization factors can be written as

$$\begin{aligned} d_1 &= d_4 \cong 1 \, , \qquad d_2 = d_5 = 0 \, , \\ d_3 &\cong \cos 2\theta_L \, , \qquad d_6 \cong \cos 2\theta_{G\text{–}L} \, , \end{aligned} \tag{8.78}$$

where $\theta_{G\text{–}L}$ is the Bragg angle of the coupling G–L reflection. With this small-angle approximation, the fundamental equation can be decomposed into two sets of matrix equations of the form

$$\begin{bmatrix} \chi_O + \phi_O^2 - u_O^2 & p_G\chi_{O\text{–}G} & p_L\chi_{O\text{–}L} \\ p_G\chi_{G\text{–}O} & \chi_O + \phi_O^2 - u_O^2 - \alpha_G & p_{G\text{–}L}\chi_{G\text{–}L} \\ p_L\chi_{L\text{–}O} & p_{G\text{–}L}\chi_{L\text{–}G} & \chi_O - \phi_O^2 - u_O^2 - \alpha_L \end{bmatrix} \begin{bmatrix} E_O \\ E_G \\ E_L \end{bmatrix} = [0] \, , \tag{8.79}$$

where

$$p_G \cong p_L \cong p_{G-L} = 1 \tag{8.80}$$

for the σ polarization, and

$$p_G = \cos 2\theta_G \, , \quad p_L = \cos 2\theta_L \, , \quad p_{G\text{–}L} = \cos 2\theta_{G\text{–}L} \tag{8.81}$$

for the π polarization. The quantities ϕ_O, u_O, and α_G, which have been defined in (8.19), (8.48), and (8.46), are rewritten as

$$\begin{aligned} \phi_O &= \frac{k_{Oz}}{k} \, , \qquad u_O = \frac{K_{Oz}}{k} \, , \\ \alpha_G &= \frac{(\boldsymbol{k}_O + \boldsymbol{g})^2 - k_O^2}{k^2} \, . \end{aligned}$$

α_L takes a similar form to that of α_G, except that G is replaced by L and $\boldsymbol{g}$ by $\boldsymbol{\ell}$. k_{Oz} and K_{Oz} are the vector components of $\boldsymbol{k}_O$ and $\boldsymbol{K}_O$ normal to the crystal surface. Since the G planes always satisfy Bragg's law, $\alpha_G = 0$.

Equation (8.79) can be solved as an eigenvalue problem as usual. The determinant of the matrix of (8.79) equal to zero with $\alpha_G = 0$ is the dispersion equation:

$$\begin{aligned} W^3 - \alpha_L W^2 + pW + (&\alpha_L\chi_{G\text{–}O}\chi_{O\text{–}G} + \chi_{O\text{–}G}\chi_{L\text{–}O}\chi_{G\text{–}L} \\ &+ \chi_{G\text{–}O}\chi_{O\text{–}L}\chi_{L\text{–}G}) = 0 \, , \end{aligned} \tag{8.82}$$

where

$$W = \chi_O + \phi_O^2 - u_O^2 \, ,$$
$$p = -\chi_{G\text{–}O}\chi_{O\text{–}G} - \chi_{L\text{–}O}\chi_{O\text{–}L} - \chi_{G\text{–}L}\chi_{L\text{–}G} \, . \tag{8.83}$$

The coordinates of the tie point and the absorption coefficient are given by the real parts and imaginary parts of the eigenvalues determined from the dispersion equation. The eigenvectors provide the wavefield-amplitude ratios among the diffracted waves. The absolute amplitudes can be determined by the boundary conditions for the continuities of the normal components of the electric displacements and the tangential components of the magnetic fields at the crystal boundary:

$$\begin{aligned}
&E_O^e + E_O^s = \sum_{j=1}^{3} c_j E_O(j) \, , \\
&E_G^s = \sum_{j=1}^{3} c_j E_G(j) \, , \qquad E_L^s = \sum_{j=1}^{3} c_j E_L(j) \, , \\
&\phi_O \left(E_O^e - E_O^s\right) = \sum_{j=1}^{3} u_O(j) c_j E_O(j) \, , \\
&-\phi_G E_G^s = \sum_{j=1}^{3} u_O(j) c_j E_G(j) \, , \quad -\phi_L E_L^s = \sum_{j=1}^{3} u_O(j) c_j E_L(j) \, ,
\end{aligned} \tag{8.84}$$

E_O^e is the incident wavefield amplitude and E_O^s, E_G^s, and E_L^s are the amplitudes of the surface-reflected waves of the O, G, and L reflections. The coefficient c_j is the proportionality constant for mode j, which can be determined from (8.84). The corresponding intensities of these surface-reflected waves are

$$\begin{aligned}
P_O^s &= \left|\frac{E_O^s}{E_O^e}\right|^2 \, , \\
P_G^s &= \left|\frac{E_G^s}{E_O^e}\right|^2 \left(\frac{\phi_G}{\phi_O}\right) \, , \\
P_L^s &= \left|\frac{E_L^s}{E_O^e}\right|^2 \left(\frac{\phi_L}{\phi_O}\right) \, .
\end{aligned} \tag{8.85}$$

All these items can be calculated numerically for a general three-wave GIXD. It is, however, also desired to have analytical expressions derived to reveal the relationship between the various physical parameters, though it is difficult to do so for a general three-wave case. Alternatively, we may find analytical expressions for diffractions in special geometries. Here we consider a symmetric three-wave GIXD, in which $OG = OL = G - L$. The Bragg condition then becomes

$$\lambda = 2d_G \sin\left(\frac{\pi}{3}\right) \cos\phi_3 \,. \tag{8.86}$$

The polarization factors are given by

$$\begin{aligned} &P_\sigma = 1\,, \qquad P_\pi = d_3 = d_6 = \cos 2\theta_G\,, \\ &d_1 = d_4 = \cos(\pi - \tau)\,, \\ &d_2 = d_5 = \sin(\pi - \tau)\sin\theta_G\,, \end{aligned} \tag{8.87}$$

where the angle $\tau = \angle MEL$. The first equation of (8.19) still holds for this three-wave GIXD, i.e.,

$$-2\varepsilon_O = \sin^2\phi_C = 2\zeta\sin\phi_O - \zeta^2\,. \tag{8.88}$$

Thus,

$$\begin{aligned} &\delta = \sin\phi_O - \sqrt{\sin^2\phi_O - \sin^2\phi_C}\,, \\ &\alpha_L = \sin^2\phi_O - \sin^2\phi_L\,. \end{aligned} \tag{8.89}$$

For simplicity, let's consider only the σ polarization. The eigenvalues for modes 1, 2, and 3 can be expressed as

$$\begin{aligned} &-2\varepsilon_O(1) = \sin^2\phi_{C1} = \chi_O - \sqrt{\chi_{G-O}\chi_{O-G}}\,, \\ &-2\varepsilon_O(2) = \sin^2\phi_{C2} = \chi_O + b_-\sqrt{\chi_{G-O}\chi_{O-G}}\,, \\ &-2\varepsilon_O(3) = \sin^2\phi_{C3} = \chi_O + b_+\sqrt{\chi_{G-O}\chi_{O-G}}\,, \end{aligned} \tag{8.90}$$

where

$$\begin{aligned} &b_\mp = \frac{1 + \Delta_L \mp \sqrt{9 - 2\Delta_L + \Delta_L^2}}{2}\,, \\ &\Delta_L = \frac{\alpha_L}{\sqrt{\chi_{G-O}\chi_{O-G}}}\,. \end{aligned} \tag{8.91}$$

As α_L approaches zero,

$$\begin{aligned} &\sin^2\phi_{C1} = \chi_O - \sqrt{\chi_{G-O}\chi_{O-G}}\,, \\ &\sin^2\phi_{C2} = \chi_O + \sqrt{\chi_{G-O}\chi_{O-G}}\,. \end{aligned} \tag{8.92}$$

The eigenvectors of the three modes are given by

$$\begin{aligned} &E_G(1) = E_O(1)\,, \qquad E_L(1) = 0\,, \\ &E_G(2) = -E_O(2)\,, \qquad E_L(2) = \sqrt{2}\Delta_- E_O(2)\,, \\ &E_G(3) = -E_O(3)\,, \qquad E_L(3) = \sqrt{2}\Delta_+ E_O(3)\,, \end{aligned} \tag{8.93}$$

where

$$\begin{aligned} &\Delta_- = \Delta\,, \qquad \Delta_+ = -\frac{1}{\Delta}\,, \\ &\Delta = \frac{\dfrac{\sqrt{\chi_{G-O}\chi_{O-G}}}{2} - \alpha_L - \dfrac{\sqrt{9\chi_{G-O}\chi_{O-G} - 2\alpha_L + \alpha_L^2}}{2}}{\sqrt{2\chi_{G-O}\chi_{O-G}}}\,. \end{aligned} \tag{8.94}$$

When α_L approaches zero, Δ is close to 2. The wavefield amplitudes inside the crystal are

$$E_O(1) = \left(\frac{\sin\phi_O}{C_1}\right) E_O^e ,$$

$$E_O(2) = \left(\frac{a_- \sin\phi_O}{C_2}\right) E_O^e ,$$

$$E_O(3) = \left(\frac{a_+ \sin\phi_O}{C_3}\right) E_O^e , \tag{8.95}$$

where E_O^e is the amplitude of the incident wave, C_j is given in (8.55), and

$$a_- = \frac{1}{1+\gamma} , \qquad a_+ = \frac{1}{1+\frac{1}{\gamma}} , \tag{8.96}$$

with

$$\gamma = \Delta^2 \left(\frac{C_3}{C_2}\right) \frac{\sin\phi_L + \sqrt{\sin^2\phi_O - \sin^2\phi_{C2} + \alpha_L}}{\sin\phi_L + \sqrt{\sin^2\phi_O - \sin^2\phi_{C3} + \alpha_L}} . \tag{8.97}$$

As ϕ_O approaches zero, γ is close to 2.

The amplitudes of the surface-reflected waves are

$$E_O^s = -E_O^e + E_O^e \left(\frac{1}{C_1} + \frac{a_-}{C_2} + \frac{a_+}{C_3}\right) \sin\phi_O ,$$

$$E_G^s = E_O^e \left(\frac{1}{C_1} - \frac{a_-}{C_2} + \frac{a_+}{C_3}\right) \sin\phi_O ,$$

$$E_L^s = \sqrt{2} E_O^e \left(\frac{\Delta_- a_-}{C_2} + \frac{\Delta_+ a_+}{C_3}\right) \sin\phi_O . \tag{8.98}$$

The corresponding intensities can be calculated by using (8.85).

The three-wave (000)(220)(202) case of germanium is used to illustrate the dynamical calculation for a three-wave GIXD. The σ polarization is chosen. It is known in the two-wave (000)(220) GIXD that the critical angles are $\phi_{C1} = 6.5419$ mrad and $\phi_{C2} = 16.9616$ mrad for $\lambda = 3.463683$ Å. The number of excited tie points for an incident angle ϕ_O in different angular regions with respect to the two critical angles is different. These characteristics remain for the three-wave GIXD. Moreover, according to (8.72) and (8.86), the incident angle ϕ_3 for the three-wave GIXD depends on the X-ray wavelength used. The three wavelengths, 3.463683 Å, 3.464341 Å, and 3.464653 Å, calculated from (8.86) for ϕ_3= 24, 14, and 4 mrad are chosen to illustrate the diffraction situations for $\phi_3 < \phi_{C1}$, $\phi_{C1} < \phi_3 < \phi_{C2}$, and $\phi_{C2} < \phi_3$, respectively. The corresponding ϕ_Δ are 29.393, 17.146, and 4.899 mrad. Figures 8.34, 8.35, and 8.36 show the calculated intensities, section of dispersion surface, mode excitation and penetration depth versus the incident angle ϕ_O for $\lambda = 3.463683$ Å, 3.464341 Å, and 3.464653 Å, respectively.

The calculated P_{000}^s, P_{220}^s, P_{202}^s, and ΣP^s versus ϕ_O are shown in Fig. 8.34 for 3.463683 Å. Since ϕ_3 (= 24 mrad) is far from the critical angles ϕ_{C1} and

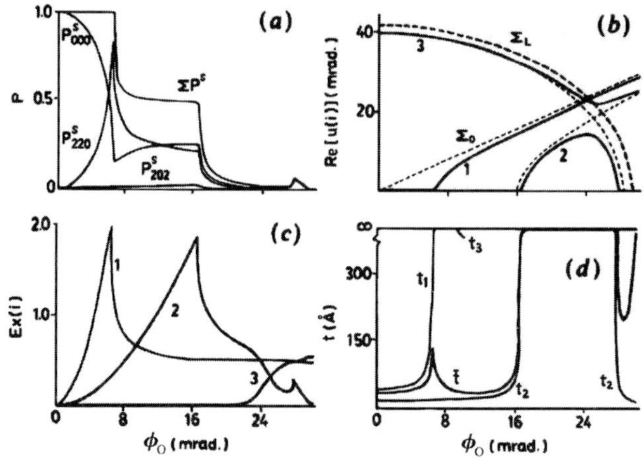

Fig. 8.34. (**a**) Calculated intensities, (**b**) dispersion surface, (**c**) mode excitation, and (**d**) penetration depth for Ge (000)(220)(202) GIXD and 3.463683 Å [566]

ϕ_{C2}, the presence of the (202) reflection has little effect on the (000) and (220). Thus, P^s_{000}, P^s_{220}, and ΣP^s behave as if they were two-wave GIXDs. The intensity distributions, therefore, resemble that shown in Fig. 8.13(a). Two peaks with the intensities of 0.056 and 0.019 for P^s_{202} occur at $\phi_O =$ 27.75 mrad and 16.5 mrad.

Three dispersion curves are shown in Fig. 8.34(b). The curve of mode 1 in the three-wave case is exactly the same as that in the two-wave (000)(220) case, which can be calculated by the first equation of (8.90). Curves 2 and 3 maintain the two-wave characteristics of the (220) and (202) reflections when ϕ_O is far away from ϕ_3(= 24 mrad). As ϕ_O approaches ϕ_3, the hyperbolic curves result. At $\phi_O = 24.5$ mrad, the distance between curves 2 and 3 is minimal, indicating that a strong interaction between modes 2 and 3 takes place. The intersection of curves 1 and 2 at ϕ_3 implies that the two modes are degenerate. The curves denoted as Σ_O and Σ_L are the wavefronts of the one-wave (000) and (202) reflections. Other dashed curves represent the two-wave dispersion curves of the (220) and (202) reflections; the curve close to mode 2 represents the former and the curve extending from mode 3 the latter. They serve as the asymptotes for the three-wave dispersion hyperbola just mentioned.

The excitation of mode 1 in the three-wave (000)(220)(202) case is the same as that in the two-wave (000)(220) GIXD (see Fig. 8.13(c)). The behavior of mode 2 is similar to that in the two-wave case, except that near ϕ_3 the effect due to the presence of the (202) reflection becomes appreciable. The small kink at $\phi_O = 27.75$ mrad is the result of the total reflection near the edge of the dispersion curve of mode 2. The excitation of mode 3 is almost zero for $\phi_O < \phi_3$, increases gradually for ϕ_O close to ϕ_3, and reaches its two-wave value of 50% for large ϕ_O.

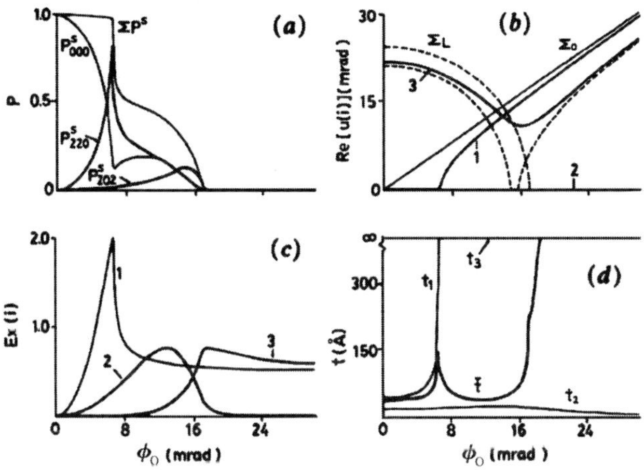

Fig. 8.35. (**a**) Calculated intensities, (**b**) dispersion surface, (**c**) mode excitation, and (**d**) penetration depth for Ge (000)(220)(202) GIXD and 3.464341 Å [566]

The penetration depths t_1, t_2 and the average depth $\bar{t}$ are similar to that of the two-wave (000)(220) case for $\phi_O < \phi_{C1}$. t_2 and t are infinite for $\phi_{C2} < \phi_O < \phi_{C1}$ because of the total transmission of mode 2. As ϕ_O increases, t_2 decreases to the two-wave value, 17 Å. The average $\bar{t}$ has a minimum of 199 Å near $\phi_O = 28.30$ mrad and t_3 is always infinite. This total transmission seems to be associated with the Poynting vectors normal to the dispersion curves of modes 1, 2, and 3.

Figure 8.35 shows the calculated results for $\phi_3 = 14$ mrad and $\lambda = 3.464341$ Å. Since the two-wave dispersion curves of the (220) and (202) reflections, represented by the dashed curves in Fig. 8.35(b), do not intersect, the dispersion curve of mode 2 is along the abscissa, compared to the hyperbola shown in Fig. 8.34(b) for $\lambda = 3.463683$ Å. The surface-reflected intensities P^s_{000}, P^s_{220}, and ΣP^s in Fig. 8.35(a) show no sharp steps near $\phi_{C2} = 15.9616$ mrad. P^s_{202} increases from zero at $\phi_O = 0$ to a maximum near $\phi_O \simeq 15$ mrad due to the total reflection in the region around $\phi_O \simeq 15$ mrad between the two dashed curves shown in Fig. 8.35(b). It decreases abruptly at $\phi_O \simeq 16.5$ mrad, the point where Σ_L touches the crystal surface, and gradually approaches zero for large ϕ_O. The dispersion curves of modes 1 and 3 are similar to those for $\lambda = 3.463683$ Å shown in Fig. 8.34(b). The degeneracy occurs at ϕ_3 for modes 1 and 3.

The excitation of mode 1 remains unchanged (Fig. 8.35(c)), while the excitation of mode 2 has a maximum at ϕ_3, weaker than that for $\lambda = 3.463683$ Å. In Fig. 8.35(d), the penetration depths t_1, t_3, and $\bar{t}$ behave like that shown in Fig. 8.34(d), except that $\bar{t}$ does not have a minimum at $\phi_O = 28$ mrad. t_2 is a slowly varying curve, starting with 16.3 Å at $\phi_O = 0$, reaching a maximum of 21.5 Å at $\phi_O = \phi_3$, and approaching 6 Å for large ϕ_O.

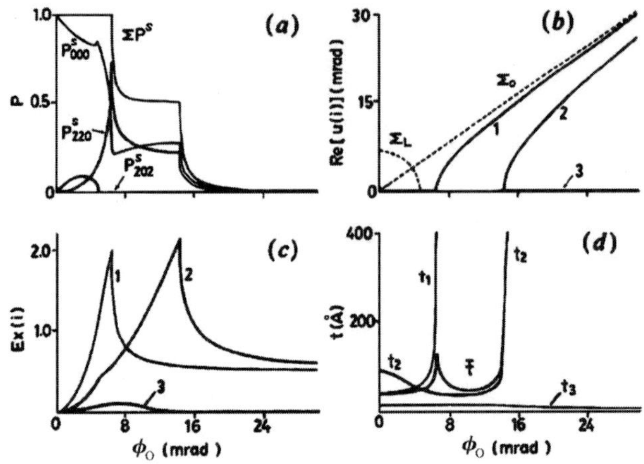

Fig. 8.36. (**a**) Calculated intensities, (**b**) dispersion surface, (**c**) mode excitation, and (**d**) penetration depth for Ge (000)(220)(202) GIXD and 3.464653 Å [566]

The incident angle for the three-wave diffraction is $\phi_3 = 4$ mrad at $\lambda = 3.464653$ Å. As shown in Fig. 8.36(b), the dispersion curves of modes 1 and 2 resemble that in the two-wave case, except that the step of curve 2 does not occur at ϕ_{C1} (= 15.96 mrad) but at 14.4 mrad. The dispersion curve of mode 3 is a straight line along the abscissa. Since the wavefront Σ_L of the (202) reflection does not intersect the dispersion curves 1 and 2, total reflection is expected to occur in the angular range from $\phi_O = 0$ to $\phi_\Delta = 4.9$ mrad, where the Σ_L cuts the crystal surface. Indeed, the intensity curve P^s_{202} in Fig. 8.36(a) shows the reflectivity of the total reflection as a bump in the same angular range. Accordingly, P^s_{000} exhibits a dip in this range. Again, P^s_{220} and ΣP^s behave similarly to that for the two-wave (220) case.

In Fig. 8.36(c), the excitation of mode 1 remains the same as that for two-wave diffraction. The curve of the excitation of mode 2 is similar to the two-wave excitation curve, with, however, a small kink at $\phi_\Delta = 4.9$ mrad and a maximum of 220% at $\phi_O = 14.5$ mrad. The maximum value exceeding 200% is due to the definition of the excitation given in (8.58), where the directions of energy flow are not considered. Mode 3 is excited in the angular region from 0 to 14.5 mrad, the total reflection ranges of the (220) and (202) reflections.

The penetration depth t_2 in the range of total reflection decreases from 90.1 Å at $\phi_O = 0$ towards the minimum value of 35 Å. The other penetration depths behave similarly to that for $\lambda = 3.464341$ Å shown in Fig. 8.35(d).

Three-wave GIXD is, in principle, a wavevector-coplanar multiple-wave diffraction for $\phi_O = 0$. The surface-reflected intensities other than the incident reflection are null. The above demonstrated calculations are for the situation in which the angle of incidence is slightly off zero and the involved wavevectors are non-coplanar. In this situation, surface-reflected waves have appreciable intensities. Because of $\phi_O > 0$, the dispersion surface lies above the crystal surface as shown in Figs. 8.34, 8.35, 8.36.

The effect of the coupling between the σ- and π-polarized wavefields in this non-coplanar diffraction can be estimated from the second equation of (8.75). The order of magnitude of the π-polarized waves compared with that of the σ-polarized ones ranges from 0.001 to 0.05, which is so small that can be neglected. In the calculations, we have purposely ignored the absorption to bring out the information about the extinction, though the crystal and wavelengths chosen heavily involve the absorption. As can be seen from the calculated penetration depths, the average extinction lengths for these particular GIXDs are about 15–50 Å. The maximum depths at the critical angle ϕ_{C1} are around 100–300 Å. Phase is another important factor affecting the reflected intensities, which will be discussed in Chap. 9 as the X-ray phase problem is concerned.

8.2.6 Three-Wave Coplanar Diffraction

Three-wave coplanar diffraction means three-wave wavevector-coplanar diffraction [32,50,55]. Consider a three-wave (O, G, L) coplanar diffraction, where all the wavevectors $\boldsymbol{K}$'s lie in the crystal surface. All the $\hat{\boldsymbol{\sigma}}$ polarization unit vectors are perpendicular to the crystal surface and all the $\hat{\boldsymbol{\pi}}$ unit vectors lie in that surface. This situation can be illustrated by moving point E_3 to coincide with point C in Fig. 8.33. Consequently, the corresponding polarization factors are exactly equal to those given in (8.78). So are the fundamental equation (8.79), the boundary conditions (8.84), and the diffraction intensities (8.85).

8.2.7 Crystal Truncation Rod Diffraction

A crystal surface is considered as the result of the truncation of a three-dimensional crystal. Although it is a two-dimensional arrangement of atoms, the relative positions of the surface atoms with respect to the crystal bulk involve the third dimension along the crystal surface normal. In using X-ray scattering and diffraction to probe this three-dimensional structure, in-plane scan and surface-normal scan are usually employed to gather the structural information in directions parallel and perpendicular to the crystal surface, respectively. Surface-normal scan is sometimes called the 'crystal-truncation-rod (CTR) scan' in the literature [168, 169] because the scan is along the reciprocal rods which are the Fourier transform of a surface-truncated crystal in the reciprocal space (Fig. 8.37). This CTR diffraction is also a powerful tool to help solve interface structures on the atomic scale [597–600]. In addition, together with in-situ experimental techniques, CTR and GIXD provide a means of investigating surface order-disorder, melting and roughening transitions [601–604].

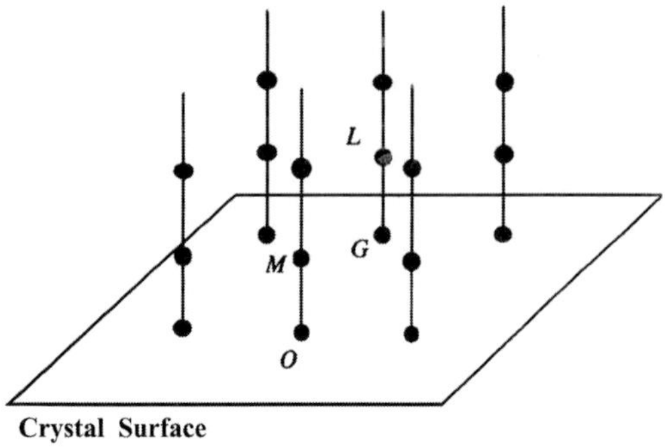

Fig. 8.37. Crystal truncation rods and reciprocal lattice [609]

There are two types of crystal truncation rods: non-specular rods with and specular rods without the in-plane components of momentum transfer with respect to the incident wavevector. A specular rod starts from the origin O of the reciprocal lattice to a given reciprocal lattice point M, which is solely normal to the crystal surface (Fig. 8.37). The scan is equivalent to the usual $\theta - 2\theta$ scan. Most CTR rods displaced from the origin O are non-specular rods, because the displacement vector, for example, $\boldsymbol{OG}$ for the rod GL, introduces an in-plane momentum transfer for the incident X-rays. Dynamical calculations for the intensity distribution of specular CTR [167,583,605–607] and non-specular CTR rods [608–610] have been reported in the literature. In fact, the specular rod can be easily dealt with by applying the theories of two-wave GIXD [167] or the modified Darwin approach [583, 607]. Here we shall only concentrate on the rather difficult situation, the non-specular rod diffraction, and introduce the multiple-wave diffraction approach [608–610]. Above all, a specular rod is a special case of the non-specular rod. On the other hand, as the non-specular rods carry structural information in both normal and in-plane directions, the interpretation of these rods would be more desired.

The diffraction geometry of a non-specular rod GL involving the reciprocal lattice points O, G, and L is schematically shown in Fig. 8.38, where the reciprocal lattice vector $\boldsymbol{OG}$ is parallel to the crystal surface and $\boldsymbol{GL}$ is normal to that surface. A Cartesian coordinates system is chosen, where the x-axis is along $\boldsymbol{OG}$, the z-axis along the inward surface normal $\hat{\boldsymbol{n}}_e$, and the y-axis perpendicular to the x and z axes. The crystal surface is defined by $z = 0$. The coordinates of the O, G, and L are $(0, 0, 0)$, $(g, 0, 0)$, and $(g, 0, -\ell)$, respectively, where $g = |OG|$ and $\ell = |GL|$. $\boldsymbol{k}_O$ and $\boldsymbol{k}_O^s$ are the wavevectors of the incident and specularly reflected waves outside the crystal, while $\boldsymbol{k}_G$ and $\boldsymbol{k}_L$ are the diffracted wavevectors at the entrance surface of the crystal. Points C_O, C_{Os}, C_G, and C_L are the starting points of the $\boldsymbol{k}_O$, $\boldsymbol{k}_O^s$, $\boldsymbol{k}_G$, and $\boldsymbol{k}_L$, respectively. Since $\boldsymbol{GL}$ is parallel to $\hat{\boldsymbol{n}}_e$, $\boldsymbol{k}_G$ and $\boldsymbol{k}_L$ are identical. Point

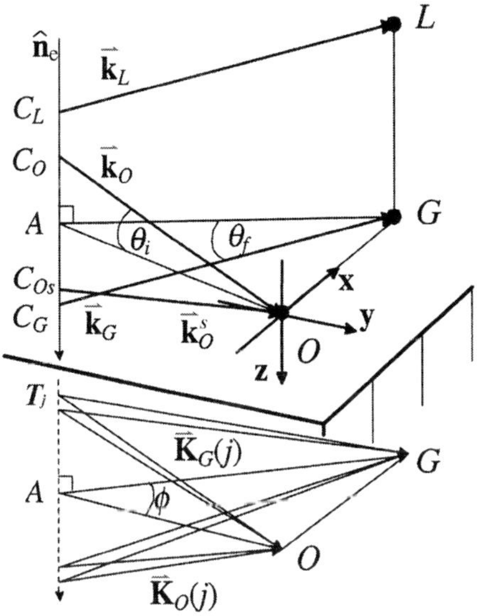

Fig. 8.38. Geometrical relation between the wavevectors and crystal surface in a non-specular surface normal scan [610]

A, located at $(x_n, y_n, 0)$, lies along the normal $\hat{\boldsymbol{n}}_e$, so that the triangle AGO is parallel to the crystal surface. AO and AG are the in-plane components of $\boldsymbol{k}_O$ (and also $\boldsymbol{k}_O^s$) and $\boldsymbol{k}_G$ (and also $\boldsymbol{k}_L$), respectively, and the angle between $\boldsymbol{AO}$ and $\boldsymbol{AG}$ is ϕ. θ_i is the angle of incidence of $\boldsymbol{k}_O$ with respect to the surface and θ_f the angle of emergence of $\boldsymbol{k}_G$ (and also $\boldsymbol{k}_L$) with respect to the surface. The tic point T_j for mode j, located at (x_n, y_n, z_j), is the starting point of the wavevectors $\boldsymbol{K}_O(j)$, $\boldsymbol{K}_G(j)$, and $\boldsymbol{K}_L(j)$ (not shown in Fig. 8.38) inside the crystal. For a CTR scan, the crystal is oriented so that the θ_i and θ_f vary. Namely, ϕ and θ vary, where θ (not shown in Fig. 8.38) is the angle between $\boldsymbol{OG}$ and the tangent of the Ewald sphere at O. The relationships between θ_i, θ_f, ϕ, and θ are [611]

$$\sin^2\theta_f = \sin^2\theta_i + 4\cos^2\theta_i \sin\theta_G\,(\sin\theta - \sin\theta_G)\ ,$$
$$\sin\phi = \frac{2\cos\theta\sin\theta_G}{\sqrt{1 - 4\sin\theta_G\,(\sin\theta - \sin\theta_G)}}\ , \tag{8.99}$$

where the Bragg angle θ_G of the G reflection also equals to $\sin^{-1}[g/(2k\cos\theta_i)]$. From (8.99), the momentum transfer during the rod scan is varying from $\boldsymbol{OG}$ to $\boldsymbol{OL}$ while the in-plane component is kept equal to $\boldsymbol{OG}$. If $\theta_i = \theta_f = 0$, then $\theta = \theta_G$ and $\phi = 2\theta$. Thus, the Bragg condition for the in-plane G reflection is satisfied.

There are two ways of carrying out the rod scan experimentally: one by varying θ_f for fixed θ_i and the other by varying θ_i and θ_f so that $\theta_i = \theta_f$ [611, 612]. The former uses a so-called z-axis diffractometer, where the

z direction is along the surface normal. The latter employs a conventional four-circle diffractometer. The two different scan schemes lead to different intensity distributions for the rods [611].

The algorithm in the representation of a Cartesian coordinates system, described in Chap. 7, is most suitable for the dynamical calculation for CTR's, because the equation of dispersion involves a high-order polynomial of the accommodation $k\zeta$, which cannot be solved as a linear eigenvalue problem. By considering the x-, y-, and z-polarized wavefields and following the fundamental equations given in (7.22) or (7.29), the eigenvalues and eigenvectors can be determined in a straightforward manner. Similar to GIXD, the real and the imaginary parts of the eigenvalues yield the coordinates of the tie points and the associated linear absorption coefficients for all the modes of propagation. The eigenvectors give the wavefield-amplitude ratios among the diffracted waves. The boundary conditions for wavefield-amplitudes, according to (7.36), are:

(i) Continuity of the x and y components of the $\boldsymbol{E}$ fields at the boundary:

$$\sum_j c_j E_O^x(j) = e_{O(e)}^x + e_O^{sx} ,$$

$$\sum_j c_j E_G^x(j) = e_G^x , \qquad \sum_j c_j E_L^x(j) = e_L^x ,$$

$$\sum_j c_j E_O^y(j) = e_{O(e)}^y + e_O^{sy} ,$$

$$\sum_j c_j E_G^y(j) = e_G^y , \qquad \sum_j c_j E_L^y(j) = e_L^y , \tag{8.100}$$

where e's and E's are the wavefield amplitudes outside and inside the crystal, respectively. $e_{O(e)}$ is the amplitude of the incident wave, and the superscript s in e_O^{sx} indicates the surface specularly reflected wave. c_j is the usual proportionality constant of wavefield-amplitude. The summation is taken over all the permitted modes. For a semi-infinitely thick crystal, there are six $(= 2 \times N$, N=3) permitted modes. However, the number of permitted modes is twelve for a thin crystal plate.

(ii) Continuity of the z component of the electric displacement $\boldsymbol{D}$:

$$\sum_j c_j D_O^z(j) = \sum_j c_j \left[E_O^z(j) + \chi_O E_O^z(j) + \chi_{O\text{-}G} E_G^z(j) + \chi_{O\text{-}L} E_L^z(j)\right]$$
$$= e_{O(e)}^z + e_O^{sz} ,$$

$$\sum_j c_j D_G^z(j) = \sum_j c_j \left[E_G^z(j) + \chi_{G\text{-}O} E_O^z(j) + \chi_O E_G^z(j) + \chi_{G\text{-}L} E_L^z(j)\right]$$
$$= e_G^z ,$$

$$\sum_j c_j D_L^z(j) = \sum_j c_j \left[E_L^z(j) + \chi_{L\text{-}O} E_O^z(j) + \chi_{L\text{-}G} E_G^z(j) + \chi_O E_L^z(j)\right]$$
$$= e_L^z \,. \tag{8.101}$$

(iii) Continuity of the x and y components of the magnetic induction $\boldsymbol{H}$:

$$\begin{aligned}
&\sum_j c_j H_O^x(j) = h_{O(e)}^x + h_O^{sx} \,, \\
&\sum_j c_j H_G^x(j) = h_G^x \,, \qquad \sum_j c_j H_L^x(j) = h_L^x \,, \\
&\sum_j c_j H_O^y(j) = h_{O(e)}^y + h_O^{sy} \,, \\
&\sum_j c_j H_G^y(j) = h_G^y \,, \qquad \sum_j c_j H_L^y(j) = h_L^y \,.
\end{aligned} \tag{8.102}$$

With $\boldsymbol{H}_M = (\boldsymbol{K}_M \times \boldsymbol{E}_M)/K_M$, (8.102) becomes

$$\begin{aligned}
\sum_j c_j \frac{K_O^y(j)E_O^z(j) - K_O^z(j)E_O^y(j)}{K_O(j)} &= -e_{O(e)}^z \frac{y_n}{k} - e_{O(e)}^y \sin\theta_i - e_O^{sz}\frac{y_n}{k} \\
&\quad + e_O^{sy} \sin\theta_i \,, \\
\sum_j c_j \frac{K_G^y(j)E_G^z(j) - K_G^z(j)E_G^y(j)}{K_G(j)} &= -e_G^z \frac{y_n}{k} + e_G^y \sin\theta_f \,, \\
\sum_j c_j \frac{K_L^y(j)E_L^z(j) - K_L^z(j)E_L^y(j)}{K_L(j)} &= -e_L^z \frac{y_n}{k} + e_L^y \sin\theta_f \,, \\
\sum_j c_j \frac{K_O^z(j)E_O^x(j) - K_O^x(j)E_O^z(j)}{K_O(j)} &= e_{O(e)}^x \sin\theta_i + e_{O(e)}^z \frac{x_n}{k} - e_O^{sx} \sin\theta_i \\
&\quad + e_O^{sz}\frac{x_n}{k} \,, \\
\sum_j c_j \frac{K_G^z(j)E_G^x(j) - K_G^x(j)E_G^z(j)}{K_G(j)} &= -e_G^x \sin\theta_f - e_G^z \frac{g - x_n}{k} \,, \\
\sum_j c_j \frac{K_L^z(j)E_L^x(j) - K_L^x(j)E_L^z(j)}{K_L(j)} &= -e_L^x \sin\theta_f - e_L^z \frac{g - x_n}{k} \,.
\end{aligned} \tag{8.103}$$

(iv) Continuity of the z components of the magnetic field $\boldsymbol{B}$: Because $\boldsymbol{B} \simeq \boldsymbol{H}$ for X-rays, the condition of continuity of the z components of the magnetic field $\boldsymbol{B}$ is equivalent to the condition of the x and y components of the $\boldsymbol{E}$ fields.

There are 15 equations for 15 unknowns, i.e., the x, y, and z components of $\boldsymbol{e}_{O(e)}$, $\boldsymbol{e}_G$, and $\boldsymbol{e}_L$, together with the c_j of the six modes for a semi-infinitely thick crystal. For a thin crystal, all the twelve modes need to be considered.

Consequently, the intensity of the rod for a given θ_i and θ_f can be calculated as

$$I\left(\theta_i, \theta_f\right) = \frac{\left(|e_G^x + e_L^x|^2 + |e_G^y + e_L^y|^2 + |e_G^z + e_L^z|^2\right) \sin\theta_f}{\left(\left|e_{O(e)}^x\right|^2 + \left|e_{O(e)}^y\right|^2 + \left|e_{O(e)}^z\right|^2\right) \sin\theta_i} , \tag{8.104}$$

where the coherent interaction between the diffracted G and L waves is considered because the wavevectors $\boldsymbol{K}_G$ and $\boldsymbol{K}_L$ are identical. The corresponding excitation of mode $Ex(j)$ of mode j and excitation of wave Ex_M of wave M are

$$Ex(j) = \sum_{M=O,G,L} \frac{|E_M(j)|^2 \sin\theta_f}{|E_O|^2 \sin\theta_i} ,$$

$$Ex_M = \sum_j \frac{|E_M(j)|^2 \sin\theta_f}{|E_O|^2 \sin\theta_i} . \tag{8.105}$$

Figure 8.39 is a schematic representation of an overall view (Fig. 8.39(a)) and a cross section (Fig. 8.39(b)) of the dispersion surface of a non-specular CTR. In Fig. 8.39(a), the three Ewald spheres centered at the reciprocal lattice points O, G, and L intersect. The G reflection occurs at the tie point B, at which the corresponding Ewald spheres centered at O and G intersect. Similarly, the L reflection takes place at the tie point F. A close-up of the dispersion surface is given in Fig. 8.39(b), where La is the Laue point and S an intermediate scanned point along GL. Point A is an entrance point at the crystal surface, previously defined for a given azimuthal angle ψ measured from LaO, so that

$$\begin{aligned}\psi &= \angle LaOG - \angle AOG \\ &= \arccos\left(\frac{g}{2a}\right) - \arccos\left(\frac{a^2+g^2-b^2}{2ag}\right) ,\end{aligned} \tag{8.106}$$

where $a = \cos\theta_i/\lambda$ and $b = \cos\theta_f/\lambda$. For a z-axis diffractometer at a small θ,

$$\psi = \arccos\left(\frac{\lambda g}{2}\right) - \arccos\left\{\frac{\left[\left(\frac{q}{d}\right)^2 + g^2\right]\lambda}{2g}\right\} , \tag{8.107}$$

and for a four-circle diffractometer ($\theta_i = \theta_f$),

$$\psi = \arccos\left(\frac{\lambda g}{2}\right) - \arccos\left(\frac{g}{2\sqrt{1-\left(\frac{q}{2d}\right)^2}}\right) , \tag{8.108}$$

where q and d are the normal component of the momentum transfer and the lattice spacing perpendicular to the crystal surface, respectively. As shown in Fig. 8.39(b), the loci of the tie points of the six modes for a thick crystal

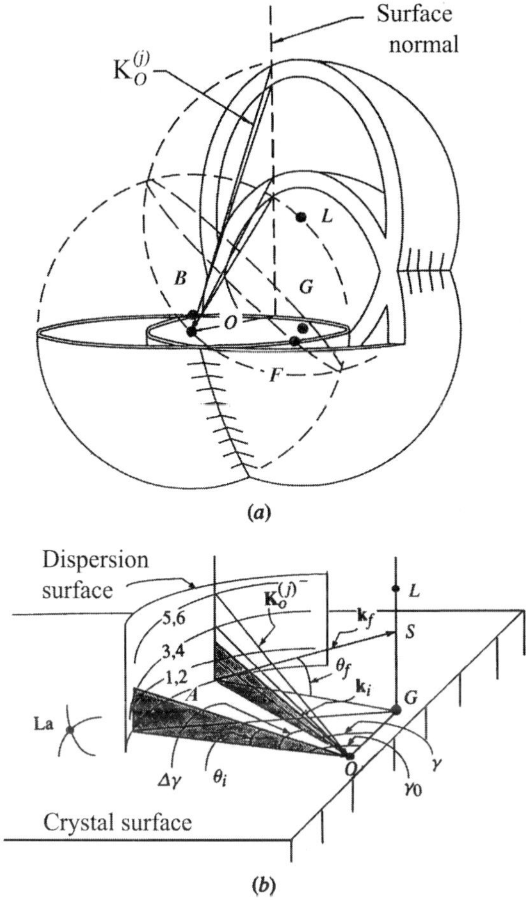

Fig. 8.39. (**a**) An overall view and (**b**) a close-up of the dispersion surface involved in a non-specular rod scan [609]

are indicated schematically as the curves on the cylindrical surface defined by the $\boldsymbol{K}_O(j)$ centered at the point O.

The non-specular rod from (200) to (202) of a tungsten [001] cut crystal for CuKα_1 is subjected to the dynamical calculation for illustration. Figure 8.40 shows the calculated dispersion curves, intensity distributions, excitation of mode, and excitation of wave versus ψ for a rod scan at $\theta_i = 0.7°$ with varying θ_f. Points A and B in Fig. 8.40(a), i.e., $\psi = 0$ and $45°$, are the positions at which the (200) and (202) reflections occur, respectively. The dispersion curves of modes 1 and 2 are so close that a single curve is plotted to represent the dispersion surface. It is similar for modes 3 and 4 and modes 5 and 6. The intensity peaks are at points A and B for the two respective reflections (Fig. 8.40(b)). The corresponding excitations of waves shown in

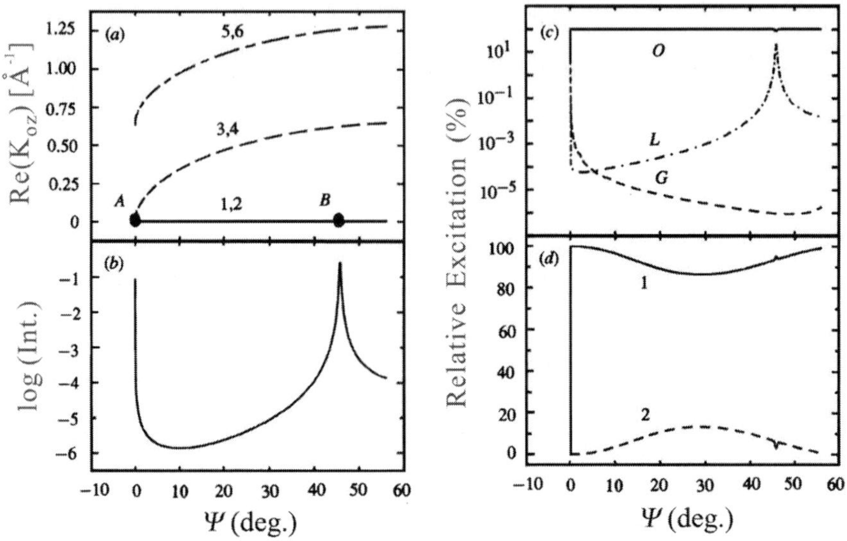

Fig. 8.40. (**a**) Calculated dispersion curves, (**b**) intensity distribution, (**c**) excitation of wave, and (**d**) excitation of mode of the W (200) rod for 1.560562 Å [609]

Fig. 8.40(c) reveal that the incident wave is always excited, while the $G =$ (200) and $L =$ (202) reflections have their maximum excitation at positions A and B, respectively. Since the excitation of mode for modes 3, 4, 5, and 6 are almost one order of magnitude smaller than that for modes 1 and 2, only the strongly excited modes are shown in Fig. 8.40(d). The excitations of modes 1 and 2 have, respectively, a minimum and a maximum at $\psi = 30°$ in between the (200) and (202) positions.

The most interesting diffraction situations are those in the vicinity of points A and B. Figure 8.41 shows the dispersion curves for ψ near 0 and 45°. In Fig. 8.41(a), near the (200) reflection position, the dispersion curves exhibit

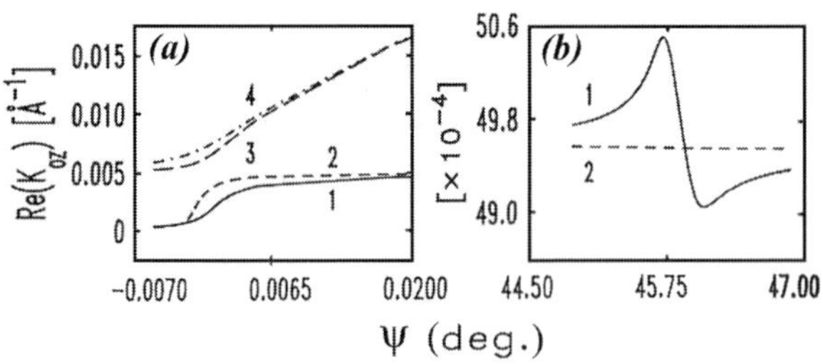

Fig. 8.41. Calculated dispersion curves near (**a**) the (200) reflection and (**b**) the (202) reflection of the W (200) rod for 1.540562 Å [609]

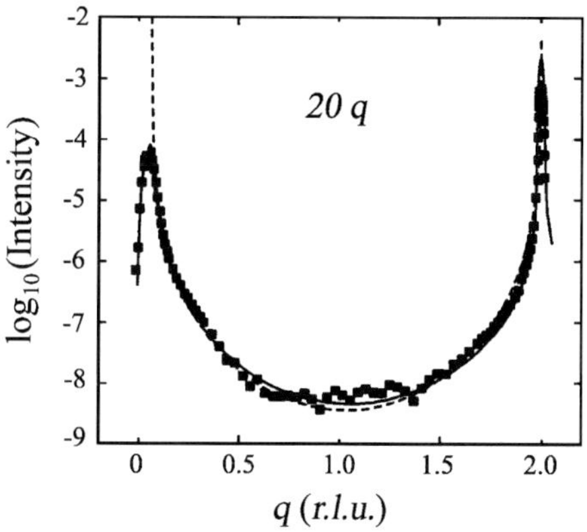

Fig. 8.42. Calculated and measured intensity distributions of the W (200) rod scan for 1.540562 Å [609]

a distorted hyperbola for the transmitted (200), where the distortion is due to the interception of the dispersion surface by the plane perpendicular to the crystal surface. However, the dispersion curve of mode 1 in Fig. 8.41(b) shows a resonance-like curve, which is a characteristic of a Bragg-type reflection. Namely, the two branches of the hyperbola are connected by the 'solution surface' [613] due to the (202) total reflection. From the calculated dispersion surface, related linear absorption coefficient, and excitation (not shown here), it is clear that the CTR diffraction is nothing more than a coalescence of a glancing Laue-type diffraction and an asymmetric Bragg-type reflection.

The calculated intensity distribution along the (200) to (202) rod with a surface roughness of 1.05 Å for CuKα_1 radiation is shown in Fig. 8.42, where the measured intensity profile [169] is displayed for comparison. The calculated intensity profile has been convoluted with the instrumental broadening and the crystal mosaic spread, which are assumed to be a Gaussian. The models for surface roughness proposed by Robinson [169] and Harada [614] can be used in calculating the integrated intensity. The calculated intensities and the experimental ones are in good agreement. Similar calculation by using the kinematical theory fails to reproduce the intensity distribution, especially at the (200) and (202) peak positions. The kinematical calculation diverges at these two positions. Clearly, the dynamical calculation by using the geometry of three-wave diffraction is indispensable in accounting for the intensity distribution of the CTR, especially of the non-specular one. More importantly, in contrast to the kinematical calculation, where the absolute zero point for momentum transfer is undefined due to grazing incidence, the dynamical calculation scheme provides an absolute scale for the diffracted intensity versus momentum-transfer.

Dealing with the surface roughness or interface roughness in the dynamical calculation is not an easy task. In an ideal approach, it would be better to consider the surface roughness in the boundary conditions for wavevectors and wavefield amplitudes. With this consideration, a detailed microscopic picture of the diffraction from a rough surface can be brought out.

8.3 Polarization Aspects of Three-Wave Diffraction

Polarization is an important factor affecting the intensity of a multiple-wave diffraction, as described in the kinematical and dynamical theories given in Chaps. 5 and 6. Many experimental investigations on the effects of polarization on multiple-wave diffraction have been reported [26,228,232,234,305,322, 324,330,347,376,514,615–617]. Among them the indirect excitation of three-wave diffraction with a polarization-forbidden reflection, the suppression of multiply diffracted intensity by polarization tuning, and the polarization-resolved multiple-wave diffraction are of fundamental and practical importance for X-ray diffraction in crystals. Although there are several ways to describe these three polarization aspects, for simplicity we shall start with the expression for the wavefield $\boldsymbol{D}_G^{(3)}$ of a three-wave (O, G, L) diffraction given in (7.294) and (7.296),

$$\begin{aligned}\boldsymbol{D}_G^{(3)} &\cong A_G\chi_{G-O}\boldsymbol{D}^{(2)}_{O[G]}+A_GA_L\chi_{L-O}\chi_{G-L}\boldsymbol{D}^{(2)}_{O[L][G]}-A_GA_OA_L\\&\times\chi_{G-O}\chi_{L-O}\chi_{O-L}\boldsymbol{D}^{(2)}_{O[L][O][G]}\\&\quad -A_G^2A_L\chi_{G-O}\chi_{L-G}\chi_{G-L}\boldsymbol{D}^{(2)}_{O[G][L][G]}\\&\equiv \boldsymbol{D}_2+\boldsymbol{D}_{\mathrm{Um}}+\boldsymbol{D}_{\mathrm{Au1}}+\boldsymbol{D}_{\mathrm{Au2}}\ .\end{aligned}\tag{8.109}$$

Assume that the incident wavefield $\boldsymbol{D}_O$ with an arbitrary polarization unit vector $\boldsymbol{p}_O$ is given by

$$\boldsymbol{D}_O = D_O\boldsymbol{p}_O\ ,\tag{8.110}$$

with

$$\boldsymbol{p}_O = \alpha\hat{\boldsymbol{\sigma}}_O+\beta\hat{\boldsymbol{\pi}}_O = \alpha\hat{\boldsymbol{\sigma}}+\beta\hat{\boldsymbol{\pi}}_O\ ,\tag{8.111}$$

where $\alpha = \cos\omega_O$ and $\beta = \sin\omega_O$, ω_O being the angle between $\boldsymbol{p}_O$ and the σ-polarized vector $\boldsymbol{\sigma}$. The polarization unit vectors of the incident wave O and the diffracted wave G are defined as

$$\begin{aligned}\hat{\boldsymbol{\sigma}} &= \hat{\boldsymbol{\sigma}}_O = \hat{\boldsymbol{\sigma}}_G = \hat{\boldsymbol{K}}_G\times\hat{\boldsymbol{K}}_O\ ,\\ \hat{\boldsymbol{\pi}}_O &= \hat{\boldsymbol{K}}_O\times\hat{\boldsymbol{\sigma}}_O = \hat{\boldsymbol{K}}_O\times\hat{\boldsymbol{\sigma}}\ ,\\ \hat{\boldsymbol{\pi}}_G &= \hat{\boldsymbol{K}}_G\times\hat{\boldsymbol{\sigma}}_G = \hat{\boldsymbol{K}}_G\times\hat{\boldsymbol{\sigma}}\ .\end{aligned}\tag{8.112}$$

The vector components in (8.109) are given by

$$\begin{aligned}\boldsymbol{D}^{(2)}_{O[G]} &= \left(\alpha\hat{\boldsymbol{\sigma}}_{O[G]}+\beta\hat{\boldsymbol{\pi}}_{O[G]}\right)D_O = \boldsymbol{p}_2D_O\ ,\\ \boldsymbol{D}^{(2)}_{O[L][G]} &= \left(\alpha\hat{\boldsymbol{\sigma}}_{O[L][G]}+\beta\hat{\boldsymbol{\pi}}_{O[L][G]}\right)D_O = \boldsymbol{p}_{\mathrm{Um}}D_O\ ,\end{aligned}$$

$$\boldsymbol{D}^{(2)}_{O[L][O][G]} = \left(\alpha\hat{\boldsymbol{\sigma}}_{O[L][O][G]} + \beta\hat{\boldsymbol{\pi}}_{O[L][O][G]}\right) D_O = \boldsymbol{p}_{\mathrm{Au1}} D_O \ ,$$
$$\boldsymbol{D}^{(2)}_{O[G][L][G]} = \left(\alpha\hat{\boldsymbol{\sigma}}_{O[G][L][G]} + \beta\hat{\boldsymbol{\pi}}_{O[G][L][G]}\right) D_O = \boldsymbol{p}_{\mathrm{Au2}} D_O \ , \qquad (8.113)$$

where $\hat{\boldsymbol{\sigma}}_{O[G]} = \hat{\boldsymbol{\sigma}}$ and $\hat{\boldsymbol{\pi}}_{O[G]} = \hat{\boldsymbol{\pi}}_O - (\hat{\boldsymbol{K}}_G \cdot \hat{\boldsymbol{\pi}}_O)\hat{\boldsymbol{K}}_G$. Since all the vectors in (8.113) are perpendicular to $\boldsymbol{K}_G$, they can be expressed in terms of $\hat{\boldsymbol{\sigma}}_G$ and $\hat{\boldsymbol{\pi}}_G$, namely, $\hat{\boldsymbol{\sigma}}$ and $\hat{\boldsymbol{\pi}}_{\boldsymbol{G}}$. The scalar products of the polarization vectors in (8.113) multiplied by $\hat{\boldsymbol{\sigma}}_G$ and $\hat{\boldsymbol{\pi}}_G$, respectively, give the following polarization factors:

$$p_2^{\sigma}(\sigma) = \hat{\boldsymbol{\sigma}}_{O[G]} \cdot \hat{\boldsymbol{\sigma}}_G = \hat{\boldsymbol{\sigma}} \cdot \hat{\boldsymbol{\sigma}} = 1 \ ,$$
$$p_2^{\pi}(\sigma) = \hat{\boldsymbol{\sigma}}_{O[G]} \cdot \hat{\boldsymbol{\pi}}_G = \hat{\boldsymbol{\sigma}} \cdot \hat{\boldsymbol{\pi}}_G = 0 \ ,$$
$$p_2^{\sigma}(\pi) = \hat{\boldsymbol{\pi}}_{O[G]} \cdot \hat{\boldsymbol{\sigma}}_G = 0 \ ,$$
$$p_2^{\pi}(\pi) = \hat{\boldsymbol{\pi}}_{O[G]} \cdot \hat{\boldsymbol{\pi}}_G = \hat{\boldsymbol{\pi}}_O \cdot \hat{\boldsymbol{\pi}}_G = \cos 2\theta_G \ , \qquad (8.114)$$

for the direct two-wave diffracted $\boldsymbol{D}^{(2)}_{O[G]}$;

$$p_{\mathrm{Um}}^{\sigma}(\sigma) = \hat{\boldsymbol{\sigma}}_{O[L][G]} \cdot \hat{\boldsymbol{\sigma}}_G = -\left(\hat{\boldsymbol{K}}_G \times \hat{\boldsymbol{K}}_G \times \hat{\boldsymbol{\sigma}}_{O[L]}\right) \cdot \hat{\boldsymbol{\sigma}}_G$$
$$= p_2^{\sigma}(\sigma) - \left(\hat{\boldsymbol{K}}_L \cdot \hat{\boldsymbol{\sigma}}\right)^2 \ ,$$
$$p_{\mathrm{Um}}^{\pi}(\sigma) = \hat{\boldsymbol{\sigma}}_{O[L][G]} \cdot \hat{\boldsymbol{\pi}}_G = p_2^{\pi}(\sigma) - \left(\hat{\boldsymbol{K}}_L \cdot \hat{\boldsymbol{\sigma}}\right)\left(\hat{\boldsymbol{K}}_L \cdot \hat{\boldsymbol{\pi}}_G\right) \ ,$$
$$p_{\mathrm{Um}}^{\sigma}(\pi) = \hat{\boldsymbol{\pi}}_{O[L][G]} \cdot \hat{\boldsymbol{\sigma}}_G = p_2^{\sigma}(\pi) - \left(\hat{\boldsymbol{K}}_L \cdot \hat{\boldsymbol{\pi}}_O\right)\left(\hat{\boldsymbol{K}}_L \cdot \hat{\boldsymbol{\sigma}}\right) \ ,$$
$$p_{\mathrm{Um}}^{\pi}(\pi) = \hat{\boldsymbol{\pi}}_{O[L][G]} \cdot \hat{\boldsymbol{\pi}}_G = p_2^{\pi}(\pi) - \left(\hat{\boldsymbol{K}}_L \cdot \hat{\boldsymbol{\pi}}_O\right)\left(\hat{\boldsymbol{K}}_L \cdot \hat{\boldsymbol{\pi}}_G\right) \ , \qquad (8.115)$$

for the Umweganregung excited wave $\boldsymbol{D}^{(2)}_{O[L][G]}$;

$$p_{\mathrm{Au1}}^{\sigma}(\sigma) = \hat{\boldsymbol{\sigma}}_{O[L][O][G]} \cdot \hat{\boldsymbol{\sigma}}_G = p_{\mathrm{Um}}^{\sigma}(\sigma) \ ,$$
$$p_{\mathrm{Au1}}^{\pi}(\sigma) = \hat{\boldsymbol{\sigma}}_{O[L][O][G]} \cdot \hat{\boldsymbol{\pi}}_G$$
$$= p_{\mathrm{Um}}^{\pi}(\sigma) + \left(\hat{\boldsymbol{K}}_L \cdot \hat{\boldsymbol{\sigma}}\right)\left(\hat{\boldsymbol{K}}_O \cdot \hat{\boldsymbol{\pi}}_G\right)\left(\hat{\boldsymbol{K}}_L \cdot \hat{\boldsymbol{K}}_O\right) \ ,$$
$$p_{\mathrm{Au1}}^{\sigma}(\pi) = \hat{\boldsymbol{\pi}}_{O[L][O][G]} \cdot \hat{\boldsymbol{\sigma}}_G = p_{\mathrm{Um}}^{\sigma}(\pi) \ ,$$
$$p_{\mathrm{Au1}}^{\pi}(\pi) = \hat{\boldsymbol{\pi}}_{O[L][O][G]} \cdot \hat{\boldsymbol{\pi}}_G$$
$$= p_{\mathrm{Um}}^{\pi}(\pi) + \left(\hat{\boldsymbol{K}}_L \cdot \hat{\boldsymbol{\pi}}_O\right)\left(\hat{\boldsymbol{K}}_O \cdot \hat{\boldsymbol{\pi}}_G\right)\left(\hat{\boldsymbol{K}}_L \cdot \hat{\boldsymbol{K}}_O\right) \ ,$$
$$p_{\mathrm{Au2}}^{\sigma}(\sigma) = \hat{\boldsymbol{\sigma}}_{O[G][L][G]} \cdot \hat{\boldsymbol{\sigma}}_G = p_{\mathrm{Um}}^{\sigma}(\sigma) \ ,$$
$$p_{\mathrm{Au2}}^{\pi}(\sigma) = \hat{\boldsymbol{\sigma}}_{O[G][L][G]} \cdot \hat{\boldsymbol{\pi}}_G - p_{\mathrm{Um}}^{\pi}(\sigma) \ ,$$
$$p_{\mathrm{Au2}}^{\sigma}(\pi) = \hat{\boldsymbol{\pi}}_{O[G][L][G]} \cdot \hat{\boldsymbol{\sigma}}_G$$
$$= p_{\mathrm{Um}}^{\sigma}(\pi) + \left(\hat{\boldsymbol{K}}_G \cdot \hat{\boldsymbol{\pi}}_O\right)\left(\hat{\boldsymbol{K}}_L \cdot \hat{\boldsymbol{\sigma}}\right)\left(\hat{\boldsymbol{K}}_L \cdot \hat{\boldsymbol{K}}_G\right) \ ,$$
$$p_{\mathrm{Au2}}^{\pi}(\pi) = \hat{\boldsymbol{\pi}}_{O[G][L][G]} \cdot \hat{\boldsymbol{\pi}}_G$$
$$= p_{\mathrm{Um}}^{\pi}(\pi) + \left(\hat{\boldsymbol{K}}_G \cdot \hat{\boldsymbol{\pi}}_O\right)\left(\hat{\boldsymbol{K}}_L \cdot \hat{\boldsymbol{\pi}}_G\right)\left(\hat{\boldsymbol{K}}_L \cdot \hat{\boldsymbol{K}}_G\right) \ , \qquad (8.116)$$

for the Aufhellung excited waves $\boldsymbol{D}^{(2)}_{O[L][O][G]}$ and $\boldsymbol{D}^{(2)}_{O[G][L][G]}$. These polarization factors play an important role in the polarization related phenomena described in the following sections.

8.3.1 Indirect Excitation of Three-wave Diffraction with a Polarization-Forbidden Reflection

Indirect excitation for X-ray diffraction is a special case in multiple-wave diffraction, where the primary reflection is a forbidden reflection by space group symmetry. The diffracted intensity of the primary reflection is usually zero when the condition of multiple-wave diffraction is not fulfilled. However, the primary wave could have appreciable intensity under a multiple-wave diffraction situation due to the indirect excitation of the primary reflection via a detoured route, the so-called Umweganregung process [4].

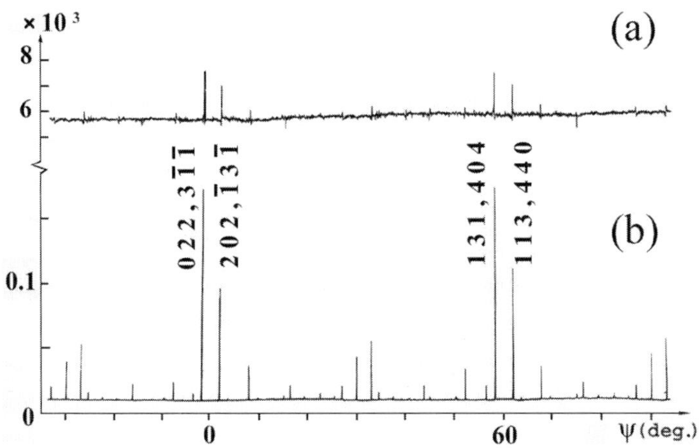

Fig. 8.43. Renninger scans of Ge (333) and Cu$K\alpha_1$ for (**a**) a σ-polarized incident wave and (**b**) a π-polarized incident wave [228]

Similar indirect excitation for multiple-wave diffraction can occur as well for a primary reflection which is forbidden by polarization. Figures 8.43(a) and (b) show the Renninger scans of germanium (333) and Cu$K\alpha_1$ with a σ linear polarized incident wave and a π-polarized one, respectively [228]. The polarization of the incident wave is controlled by the relative position of the sample Ge crystal with respect to a Ge (333) monochromator. For the σ and π polarization arrangements, the plane of incidence of the monochrmator is parallel and perpendicular to that of the sample crystal, respectively. Since the Bragg angle of (333) is very close to 45°, the (333) reflected wave from a laboratory Cu$K\alpha_1$ source is linearly polarized in the direction normal to the plane of incidence. The primary reflection is the (333) of the sample crystal, which is a strong reflection for the σ-polarized incident wave and is a forbidden reflection for the π-polarized incident wave. The Miller indices of the secondary reflections of a few strong four-wave diffractions are also given. Many multiple-wave cases, having nearly zero intensity in the Renninger scan around [333] shown in Fig. 8.43(a) for the σ-polarization, have notable intensity distributions for the π-polarized incident wave. This indirect excitation of the (333) reflection takes place through the detoured channels: first the corresponding secondary reflection and then the coupling reflection. All the secondary and coupling reflections involved are not forbidden by the space group symmetry nor by the polarization. This excitation can be understood to a first-order approximation by considering the interference of the direct-diffracted wavefield $\boldsymbol{D}_2$ and the Umweg-excited $\boldsymbol{D}_{\mathrm{Um}}$, i.e., the first two terms on the right of (8.109),

$$\begin{aligned} \boldsymbol{D}_2 &= A_G \chi_{G-O} \boldsymbol{D}^{(2)}_{O[G]} = A_G \chi_{G-O} \boldsymbol{p}_2 D_O \\ &= A_G \chi_{G-O} \left(\alpha p^{\sigma}_G(\sigma) \hat{\boldsymbol{\sigma}} + \beta p^{\pi}_G(\pi) \hat{\boldsymbol{\pi}}_G \right) D_O \ , \\ \boldsymbol{D}_{\mathrm{Um}} &= A_G A_L \chi_{L-O} \chi_{G-L} \boldsymbol{D}^{(2)}_{O[L][G]} = A_G A_L \chi_{L-O} \chi_{G-L} \boldsymbol{p}_{\mathrm{Um}} D_O \\ &= A_G A_L \chi_{L-O} \chi_{G-L} \left[p^{\sigma}_{\mathrm{Um}}(\boldsymbol{p}_O) \hat{\boldsymbol{\sigma}} + p^{\pi}_{\mathrm{Um}}(\boldsymbol{p}_O) \hat{\boldsymbol{\pi}} \right] D_O \ , \end{aligned} \tag{8.117}$$

with

$$\begin{aligned} p^{\sigma}_{\mathrm{Um}}(\boldsymbol{p}_O) &= \alpha p^{\sigma}_{\mathrm{Um}}(\sigma) + \beta p^{\sigma}_{\mathrm{Um}}(\pi) \ , \\ p^{\pi}_{\mathrm{Um}}(\boldsymbol{p}_O) &= \alpha p^{\pi}_{\mathrm{Um}}(\sigma) + \beta p^{\pi}_{\mathrm{Um}}(\pi) \ , \end{aligned} \tag{8.118}$$

where (8.113)–(8.116) have been employed.

For the σ polarization, $\omega_O = 0$ and $\alpha = 1$, $\beta = 0$. The $\boldsymbol{D}_2$ and $\boldsymbol{D}_{\mathrm{Um}}$ become

$$\begin{aligned} \boldsymbol{D}_2(\sigma) &= A_G \chi_{G-O} \hat{\boldsymbol{\sigma}} \ , \\ \boldsymbol{D}_{\mathrm{Um}}(\sigma) &= A_G A_L \chi_{L-O} \chi_{G-L} \left\{ \left[1 - \left(\hat{\boldsymbol{K}}_L \cdot \hat{\boldsymbol{\sigma}} \right)^2 \right] \hat{\boldsymbol{\sigma}} \right. \\ &\qquad \left. - \left(\hat{\boldsymbol{K}}_L \cdot \hat{\boldsymbol{\sigma}} \right) \left(\hat{\boldsymbol{K}}_L \cdot \hat{\boldsymbol{\pi}}_G \right) \hat{\boldsymbol{\pi}}_G \right\} . \end{aligned} \tag{8.119}$$

Since (333) is a strong reflection and the polarization factor in the braces of the second equation in (8.119) is very small, the three-wave diffracted intensity is dominated by the $\boldsymbol{D}_2$, forming the high intensity background in the ψ scan. Hence the contribution of the Umweg process to the multiple-wave diffraction is small. If the Aufhellung is taken into account as in (8.109), small dips on the high intensity background are expected to be observable (Fig. 8.43(a)).

For the π polarization, $\omega_O = 90°$ and $\alpha = 0$, $\beta = 1$. The amplitude of $\boldsymbol{D}_2$ is zero, resulting in the zero intensity background for the (333) reflection. The Umweg-excited wave with a non-zero amplitude contributes to the three-wave diffraction intensity as shown in Fig. 8.43(b). This qualitative analysis based on polarization factors gives a reasonable explanation for the indirect excitation of the reflection forbidden by polarization. Similar argument may be used to account for the indirect excitation of four- and five-wave diffractions shown in the same figure. An experiment similar to that reported by Kshevetskii, Stetsko, and Shelud'ko has been recently investigated by using synchrotron radiation [616].

8.3.2 Polarization Suppression of Multiple-Wave Interaction

The coherent dynamical interaction in multiple-wave diffraction, as described in the preceding section, depends on the interference between the direct-diffracted wave $\boldsymbol{D}_2$ and the Umweg-excited wave $\boldsymbol{D}_{\mathrm{Um}}$. The visibility of intensity distribution due to that interaction is then governed by the comparability of the amplitudes of those two waves. The ideal condition for revealing the interference effect is that both waves have equal amplitudes. In reality, this desired situation may not be encountered. For example, sometimes the amplitude of the Umweg-diffracted wave is very large compared to that of the direct-diffracted wave. Under this circumstance, the way of observing the dynamical interaction is to lower the amplitude of the Umweg wave so as to have a comparable amplitude with that of the directly diffracted wave. According to (8.117), to weaken $\boldsymbol{D}_{\mathrm{Um}}$ by adjusting the polarization factors, the two equations in (8.118) should be close or equal to zero, i.e.,

$$p^{\sigma}_{\mathrm{Um}}(\sigma)p^{\pi}_{\mathrm{Um}}(\pi) - p^{\pi}_{\mathrm{Um}}(\sigma)p^{\sigma}_{\mathrm{Um}}(\pi) = \left(\hat{\boldsymbol{K}}_O \cdot \hat{\boldsymbol{K}}_L\right)\left(\hat{\boldsymbol{K}}_G \cdot \hat{\boldsymbol{K}}_L\right) = 0\,. \tag{8.120}$$

To fulfill this condition for polarization suppression, the Bragg angle of either the secondary L or the coupling $G-L$ reflections is close or equal to 45°. In other words, setting the first and second equations of (8.118) equal to zero leads to the polarization angle ω_s for polarization suppression,

$$\omega_s = -\arctan\left[\frac{p^{\sigma}_{\mathrm{Um}}(\sigma)}{p^{\sigma}_{\mathrm{Um}}(\pi)}\right]\,. \tag{8.121}$$

This suggests that with an appropriate experimental setup, the Umweg-excited wave for a given multiple-wave case be suppressed by setting $\omega_O = \omega_s$. Such an experiment has recently been realized [322]. Figure 8.44 is the

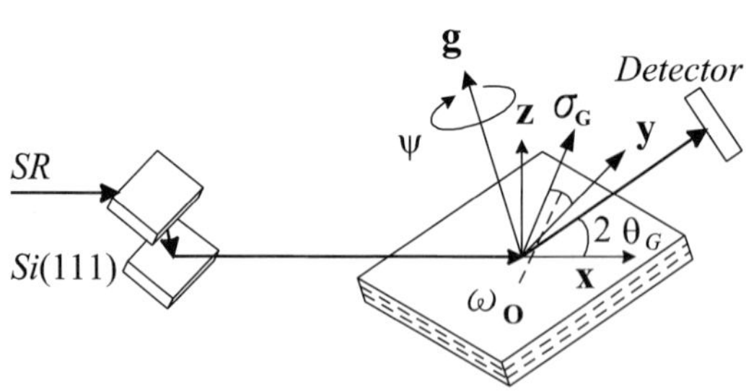

Fig. 8.44. Schematic of the experimental setup for polarization suppression [322]

schematic representation of the experiment setup, where the incident synchrotron radiation monochromatized by a silicon (111) cut double-crystal is incident on a sample crystal of (111) cut silicon. The sample crystal is mounted on a κ-goniostat of the six-circle soft X-ray diffractometer described in Chap. 4. The linearly polarized incident wave is along the x-axis and normal to the z-axis. The y-axis is perpendicular to the plane of incidence. The incident radiation of $\lambda = 1.762$ Å is used. The vector $\boldsymbol{\sigma}_G$ is perpendicular to the plane of incidence of the primary G reflection, whose reciprocal lattice vector is $\boldsymbol{g}$. The angle between the y-axis and $\boldsymbol{\sigma}_G$ is the polarization angle ω_O of the incident wave. Various polarization states for the incident wave can be chosen by changing ω_O, namely by rotating the yz-plane of the crystal with respect to the x-axis with the κ goniostat. The vertical and horizontal beam divergences are 0.010° and 0.025°, respectively. Multiple-wave diffraction patterns are recorded during the ψ scan around the $\boldsymbol{g}$. A solid-state pin diode, which can be moved along the 2θ and γ circles, is employed as a detector.

Figures 8.45(a) and (b) show the same portions of the ψ scans of Si symmetric (222) Bragg reflection for a π-polarized incident wave and an intermediate-polarized one, where $\omega_O = 90°$ and 65°, respectively. To satisfy the condition of (8.120), the wavelength of 1.762 Å is chosen so that the Bragg

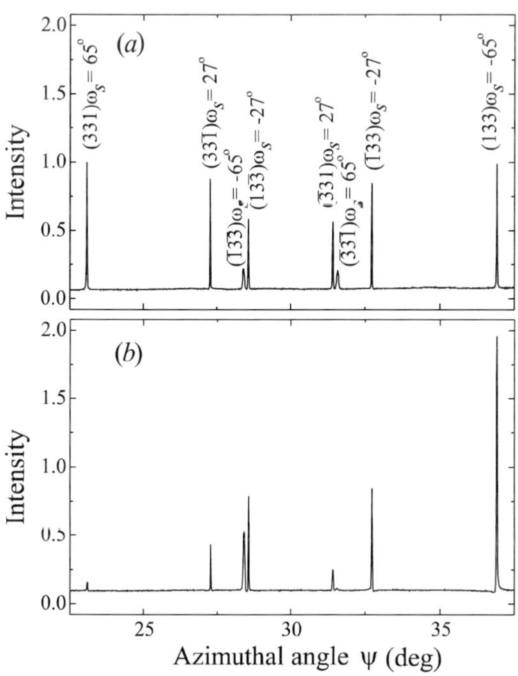

Fig. 8.45. Renninger scans of Si (222) and 1.762 Å for (**a**) a π-polarized ($\omega_O = 90°$) incident wave and (**b**) an intermediate-polarized ($\omega_O = 65°$) incident wave [322]

angle of a secondary {331} reflection is close to 45°. In Fig. 8.45(a), the background is the intensity of the (222) reflection. The multiple diffraction peaks are labelled with the secondary reflection together with the corresponding ω_s for polarization suppression. For the intermediate polarization ($\omega_O = 65°$), a drastic decrease in intensity occurs for the (331) and $(3\bar{3}\bar{1})$ peaks because $\omega_O = \omega_s$ (Fig. 8.45(b)). An intensity increase also occurs for the (133) and $(\bar{1}3\bar{3})$ due to $\omega_O = -\omega_s$. A small intensity variation is observed for those diffractions with $\omega_s = 27°$ and $-27°$.

The same experimental arrangement can be used to improve the visibility of the phase effect on multiple-wave interaction. Related topics to the phase of X-ray reflection will be discussed in Chap. 9.

8.3.3 Polarization-Resolved Multiple-Wave Diffraction

The partial polarization suppression for the enhancement of multiple-wave interaction inside a sample crystal relies on the polarization of the incident radiation and a properly chosen photon energy. This prerequisite condition for the photon energy may not be necessary if the polarization of both the incident wave and the diffracted wave can be modified simultaneously [618]. Following this idea, a recent experiment has demonstrated the enhancement of the visibility of multiple-wave interference by using an analyzing perfect crystal as a polarizer for the diffracted wave [324]. The diffraction geometry is shown in Fig. 8.46, where the arrangement of the X-ray source and the sample crystal is the same as that shown in Fig. 8.44. A polarizer crystal is placed along the direction of the diffracted wave $\boldsymbol{K}_G$ and is capable of being rotated around $\boldsymbol{K}_G$. The incident wave along the x direction is linearly polarized in the direction of the unit vector $\boldsymbol{p}_O = \hat{\boldsymbol{y}}$.

The polarization unit vectors $\boldsymbol{p}_2$ and $\boldsymbol{p}_{\mathrm{Um}}$ of the directly diffracted wave $\boldsymbol{D}_2$ and the detoured excited wave $\boldsymbol{D}_{\mathrm{Um}}$ defined in (8.117) are schematically shown in Fig. 8.47. The angle between the $\boldsymbol{p}_2$ and $\boldsymbol{p}_{\mathrm{Um}}$ is γ. ω_O is the polarization angle of the incident wave.

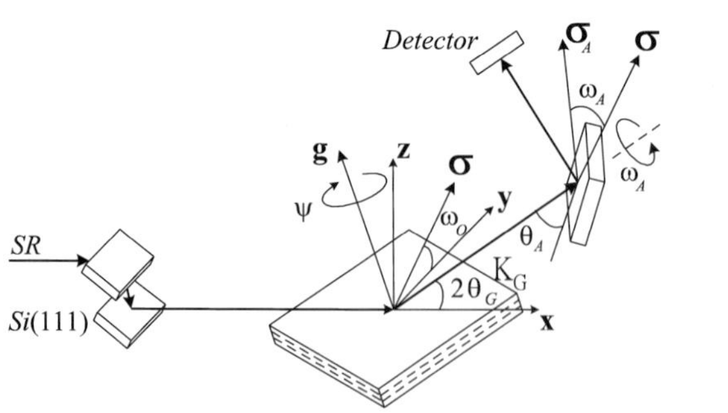

Fig. 8.46. Experimental setup for polarization-resolved multiple-wave experiments [324]

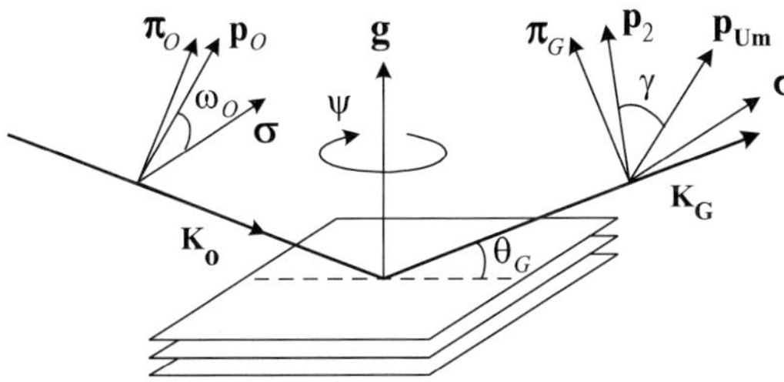

Fig. 8.47. Geometrical relation between the polarization vectors for the G reflection shown in Fig. 8.46 [324]

Consider then the reflection of the three-wave diffracted $\boldsymbol{D}_G^{(3)}$ wave by the polarizer. The diffracted wave with $\boldsymbol{K}_G$ from the sample crystal becomes the incident wave for the polarizer. The polarization of the reflected wave from the polarizer depends on the angle ω_A between the plane of incidence of the reflection A of the polarizer and that of the G reflection of the sample crystal. Assume that $\hat{\boldsymbol{\sigma}}_A$ is the σ-polarization vector perpendicular to the plane of incidence of the polarizer. The polarization angle ω_A is then the angle between $\hat{\boldsymbol{\sigma}}_A$ and the incident polarization vector $\hat{\boldsymbol{\sigma}}$ for the sample crystal (Fig. 8.46). The Bragg angle of reflection A for the polarizer is denoted by θ_A, which is set to be very close to 45° to suppress the π component. The direction of the polarization of the polarizer can be varied by rotating the polarizer around $\boldsymbol{K}_G$.

The resultant wavefield of a three-wave diffraction after the polarizer, composed of the directly diffracted and Umweg-excited waves, can be written as

$$\begin{aligned} \boldsymbol{D}_G^{(3)A} &\cong \boldsymbol{D}_2^A + \boldsymbol{D}_{\mathrm{Um}}^A \\ &= A_A A_G \chi_{A\text{–}O} \left(\chi_{G\text{–}O} \boldsymbol{p}_2^A + A_L \chi_{G\text{–}L} \chi_{L\text{–}O} \boldsymbol{p}_{\mathrm{Um}}^A \right) D_O \ , \end{aligned} \tag{8.122}$$

where A_A and $\chi_{A\text{–}O}$ are the resonance term and the Fourier component of the reflection A of the polarizer, respectively;

$$\boldsymbol{p}_2^A = (\boldsymbol{p}_2 \cdot \hat{\boldsymbol{\sigma}}_A)\,\hat{\boldsymbol{\sigma}}_A \quad \text{and} \quad \boldsymbol{p}_{\mathrm{Um}}^A = (\boldsymbol{p}_{\mathrm{Um}} \cdot \hat{\boldsymbol{\sigma}}_A)\,\hat{\boldsymbol{\sigma}}_A \tag{8.123}$$

are the projections of the polarization vectors $\boldsymbol{p}_2$ and $\boldsymbol{p}_{\mathrm{Um}}$ onto $\hat{\boldsymbol{\sigma}}_A$. The projected $\boldsymbol{p}_2^A$ and $\boldsymbol{p}_{\mathrm{Um}}^A$ can be either along the same direction or opposite to each other, depending on the polarization angle ω_A. The two situations, denoted as case I and case II, are depicted respectively in Fig. 8.48 for two different polarization angles ω_A^{I} and ω_A^{II}, namely, two different $\hat{\boldsymbol{\sigma}}_A^{\mathrm{I}}$ and $\hat{\boldsymbol{\sigma}}_A^{\mathrm{II}}$. The former leads to polarization enhancement and the latter polarization suppression.

Figure 8.49 shows the measured Renninger scans around the GaAs $(\bar{2}\bar{2}2)$ of a GaAs $(000)(\bar{2}\bar{2}2)(\bar{3}\bar{1}3)$ three-wave diffraction for $\lambda = 1.5399$ Å with a π-polarized incident radiation ($\omega_O = 90°$). The primary reflection $(\bar{2}\bar{2}2)$ is a weak reflection and the secondary $(\bar{3}\bar{1}3)$ a medium strong reflection. The ψ rotation from negative angles to positive angles corresponds to the movement

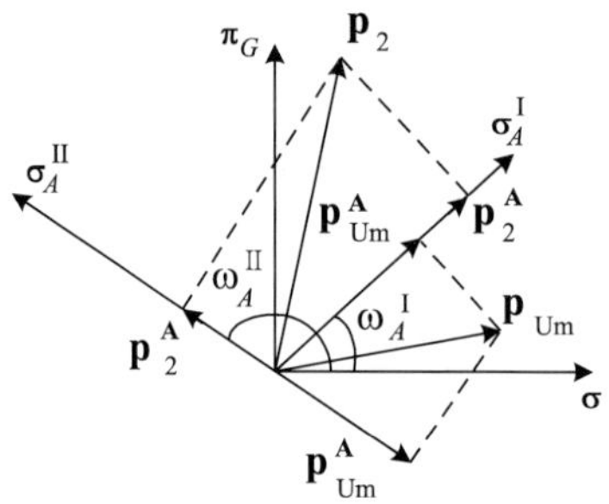

Fig. 8.48. Projection of the polarization vectors for the polarizer shown in Fig. 8.46 at the two polarization angles ω_A^{I} and ω_A^{II} [324]

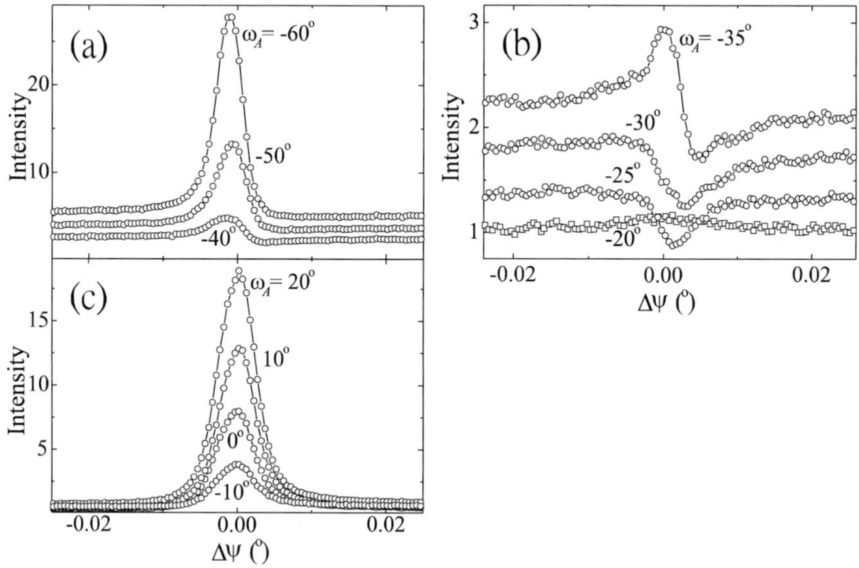

Fig. 8.49. Measured $I_{\bar{2}\bar{2}\bar{2}}$ of GaAs (000)($\bar{2}\bar{2}\bar{2}$)($\bar{3}\bar{1}3$) three-wave diffraction for $\lambda = 1.5399$ Å and a π-polarized incident radiation ($\omega_O = 90°$) with different polarization angles ω_A for a Ge (333) polarizer: (**a**) $-60° \leq \omega_A \leq -40°$, (**b**) $-35° \leq \omega_A \leq -20°$, (**c**) $-10° \leq \omega_A \leq 20°$ [324]

of the reciprocal lattice point of the secondary ($\bar{3}\bar{1}3$) reflection towards the interior of the Ewald sphere, i.e., the 'IN' situation. The (333) reflection of a perfect germanium crystal is used for polarization analysis. Under this experimental condition, the angle γ is 21°, and the angles between $\hat{\boldsymbol{\sigma}}$ and $\boldsymbol{p}_2$ and between $\hat{\boldsymbol{\sigma}}$ and $\boldsymbol{p}_{\mathrm{Um}}$ are 90° and 69°, respectively. Various polarization angles ω_A for the Ge (333) polarizer are chosen to give different polarizations for the diffracted wave. Only the intensity profiles for $-60° \leq \omega_A \leq 20°$ are shown in the figure. The profiles for $-90° \leq \omega_A \leq -60°$ and $20° \leq \omega_A \leq 90°$ are qualitatively the same as those in Fig. 8.49(a) and (b).

The vectors $\hat{\boldsymbol{\sigma}}_A$ and $\boldsymbol{p}_2^A$ are collinear at $\omega_A = \pm 90°$. Hence the weak primary ($\bar{2}\bar{2}\bar{2}$) reflected wave is not suppressed at all by the polarizer, because

$|\boldsymbol{p}_2^A|/|\boldsymbol{p}_2| = 1.0$. However, the strong Umweg $(\bar{3}\bar{1}3)$ diffracted wave is slightly suppressed with $|\boldsymbol{p}_{Um}^A|/|\boldsymbol{p}_{Um}| = 0.93$. When ω_A is increased from $-90°$ to $-20°$, the Umweg wave is gradually suppressed (see Figs. 8.49(a) and (b)). The total suppression of the Umweg reflection occurs when ω_A is around $-20°$, where the primary reflected wave still has appreciable intensity. A further increase in ω_A tends to increase the Umweg peak intensity (Fig. 8.49(c)), and the primary reflected wave is totally suppressed at $\omega_A = 0$.

The same polarization suppression phenomena also occur for three-wave diffractions with strong primary reflections and weak secondary reflections as well as those with other combinations.

8.4 High-Order Multiple-Wave Diffraction

A high-order multiple-wave diffraction involves more than three waves. It usually occurs in the situation with a large d/λ. That means the radius of Ewald sphere is relatively large compared with the magnitude of the reciprocal lattice basis vectors. This kind of multiple-wave diffraction are encountered frequently for a short X-ray wavelength and for large crystal lattice constants. When the number of waves increases, the interaction among the diffracted waves, i.e., the dynamical diffraction effect, becomes much more complicated than in a three-wave diffraction. In the following, the dynamical aspects of four-, six-, twenty-six-wave cases are described.

8.4.1 Four-Wave Diffraction

The reciprocal lattice points of a four-wave case usually form a square, rectangle, diamond, or trapezoid. With respect to a crystal surface, a four-wave case can be of a Laue transmission, a Bragg-Laue, or a Bragg-surface type, depending on whether the diffracted wave is of a Laue, a Bragg, or a surface type. For the Ge four-wave $(000)(1\bar{1}1)(20\bar{2})(3\bar{1}\bar{1})$ diffraction and $\mathrm{CuK}\alpha_1$ radiation, the four reciprocal lattice points are at the corners of a rectangle. If the crystal surface normal is [121], then the case is of a transmission type. Since $(1\bar{1}1)$, $(20\bar{2})$, $(3\bar{1}\bar{1})$ and their coupling reflections are relatively strong, the dispersion surface, the excitation of mode, and the linear absorption coefficient along the ψ rotation for the $(1\bar{1}1)$ and $(20\bar{2})$ reflection lines are similar to those for the (111) reflection in the three-wave $(000)(111)(\bar{1}11)$ Laue case, except that now two more modes need to be considered. The azimuthal rotation around $[3\bar{1}\bar{1}]$ is symmetric about $\Delta\psi = 0$, because [311] lies in between $[1\bar{1}1]$ and $[20\bar{2}]$ and the reciprocal lattice points of $(1\bar{1}1)$ and $(20\bar{2})$ is symmetric about $[3\bar{1}\bar{1}]$. An anomalous transmission is expected to occur for thick crystals with $\mu T \gg 1$. The calculated transmission intensities for $(1\bar{1}1)$, $(20\bar{2})$, and $(3\bar{1}\bar{1})$ shown in Fig. 8.50 exhibit such effects. This four-wave case has been studied extensively [219,619–622]. Details about the dynamical calculation and divergent-beam experiment can be found in the references [53, 219].

The Bragg-Laue four-wave (000)(004)(111)(113) diffraction of germanium occurs when the crystal surface normal is along [001], where (004) is a symmetric Bragg reflection, (113) an asymmetric Bragg reflection, and (111) a transmission (Laue) diffraction. Again, all the reflections involved, including the coupling reflections, are strong. The dynamical interaction among the diffracted waves is similar to that of the Bragg-Laue $(000)(111)(\bar{1}\bar{1}1)$ and Bragg-Bragg (000)(111)(220) cases. The number of permitted modes for a thick crystal with both σ and π polarizations is $2 \times (4-2) = 4$. Therefore, the dispersion surface, excitation of mode, linear absorption coefficient, and intensity distribution resemble those for the Bragg-Laue and Bragg-Bragg cases. The diffraction profile obtained with the Renninger scan is shown in Fig. 4.7. Detailed calculations and measurements are given in the references [53].

The dynamical interaction in multiple-wave cases is governed by the fundamental equation of wavefield and the dispersion relation. Many structure-factor multiplets are involved in the dispersion equation. They play an important role in affecting the dispersion surface, wavefiled-amplitude ratio, and diffracted intensities. In a four-wave (O, G, L, M) case, four structure-factor triplets and three quartets are involved, which are displayed in Fig. 8.51, where A's represent the multiplets. Both the magnitudes and phases of these multiplets contribute to the diffraction intensity. We shall come back to this point in Chap. 9 when we discuss the X-ray phase problem.

The four-wave diffracted intensity can also be described by the kinematical theory. In the quantum field approach given in Chap. 7, (7.283) also gives a kinematical expression. The involved number of structure-factor multiplets, as shown in Fig. 8.52, is more than seven. This is due mainly to the

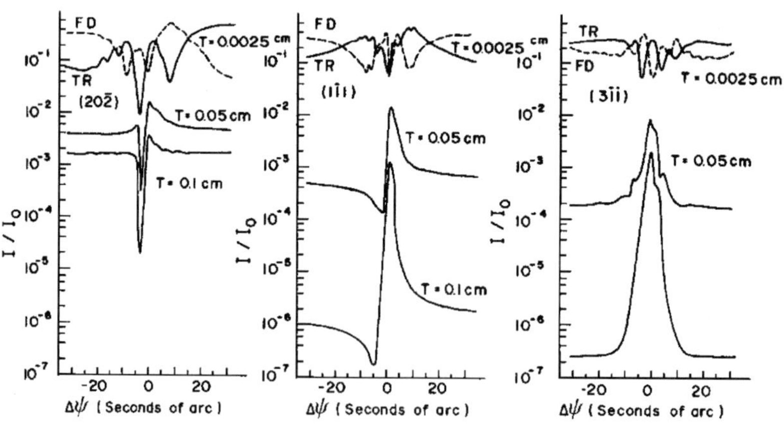

Fig. 8.50. Transmitted intensities calculated for the Ge four-wave $(000)(1\bar{1}1)$ $(20\bar{2})(3\bar{1}\bar{1})$ Laue diffraction with an unpolarized incident wave and CuKα_1: (**a**) $(20\bar{2})$, (**b**) $(1\bar{1}1)$, (**c**) $(3\bar{1}\bar{1})$ [219]

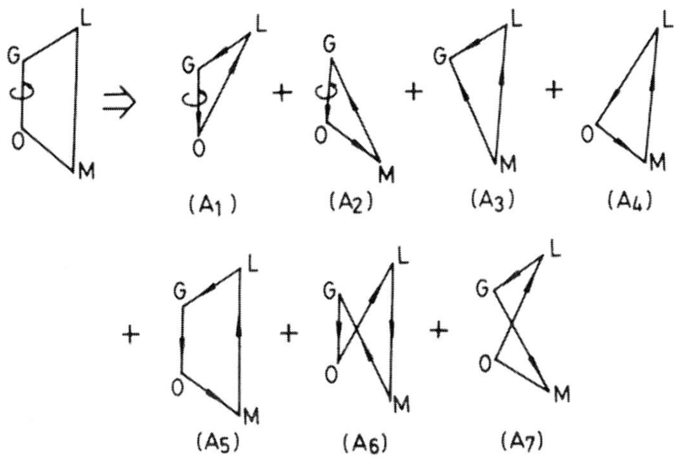

Fig. 8.51. Structure-factor multiplets involved in a four-wave (O, G, L, M) dynamical interaction [636, 638]

series expansion adopted for the intensity. The close-loop polygons are the structure-factor invariants, which are phase-dependent. The double dashed lines indicate the phase-independent interactions.

8.4.2 Coincidental Twenty-Six-Wave Diffraction

Coincidental multiple-wave diffraction occurs only for a specific X-ray wavelength. The involved reciprocal lattice points are not confined to a plane but to the surface of the corresponding Ewald sphere of the specific wavelength. A coincidental twenty-six-wave diffraction takes place for silicon at λ_c defined as

$$\lambda_c = \frac{2a_{\text{Si}}}{9\sqrt{3}n} , \tag{8.124}$$

where n is the refractive index equal to 0.9999985 and $a_{\text{Si}} = 5.4309\,\text{Å}$. The involved reflections are $(000)(\bar{3}33)(3\bar{3}3)(33\bar{3})(5\bar{1}\bar{1})(\bar{1}5\bar{1})(\bar{1}\bar{1}5)(\bar{2}28)(\bar{2}82)(2\bar{2}8)$ $(28\bar{2})(8\bar{2}2)(82\bar{2})(1\ 7\ 11)(1\ 11\ 7)(7\ 1\ 11)(7\ 11\ 1)(11\ 1\ 7)(11\ 7\ 1)(4\ 10\ 10)(10\ 4\ 10)(10\ 10\ 4)(6\ 6\ 12)(6\ 12\ 6)(12\ 6\ 6)(999)$. For simplicity, let's consider this case a twenty-six-wave Laue transmission through a plane-parallel crystal plate with the surface normal along [111]. The reciprocal lattice points of (000) and (999) lie on the diameter along the incident-wave direction. In other words, (999) is a backward diffraction. The rest 24 reciprocal lattice points are distributed on the surface of the Ewald sphere with an appropriate symmetry with respect to the incident wave (see Fig. 8.53). That is, there are 12 points on each side of the equatorial plane. Below the equatorial plane, the six reciprocal lattice points $(\bar{3}33), (3\bar{3}3), (33\bar{3}), (5\bar{1}\bar{1}), (\bar{1}5\bar{1})$, and $(\bar{1}\bar{1}5)$ are on the surface of the Ewald sphere at the level of distance

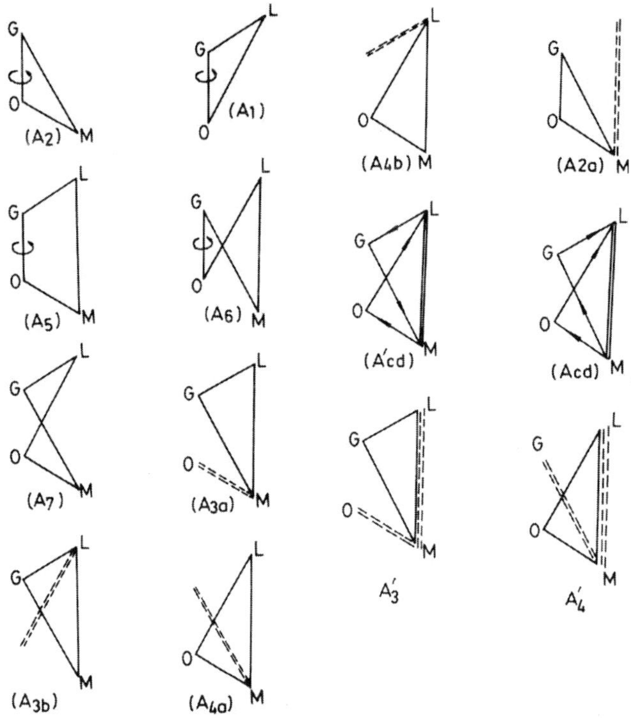

Fig. 8.52. Structure-factor multiplets involved in a four-wave (O, G, L, M) kinematical diffraction [638]

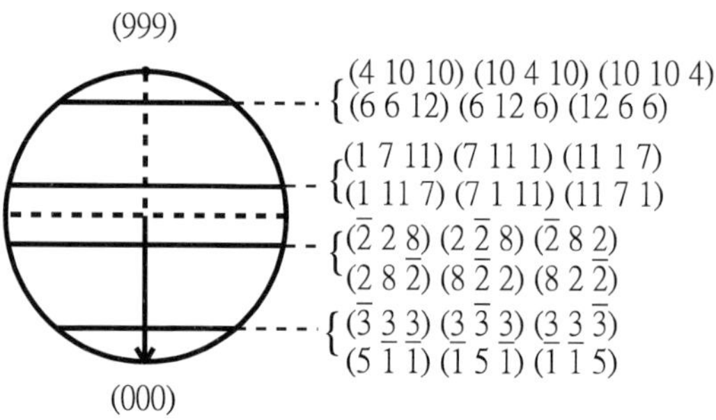

Fig. 8.53. The geometry for the twenty-six wave case in the reciprocal space

$\sqrt{3}/(a_{\mathrm{Si}})$ from the origin (000) perpendicular to [111]. Both $(\bar{3}33)(3\bar{3}3)(33\bar{3})$ and $(5\bar{1}\bar{1})(\bar{1}5\bar{1})(\bar{1}\bar{1}5)$ form two groups of three-fold symmetry with respect to [111]. The angular separation between the two is 60 degrees. The other six points $(\bar{2}28)$, $(\bar{2}82)$, $(2\bar{2}8)$, $(28\bar{2})$, $(8\bar{2}2)$, and $(82\bar{2})$ situated at the level of distance $8/(\sqrt{3}a_{\mathrm{Si}})$ from the origin remain a six-fold symmetry. Similar arrangements are for the rest twelve points above the equatorial plane. Since

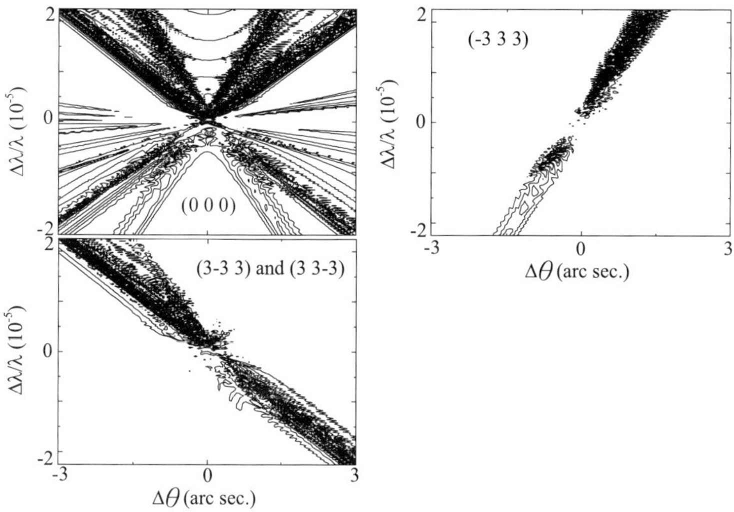

Fig. 8.54. Calculated intensity distributions versus $\Delta\lambda/\lambda$ and $\Delta\theta$ for the twenty-six wave case

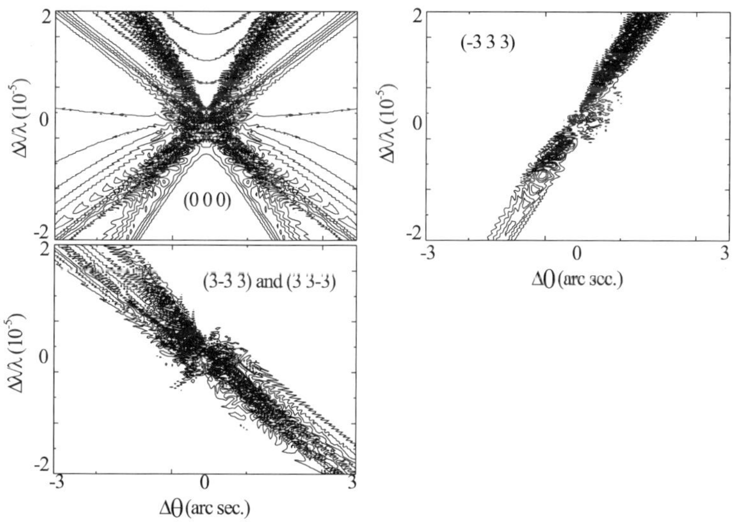

Fig. 8.55. Calculated intensity distributions versus $\Delta\lambda/\lambda$ and $\Delta\theta$ for the seven-wave case

the coincidental multiple-wave diffraction is very sensitive to small variation in wavelength, we shall demonstrate the diffraction intensity distribution in $\Delta\theta$ versus $S = \Delta\lambda/\lambda$, $\Delta\theta$ is the angular deviation from the Bragg angle of $\{\bar{3}33\}$.

Figures 8.54 and 8.55 are the intensity distributions of the forward diffracted (000) reflection and the transmitted ($\bar{3}33$) and ($3\bar{3}3$) reflections for the twenty-six-wave and the hypothetical seven-wave (000)($\bar{3}33$)($3\bar{3}3$)($33\bar{3}$) ($5\bar{1}\bar{1}$)($\bar{1}5\bar{1}$)($\bar{1}\bar{1}5$) cases, respectively. The crystal thickness is 1.97 mm.

The intensity distribution of (000) in the 26-wave case is much more complicated than that in the 7-wave case. As can be seen, there are more reflection bands with appreciable intensity radiated from the origin, where $\Delta\theta = 0$ and $\Delta\lambda/\lambda = 0$. The darker the band, the stronger the reflection. Each reflection ban corresponds to the intensity distribution along the reflection line of a reflection involved. The relative positions of these diffraction bans can be identified from the transmitted reflected intensity distributions of ($\bar{3}33$) and ($3\bar{3}3$). Intensity variation due to the spectral change near the origin is much more severe in the 26-wave case than in the 7-wave case. For example, the intensity of ($\bar{3}33$) in the vicinity of the origin is nearly zero in the 26-wave case, while there is still some intensity around the origin for the same reflection in the 7-wave case. This is due to the difference in the multiple-wave interaction between the two cases. Moreover, the spectral condition for the 26-wave diffraction is more stringent than that for the 7-wave case.

8.4.3 Multiple-Wave Back Diffraction

The experimental investigation on the multiple-wave back reflection, which is similar to the coincidental twenty-six-wave case, has been pursued by using a conventional Co radiation [584]. As mentioned in the previous section, the coincidental multiple-wave diffraction with a back reflection is very sensitive to spectral variation. An interesting way of generating wavelength-dispersion situations is using a coupling crystal plate to tune the incident spectrum to the energy for a back reflection. Figure 8.56 is the schematic representation of the optics. The incident radiation from a cobalt target (labelled by '1') goes

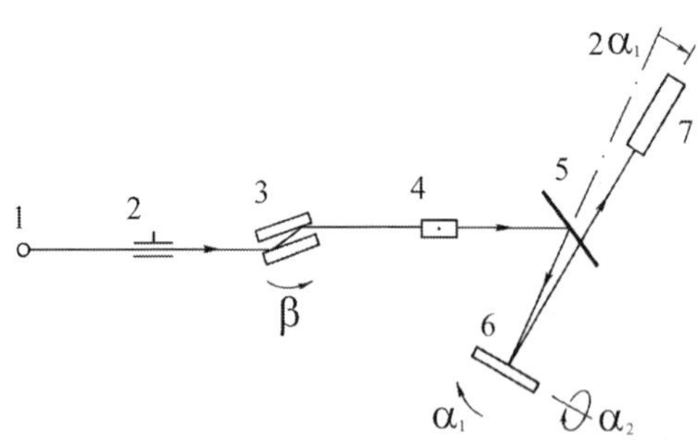

Fig. 8.56. Schematic of the optics setup for the four-wave (000)(440)($2\bar{2}0$)(620) case involving a $\pi/2$ (620) reflection for Co$K\alpha_1$ [584]

through a vertical slit (labelled by '2'), a Ge (111) channel-cut monochromator (labelled by '3'), and a horizontal slit (labelled by '4'). Then it is reflected by the (333) coupling plate (labelled by '5') of a thin Ge crystal and impinges on a Ge sample crystal (labelled by '6') for the (620) reflection. The Bragg angle θ_B is close to $\pi/2$ for the (620) reflection with CoKα_1 radiation ($\lambda = 1.78912$ Å). The (620) diffracted wave after going through the coupling plate is monitored by a scintillation counter (labelled by '7').

The second leaf of the channel-cut monochromator and the coupling plate form a (+, +) dispersive arrangement. Due to the special feature of the DuMont diagram for this (+, +) arrangement with a Bragg angle near $\pi/2$, a high-energy resolution tuning with a rather large angular steps in the Bragg angle is achieved. The spectrum of the CoKα_1 radiation reflected by (620) is widened considerably with respect to the θ by tuning the coupling plate. The sample crystal can be tilted with respect to the normal-incident wave

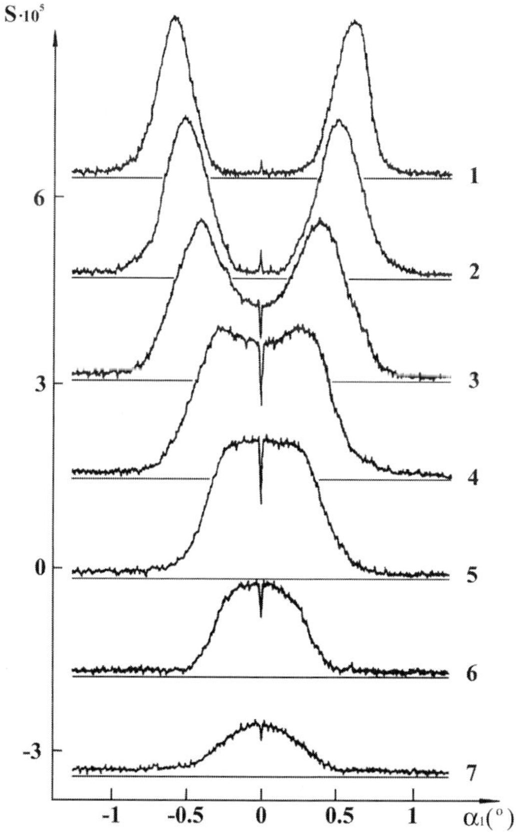

Fig. 8.57. Measured intensity distribution near the four-wave (000)(440)($2\bar{2}0$)(620) diffraction position for various S close to CoKα_1 [584]

by the angle α_1 and can be rotated by the angle α_2 around the [001] direction to bring the four-wave $(000)(440)(2\bar{2}0)(620)$ case to position for all the wavelengths near $\lambda = 1.78912$ Å. This is because the four-wave case is a systematic multiple-wave diffraction, which occurs for all wavelengths. As discussed for a two-wave back diffraction, the pair of Darwin curves form a doughnut shape intensity distribution versus the azimuth angle around the reciprocal lattice vector of the back reflection for photon energies higher than the energy E_c for the exact back diffraction (see Fig. 8.15). When the photon energy E gradually decreases to approach E_c, the diameter of the doughnut ring decreases and the two Darwin curves eventually become one single curve. When the photon energy $E < E_c$, the intensity of this single Darwin curve gradually diminishes. Since this four-wave case is a systematic diffraction, it takes place when the direction [620] coincides with the direction of the incident wave for all the photon energies close to E_c. This angular position is defined as $\alpha_1 = 0$, which also corresponds to the center of the doughnut ring for $E \geq E_c$. Figure 8.57 shows the $\alpha_1 - 2\alpha_1$ scans of the measured intensity of (620) for various wavelengths, i.e., $\lambda - \lambda S$, where $S = \Delta\lambda/\lambda$, in the vicinity of 1.78912 Å. The upper two curves, which are for short wavelengths, show a pair of Darwin peaks, and the tiny peak intensity at $\alpha_1 = 0$ is due to the four-wave case. When the two Darwin peaks gradually merge, forming a nonzero intensity background, the four-wave case appears as a dip, similar to the Aufhellung in multiple-wave diffraction. At $\lambda = 1.78912$ Å, a four-wave dip appears in the middle of a broad peak. When the wavelength increases, the broad peak gradually disappears.

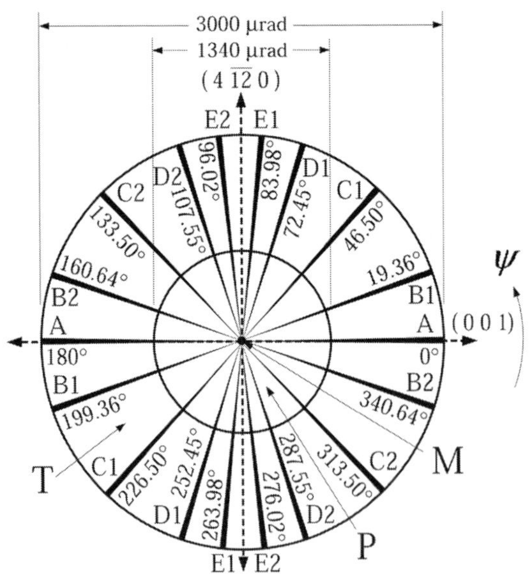

Fig. 8.58. Polar plot of the twenty-four wave case normal to [12 4 0] [590]

A different kind of back diffraction involving twenty-four waves has recently been investigated by using a synchrotron undulator source [590]. The (12 4 0) reflection of silicon is a back diffraction at $\lambda = 0.85872946$ Å and $E = 14.438103$ keV. At this wavelength, twenty-two additional reflections take part in the diffraction to form a coincidental twenty-four-wave diffraction, where an eight-wave (000)(12 4 0)($8\bar{4}0$)(4 8 0)(8 8 0)($4\bar{4}0$)(12 0 0)(040) case, denoted as 'A', and eight four-wave cases are involved. The latter include the 'B1' (000)(12 4 0)(6 8 2)($6\bar{4}\bar{2}$), the 'C1' (000)(12 4 0)(12 2 $\bar{2}$)(022), the 'D1' (000)(12 4 0)(6 4 6)($60\bar{6}$), the 'E1' (000)(12 4 0)($82\bar{6}$), the 'E2' (000)(12 4 0)(826)($42\bar{6}$), the 'D2' (000)(12 4 0)($64\bar{6}$), the 'C2' (000)(12 4 0)(12 2 2), and the 'B2' (000)(12 4 0)($68\bar{2}$)($6\bar{4}2$) case. Figure 8.58 shows the polar plot of the multiple-wave cases involved in this twenty-four wave diffraction. The [12 4 0] direction is perpendicular to the paper. So is the back diffracted wave. The reciprocal lattice points of each of the multiple-wave cases lie in a section normal to the page. At the center, the twenty-four wave diffraction condition is completely fulfilled. The two directions [001] and [4 $\overline{12}$ 0] define the plane perpendicular to the (12 4 0) diffracted wave. The incident angular deviations from the center along [001] and [4 $\overline{12}$ 0] are denoted by ε and η, respectively. The azimuth angle around [12 4 0] is ψ. Since the angular width of a $\pi/2$ reflection is about $\sqrt{\chi}$, the width of the (12 4 0) reflection is rather wide, about 1340 µrad, which is also indicated in the figure.

The experimental setup is schematically shown in Fig. 8.59, where WBS is the tungsten white beam slits, FS and GS the front and guard slits, IC1 and IC2 the first and second N_2-filled ionization chambers, and GO a four-circle diffractometer. The incident wave divergence at the fundamental energy of 14.438 keV of the undulator used is 17 µrad vertical and 42 µrad horizontal. The key optical components are the high-heat-load monochromator (HHLM) and the high-resolution monochromator (HRSM). The former consists of two parallel (111) diamond crystals and selects a bandpass of about 1.1 eV. The latter is a dispersive four-crystal monochromator, composed of two channel-cut perfect silicon crystals with (440) and (975) as the outer and inner reflections [590]. The asymmetry angles of the crystal surface with respect to (440) and (975) are $-24.7°$ and $71.0°$, respectively, leading to the energy bandpass of about 2.3 meV.

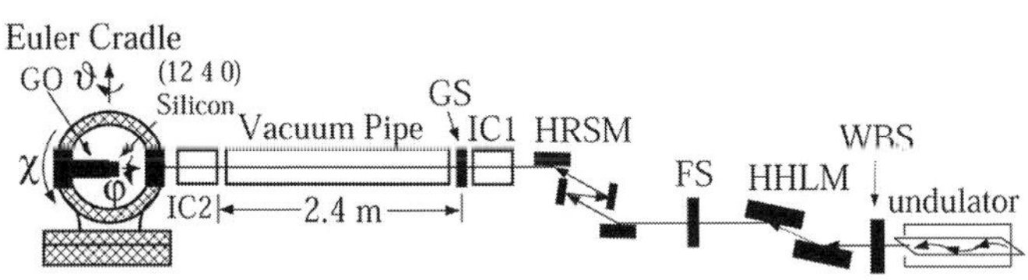

Fig. 8.59. Experimental setup for the twenty-four wave back diffraction [590]

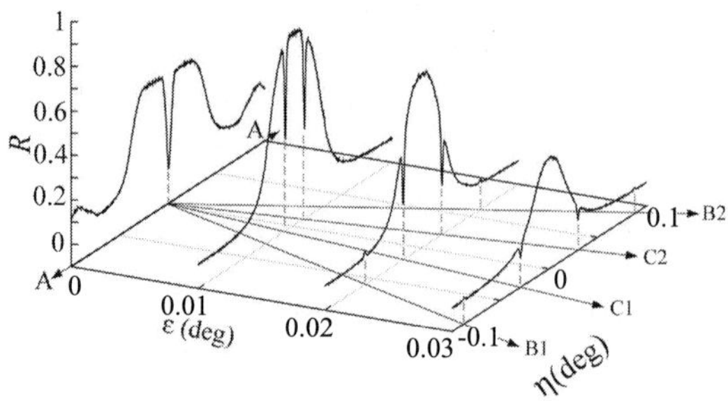

Fig. 8.60. Two-dimensional intensity of (12 4 0) near twenty-four wave diffraction point for $\Delta E = -1.0\,\mathrm{meV}$ [590]

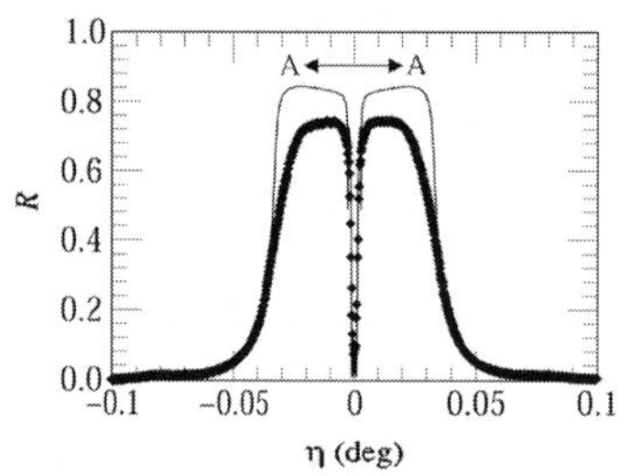

Fig. 8.61. Calculated and measured intensities of (12 4 0) near the twenty-four wave diffraction point with $\varepsilon = 0.001°$ [590]

Figure 8.60 shows the two-dimensional intensity distribution of (12 4 0), which is composed of one-dimensional scan over η at different ε for $\Delta E = -1.0\,\mathrm{meV}$ from $E = 14.438103\,\mathrm{keV}$. The polarization of the incident wave is chosen to be along [001]. The ten-wave diffraction, consisting of the four-wave cases B1, B2, C1, and C2, appears as a dip on the (12 4 0) reflection plateau at the center position where $\varepsilon = 0$ and $\eta = 0$. As the incident wave deviates from $\varepsilon = 0$ along [001], the ten-wave case is decomposed into four four-wave cases.

The dynamical calculation also shows similar features, which are in a reasonable agreement with the measured intensity distribution in the vicinity of the twenty-four wave case. Figure 8.61 gives such an example for the calculated and measured back reflected intensities R over η for $\Delta E = -0.5\,\mathrm{meV}$ at $\varepsilon = 0.001°$.

9. Applications

X-ray multiple-wave diffraction results from simultaneous reflections from two-, three-, and higher-dimensional periodic atomic structures, termed as 'crystals'. The intensity distribution as a function of momentum transfer due to scattering for a given photon energy provides structural information about the crystals, including the crystal lattice constants, the strain field, the structure factors, and then the electron density maps. If a crystal with known lattice constants is used, the diffraction intensity versus photon energy will give accurate spectra of the radiation employed. Since the diffraction conditions for multiple-wave diffraction is more stringent than for two-wave cases, precise measurements of lattice constants and radiation spectra as well as the design of high-resolution X-ray optical components are feasible by using multiple-wave diffraction.

If the X-ray photon energies are tuned to the absorption edges of atoms, resonance multiple-wave diffraction will take place. The diffracted intensity versus momentum transfer as well as photon energy should, in principle, provide both the atomic and electronic structures of the matter investigated. Moreover, multiple-wave diffraction, being phase sensitive, lends itself as a technique for measuring the phases of diffracted waves from a variety of materials which retain a certain degree of structure coherence, namely periodic ordering. In this chapter, various applications derived from the unique characteristics of multiple-wave diffraction are described. Prospects of future development of this diffraction technique for material characterization and condensed matter research will also be tentatively given.

9.1 Determination of X-Ray Reflection Phase

Phase, like potential, is a relative physical quantity, which plays an important role in many branches of research. The so-called Aharonov-Bohm effects, Aharonov-Casher effects, magnetic flux quantization, Berry phase in parameter space, gauge-field theory, etc., involve dominant phase factors one way or another. In the optics of the visible spectrum, electrons, atoms, and molecules, interference between various waves is a proper means of extracting phase information from the intensity of photons or particles scattered from an object of interest. In the case of X-ray diffraction in crystals, the magnitude and the phase of the diffracted wave from a crystal unit cell, namely, the structure factor, are the necessary and sufficient conditions for the determination of the relative positions of atoms in the unit cell. However, the measured intensity of a reflection G is only proportional to the product of the structure factor F_G and its complex conjugate. Information about the phase is therefore lost. This fact constitutes the so-called 'X-ray phase problem'. Many methods have been developed to solve this problem, such as direct methods and methods utilizing anomalous dispersion, heavy atoms, isomorphous and molecular replacement, maxima entropy, and thermal annealing. Nevertheless, in terms of physics, the technique of using multiple-wave diffraction is another potential method for direct phase determination from measured intensities.

The coherent interaction among the diffracted waves inside a crystal for a multiple-wave diffraction has long been thought to provide a means for phase determination. In analogy with optical holography, one of the diffracted waves in multiple-wave diffraction serves as a reference to the others. The interference between the two therefore modifies the intensity of each of the diffracted waves. The phase information can then be extracted from the intensity variation. The development of this diffraction method for phasing structure factors from the measured multiply diffracted intensity variation, as already described in Chap. 1, results from the efforts of many investigators in this field of research. In the following we shall first present the phasing method for the qualitative and the quantitative analysis of structure factors and then describe the various applications in relation to phase determination.

9.1.1 Qualitative and Quantitative Determination of Phase

General Consideration for Qualitative and Quantitative Phase Determination. The phase of an X-ray diffraction from a single-crystal depends on whether the crystal is centrosymmetric or noncentrosymmetric, i.e., whether the crystal is with or without a center of inversion. From the definition of the structure factor, (2.27), the reflection phase is either 0 or π for centrosymmetric crystals and can be any value between 0 and 2π for noncentrosymmetric crystals. In addition, the reflections with phase values of 0 and π are centric reflections and the others are acentric reflections. Obviously, if a reflection G with a phase δ_G is a centric reflection, then the cosine of δ_G, either $+1$ or -1, is important; if it is an acentric reflection, both the sine and cosine of δ_G need to be considered. For the former, if the sign of $\cos\delta_G$ can be qualitatively distinguished, then the associated phase can be known. For the latter, the signs of both $\cos\delta_G$ and $\sin\delta_G$ are important, therefore the qualitative determination of the signs and even quantitative determination of δ_G become necessary. In any case, how to extract phase information from measured intensity distributions is the key issue in solving the X-ray phase problem.

The approximate expressions for the multiply diffracted intensity as a function of phase and some geometric factors described in Chap. 7 provide a clue to solving this problem. In the Bethe approximation, the intensity of a three-wave (O, G, L) diffraction for a σ polarization depends on the interference between the primary diffracted wave G and the detoured diffracted wave via the secondary L and the coupling G–L reflection. Therefore, the diffracted intensity is mainly governed by the modified polarizability $|\chi^{\sigma}_{GO,LO}|$ given in (7.66) or (7.88) as

$$|\chi^{\sigma}_{GO,LO}| = \left\{ P_{\sigma}^2 |\chi_{G\text{–}O}|^2 + \frac{P_1^2 |\chi_{L\text{–}O}\chi_{G\text{–}L}|^2}{4|\xi_L|^2} \right.$$
$$\left. - \frac{1}{2\xi_L}\left(P_1 P_{\sigma}\chi_{L\text{–}O}\chi_{G\text{–}L}\chi^{*}_{G\text{–}O}\right) - \text{c.c.} \right\}^{\frac{1}{2}}, \tag{9.1}$$

where $1/2\xi_L$ defined in (7.55) can be represented by the resonance function [279] defined in (7.286) and (7.288) as

$$\frac{1}{2\xi_L} = -A_L = -\frac{K'^2}{K_L^2 - K'^2} = -|A_L|\,\mathrm{e}^{\mathrm{i}\Delta}, \tag{9.2}$$

and c.c. stands for the complex conjugate of its preceding term. With the consideration of the absorption for the wavevector $\boldsymbol{K}_L$, A_L takes the approximate form given in (7.301). The magnitude $|A_L|$ is a symmetric function of $K_L - K'$ with the maximum at $K_L = K'$, the exact three-wave diffraction position. K' is the distance between the three-wave Lorentz point and the reciprocal lattice point O. The geometric phase Δ due to the crossing of the surface of the Ewald sphere by the reciprocal lattice point L is a smooth step function, with a phase of 0° for $K_L - K' < 0$ and 180° for $K_L - K' > 0$. At and near $K_L = K'$, Δ changes linearly and abruptly from 0° to 180° over an angular range proportional to the diffraction peak width. The change in $|K_L|$ from $K_L < K'$ to $K_L > K'$ corresponds to the motion of point L from the inside to the outside of the Ewald sphere. This is equivalent to the outgoing (OUT) situation in the Renninger scan and the reference-beam/stereoscopic imaging [309, 310, 318, 319, 326, 623]. Hence, for the incoming (IN) situation for point L, Δ changes from 180° to 0° when point L crosses the three-wave position. Hereafter, the IN situation will be adopted in the discussion for phase determination, because in multiple diffraction experiments, diffraction profiles are represented in an increasing order of the diffraction angle, either θ or ψ, that corresponds to the IN situation. With this geometric phase shift Δ due to the L diffraction, the three-wave diffraction intensity $I_G(3)$ is related to

$$I_G^{(3)} \sim \left\{ P_{\sigma}^2 |\chi_{G\text{–}O}|^2 + \frac{P_1^2 |\chi_{L\text{–}O}\chi_{G\text{–}L}|^2}{4|\xi_L|^2} \right.$$
$$\left. + |A_L| P_1 P_{\sigma} |\Gamma|^3 |F_{L\text{–}O}F_{G\text{–}L}F_{G\text{–}O}| \cos\left(\delta'_{3+} + \Delta\right) \right\}. \tag{9.3}$$

Hence, the interaction among the diffracted waves is governed by the cosine of the effective phase sums, both $\pm(\delta'_{3+} + \Delta)$ and $\mp(\delta'_{3-} - \Delta)$, where

$$\begin{aligned} \delta'_{3+} &= \delta_{L\text{–}O} + \delta_{G\text{–}L} + \delta_{O\text{–}G} \qquad \text{and} \\ \delta'_{3-} &= \delta_{O\text{–}L} + \delta_{L\text{–}G} + \delta_{G\text{–}O} \,. \end{aligned} \tag{9.4}$$

This result, included in the intensity expressions (7.307), (7.337), and (7.355), resembles that of the two-slit interference experiment. That is, the intensity variation of the three-wave (O, G, L) diffraction relative to the two-wave (O, G) case for a given primary reflection G is composed of the phase-dependent part I_D and the phase-independent part I_K, which can be rewritten as

$$\begin{aligned} I'_G(\psi_u) &= \frac{I_G^{(3)} - I_G^{(2)}}{I_G^{(2)}} = I_D + I_K \\ &= \frac{A\psi_u \cos\delta_3 - B\sin\delta_3}{d_2^2(\psi_u^2+1)} + \frac{C}{d_2^2(\psi_u^2+1)} \,, \end{aligned} \tag{9.5}$$

where the parameters A, B, C, and d_2 have been defined in (7.307) and B is usually positive. The triplet phases δ_3 and $\delta_{3\pm}$ are now defined as

$$\begin{aligned} \delta_3 = \delta_{3+} &= \delta_{L\text{–}O} + \delta_{G\text{–}L} - \delta_{G\text{–}O} \,, \qquad \text{and} \\ \delta_{3-} &= \delta_{O\text{–}L} + \delta_{L\text{–}G} - \delta_{O\text{–}G} \,, \end{aligned} \tag{9.6}$$

which are equal to $\delta'_{3\pm}$ of (9.4) when anomalous dispersion effects are negligibly small. As clearly seen, the phase-independent I_K is a symmetric function of the reduced azimuth angular parameter ψ_u, defined as $\psi_u = 2\psi/\eta_\psi$, where η_ψ is the FWHM of the measured profile. The effective phase sum u can also be expressed in terms of the reduced parameter ψ_u as

$$u = \delta_{3+} + \Delta = \delta_{3+} + \arctan\left(\frac{1}{\psi_u}\right) \,. \tag{9.7}$$

Consequently, the phase-dependent I_D is related to the real and imaginary parts of the resonance function A_L given in (9.2). Furthermore, the real part and imaginary parts of I_D satisfy the Kramers-Krönig relations in the momentum space [329].

Similarly, (9.5) is also valid for the reduced θ angle, θ_u, in the reference-beam/stereoscopic imaging [624]. For a centrosymmetric crystal, the parameter $B = 0$ and the asymmetry of the intensity profile depends solely on $(A\psi_u \cos\delta_3 + C)/({\psi_u}^2 + 1)$. During the azimuth rotation, $\psi_u = 0$ corresponds to the exact three-wave diffraction position. For the 'IN' situation, the angular position of ψ_u is changed from negative to positive when the reciprocal lattice point L moves towards the interior of the Ewald sphere. If $\delta_3 = 0°$, the effective phase sum u changes from 180° to 0°. Hence, the diffraction profile shows first a decrease then an increase in intensity. If $\delta_3 = 180°$, a reversed profile asymmetry results. The same conclusion can also be drawn from (9.3).

Although the phase effect on the intensity profile of the three-wave diffraction is clear from the approximate analytical expressions, the observations of the asymmetry of the intensity distribution due to phase effect in centrosymmetric crystals, as mentioned in Chap. 1, were reported in the literature [224, 251, 256–262, 287] earlier than the theoretical explanation. The details of these studies can be found in the review article [453]. As an example, the diffraction profiles indexed as $\bar{1}11$(OUT) and 311(IN) show opposite asymmetry in Fig. 4.2. The corresponding three-wave cases are $(000)(222)(\bar{1}11)$ and (000)(222)(311) for silicon. For phase determination, the signs S_L and S_R are defined for the experimental line-profile asymmetry and the sense of the azimuthal rotation, respectively. S_L is positive for the diffracted intensity first decreasing then increasing during the azimuth scan and negative for the opposite asymmetry. The sign S_R is positive for the 'IN' and negative for the 'OUT' situation. Then the sign of $\cos\delta_3$ follows the relationship [260, 261]

$$S(\cos\delta_3) = S_L \cdot S_R \tag{9.8}$$

for these two three-wave cases. Care must be taken in distinguishing the IN and OUT situations when $\ell^2 - \boldsymbol{\ell}\cdot\boldsymbol{g} < 0$, because the sense of rotation is opposite to that for $\ell^2 - \boldsymbol{\ell}\cdot\boldsymbol{g} > 0$ [286]. Following the sign equation (9.8), the determined triplet phase δ_3 is 180° for (000)(222)(311) and 0° for $(000)(222)(\bar{1}11)$.

Consider an ideal phase-sensitive three-wave diffraction of a noncentrosymmetric crystal with $A > 0$, $B > 0$ and $C = 0$, where the phase-independent part is null. Then the expected profiles for $\delta_3 = 0°$, 90°, 180°, 270° can be drawn as shown in the first row of Fig. 9.1 [279,305,328]. The profiles for $\delta_3 = 45°$, 135°, 225°, 315° can also be deduced. Therefore, qualitative and semi-quantitative phase determination can in principle be achieved by comparing the experimental profiles with those shown in Fig. 9.1. However, in practice, the phase-independent contribution I_K may not be zero and can alter the maximum (or minimum) of the profile for $C > 0$ (or $C < 0$), especially when $\delta_3 = \mp 90°$. The case with $C > 0$ corresponds to the Umweganregung situation and the one with $C < 0$ the Aufhellung situation. In other words, the most convenient way of distinguishing the case with $\delta_3 = 90°$ from that with $\delta_3 = -90°$ is to compare the intensity profiles of two centrosymmetrically related three-wave cases, say, (O, G, L) and $(O, -G, -L)$, in which the triplet phase of the former has an opposite sign to that of the latter, i.e., $\delta_{3+} = -\delta_{3-}$ [279, 328]. The sign '+' in the subscript also refers to the (O, G, L) case and the '−' the $(O, -G, -L)$ case. As shown in Fig. 9.1, the peak intensity of the former with $\delta_3 = 90°$ is weaker than that of the latter with $\delta_3 = -90°$ for the ideal, Umweganregung, and Aulfhellung situations. This semi-quantitative phase determination technique by profile inspection was developed by Hümmer and coworkers in 1986 [279, 305, 328].

Figure 9.2 shows the three-wave profiles for the centrosymmetric pair $(000)(4\ 15\ \bar{4})(2\bar{1}0)$ and $(000)(\bar{4}\ \overline{15}\ 4)(\bar{2}10)$ of the noncentrosymmetric crystal of tetragonal lysozyme for $\lambda = 1.2438$ Å at the IN position [305]. According to Fig. 9.1, the triplet phases determined by inspection are about 45° for

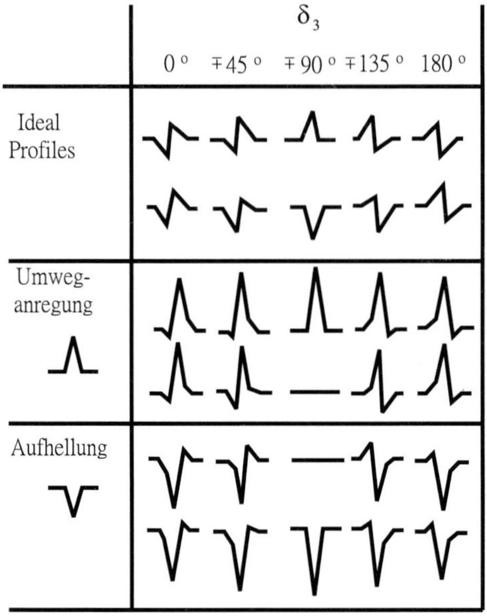

Fig. 9.1. Schematic diagram of three-wave diffraction profile, the ψ-scan, in relation to δ_3. The ψ rotation is from left to right. '+': the (O, G, L) case; '−': the $(O, -G, -L)$ case [328]

the '+' case and $-45°$ for the '−' case. The error in the determined phase is $\pm 45°$.

The quantitative phase determination technique was developed mainly based on (9.5) [291, 301]. The idea is the following. If the phase-independent I_K can be estimated and then subtracted from the measured I'_G, then the remainder is the phase-dependent I_D, which should in principle exhibits the phase-sensitive characteristics. Figure 9.3 depicts the profiles I_D and the result of the subtraction of the calculated I_K from the intensity profiles I'_G for various δ_3. The Kramers-Krönig transformed I_D, denoted as $K[I_D]$, and the effective phase sum u are also shown in this figure. As indicated in the figure, the value of u for a large $\Delta\psi$ is the corresponding triplet phase δ_3. Moreover, the relation $K[I_D] = I_D(\delta_3 - 90°)$, indicating the connection between the real and imaginary parts of the I_D profile, is also revealed.

The I_D profiles do show in Fig. 9.3 the difference due to different δ_3. Hence, experimentally, the $\cos\delta_3$ and $\sin\delta_3$ can be calculated directly from the intensities I_D at the angular positions of half width, i.e, $\psi_u = \mp 1$. Namely,

$$I_D\,(\psi_u = -1) = \frac{-A\cos\delta_3 - B\sin\delta_3}{2d_2^2}\,,$$

$$I_D\,(\psi_u = +1) = \frac{A\cos\delta_3 - B\sin\delta_3}{2d_2^2}\,. \tag{9.9}$$

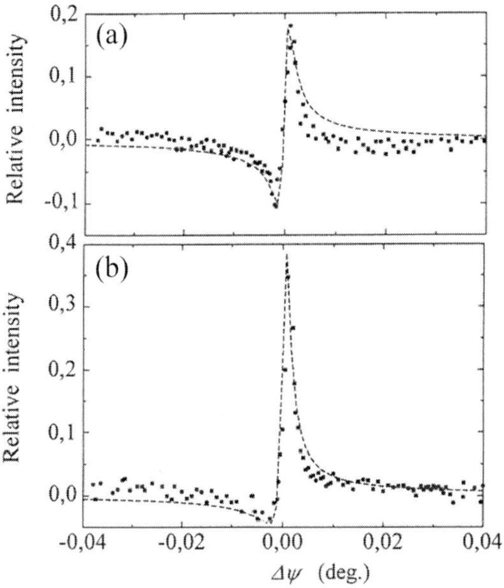

Fig. 9.2. Renninger scans of three-wave pairs (**a**) (000)(4 15 $\bar{4}$)(2$\bar{1}$0) and (**b**) (000)($\bar{4}$ $\overline{15}$ 4)($\bar{2}$10) of tetragonal lysozyme for $\lambda = 1.2438$ Å at the IN position [305]

Note that $\psi_u = 2\psi/\eta_\psi$, where η_ψ is the FWHM of the measured profile.

The estimation of I_K can be made by taking an average of intensity over all the symmetry-related three-wave cases, so that the averaged I_D approaches zero due to the cancellation of the contributions from cases with $\perp\delta_3$ [292, 297]. If two centrosymmetrically related three wave diffraction, (O, G, L) and $(O, -G, -L)$, are used for the estimation of I_K, then at $\psi_u = 0$, $C/(d_2^2) = (I'_G(+) + I'_G(-))/2$ because $\delta_{3+} = -\delta_{3-}$ and $I_D(+) = -I_D(-)$. With the parameter $C/(d_2{}^2)$ and η_ψ, $I_K(\psi_u)$ is then known. The quantitative phase determination becomes straightforward [301] via

$$\tan\delta_3 = \frac{-I_+ - I_-}{I_+ - I_-} , \tag{9.10}$$

where

$$I_\pm = I_D(\psi_u = \pm 1) = I'_G(\psi_u = \pm 1) - I_K(\psi_u = \pm 1) . \tag{9.11}$$

The quadrant to which δ_3 belongs is determined by the signs of the numerator and denominator of (9.10). The $I_\pm$ should be taken at the IN position. At the OUT position, $I_\pm$ has to be replaced by $I_\mp$.

Equation (9.10) is valid only when the exact three-wave diffraction position, $\psi_u = 0$, is experimentally determined. Since multiple-wave diffraction

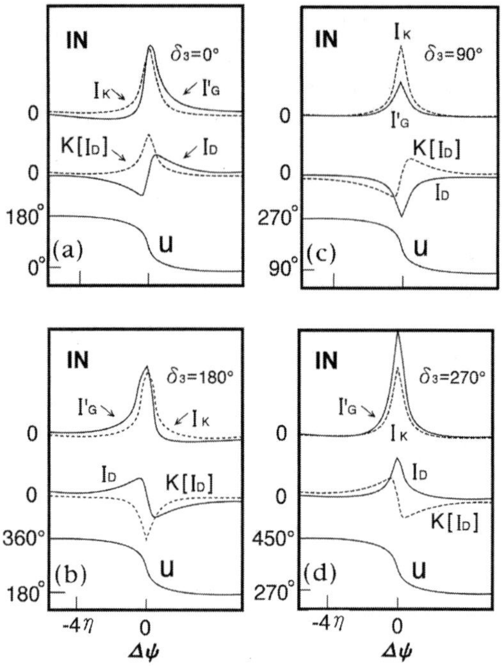

Fig. 9.3. Calculated intensity profiles I'_G, I_K, I_D, $K[I_D]$, and u for $\delta_3 = 0°$, $90°$, $180°$, $270°$ [291]

profiles are rather sharp and span in a very small angular range, sometimes it is not easy to determine precisely the position corresponding to $\psi_u = 0$. An alternative is to use a quasi-universal function $I_U(\psi_u)$ to describe the measured intensity profiles for both '+' and '−' cases so that

$$I'_{G\pm}(\psi_u) = sI_{U\pm}(\psi_u) , \tag{9.12}$$

with

$$I_{U\pm}(\psi_u) = \frac{\psi_u - 1/2\left(\sqrt{K_\pm} - \frac{1}{\sqrt{K_\pm}}\right)}{(\psi_u)^2 + 1} , \tag{9.13}$$

where the scaling factor $s = A\cos\delta_3/(d_2{}^2)$. The quantities $K_\pm = |I^\pm_{\max}/I^\pm_{\min}|$ are the intensity ratios of the maximum and minimum of the profiles for the '+' and '−' cases. They satisfy the relation

$$\frac{1}{2}\left(\sqrt{K_\pm} - \frac{1}{\sqrt{K_\pm}}\right) = \frac{C}{A\cos\delta_{3\pm}} - \frac{B\tan\delta_{3\pm}}{A} . \tag{9.14}$$

A simple intensity-ratio formula [316]

$$\tan\delta_3 = \frac{1}{4}\left[\sqrt{K_-} - \frac{1}{\sqrt{K_-}} - \left(\sqrt{K_+} - \frac{1}{\sqrt{K_+}}\right)\right] \tag{9.15}$$

can be derived to find δ_3 with the reliabilities r_1 and r_2 defined as

$$r_1 = \frac{I_+(\max)\sqrt{K_-}}{I_-(\max)\sqrt{K_+}}\,,$$
$$r_2 = \frac{[I_+(\max) - I_+(\min)]\,[\psi_{u+}(\max) - \psi_{u+}(\min)]}{[I_-(\max) - I_-(\min)]\,[\psi_{u-}(\max) - \psi_{u-}(\min)]}\,, \tag{9.16}$$

where $\psi_{u\pm}(\max)$ and $\psi_{u\pm}(\min)$ are the positions of ψ_u with the maximum and minimum intensities for the '+' and '−' cases, respectively. For a reliable phase determination, both r_1 and r_2 are close to unity.

The scaling factor s can be either positive or negative, depending on $\cos\delta_3$ and the sign of the coefficient A. The latter is governed by the polarization and geometrical factors. For the intensity profiles shown in Fig. 9.2 at the IN position, $s > 0$, then $|\delta_3| < 90°$ for $A > 0$ and $90° < |\delta_3| < 180°$ for $A < 0$. For the profiles with opposite asymmetry, $s < 0$, then $|\delta_3| < 90°$ for $A < 0$ and $90° < |\delta_3| < 180°$ for $A > 0$. When the intensity ratios $K_\pm$ approach zero or infinity, $\delta_{3\pm}$ tend to $\pm 90°$. If $K_\pm$ approach zero, then $\delta_{3\pm}$ are close to $90°$. If $K_\pm$ approach infinity, then $\delta_{3\pm}$ are close to $-90°$. As an example, the universal curves calculated for the measured values of $K_+ = 1.9$ and $K_- = 8.7$ of the profiles shown in Fig. 9.2 are indicated as dashed curves with $s = 0.13$, $\eta = 3.8''$, and $A = B$ in the same figure. The determined phases are $\delta_{3\pm} = \pm 26°$, which are very close to the calculated values of $\pm 35°$ from the known structure. The reliability indices r_1 and r_2 range from 0.75 to 0.85.

A much simpler way of estimating δ_3 is to fit the three-wave profiles of the two centrosymmetrically related (O, G, L) and $(O, -G, -L)$ by the asymmetric Lorentzian functions defined as

$$L_\pm(\psi_u) = \frac{a_\pm(\psi_u - \psi_{uo\pm}) - b_\pm}{(\psi_u - \psi_{uo\pm})^2 + c_\pm^2}\,, \tag{9.17}$$

where $\psi_{uo\pm}$ are the zero-point of the Lorentzians. The adjustable parameters are $a_\pm, b_\pm, c_\pm$, and $\psi_{uo\pm}$. From (9.5) and the relation $\delta_{3+} = -\delta_{3-}$, it follows that $a_+ = a_-(\equiv a)$ and $c_+ = c_-(\equiv c)$. The best fit parameters give

$$\tan\delta_3 = \frac{b_- - b_+}{2ac} \tag{9.18}$$

for the quantitative estimation of δ_3. In practice, owing to experimental errors and theoretical approximations, the fitting parameters a_+ and c_+ may differ from a_- and c_-. Under this circumstance, a and c serve as the weight average for a_+ and a_- and for c_+ and c_-, respectively. Hence, the reliability indices of fitting can be defined as

$$r_a = 1 - \frac{|a_+ - a_-|}{a} ,$$
$$r_c = 1 - \frac{|c_+ - c_-|}{c} . \tag{9.19}$$

For a reliable phase determination, the values r_a and r_c should be close to unity. Although the asymmetric Lorentzian given in (9.17) is a function of ψ_u, (9.17) is equally valid for intensity profiles in terms of θ_u. Figure 9.4 shows the best fit of Lorentzians to the experimental profiles of the four-wave $(000)(200)(12\ 2\ \bar{1})(\overline{10}\ 2\ \bar{1})$ and $(000)(\bar{2}00)(\overline{12}\ \bar{2}\ 1)(10\ \bar{2}\ 1)$ diffractions of a tetragonal lysozyme for $\lambda = 1.24$ Å, obtained with the reference-beam/stereoscopic imaging technique. Due to the two-fold rotation symmetry of the (200) and ($\bar{2}00$), (9.15) and (9.18) can also be applied to the quantitative phase determination of the above four-wave cases (see the later section for four-wave and higher-order diffractions). Compared to the calculated phase values, $-126°$ for the triplet phase of $F(12\ 2\ \bar{1})F(\overline{10}\ \bar{2}\ 1)/F(200)$ and $126°$ for that of $F(\overline{12}\ \bar{2}\ 1)F(10\ 2\ \bar{1})/F(\bar{2}00)$, the determined phases are $-145°$ and $145°$, respectively. The reliability indices are $r_a = 0.8$ and $r_c = 0.75$.

For a quantitative phase determination, other asymmetric functions like those derived from the Takagi-Taupin equations [336] and the hyperbolic tangent [318] could be used for profile fitting. The following fitting function is derived from a distorted-wave approximation for triplet phase estimation in reference-beam experiments [318]:

$$\frac{I_G^{(3)}}{I_G^{(2)}} = 1 - A_o \sin\delta_3 \left(\frac{\sin^2(\Delta\theta_u/2)}{(\Delta\theta_u/2)^2}\right) + \frac{A_o}{\Delta\theta_u/2}\left(\cos\delta_3 + \frac{A_o}{\Delta\theta_u}\right)$$
$$\times \left[1 - \frac{\sin(\Delta\theta_u)}{\Delta\theta_u}\right] , \tag{9.20}$$

where the reduced angle is given by

$$\Delta\theta_u = \frac{2(\theta - \theta_G)}{\eta_\theta} , \tag{9.21}$$

with η_θ being the width of the primary G reflection. The fitting parameters are the two-wave diffraction background $I_G^{(2)}$, the Bragg angle θ_G, the amplitude A_o, and the triplet phase δ_3.

For a qualitative phase analysis for $\cos\delta_3$, the intensity differences between the right and left tails of a profile at equal distances from the peak position versus the azimuth angle ψ are sometimes plotted for the measured profile and theoretical profiles calculated for various δ_3. By comparing with the theoretical plot, the asymmetry of the measured profile can be distinguished and the value of $\cos\delta_3$ can be determined [286].

An attempt to deal with non-perfect crystals for phase determination by using three-wave diffraction has recently been made [335], where a general parametric function was employed to describe diffraction intensity profiles.

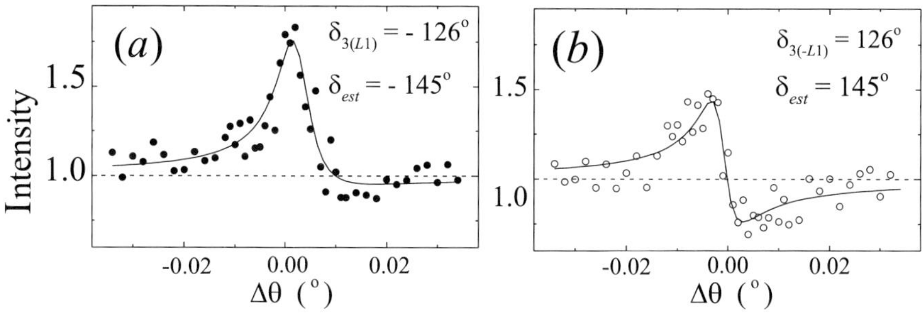

Fig. 9.4. $\Delta\theta$ distribution of measured semi-integrated relative intensities for (**a**) L_1 $(12\ 2\ \bar{1})$ and (**b**) $-L_1$ $(\overline{12}\ \bar{2}\ 1)$ reflections of the two inversion-symmetry related $(000)(200)(12\ 2\ \bar{1})(\overline{10}\ 2\ \bar{1})$ and $(000)(\bar{2}00)(\overline{12}\ \bar{2}\ 1)(10\ \bar{2}\ 1)$ for $\lambda = 1.24\,\text{Å}$ [326]

Polarization and Crystal-Thickness Effects on Phase Determination. The polarization of the incident and diffracted waves is another important factor affecting the intensity distribution of a multiple-wave diffraction, as has already been described in Chap. 8 for linear polarization. Consequently, the phase determination based on the diffraction intensity profile is also closely related to polarization factors.

An elliptically polarized incident wave is usually represented by

$$\boldsymbol{D}_O = \boldsymbol{D}_{\sigma O}\left(\hat{\boldsymbol{\sigma}}_\sigma + \mathrm{i}b_e\hat{\boldsymbol{\pi}}\right) , \tag{9.22}$$

where b_e is the ellipticity parameter defined as

$$b_e = \frac{D_{\pi O}}{D_{\sigma O}} . \tag{9.23}$$

More generally, the polarization vector of an elliptically polarized radiation can be expressed for any linear polarization vector $\boldsymbol{p'}_O$ as

$$\boldsymbol{p'}_O = \boldsymbol{p}_O + \mathrm{i}b_e\boldsymbol{p}_O^{\perp} = (\alpha - \mathrm{i}b_e\beta)\,\hat{\boldsymbol{\sigma}}_O + (\beta + \mathrm{i}b_e\alpha)\,\hat{\boldsymbol{\pi}}_O , \tag{9.24}$$

where $\boldsymbol{p}_O = \alpha\hat{\boldsymbol{\sigma}}_O + \beta\hat{\boldsymbol{\pi}}_O$ with $\alpha = \cos\omega$ and $\beta = \sin\omega$, as defined in (8.111). $\boldsymbol{p}_O^{\perp}$ is normal to $\boldsymbol{p}_O$. By replacing the α and β in the $\boldsymbol{D}_2$ and $\boldsymbol{D}_{\mathrm{Um}}$ of (8.119) with $\alpha - \mathrm{i}b_e\beta$ and $\beta + \mathrm{i}b_e\alpha$, respectively, and substituting the $\boldsymbol{D}_2$ and $\boldsymbol{D}_{\mathrm{Um}}$ into the $\boldsymbol{D}_G^{(3)}$ of (7.294), the relative three-wave intensity $I'_G = \boldsymbol{D}_G^{(3)}\boldsymbol{D}_G^{(3)*}/(\boldsymbol{D}_G^{(2)}\boldsymbol{D}_G^{(2)*}) - 1$ in terms of the reduced angular parameter ψ_u becomes

$$I'_G = \frac{A_e^{-1}F_e\left\{B_e\left[\psi_u\cos\left(\delta_3 - \delta_e\right) - \sin\left(\delta_3 - \delta_e\right)\right] + F_eC_e\right\}}{\psi_u^2 + 1} , \tag{9.25}$$

where

$$A_e = a_\sigma^2 + a_\pi^2 + b_e^2\left(b_\sigma^2 + b_\pi^2\right) ,$$

$$F_e = \frac{|F_{G\text{-}L}|\,|F_{L\text{-}O}|}{(|F_O|\,|F_{G\text{-}O}|)} ,$$

$$
\begin{aligned}
B_e &= \sqrt{B_{e1}^2 + B_{e2}^2} \,, \\
B_{e1} &= b_e \left(a_\sigma d_\sigma + a_\pi d_\pi - b_\sigma c_\sigma - b_\pi c_\pi \right) \,, \\
B_{e2} &= a_\sigma c_\sigma + a_\pi c_\pi + b_e^2 \left(b_\sigma d_\sigma + b_\pi d_\pi \right) \,, \\
C_e &= c_\sigma^2 + c_\pi^2 + b_e^2 \left(d_\sigma^2 + d_\pi^2 \right) \,, \\
a_\sigma &= \alpha P_G^\sigma \,, \qquad a_\pi = \beta P_G^\pi \,, \qquad b_\sigma = -\beta P_G^\sigma \,, \qquad b_\pi = \alpha P_G^\pi \,, \\
c_p &= p_{\mathrm{Um}}^p \left(\boldsymbol{p}_O \right) \,, \qquad d_p = -\beta p_{\mathrm{Um}}^p(\sigma) + \alpha p_{\mathrm{Um}}^p \left(\pi_O \right) \,, \\
\delta_e &= \arctan \left(\frac{B_{e1}}{B_{e2}} \right) \,.
\end{aligned} \tag{9.26}
$$

The superscript p can be either σ or π. Thus, the use of an elliptically polarized incident radiation introduces the phase shift δ_e. The experimental and theoretical demonstration of this phase shift due to elliptical polarization in three-wave diffraction was first reported by Shen and Finkelstein [330]. Figure 9.5 shows the diffraction profiles for the GaAs (000)(442)(151) three-wave case in the OUT situation and $\lambda = 1.3$ Å with the left-handed and the right-handed elliptically polarized incident waves of the ellipticity parameter $b_e = 0.45$. The primary reflection is (442). The right-handed elliptical polarization was obtained at 0.11 mrad above the electron orbital plane of the Cornell High Energy Synchrotron Source, while the left-handed polarization was obtained at 0.11 mrad below the orbital plane. The calculated profiles for $\delta_3 = 0°$, $90°$, $180°$, and $-90°$ are also shown. By comparing the measured profiles with the calculated ones, the triplet phase δ_3 is $90°$. The difference in the profile asymmetry for the left- and right-handed polarizations is due to the additional phase shift δ_e. Because of this phase shift, the intensity profiles of multiple-wave diffraction can be tuned to exhibit phase sensitivity. If one of the strong component waves, either the primary reflected or the detoured diffracted one, could be partially suppressed so as to have comparable amplitudes for interference, use of an elliptically polarized incident wave may facilitate phase determination. Figure 9.6 demonstrates the increase in the phase sensitivity of diffraction intensities by using an elliptically polarized incident radiation under a polarization suppression condition. The intensity profiles of a GaAs three-wave (000)(222)(311) diffraction and $\lambda = 2.4108$ Å for a σ-, a π-, and an intermediate polarization with the polarization angle $\omega_O = 47°$ of an elliptically polarized incident radiation ($b_e = 0.15$) were measured with the soft X-ray diffractometer mentioned in Chap. 4 at the National Synchrotron Radiation Research Center (NSRRC). This case involves a weak primary reflection (222), a strong secondary (311), and a coupling ($\bar{1}11$) reflection. Since the Bragg angle of the (311) reflection is close to 45° for $\lambda = 2.4108$ Å, the condition for polarization suppression, (8.121), of the Umweg peak is fulfilled at the polarization angle $\omega_O = 47°$. As shown in Fig. 9.6(a), the intensity profiles of the σ- and the π-polarization exhibit no phase sensitive asymmetry. While strong intensity asymmetry is observed with the elliptically polarized incident wave, the associated phase δ_3 is determined to be $-90°$ according to

(9.25). The corresponding phase sensitivity S is governed by $S = F_e C_e / B_e$. According to the calculation based on (9.25), the best phase sensitivity is $S \leq 1$. It should be noted that the intensity shown in Fig. 9.6 is normalized with the two-wave (222) intensity in Fig. 9.6(b), which is much weaker than the peak intensities in Fig. 9.6(a).

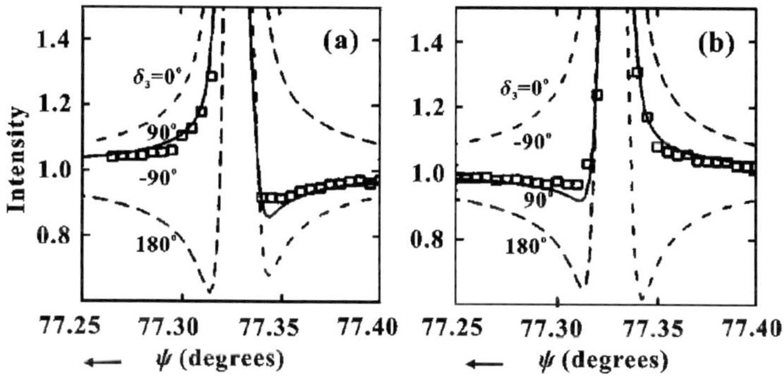

Fig. 9.5. Renninger scans of the three-wave (000)(442)(151) case of GaAs and 1.3 Å in the OUT situation with (**a**) left-handed and (**b**) right-handed elliptical polarization ($b_e = 0.45$) [330]

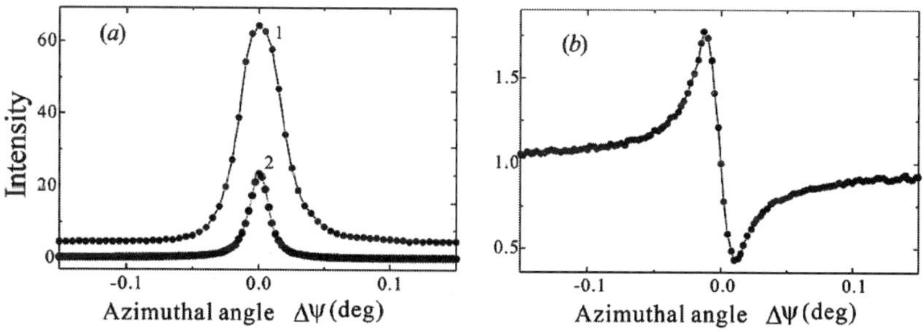

Fig. 9.6. Renninger scans of GaAs three-wave (000)(222)(311) diffraction for 2.4108 Å with (**a**) σ (*curve 1*) and π (*curve 2*) and (**b**) intermediate polarization ($\omega_O = 47°$) elliptical polarized incident wave ($b_e = 0.15$) [322]

The qualitative increase of phase sensitivity of a three-wave diffraction profile can also be achieved for a linearly polarized incident wave by partial polarization suppression of the strong component wave with a proper analyzer for the diffracted wave. Referring to the description of polarization-resolved analysis for multiple-wave diffraction given in Chap. 8 and the iterative Born approximation of Chap. 7 [291,296,322], we can describe the relative intensity I'_G after an analyzer by the same equation as (9.25) with the parameters defined as

$$A_e = \left|\boldsymbol{p}_2^A\right|^2 , \qquad B_e = \boldsymbol{p}_2^A \cdot \boldsymbol{p}_{\mathrm{Um}}^A , \qquad C_e = \left|\boldsymbol{p}_{\mathrm{Um}}^A\right|^2 , \tag{9.27}$$

where $\boldsymbol{p}_2^A$ and $\boldsymbol{p}_{\mathrm{Um}}^A$ have been given in (8.123). Again, for the Umweg type three-wave diffraction, the most favorable condition for achieving high phase sensitivity is $S = F_e C_e/|B_e| \leq 1$. For the Aufhellung type three-wave diffraction, which usually involves three strong reflections, the phase sensitivity is expected to be high when the modulus $|\boldsymbol{p}_2^A|$ is comparable with the $|\boldsymbol{p}_{\mathrm{Um}}^A|$. For example, let's consider two three-wave cases: the '+' case for GaAs $(000)(222)(31\bar{3})$, and the '−' case for GaAs $(000)(\bar{2}\bar{2}\bar{2})(\bar{3}\bar{1}3)$, with $\lambda = 1.5399$ Å. Figure 9.7 shows the measured intensity profiles of these two cases for a π-polarized incident wave and the polarization states specified by $\omega_A = -90°$ and $\omega_A = -25°$ of the analyzing Ge (333) crystal. The phase sensitivity S is 4.4 for $\omega_A = -90°$ and 0.79 for $\omega_A = -25°$. Hence, the two profiles shown in Fig. 9.7(a) exhibit low phase sensitivity, while the profiles in Fig. 9.7(b) display distinct phase sensitive features. The phase δ_{3+} of the upper profile is $-90°$ and that of the lower curve is $90°$, which are in a good correspondence with the characteristic profiles shown in Fig. 9.1 for an ideal situation.

The inverted asymmetry, the so-called anomalous asymmetry, of three-wave intensity profiles for a π-polarized incident wave sometimes occurs when the corresponding polarization factors change their signs [281,305,312,514]. According to the polarization factors defined in (8.114) and (8.115), no anomalous asymmetries occur for the σ-polarized incident wave, because $p_2^\pi(\sigma)$ $(= 1)$ and $p_{\mathrm{Um}}^\pi(\sigma)$ (> 0) have the same sign. For a π-polarized incident wave, the inversion of profile asymmetry takes place when the signs of $p_{\mathrm{Um}}^\pi(\pi)$ and $p_2^\pi(\pi)$ are different. This means an additional 180° phase shift is introduced in the diffraction process. Referring to the last equation of (8.115), there are two situations in which $p_2^\pi(\pi)$ and $p_{\mathrm{Um}}^\pi(\pi)$ may have different signs: (i) when $p_2^\pi(\pi) > 0$, i.e., $\cos\theta_G < 45°$, the wavevector $\boldsymbol{K}_L$ satisfies the condition $(\hat{\boldsymbol{K}}_L \cdot \hat{\boldsymbol{\pi}}_O)(\hat{\boldsymbol{K}}_L \cdot \hat{\boldsymbol{\pi}}_G) > p_2^\pi(\pi)$; (ii) when $p_2^\pi(\pi) < 0$, i.e., $\cos\theta_G > 45°$, the condition $|(\hat{\boldsymbol{K}}_L \cdot \hat{\boldsymbol{\pi}}_O)(\hat{\boldsymbol{K}}_L \cdot \hat{\boldsymbol{\pi}}_G)| > |p_2^\pi(\pi)|$ is fulfilled. In other words, the change of sign takes place when (i) the secondary reflection L with $90° < 2\theta_L < 90° + 2\theta_G$ or $270° < 2\theta_L < 270° + 2\theta_G$ for $\cos\theta_G < 0$, and (ii) the secondary reflection L with $2\theta_G - 90° < 2\theta_L < 90°$ or $90° + 2\theta_G < 2\theta_L < 270°$ for $\cos\theta_G > 0$. Figure 9.8 shows the two shaded regions of the Ewald sphere in which the inversion of profile asymmetry occurs. The secondary reciprocal lattice point L is projected onto the plane of incidence of the primary reflection. The shaded regions are bounded by the branches of hyperbola with the asymptotes coincident with $\hat{\boldsymbol{K}}_O$ and $\hat{\boldsymbol{K}}_G$ according to the last equation of (8.115).

The inversion of profile asymmetry also occurs in the experiments in which the π-polarized incident wave together with an analyzing crystal for the primary reflected wave are used. Based on (8.123) and Fig. 8.48, a polarization-enhanced (case I) or a polarization suppressed (case II) three-wave diffraction situation takes place depending on whether $\boldsymbol{p}_2^A \cdot \boldsymbol{p}_{\mathrm{Um}}^A$ is greater or smaller than

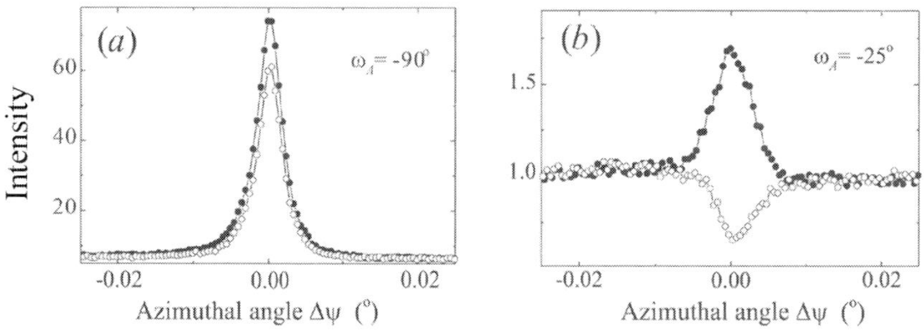

Fig. 9.7. Intensity profiles of GaAs $(000)(222)(31\bar{3})$ (*solid circles*) and GaAs $(000)(\bar{2}\bar{2}\bar{2})(\bar{3}\bar{1}3)$ (*open circles*) three-wave diffraction and 1.5399 Å for a π-polarized incident wave with (**a**) $\omega_A = -90°$ and (**b**) $\omega_A = -25°$ for the Ge (333) analyzer [324]

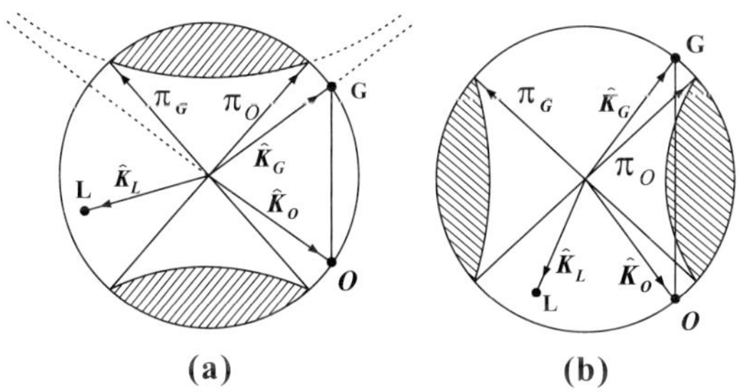

Fig. 9.8. Projection of the Ewald sphere onto the primary two-wave diffraction plane: (**a**) $\theta_G < 45°$, and (**b**) $\theta_G > 45°$ [312]

zero, respectively. Moreover, according to (9.25) and (9.27), the asymmetry of the intensity profile of case I is the same as that obtained for a σ-polarized incident wave for a given positive B_e parameter. However, the profile asymmetry is inverted for case II because of a negative B_e value. This inversion happens every time when the polarization state ω_A of the analyzer crosses the polarization of the total suppression of the primary or the Umweg wave. Figure 9.9 shows the experimental intensity profiles of the GaAs (000)(111) (220) three-wave diffraction for a π-polarized incident wave and various polarization states ω_A of a Ge (333) analyzer at $\lambda = 1.5399$ Å. All the primary (111), the secondary (220), and the coupling $(\bar{1}\bar{1}1)$ reflections are strong reflections. The intensity profiles (not shown here) for $-90° \leq \omega_A < -10°$ and $30° < \omega_A \leq 90°$ are qualitatively the same as that in Fig. 9.9(a). The γ angle defined in Fig. 8.47 is 17.1°. When $0° \leq \omega_A < 17.1°$, an inverted asymmetry (Fig. 9.9(b)) compared with those (Fig. 9.9(a) and (c)) outside this analyzer angular range is observed. This is the situation for case II.

Although the polarization can alter the profile asymmetry, the involved polarization factors of the various situations discussed can be easily calcu-

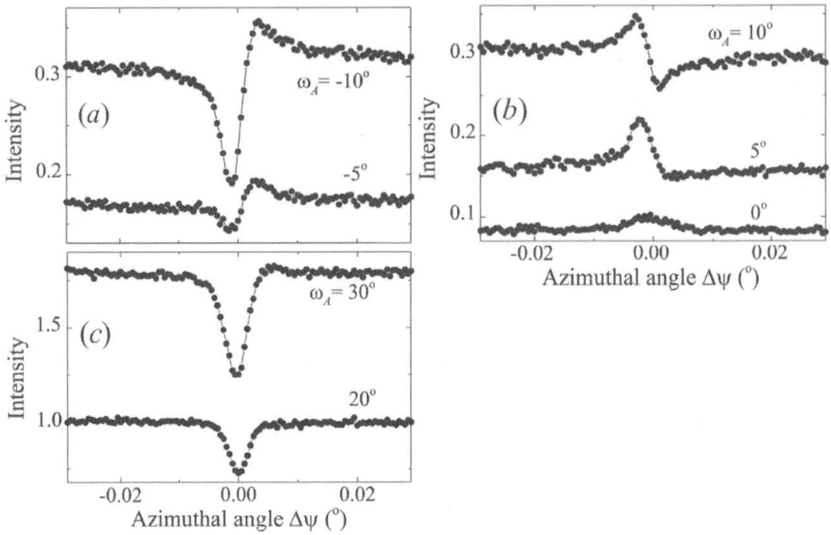

Fig. 9.9. Intensity profiles of GaAs (000)(111)(220) three-wave diffraction at 1.5399 Å for a π-polarized incident wave and various polarization states ω_A of a Ge (333) analyzer. Intensities are normalized with the two-wave intensity of (111) at $\omega_A = 20°$ [324]

lated. Therefore, a suitable three-wave diffraction can be identified before experimental investigation and the ambiguity due to asymmetry inversion can be avoided. If the experiment is for macromolecular crystals, anomalous asymmetry will not occur, because the Bragg angles are relatively small due to large unit cell dimensions.

Diffraction intensity is also a function of crystal thickness for two-wave and multiple-wave diffractions. Consequently, the accuracy of phase determination also depends on the crystal thickness. As described in (7.132) and (7.166) for Bragg type three-wave diffractions, the integrated intensity profile or reflectivity is less affected by the crystal thickness, mainly due to the effect of total reflection. The total reflection range as well as the intensity can be widened and increased for a constructive interference with the effective phase sum $\varDelta + \delta_3$ equal to 0°, while the angular width and intensity decrease for a destructive interference with the phase sum equal to 180°. The overall profile asymmetry therefore remains almost unchanged, irrespective of the crystal thickness.

The situation for a Laue type three-wave diffraction is different from that in a Bragg type diffraction. For the crystal thickness less than the three-wave extinction length, the profile asymmetry is in good correspondence with those shown in Fig. 9.1 for the given phases. However, unambiguous phase determination from the profile asymmetry is not possible when the crystal thickness is larger than the three-wave extinction length. This is because the profile asymmetry changes as the thickness varies [251, 305, 516].

Visibility and Phase Sensitivity of Three-Wave Interaction. The relative intensity variation I'_G given in (9.5) is composed of the phase-dependent I_D and phase-independent I_K. If $I_K \gg I_D$, then the visibility of the profile asymmetry of the two centrosymmetrically related three-wave cases due to I_D and the phase-sensitivity of the three-wave diffractions are very low. Under this circumstance, phase determination from the diffraction profiles is nearly impossible. This implies that the three-wave cases studied are not suited for phase determination. In order to select appropriate three-wave cases for phase determination, it is desired to establish some kind of criteria. An empirical ratio Q has thus been defined as [298]

$$Q = \frac{\left|F'_{L\text{–}O} F'_{G\text{–}L}\right|}{\left|F'_{G\text{–}O}\right|^2} , \tag{9.28}$$

for a given centrosymmetrically related three-wave pair, where F' is the structure factor corrected for the geometrical polarization factors. The appropriate cases for phase determination are those with $2 < Q < 6$. Analytical expressions for visibility and phase-sensitivity may be derived from (9.5). Since the phase-dependent I_D is the key in the phase determination procedure, the phase sensitivity can be considered as the ratio between the phase dependent I_D and the phase-independent I_K. It is also known from (9.5) that the peak intensities and the profile tails are governed by the sine and cosine functions of the triplet phase, respectively. In addition, the major difference in peak intensity between I_D and I_K happens when $\delta_3 = \pm 90°$. Therefore, the cases involving $\delta_3 = \pm 90°$ are the most phase-sensitive situation. Under this consideration, intensity extremes, $I_{\text{max}} = I'_G(\delta_3 = -90°)$ and $I_{\text{min}} = I'_G(\delta_3 = 90°)$ for $\delta_3 = \pm 90°$, usually occur when ψ_u is very close to 0. However, the phase sensitive cases may not have high visibility of profile asymmetry for the pair of diffractions. To estimate both sensitivity and visibility for the cases with $\delta_3 = \pm 90°$, the sensitivity S_n and visibility V_b can be defined as [624]

$$\begin{aligned} S_n &= \left| \frac{I_D\left(\psi_u = 0\right)}{I_K\left(\psi_u = 0\right) + I_D\left(\psi_u = 0\right)} \right|_{\delta_3 = \pm 90°} \cong \left| \frac{B}{B + C} \right| , \\ V_b &= \frac{\left|I_{\text{min}} - I_{\text{max}}\right|}{\left|I_{\text{min}} + I_{\text{max}}\right|} = \left| \frac{2B}{C} \right| , \end{aligned} \tag{9.29}$$

which can be calculated based on the expressions derived from the Bethe or Born approximations. Although (9.5) is valid for $\delta_3 = \pm 90°$, it can also be used as a guide to estimate which multiple diffraction is worth being measured before we carry out the experiment.

The inelastic coherent scattering by phonons, apart from the purely elastic scattering, in a three-wave diffraction may affect considerably the diffraction intensity and the visibility and phase sensitivity of the three-wave interaction. Figure 9.10 shows the experimental rocking curves of the primary

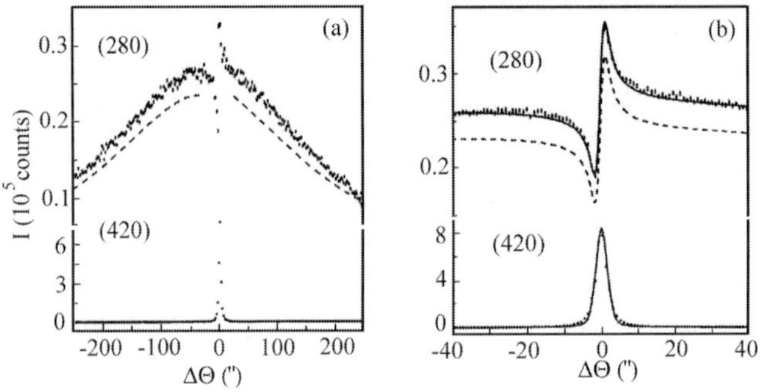

Fig. 9.10. Rocking curves of KDP (420) and (280) reflections in the vicinity of the three-wave (000)(420)(280) diffraction position for $\lambda = 1.789$ Å in (**a**) wide-angle scan and (**b**) detailed scans near the three-wave diffraction positions: experimental (*vertical short line*); calculation with (*solid line*) and without (*dashed line*) inelastic coherent scattering components [625]

(420) and the secondary (280) reflection at the three-wave (000)(420)(280) coplanar diffraction position for a KDP crystal with CoKα_1 radiation ($\lambda = 1.789$ Å) [625]. The calculated profiles for the purely elastic scattering are also shown in the figure by dashed curves. The calculation with the phonon contribution is represented by the solid curve, which matches well with the measured one. Although the profile asymmetry due to the phase effect is clearly revealed, the difference in intensity between the elastic and inelastic scattering contributions, about 12%, is still evident. This indicates that the phonon induced inelastic coherent scattering cannot be totally ignored in a multiple-wave diffraction process. In fact, the interaction between X-ray photons and phonons in a crystal is a well-known phenomenon. Theoretical and experimental studies on this topic are well documented in the literature [31, 48, 53, 83, 626–635].

Phase Determination with Four-Wave and Higher-Order Diffraction. The phases involved in a four-wave diffraction is also related to the structure-factor multiplets given in the relative intensity variation I'_G. For a four-wave (O, G, L, M) diffraction, I'_G can be described under either the Born approximation or the Bethe approximation discussed in Chap. 7 [267, 280, 281, 295, 331, 636–638] as

$$
\begin{aligned}
I'_G &= \frac{I_G^{(4)} - I_G^{(2)}}{I_G^{(2)}} \qquad (9.30)\\
&= -\sum_{Q=L,M} p_3(Q)\left(\frac{F_{G-Q}F_{Q-O}}{F_{G-O}} + \frac{F_{Q-G}F_{O-Q}}{F_{O-G}}\right)\frac{1}{2\xi_Q}\\
&\quad + \sum_{Q=L,M}\sum_{R=L,M} p_4(Q,R)\frac{F_{O-Q}F_{Q-G}F_{G-R}F_{R-O}}{F_{G-O}F_{O-G}}\frac{1}{4\xi_Q\xi_R} + O\left(\xi^{-n}\right),
\end{aligned}
$$

where $I_G(2)$ and $I_G(4)$ are the two-wave and four-wave intensities of the primary reflection G. The function $O(\xi^{-n})$ represents the higher-order terms in $1/\xi$ with $n > 2$. The F's are the structure factors and p_3 and p_4 are the polarization factors of the three-wave and four-wave interactions normalized by the two-wave polarization factor of the G-reflection. ξ_Q is the resonance failure defined by (7.55). In (9.30), there are four triplets associated with the first-order term in ξ^{-1} and three quartets associated with the second-order ξ^{-2}. If the diffraction is a coincidental four-wave case, it is not possible to single out the connection among the multiplets. For a systematic four-wave diffraction, as described in Sect. 8.4.1, the dominant structure-factor multiplets, which are structure invariants, are the four triplets and three quartets shown in Fig. 8.51. Among these, the triplets $F_{L\text{–}O}F_{G\text{–}L}/F_{G\text{–}O}$ and $F_{M\text{–}O}F_{G\text{–}M}/F_{G\text{–}O}$ denoted as A_1 and A_2 in Fig. 8.51 are the dominant triplets if the ψ rotation axis is along $\boldsymbol{OG} = \boldsymbol{g}$. Moreover, the reciprocal lattice vectors of the reflections $L-O$ and $G-L$ are symmetrically related to those of $G-M$ and $M-O$, respectively. If the rotation axis $\boldsymbol{g}$ is a two-fold axis or two-fold screw axis 2_1, then the phase relations of individual reflections are $\delta_{L\text{–}O} = \delta_{G\text{–}M}$ and $\delta_{G-L} = \delta_{M\text{–}O}$. This implies that $\delta_{3a} = \delta_{3b}$, where δ_{3a} and δ_{3b} are the triplet phases of A_1 and A_2, respectively. This triplet-phase equality holds for space groups involving rotational symmetry elements 2, 4, $\bar{4}$, and 6. Under this condition, the dominant triplet phase of $F_{L\text{–}O}F_{G\text{–}L}/F_{G\text{–}O}$ can be determined in the same way as for three-wave diffraction. For example, the four-wave (000)(004)($\bar{1}\bar{1}1$)($\bar{1}\bar{1}3$) diffraction of germanium involves the ψ rotation around [004] to produce the Renninger scan shown in Fig. 4.7(b) at the 'OUT' position. The dominant triplets $F_{\bar{1}\bar{1}1}F_{113}/F_{004}$ and $F_{\bar{1}\bar{1}3}F_{111}/F_{004}$ are nearly equal. The associated triplet phases, which are equal, determined from the Renninger scan are 0° according to (9.8) for the centrosymmetric Ge crystal. Similarly, if a higher-order multiple-wave diffraction involves forbidden or very weak reflections so that the dominant structure-factor multiplets are still triplets, which follow the triplet-phase equality, then this dominant triplet phase can be estimated by the same token [636]. In contrast, the space groups with the symmetry elements $\bar{1}$, m, $\bar{3}$, $\bar{6}$, 3_1, 3_2, and glide planes do not satisfy the triplet phase equality [305,331], and thus the phase determination based on multiple-wave intensities is not possible.

9.1.2 Determination of Absolute Configuration of Crystals

Distinguishing the enantiomorphism, i.e., the handedness, is important for the absolute structure determination of non-centrosymmetric crystals. In an ordinary two-wave X-ray diffraction where Friedel's law holds, it is impossible to differentiate the two enantiomorphic forms, the right- and the left-handedness, related by the center of symmetry for a given non-centrosymmetric crystal. Although the enantiomorphism may be associated with specific structural features of different point groups, such as polar directions, absolute axes for roto-inversion, absolute configuration of chiral and achiral species [639], the absolute structure can be determined by distinguishing the signs of the phases of structure-factors. This is because the phase of a given reflection changes its sign when the structure handedness is inverted from the left to the right, or vice versa.

Anomalous scattering is usually employed to exhibit the intensity difference of appropriate Bijvoet pairs of reflections [640], thus leading to the discrimination of the sign of a phase. An alternative, without invoking anomalous scattering, is to determine the signs of two triplets associated with the two centrosymmetrically related three-wave diffractions, the '+' case for (O, G, L) and the '−' case for $(O, -G, -L)$, where $\delta_{3+} = -\delta_{3-}$. As discussed in the previous sections, the most sensitive situations to the sign of phase are the ones with the triplet phases close to $\pm 90°$ [641]. Once the signs of the triplet phases can be distinguished according to the three-wave intensity profiles shown in Fig. 9.1, the absolute structure can be unambiguously determined. The absolute structure determination by using three-wave diffraction has been reported in the literature by several investigators [295, 297, 304, 305, 642–644].

9.1.3 Phase Determination for Charge-Density-Wave Reflections

Modulated crystal structures are usually encountered in many materials systems where displacive modulation and/or density modulation take place. The displacive modulation is related to perfect crystal structures with a periodic distortion of the atomic positions and the density modulation is related to the modification of the occupation probability of atoms. The diffraction patterns of modulated structures therefore consist of diffraction spots of principal reflections and satellite reflections. The spots of the former are due to the undistorted lattice; those of the latter, distributed around the principal reflection peaks, are due to the periodic distortion. The intensities of the satellite reflections are usually very weak compared with those of the principal reflections. The diffraction satellites can be indexed with rational or irrational numbers with respect to the undistorted lattice, depending on whether the periodic distortion is commensurate or incommensurate with the undistorted lattice.

The reciprocal lattice vector of a given reflection G for an incommensurate modulation structure is characterized by four integer indices (h, k, ℓ, m) as [448, 645]

$$\boldsymbol{g} = h\boldsymbol{a}^* + k\boldsymbol{b}^* + \ell\boldsymbol{c}^* + m\boldsymbol{Q} , \tag{9.31}$$

where the modulated wavevector $\boldsymbol{Q}$ is given by

$$\boldsymbol{Q} = u\boldsymbol{a}^* + v\boldsymbol{b}^* + w\boldsymbol{c}^* . \tag{9.32}$$

At least one of the indices u, v, and w is irrational, depending on the dimensionality of the modulation. For a one-dimensional displacive modulation structure, the position vector $\boldsymbol{r}_j$ of atom j in the unit cell is given by

$$\boldsymbol{r}_j = \boldsymbol{r}_j^0 + \boldsymbol{A}_j \sin\left[2\pi \boldsymbol{Q} \cdot \boldsymbol{r}_j^0 - \phi_j\right] , \tag{9.33}$$

where $\boldsymbol{r}_j^0$ is the equilibrium position of atom j in the undistorted lattice, $\boldsymbol{A}_j$ and ϕ_j are the amplitude and phase of the modulation. By substituting (9.33) into (2.27) and employing the Jacobi expansion

$$\exp(\mathrm{i}z \sin\alpha) = \sum_{m=-\infty}^{+\infty} \exp(-\mathrm{i}m\alpha) J_{-m}(z) , \tag{9.34}$$

the structure factor of the G reflection is expressed as

$$\begin{aligned} F_G = &\sum_{j=1}^{N} f_j(\boldsymbol{g}) \exp\left(2\pi \mathrm{i} \boldsymbol{g} \cdot \boldsymbol{r}_j^0\right) J_{-m}\left(2\pi \boldsymbol{g} \cdot \boldsymbol{A}_j\right) \\ &\times \exp\left(-2\pi \mathrm{i} \boldsymbol{Q} \cdot \boldsymbol{r}_j^0\right) \exp\left(\mathrm{i}m\phi_j\right) , \end{aligned} \tag{9.35}$$

where J_m is the Bessel function of the first kind of order m. According to the properties of the Bessel function, the average intensity of a satellite reflection decreases rapidly with increasing m.

The inhomogeneous distribution of the charge or the spin density due to electron-phonon or electron-electron interactions in solids can also produce modulations accompanying with a phase transition at a transition temperature, thus introducing additional phase factors to the electron or spin densities [646–653]. Considering a system involving the charge-density-wave (CDW) instability at low temperatures, we can write the electronic density in the ground state as

$$\rho = \rho_0 \left[1 + \sum_j P_j \cos\left(2\pi \boldsymbol{Q}_j \cdot \boldsymbol{r}_j + \phi_j\right)\right] , \tag{9.36}$$

where the ρ_0 is the undistorted electron density, P_j the distorted amplitude, $\boldsymbol{Q}$'s the wavevectors of the CDW, and ϕ_j the phases. For a one-dimensional CDW, $2\pi Q = 2k_F$, where $\boldsymbol{k}_F$ is the Fermi wavevector. The structure factor given in (9.35) is equally valid for the CDW modulation. The indices of the CDW satellite reflections may involve fractional numbers for incommensurate modulation, while the indices of the principal reflection are integers, which are referred to the undistorted lattice.

The triplet phases of a three-wave diffraction involving fractional reflections can be determined in the same way as that involving all integer reflections for crystals without any modulation. Figure 9.11 shows the Renninger scan of 2H-$NbSe_2$ (1/3 0 9) at $\lambda = 1.0$ Å and $T = 16$ K. 2H-$NbSe_2$ is a quasi-two-dimensional CDW material, having a hexagonal layer structure with space group $P6_3/mmc$ and lattice parameters a = b = 3.45 Å and c =

12.54 Å. It undergoes a phase transition to a CDW state at $T_{CDW} = 32.5\,\mathrm{K}$. Upon cooling it becomes a superconductor at $T_C = 7.2\,\mathrm{K}$, while the CDW remains incommensurate down to 5 K [656].

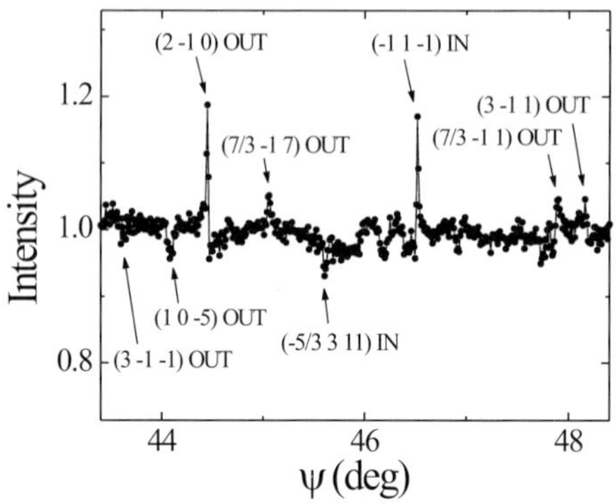

Fig. 9.11. Multiple diffraction pattern of a CDW satellite (1/3 0 9) reflection at T = 16 K. Intensity is normalized with the two-wave intensity of (1/3 0 9) [654,655]

A large number of multiple diffractions are possible to be generated, but only those with their 'strong' integer reflections being either secondary or coupling reflections have appreciable intensities. This situation is similar to that for macromolecular crystals. Figure 9.11 is the multiple diffraction pattern obtained by the ψ rotation around the reciprocal lattice vector of the primary reflection $G = (1/3\ 0\ 9)$ at 0.005° per step. The ψ angle was measured counterclockwise from the [010] direction. Only the relatively intense peaks and dips are indexed with the Miller indices of the secondary L reflection. The 'IN' and 'OUT' positions are also indicated. The rather busy background is due to the presence of many weak fractional reflections. Obviously, the peak-type (Umweganregung) profiles $(2\bar{1}0)/(\bar{5}/3\ 1\ 9)$ and $(\bar{1}1\bar{1})/(\bar{4}/3\ \bar{1}\ 10)$ as well as the dip-type (Aufhellung) profile $(\bar{5}/3\ 3\ 11)/(2\bar{3}\bar{2})$ have distinct asymmetry, where the three-wave cases are denoted by $(L)/(G-L)$. These asymmetric profiles are readily seen to show the phase effect on the multiply diffracted intensities .

The triplet phases δ_3 were analyzed based on the measured profiles for a centrosymmetric crystal, because the host structure of 2H-$NbSe_2$ at 16 K retains the same symmetry as that at 297 K [656] and the formation of static CDWs does not affect the symmetry. Since the phases δ_3's are either 0° or 180°, only the sign of $\cos\delta_3$ matters. According to (9.8), the triplet phases can be determined from the measured asymmetry of the profiles shown in Fig. 9.11 and the 'IN' and 'OUT' positions on the ψ-scan diffraction diagrams as $\delta_3 = -\delta(1/3\ 0\ 9) + \delta(2\bar{1}0) + \delta(\bar{5}/3\ 1\ 9) = 0°$, $\delta_3 = -\delta(1/3\ 0\ 9) + \delta(\bar{5}/3\ 3\ 11) + \delta(2\bar{3}\bar{2}) = 180°$, and $\delta_3 = -\delta(1/3\ 0\ 9) + \delta(\bar{1}1\bar{1}) + \delta(\bar{4}/3\ \bar{1}\ 10) = 0°$. Direct dynamical calculations give the best-fit diffraction profiles to the

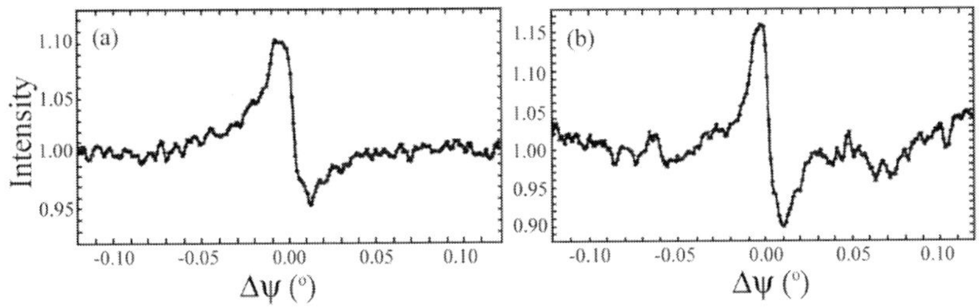

Fig. 9.12. Three-wave diffraction profiles of an icosahedral $Al_{70}Pd_{20}Mn_{10}$ in the OUT situation with $\lambda = 0.7401$ Å and a π-polarized incident wave for (**a**) $(000000)(\bar{1}31\bar{3}\bar{3}1)(\bar{1}\bar{3}\bar{1}11\bar{1})$ and (**b**)$(000000)(1\bar{3}\bar{1}33\bar{1})(131\bar{1}\bar{1}1)$ [305]

measured ones with the triplet phases as the tuning parameter. The most appropriate phase values found are in agreement with the above qualitative phase estimation.

Other multiple-wave interference experiments for phase determination have also been carried out for incommensurately modulated calaverite ($AuTe_2$) [305].

9.1.4 Phase Determination for Diffraction in Quasi-Crystals

Quasi-crystals, first discovered in rapidly quenched Al_6Mn alloy by Shechtman, Gratis, and Cahn in 1984 [657], are materials with rotational symmetry but without three-dimensional long-range translational ordering. The crystal lattice can be considered as the projection of a higher-dimensional periodic arrangement onto a three-dimensional physical space, which is aperiodic. Some two-dimensional quasi-crystals, such as octagonal, decagonal, and dodecagonal lattices, can be represented in a periodic five-dimensional space. The three-dimensional icosahedral lattice is periodic in a six-dimensional space. For the former, two of the five indices in terms of the golden mean $\tau = 2\cos 36° = (\sqrt{5}+1)/2$ are used to label diffraction spots, while for the latter, three out of the total six indices are related to the golden mean.

Three-wave diffraction, say, (O, G, L), takes place in quasi-crystals when the reciprocal lattice vectors $\boldsymbol{g}$, $\boldsymbol{\ell}$, and $\boldsymbol{g}-\boldsymbol{\ell}$ form a triangle in both three-dimensional and higher-dimensional space. In other words, two of the three vectors are not collinear in a three-dimensional space. Investigation on multiple-wave interference in quasi-crystals has been reported in the literature for an icosahedral $Al_{63.7}Cu_{23.6}Fe_{12.7}$ [658], a decagonal $Al_{70}Ni_{15}Co_{15}$ and an icosahedral $Al_{70}Pd_{20}Mn_{10}$ [305]. Figure 9.12 shows the Renninger scans of $Al_{70}Pd_{20}Mn_{10}$ with $\lambda = 0.7401$ Å and a π-polarized incident wave for the two centrosymmetrically related three-wave cases $(000000)(\bar{1}31\bar{3}\bar{3}1)(\bar{1}\bar{3}\bar{1}11\bar{1})$ and $(000000)(1\bar{3}\bar{1}33\bar{1})(131\bar{1}\bar{1}1)$. The estimated triplet phases are 0°, indicating that the quasi-crystal is centrosymmetric.

9.1.5 Determination of Phase Shift due to Resonance

Additional phase shift is introduced to X-ray scattered or diffracted waves when the X-ray photon energies are very close to the absorption edges of the constituent atoms in the crystal. Under this resonance condition, the effects of anomalous dispersion are greatly enhanced in the scattering/diffraction. In turn, this additional phase shift, the so-called resonance phase, may reflect the electronic characteristics of the excited atoms. From the diffraction point of view, the resultant phase involved in the diffraction process is then the sum of this resonance phase Δ_G and the crystallographic phase $\delta_{G\text{–}O}$ of a given reflection G, where $\Delta_G = \delta_{G\text{–}O} + \delta_{O\text{–}G}$ as defined in (7.403). Referring to the resonance perturbation Bethe approximation (RPB) for three-wave diffraction described in Chap. 7, we can see that the relative intensity variation is related to the resultant phase [323]. Hence, the information about the resonance triplet phase $\Delta = \delta_{3+} + \delta_{3-}$ may be extracted from the intensity measurements of three-wave diffractions at resonance.

Attempts to determine resonance phases and to understand the interaction of an X-ray wavefield with atoms in electronic transitions were made experimentally and numerically by using the X-ray photon energies about 500–800 eV above or below the K-absorption edges of Ga and As for a GaAs (111) crystal [278, 298]. Recent experiments of using photon energies very close, about 20 eV, to the Ga and As K-edges have demonstrated the determination of resonance phases from three-wave diffraction intensity distributions [323]. In addition, the measurement of the phases of the forbidden (600) and (222) reflections of germanium [339] by employing a three-wave resonance diffraction has also been pursued. For the (600) and (222) measurements, the hybridization of p and d atomic states in Ge has been considered theoretically the main factor resulting in an additional phase shift at the resonance of Ge atoms [659]. In the following, how to determine the resonance phase from a three-wave diffraction using the RPB approximation is demonstrated for the study involving GaAs [323].

Figures 9.13 and 9.14 display the measured three-wave profiles of GaAs $(000)(222)(33\bar{1})$ (the '+' case) and $(000)(\bar{2}\bar{2}\bar{2})(\bar{3}\bar{3}1)$ (the '−' case) at the GaK (10.369 keV) and AsK (11.869 keV) edges with a σ-polarized incident wave. The energy resolution provided by a double-crystal monochromator is about 0.2 eV.

The analysis of the intensity profiles based on $I_G^{(3)}(\Delta\psi_G)$ in (7.401) derived from the resonance Bethe (RPB) approximation is complicated. However, if we consider the ratio R of $I_G^{(3)}(\Delta\psi_G)$ between the '+' and '−' cases, then a simple relation between the intensity ratio and phases can be obtained as

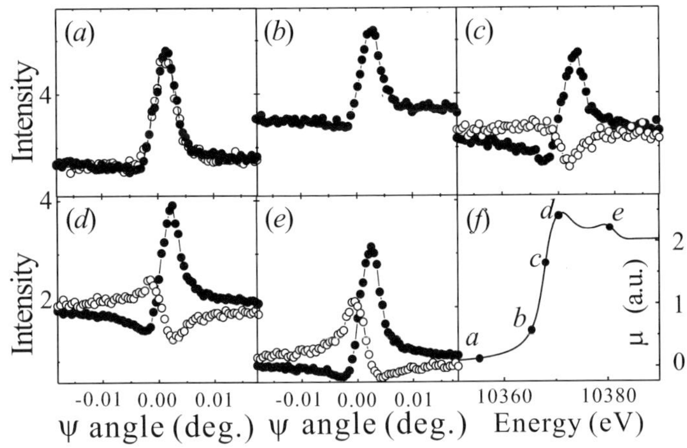

Fig. 9.13. Three-wave diffraction profiles of GaAs (000)(222)(33$\bar{1}$) (the '+' case) and (000)($\bar{2}\bar{2}\bar{2}$)($\bar{3}\bar{3}$1) (the '−' case) at the GaK (10.369 keV) edge. Profiles (**a**)–(**c**) are obtained for energies indicated in the absorption spectrum (**f**). Intensities are normalized by the two-wave intensity of (**e**). (*solid circle*): '+' case; (*open circle*): '−' case [323]

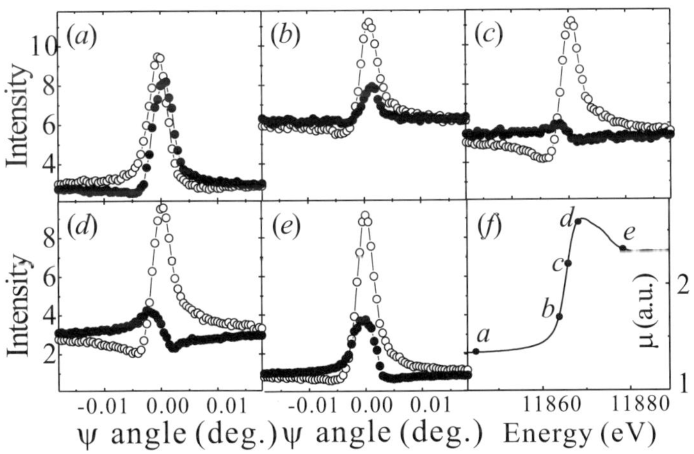

Fig. 9.14. Three-wave diffraction profiles of GaAs (000)(222)(33$\bar{1}$) (the '+' case) and (000)($\bar{2}\bar{2}\bar{2}$)($\bar{3}\bar{3}$1) (the '−' case) at the AsK (11.869 keV) edge. Profiles (**a**)–(**e**) are obtained for energies indicated in the absorption spectrum (**f**). Intensities are normalized by the two-wave intensity of (**e**). (*solid circle*): '+' case; (*open circle*): '−' case [323]

$$R(\Delta\psi_G) = \frac{I_G^{(3)}(\Delta\psi_G)_+}{I_G^{(3)}(\Delta\psi_G)_-} \cong \left|\frac{\chi_{G\text{–}O}}{\chi_{O\text{–}G}}\right|^2 \left\{1 + \frac{2(\Delta\psi_G)\,d_c - 2\eta_L d_s + d_d}{(\Delta\psi_G - \psi_-)^2 + q_-^2}\right\}, \quad (9.37)$$

where

$$
\begin{aligned}
\eta_L &= -\chi''_O \left(\frac{1+\frac{\gamma_L}{\gamma_G}}{2} \right) + \sqrt{p^2_{GL} \left|\chi_{L\text{–}O}\right| \left|\chi_{O\text{–}L}\right| \frac{\gamma_L}{\gamma_G}} \, , \\
d_c &= p_s \left(f_+ \cos\delta_{3+} - f_- \cos\delta_{3-} \right) \, , \\
d_s &= p_s \left(f_+ \sin\delta_{3+} - f_- \sin\delta_{3-} \right) \, , \\
d_d &= p_d \left(f^2_+ - f^2_- \right) \, , \\
\psi_- &= -p_s f_- \cos\delta_{3-} \, , \\
q^2_- &= \left(\eta_L - p_s f_- \cos\delta_{3-} \right)^2 + f^2_- \left(p_d - p^2_s \right) \, , \\
p_s &= \frac{\boldsymbol{p}_{G(2)} \cdot \boldsymbol{p}_{G(d)}}{p^2_{G(2)}} \, , \\
p_d &= \frac{p^2_{G(d)}}{p^2_{G(2)}} \, , \\
f_+ &= \frac{\left|\chi_{G\text{–}L}\right| \left|\chi_{L\text{–}O}\right|}{\left|\chi_{G\text{–}O}\right|} \, , \\
f_- &= \frac{\left|\chi_{L\text{–}G}\right| \left|\chi_{O\text{–}L}\right|}{\left|\chi_{O\text{–}G}\right|} \, .
\end{aligned}
\tag{9.38}
$$

p^2_{GL} is the polarization factor defined in (7.381). For simplicity, the dependence of z_L on $\Delta\theta_G$ given in (7.389) has been employed in deriving (9.37).

Figure 9.15 shows the calculated semi-integral profiles $I^{(3)}_G(\Delta\psi_G)$ based on the resonance Bethe (RPB) approximation in comparison with the exact dynamical calculations for the '+' and '−' cases. The anomalous dispersion corrections of the atomic scattering factors, shown in Figs. 9.15(c) and (f), are used for these calculations. The agreement of profiles from the resonance Bethe approximation and from the exact dynamical calculation is seen. The calculated approximate ratios $R(\Delta\psi_G)$ are also shown to reveal the different diffraction behaviors in the vicinity of the GaK and AsK edges. As can be seen from the figure, the asymmetry of the ratio remains the same when the photon energy crosses the GaK edge and is inverted when the photon energy crosses the AsK edge. This difference is, in principle, closely related to the triplet phases δ_{3+} and δ_{3-} and the resonance phases Δ's. In order to clarify the behavior of diffraction profiles, Figs. 9.16(a) and (b) show the distributions of phases δ_{3+}, δ_{3-}, $\Delta/2$, and $\Delta_d/2$ as functions of photon energy calculated with the anomalous dispersion corrections, where $\Delta_d = \delta_{3+}$-δ_{3-}. In region I, $E < E_{K(Ga)}$, Δ and d_c are close to zero, where d_c is roughly proportional to $\sin(\Delta/2)\sin(\Delta_d/2)$. Therefore, the ratio $R(\Delta\psi_G)$ is practically a symmetrical function close to a Lorentzian, and the two centrosymmetrically related three-wave profiles have the same asymmetry. This feature reveals the characteristics of the profiles far from absorption edges as usual. When the photon energy approaches the GaK edge, the ratio becomes slightly asymmetrical (see Fig. 9.15(a)), and the asymmetry increases dramatically in region II, $E_{\mathrm{GaK}} < E < E_{\mathrm{AsK}}$, with increasing Δ (see Fig. 9.15(b)). Near the AsK edge, when the energy changes from region II to region III ($E_{\mathrm{AsK}} < E$), the ratio

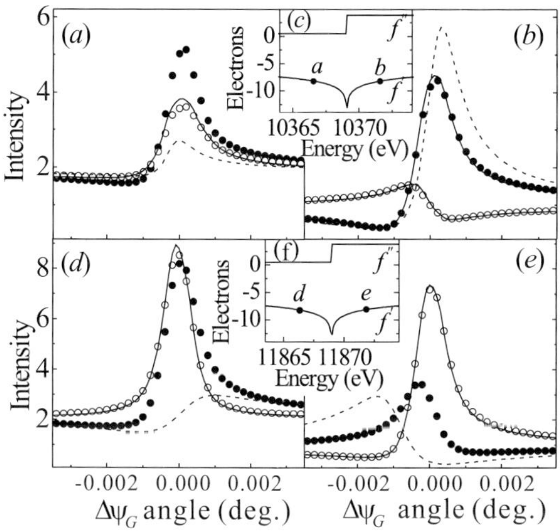

Fig. 9.15. Calculated three-wave diffraction profiles of the '+' and '−' cases near the GaK edge (**a**)–(**b**) and AsK edge (**d**)–(**e**) obtained by the RPB approach (*solid lines*) and by the direct dynamical calculation (*solid circles*: '+' case; *open circles*: '−' case). Photon energies for (**a**), (**b**), (**d**), and (**e**) are indicated in (**c**) and (**f**) of the dispersion corrections f' and f'' of the corresponding K edges. The dashed curves are the ratios $R(\Delta\psi_G)$ defined in the text. Intensities of (**a**)–(**b**) and (**d**)–(**e**) are normalized by the two-wave intensity of (**b**) and (**e**), respectively [323]

asymmetry is inverted (see Figs. 9.15(d) and (e)). This inversion of asymmetry is accompanied with the change of the sign of Δ shown in Fig. 9.16(c). This change of ratio asymmetry is independent of the change of asymmetry of the individual diffraction profiles. The inversion of the ratio asymmetry takes place when the phases δ_{3+} or δ_{3-} cross over the values around $\pm 90°$. Moreover, for a zinc blend structure like GaAs, the resonance phase shift of the primary (222) reflection is mainly due to the change of the imaginary parts f'' of the dispersion correction on the atomic scattering factors of Ga and As, while the phase shifts of the strong secondary and coupling reflections, like $(33\bar{1})$ and $(\bar{1}\bar{1}3)$, are dominated by the change in the real parts f'.

The spectral distribution of phases δ_{3+}, δ_{3-}, $\Delta/2$, and $\Delta_d/2$, determined by the curve fitting to the pair of three-wave diffraction profiles by using the RPB calculation with various values of the dispersion corrections f' and f'', are plotted in Figs. 9.16(b) and (d). The accuracy in the determined phases is better than 2°. The phases can also be determined by the fitting to the asymmetric Lorentzian functions, described in (9.17) and (9.18), with a phase accuracy of about 20°.

This phase determination procedure is most suited for the cases having large phase shift Δ due to resonance. This favorable case usually occurs when the direction of changing Δ_G is opposite to that of changing Δ_L and Δ_{G-L}.

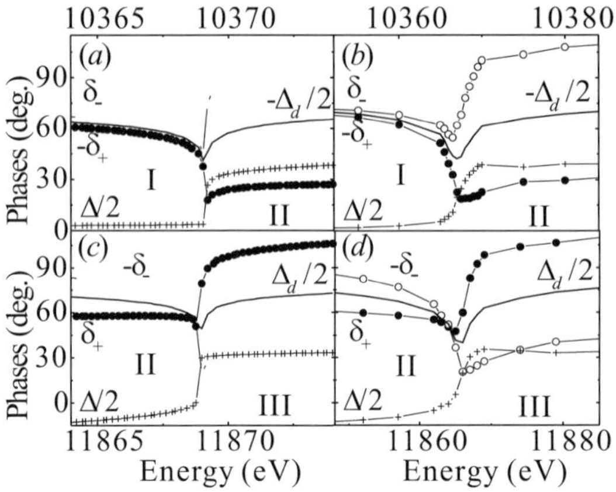

Fig. 9.16. The spectral distribution of phases δ_{3+}, δ_{3-}, $\Delta/2$, and $\Delta_d/2$: the calculated ones for (**a**) GaK edge and (**c**) AsK edge; the measured ones for (**b**) GaK edge and (**d**) AsK edge [323]

9.1.6 Phase Determination of Surface In-Plane Reflection

The phase determination methods discussed in the previous sections are used to deal with three- or higher-dimensional crystal/quasi-crystal lattices. However, for a lower-dimensional structure, such as crystal surfaces, epitaxial interfaces, Langmuir-Blodgett films, two-dimensional biological structures, and various self-assembled films [660–662], the phase determination is equally important in structure analysis. Like for a three-dimensional bulk crystal, the phases of reflections from surfaces and interfaces have been determined mainly by structural modelling via intensity matching by using the Patterson methods [612,663] or the direct methods [664–668] on a trial and error basis. Direct physical measurements of phases from surface related reflections have been reported only recently [276]. The idea is to apply the three-wave diffraction technique to surface in-plane reflections. Owing to the two-dimensional nature of surfaces and interfaces, it is not possible to adopt the Renninger scan with an wide-angle incidence for surface in-plane reflections. In order to bring simultaneously the reciprocal lattice points of surface in-plane reflections onto the surface of the Ewald sphere, only the equatorial circle of the Ewald sphere, which is parallel to the crystal surface, needs to be considered for a surface in-plane diffraction. To fulfill this requirement, the grazing incidence geometry shown in Fig. 3.3(c) has to be adopted. As indicated in Fig. 3.3(d) for a three-wave (O, G, L) grazing incidence diffraction (GIXD), the three reciprocal lattice points O, G, and L lie in an equatorial circle, whose radius has a fixed value due to the geometrical relation among the involved reciprocal lattice points. In other words, a three-wave grazing incidence diffraction occurs only for a specific wavelength and photon energy, as

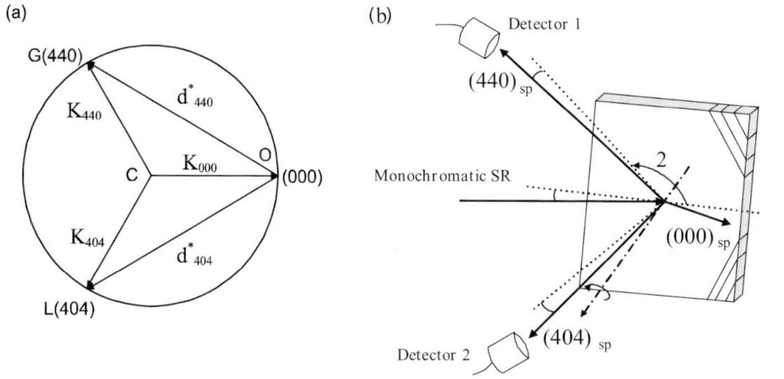

Fig. 9.17. Three-wave (000)(440)(404) GIXD of silicon at the photon energy E_M = 7.45613 keV: (**a**) geometry in the reciprocal space and (**b**) schematic of the experimental setup [276]

in a coincidental three-wave case. In contrast to the Renninger scan, where multiple-wave diffraction profiles are obtained with the ψ rotation around the reciprocal lattice vector of a given primary reflection, the energy scans are needed to obtain the diffraction profiles for the multiple-wave GIXD so as to retain the grazing incidence geometry.

The three-wave (000)(440)(404) GIXD of a silicon $(\bar{1}11)$ cut crystal is described below as an example [457]. Reflections (440) and (404) are surface in-plane reflections. Figure 9.17(a) shows the geometry in the reciprocal space, where the wavelength $\lambda_M = (2\cos 30°)/d^*_{440}$, d^*_{440} being the modulus of the reciprocal lattice vector of the (440). In this particular case, $\lambda_M = 1.66285$ Å, namely, the photon energy E_M = 7.46613 keV. The wavevectors of the involved reflections are also indicated by $\boldsymbol{K}$'s. The experimental setup is schematically shown in Fig. 9.17(b). The σ-polarized incident wave from a 1.86 tesla 25-pole wiggler with the magnet gap 22 mm was focused and monochromatized by a saggittal double-crystal monochromator (SDCM) of Si (111) when the storage ring at the National Synchrotron Radiation Research Center (NSRRC) was operating at 1.5 GeV and 200 mA. The σ-polarization vector was perpendicular to the crystal surface along the $[\bar{1}11]$ direction. An ion chamber acting as the incident wave monitor I_0 was placed in the middle of a beam path. A 1 mm thick $(\bar{1}11)$ cut Si crystal was mounted on a Huber 8-circle diffractometer. The angular divergences of the incident wave were 0.02° vertical and 0.03° horizontal. The nominal energy resolution of the SDCM was 2 eV. However, 0.2 eV per step could be achieved for one-way rotation in θ. The incident wave at the angle of incidence $\chi = 0.18°$, close to the angle of total reflection, was diffracted towards the interior of the crystal along $\boldsymbol{K}_{440}$ and $\boldsymbol{K}_{404}$, which were specularly reflected by the crystal surface at the scattering angles $\alpha = \beta = 0.4°$. Two scintillation counters were placed at $2\theta = 120°$ and $-120°$ to monitor the (440) and (404) specularly reflected waves, respectively (see Fig. 9.17(b)).

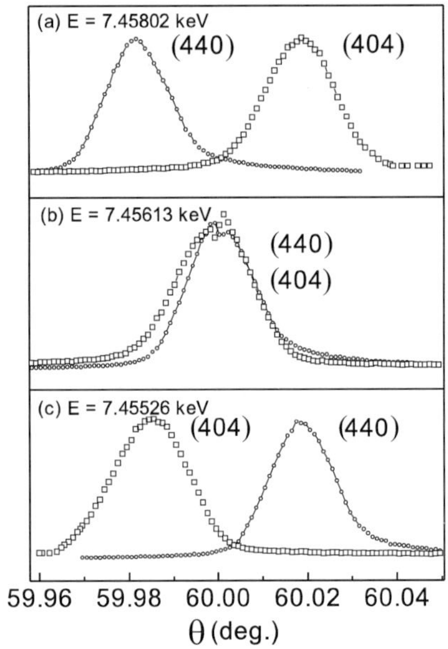

Fig. 9.18. Rocking curves of the specularly reflected (440) and (404) during photon energy scans at the nominal energies (**a**) 7.45802 keV, (**b**) 7.45613 keV, where the three-wave (000)(440)(404) GIXD occurs, and (**c**) 7.45526 keV. [276]

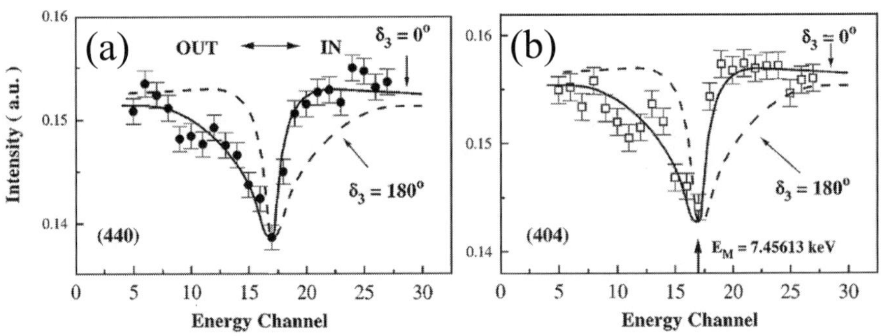

Fig. 9.19. Integrated intensities of the surface specularly reflected (**a**) (440) and (**b**) (404) versus energy channel near the three-wave (000)(440)(404) GIXD position (1 channel=0.2 eV). (*solid curve*): the calculation for $\delta_3 = 0°$; (*dashed curve*): the calculations for $\delta_3 = 180°$ [276]

The ω scan for the θ angle, the rotation of the crystal around the surface normal $[\bar{1}11]$, at each photon energy near E_M was employed and the corresponding rocking curves of the (440) and (404) were recorded (Fig. 9.18). The angular separation of the peaks between the two rocking curves gives an additional calibration of the incident energy. When the two rocking curves overlapped with each other, the three-wave GIXD was ensured to occur at E_M (Fig. 9.18(b)). The intensity modification on both rocking curves near $\theta = 60°$, which is mainly due to the three-wave interaction, is readily seen. The integrated intensities of the surface specularly reflected (440) and (404) versus the photon energies are plotted in Figs. 9.19(a) and (b). The asymmetry of the intensity distributions is clearly visible. The IN and OUT situations of the rotation of the (440) and (404) with respect to the Ewald sphere are also indicated in the same figure. Since silicon is centrosymmetric, the involved triplets in this three-wave case determined by following the sign equation (9.8) are $\delta_{3a} = \delta(\bar{4}\bar{4}0) + \delta(404) + \delta(04\bar{4}) = 0°$ and $\delta_{3b} = \delta(\bar{4}0\bar{4}) + \delta(440) + \delta(0\bar{4}4) = 0°$. The exact dynamical calculation of the intensity profiles, also shown in Fig. 9.19, for δ_{3a} and δ_{3b} equal to 0° and 180° reconfirmed the determined phases.

The X-ray penetration depth of a three-wave GIXD, similar to that of a two-wave GIXD, ranges from zero to a few hundreds of Angstrom depending on the incident angle. This characteristic makes the three-wave GIXD phasing technique applicable to reconstructed surface [337] and thin films [338, 669].

9.1.7 Phase Determination for Diffraction in Thin Films and Substrates and Reconstructed Surfaces

The possibility of phase determination for X-ray diffraction from an epitaxial Au/GaAs/AlAs heterostructure has been demonstrated by using the interference between the diffracted wave from a two-dimensional crystal and the backward reflected wave from a gold mirror layer grown on the backside of the GaAs/AlAs crystal [670, 671]. However, the applicability of that technique to common thin-film and overlayer systems is limited by the required special sample treatment.

Multiple-wave GIXD, as described in the previous section, may provide a clue to solving the phase problem for thin-film/substrate systems, provided that the interference of the diffracted waves between the thin layer and the substrate crystal can be detected. Along this direction, a substrate-mediated multiple-wave GIXD was developed for thin-film [338, 669] and surface systems [337]. In this approach, the strong in-plane reflections of the substrate act as an intermediary to enhance the three-wave interaction in the thin films/reconstructed surface layers and thus lead to direct phase determination of X-ray reflections for the thin films/reconstructed surfaces. This diffraction technique is illustrated below for thin films and reconstructed surfaces:

A 90 Å thick $Ge_{0.9}Si_{0.1}$ thin film grown on the $(\bar{1}11)$ surface of a 1 mm thick Ge substrate by organometallic chemical vapor deposition (MOCVD) is subject to the three-wave (000)(440)(404) GIXD experiment near the energy $E_M = 7.1571$ keV. The mosaic spread of the thin film is about 0.066°, much larger than that of the substrate. The geometry of the three-wave GIXD in the reciprocal space is similar to that in Fig. 9.19, except that the reciprocal lattice points of both Ge and GeSi are present. The size of the reciprocal lattice point of GeSi is larger than that of Ge, because the mosaic spread of GeSi is larger that that of Ge.

The experimental setup is the same as that shown in Fig. 9.17(b), except that the incident angle is set at 0.3°, the critical angle for the total external reflection of Ge. Since the lattice constants of Ge and GeSi are 5.6578 Å and

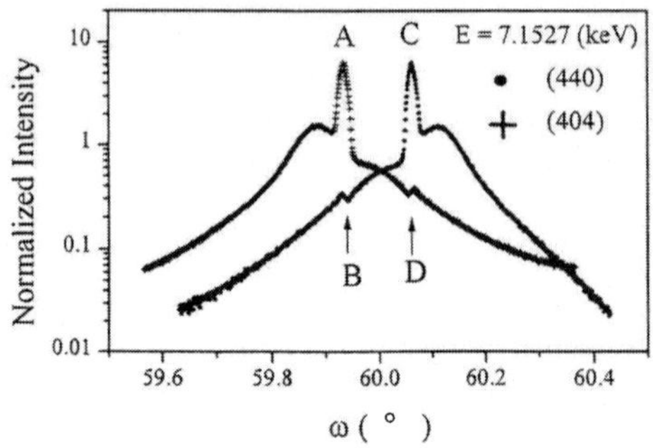

Fig. 9.20. The ω scans of the surface specularly reflected (440) and (404) at $E = 7.1527\,\text{keV}$ [338]

5.6541 Å, respectively, the difference $\Delta\theta$ in Bragg angle for (440) is 0.065° for the energy $E = 7.1527\,\text{keV}$. The ω scans of the specularly reflected (440) and (404) at this energy are shown in Fig. 9.20. The broad profiles are the diffraction intensity distributions of (440) and (404) reflections from the GeSi thin film, while the sharp peaks A and C on top of the broad profiles are the corresponding (404) and (440) reflections of the Ge substrate, respectively. The small kinks B and D are the intensity variations caused by the multiple-wave interactions of the GeSi (440) with the Ge (404) and of the GeSi (404) with the Ge (440), respectively. The positions of the peaks and kinks change as the photon energy varies: the peak position shifts towards higher ω angles for lower energies, while the kink position shifts towards the opposite direction. According to the ω rotation of the crystal, the reciprocal lattice point of (440) moves towards the equatorial circle of the Ewald sphere, which corresponds to the incoming (IN) situation. The reversal is true for the (404) reflection. The multiple-wave diffractions associated with the peaks A and C and the kinks B and D are the combined four-wave cases (A) (000),(404)F,(404)S,(440)F, (B) (000),(440)F,(404)S,(404)F, (C) (000),(440)F,(440)S,(404)F, and (D) (000),(404)F,(440)S,(440)F, where the labels S and F denote the substrate and the film, respectively.

The detailed profiles of the kink B and the peak C, after subtracting the intensity distribution of the thin film (440) from the measured raw data, are shown in Figs. 9.21(a) and (b), where the 'IN' directions of the crystal rotations are indicated. The dashed lines represent the two-wave intensities of the (440)F reflection at the corresponding ω angles. The asymmetry of the intensity profiles is clearly seen, indicating that coherent interaction takes place for the peaks and kinks mentioned. Moreover, from $\theta - 2\theta$ scans the particle-size broadening of GeSi is about 0.080° and the crystallite size is about 450 Å.

The reason why the coherent interaction retains in the diffraction process involving two different crystals in thin-film/substrate systems is due to the

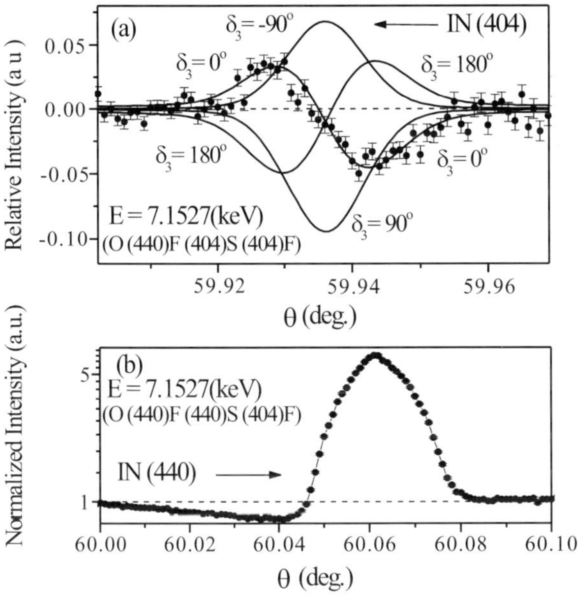

Fig. 9.21. Intensity profiles (ω scans) for (**a**) case B and (**b**) case C at E = 7.1527 keV. Calculated profiles with different δ_3 are indicated (*solid curve*) in (**a**) [338]

following facts: First, the two diffracted waves of (404)F and (404)S have the possibility of interfering with each other, because they can propagate along the same direction as long as the broad profile overlaps the sharp peak. This implies that when the separation $\Delta\theta$ between the substrate peak and the film profile is smaller than the mosaic spread $\Delta\omega$ of the thin film, i.e., $\Delta\theta \leq \Delta\omega$, the phase matching of the two (404) waves propagating in the same direction at the interface is met, that ensures the interference to occur. Second, since (404)S is much stronger than (404)F, the coupling between the two can enhance the three-wave interaction in the thin film. For example, the (404)S enhances the kink B via the interaction (000),(404)F,(440)F and the (440)S increases the peak C through the interaction (000),(440)F,(404)F in the GeSi thin film. Hence the combined four-wave case becomes actually a three-wave diffraction for the thin film. Again, GeSi and Ge are centrosymmetric. Based on (9.8), the asymmetries of the profiles shown in Figs. 9.21(a) and (b) lead to the phase determination of $\delta_{3B} = \delta(\bar{4}\bar{4}0F) + \delta(404F) + \delta(04\bar{4}F) = 0°$ and $\delta_{3C} = \delta(\bar{4}0\bar{4}F) + \delta(440F) + \delta(0\bar{4}4F) = 0°$, respectively. Similarly, the triplet phases of cases A and D can also be determined as $\delta_{3A} = \delta(\bar{4}0\bar{4}F) + \delta(440F) + \delta(0\bar{4}4F) = 0°$ and $\delta_{3D} = \delta(\bar{4}\bar{4}0F) + \delta(404F) + \delta(04\bar{4}F) = 0°$.

The required condition for three-wave GIXDs is relaxed for a single-crystal with surface reconstruction. As described in Chap. 8, the reciprocal lattice of a two-dimensional periodic structure is composed of periodically arranged

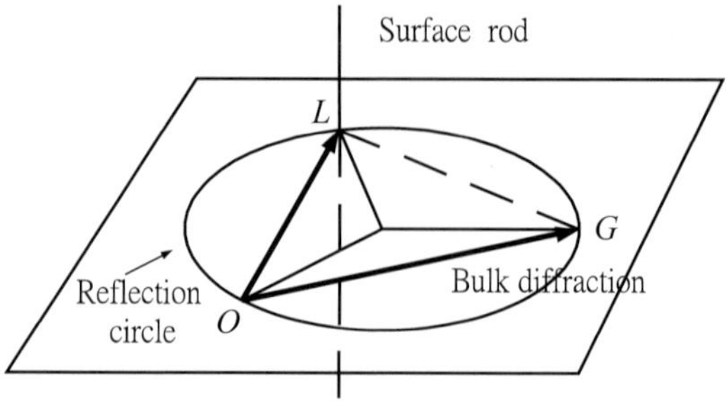

Fig. 9.22. Three-wave GIXD involving a bulk and surface reflections [337]

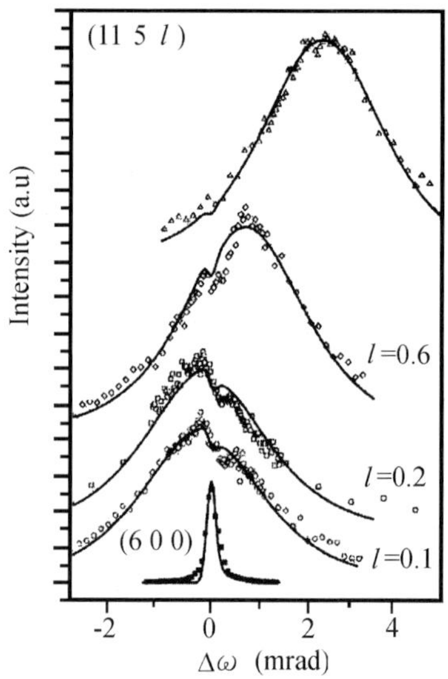

Fig. 9.23. ω scans of (600) and (11 5 ℓ) in the vicinity of the three-wave (000)(600)(11 5 ℓ) diffraction at $\lambda = 1.353$ Å for various ℓ [337]

reciprocal rods. With the aid of the reciprocal rods, a three-wave (O, G, L) diffraction can be easily found by having an in-plane reflection G of the crystal and a surface reflection L lying in a reciprocal rod of the reconstructed surface. In other words, the Ewald sphere passing through the reciprocal lattice points O and G cuts the reciprocal rod at the point L (see Fig. 9.22) [337].

Phase determination by using a three-wave GIXD is illustrated for the sample system, a Ge (113)-(3×1) reconstructed surface [673]. Figure 9.23 shows the intensity distributions of ω scans around the (600) surface reflection, equivalent to the $(2\bar{2}0)$ bulk reflection, in the vicinity of the three-wave (000)(600)(11 5 ℓ) diffraction at $\lambda = 1.353$ Å. The relations between the reciprocal basis vectors of the surface, denoted by the subscript S, and that of a cubic bulk crystal, denoted by B, are $[100]_S = 1/3[1\bar{1}0]_B$, $[010]_S = 1/11[33\bar{2}]_B$, and $[001]_S = 1/11[113]_B$. The intensity profiles are obtained using the wiggler beamline W1 of HASYLAB (DESY, Hamburg). Since the secondary reflection L is $(11\ 5\ \ell)$, the coupling reflection $G-L$ is $(\bar{5}\bar{5}\bar{\ell})$. The broad peaks are the $(11\ 5\ \ell)$ reflections with various ℓ, while the kinks are due to the three-wave interference. The profile feature for the GeSi/Ge system is similar to that in Fig. 9.17. Using the exact dynamical calculation or the two-wave Born approximation to fit the experimental profiles, we find that the phase difference between the $(11\ 5\ \ell)$ and (55ℓ) reflections are $-63°$, $-56°$, and $-22°$ for $\ell = 0.1$, 0.2, and 0.7, respectively.

9.2 Phasing Macromolecular Structures and Structure Determination with Triplet Phases

9.2.1 General Considerations

Phase determination using multiple-wave diffraction for protein crystals was first reported in 1991 for myoglobin [331] and oxyhaemoglobin [301]. Since then several studies on phasing macromolecular structures and on attempting to determine electron density distributions by the Renninger scan or the reference-beam/stereoscopic imaging have followed [303,305,307,310,311,314, 317–319,325,326,332,456,457,623,624,672,674–683]. The investigated macromolecular crystals include tetragonal lysozyme, catalase oxidoreductase, triclinic lysozyme, proteinase K, rubredoxin, T2A-Im7 protein complex, guineafowl hexagonal lysozyme, C. thermocellum endoglucanase CelA, and many others. The unit-cell volumes of these proteins range from about 66800 Å^3 to 1.2×10^6 Å^3.

The large volume of crystal unit cells for macromolecular crystals, according to (2.32), makes diffraction intensity much weaker than that from crystals of small molecules. Moreover, the reciprocal lattice points of macromolecular crystals are also much more densely populated in the reciprocal space. The possibility of having multiple-wave diffraction and the chance of having overlapped diffraction profiles are much higher for macromolecular crystals than for crystals of small molecules. For example, in a 1° ψ scan of a horse oxyhemoglobin crystal with $\lambda = 1$ Å, there are about two thousand multiple-wave cases. The population of multiple-wave diffraction in a given small angular range is enormous. In addition, macromolecular crystals usually deteriorate under radiation exposure. All these factors affect considerably the intensity of multiple-wave diffraction, thus decreasing the sensitivity and visibility of the diffraction profile in relation to the triplet phases. In the following, we shall address possible solutions to the practical issues with regard to the phase determination for macromolecular crystals by using three-wave interaction.

First, the weak diffraction effect in macromolecular crystals can be overcome by using intense X-ray sources of synchrotron radiation. Part of the profile-overlapping problem can also be resolved with the extremely parallel incident synchrotron radiation to keep the profile width small. On the other hand, although there are so many multiple-wave diffractions, the dominant strong interactions appear as a well-resolved three-wave case when the nearby overlapped diffractions are much weaker than the dominant one. An empirical criterion to classify the weak three-wave diffraction as the one that can be ignored is the ratio of the structure-factor product of the secondary and the coupling reflection between the weak and the dominant case being smaller than 0.25 [305]. For the Renninger scan technique, the angular range of the ψ scan of 2 Å resolution is about 0.03 to 0.04°. With this angular range, a well-resolved dominant three-wave diffraction can be detected for a small protein of good crystal quality, like tetragonal lysozyme, with low-emittance synchrotron radiation sources. This angular range is sufficient for measuring dominant three-wave profiles and for an unambiguous phase estimation, because it is wider than the ideal peak width of about 0.002°–0.005° for strong three-wave interference by one order of magnitude. For larger molecules, the range without overlapping for strong multiple-wave diffraction decreases.

Weak diffracted intensity also implies a large extinction length, which is of the order of millimeters for protein crystals. The diffraction from protein crystals, usually of the size of sub-millimeters, is kinematical. The multiple-wave interference effects are therefore independent of whether the diffraction geometry is of Bragg or Laue. All the Born, the Bethe approximation, and the Takagi-Taupin equation are valid for this diffraction situation.

Profile overlapping can also be eliminated by tuning the photon energy or wavelength so that two overlapped profiles are resolved. For shorter wavelengths, the crystal absorption is reduced, the radiation damage of the crystal is also decreased. However, the extinction length is increased. The diffraction is then more kinematical than dynamical. Hence, the interference effect is also decreased. For longer wavelengths, the dynamical effect is comparably stronger than for shorter wavelengths. However, the crystal is easy to deteriorate. To compromise, the wavelengths from 0.7 Å to 1.3 Å may be suited for multiple-wave diffraction experiments under various conditions.

9.2.2 Structure Determination with Triplet Phases

Aside from the measurement of triplet phases from a number of three-wave diffraction profiles, the determination of phases of individual reflections associated with the triplets and the phase extension to unmeasured reflections is another important task towards the goal of crystal-structure determination. For centrosymmetric crystals of small molecules, attempts have been made to link the experimental sign relationships of structure-factor triplets, the so-called Σ_1 and Σ_2 relations [684–686] in the direct methods, by the symmetry elements associated with the given space group so that the involved structure factors form a loop or trees. Hence the sign of each of the individual structure factors involved could be determined [264, 266]. Phase extension was then done with the aid of the direct methods [53, 266]. For macromolecular crystals, the number of non-hydrogen atoms is much larger than that for small molecules. How to obtain, in a reasonable time scale, a sufficient number of triplets which effectively link those individual reflections is one of the key issues in the attempt to use three-wave diffraction for single-crystal structure determination.

First, let's consider the way of collecting a large number of multiple-wave diffraction data. The conventional Renninger scan, as described in Chap. 4, by using a point detector can only give a multiple-wave profile one at a time. The total number of multiple-wave profiles that can be measured is about 100 to 150 per day with the synchrotron radiation under optimal conditions. This number is far less than the number that a complete set of data collection needs for structure determination. In addition, a protein crystal can stand for radiation damage only for one or two days without sample cooling, depending on the brilliance of the radiation source. This fact makes Renninger scan incapable of collecting a large set of multiple-wave diffractions. An alternative is to employ the reference-beam/stereoscopic imaging techniques, where a huge collection of centrosymmetrically related three-wave pairs can be recorded by a two-dimensional detector. Another advantage is that the overlapping of diffraction profiles can be eliminated, either by tuning the photon energies or by using a strong primary reflection and weak secondary and coupling reflections. The diffraction profiles of the latter situation (shown in Fig. 9.24) illustrate the elimination of profile overlapping. In Fig. 9.24, the '+' four-wave case (000)(004)(1 12 2)($1\ \overline{12}\ 2$) of the tetragonal lysozyme can be considered as two overlapped three-wave cases (000)(004)(1 12 2) and (000)(004) ($1\ \overline{12}\ 2$). (004) is the primary reflection; L_1 (1 12 2) and L_2 ($1\ \overline{12}\ 2$) are the corresponding secondary reflections of the two three-wave cases. The ψ scan, I_{004} versus $\Delta\psi$, of the four-wave diffraction is completely different from that of the two three-wave cases. This means that the four-wave Renninger profile does not show the characteristic asymmetries of the three-wave diffractions. The θ scans for the four-wave case, adopted by the reference beam/stereoscopic imaging, on the contrary, exhibit the same profile asymmetry as that of the two three-wave diffractions (Fig. 9.24(b)). This fact indicates the overlapped four-wave profiles obtained with the θ scan (or ω scan) retain the information of the triplet phase of the three-wave cases involved.

The drawback of the reference-beam/stereoscopic imaging techniques is that the measured intensity profile is usually weaker than that obtained with the Renninger scan and the interference of three-wave diffraction is not as pronounced as that in the Renninger scan. Moreover, the average error in the determined phases is about 15° to 20° for the Renninger scan [305] and 40° to 50° for the reference-beam techniques [682].

Second, assume that a large number of multiple-wave diffraction are collected. The next step would be the determination of individual phases and phase extension. This task depends on the measured triplet phase relationships and the connections among them. Since both the Renninger scan and the reference-beam/stereoscopic imaging techniques can only extract phase information from the three-wave cases involving reflections of comparable

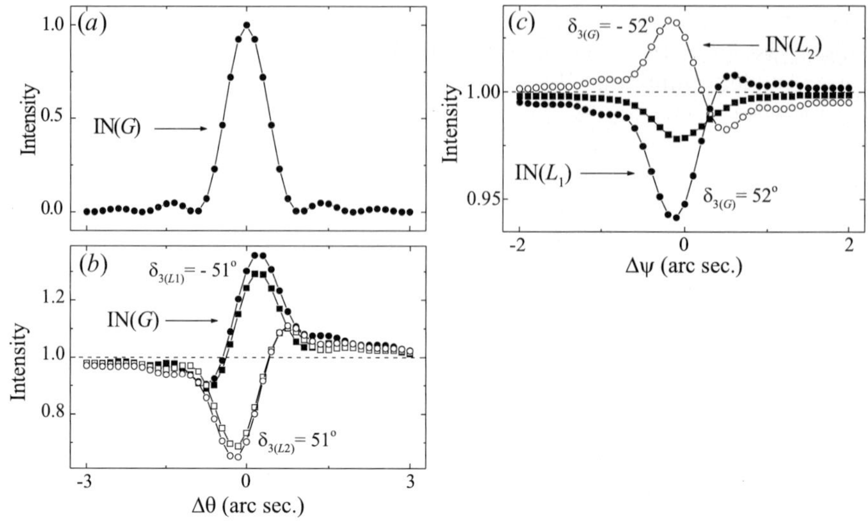

Fig. 9.24. Calculated semi-integrated relative intensity profiles for the four-wave (000)(004)(1 12 2)(1 $\overline{12}$ 2) case and $\lambda = 1.2374\,\text{Å}$: (**a**) $\Delta\theta$ scan of the G (004) reflection, (**b**) $\Delta\theta$ scan of the L_1 (1 12 2) reflection for four-wave case (*solid square*) and three-wave (000)(004)(1 12 2) case (*solid circle*) and of the L_2 (1 $\overline{12}$ 2) reflection for four-wave case (*open square*) and three-wave (000)(004)(1 $\overline{12}$ 2) case (*open circle*), and (**c**) $\Delta\psi$ scan of the G (004) reflection for four-wave case (*solid square*) and the three-wave (000)(004)(1 12 2) (*solid circle*) and (000)(004)(1 $\overline{12}$ 2) (*open circle*) cases [326]

strength, the linkage between the structure-factor triplets may not be sufficient to cover enough number of individual reflections in forming a complete data set. Under these circumstances, phase extension by employing probabilistic and/or thermodynamical methods, like in centrosymmetric crystals, may be a solution to find sufficient individual phases for structure determination. Along this direction, several investigations [305,307,311,317,325,332, 333,623,672,675–680,682] by using the direct methods [478,684–689], maximum entropy methods [690–695], simulated annealing [696,697], and others have been pursued.

As an example, Fig. 9.25 shows a portion of the electron density map of the tetragonal hen egg-white lysozyme. The electron density is calculated with maximum entropy methods. The map is drawn based on 753 individual phases deduced from 847 triplet phases measured by the Renninger scan method [317]. The structure model from the protein data bank (PDB) entry 1931 is also shown [698]. The tetragonal lysozyme crystalizes in space group $P4_32_12$ with unit-cell parameters $a = 79.1\,\text{Å}$, $c = 37.9\,\text{Å}$. It contains 129 amino acid residues and 1001 non-hydrogen atoms. The molecular weight is about 14.6 kDa. The derivation of individual phases from the measured triplets has utilized the linkage between an invariant triplet phase

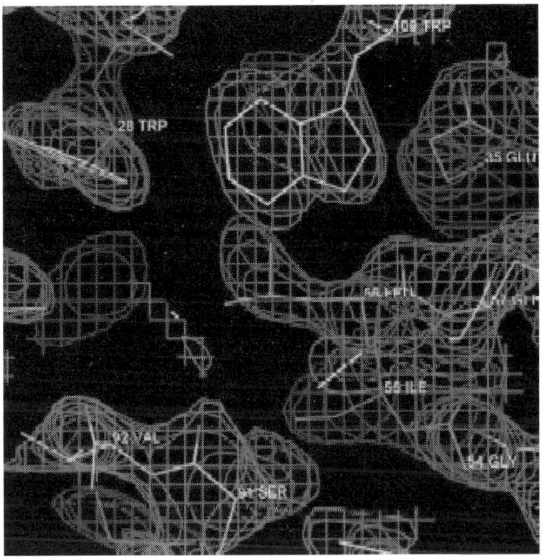

Fig. 9.25. The maximum entropy electron density map of tetragonal lysozyme calculated by using 752 structure-factor phases [317]

$$\delta_3 = \delta_{L\text{–}O} + \delta_{G\text{–}L} - \delta_{G\text{–}O} \cong \delta_{L\text{–}O} + \delta_{G\text{–}L} + \delta_{O\text{–}G} \tag{9.39}$$

and the Σ_1 relation [684–686]

$$\delta_3 = \delta_{L\text{–}O} + \delta_{\mathbf{R}[L\text{–}O]} + \delta_{O\text{–}G} = \delta_{O\text{–}G} + 2\pi\mathbf{t}[L-O] , \tag{9.40}$$

where $\mathbf{R}$ and $\mathbf{t}$ are the rotational and translational symmetry operators of the corresponding space group, and $O-G$ is a semi-invariant reflection.

A similar electron density map of the tetragonal lysozyme obtained with the reference-beam method, based on 1085 individual phases derived from 7360 measured triplet phases, is shown in Fig. 9.26(a) [682]. A map where all the calculated phases from the PDB are used for the 7360 triplets is also given in Fig. 9.26(b) for comparison. The correlation between the two maps is about 69.7%. A recursive phasing algorithm utilizing the special experimental feature of the reference-beam/stereoscopic imaging has been developed for the deduction of individual phases. That is, for a given primary (reference) reflection G, the diffraction spots of the secondary reflections L, $L+G$, $L+2G$,... lie in the same column parallel to the projection of the reciprocal lattice vector $\boldsymbol{g}$ of the G reflection on the imaging plane. The triplet phases for the three-wave cases (O, G, L), $(O, G, L+G)$, $(O, G, L+2G)$, ... in the same column are

$$\begin{aligned}
\delta_{3,0} &= \delta_{G\text{–}O} + \delta_{L\text{–}G} - \delta_{L\text{–}O} , \\
\delta_{3,1} &= \delta_{G\text{–}O} + \delta_{L\text{–}O} - \delta_{L+G} , \\
\delta_{3,2} &= \delta_{G\text{–}O} + \delta_{L+G} - \delta_{L+2G} , \\
\delta_{3,3} &= \delta_{G\text{–}O} + \delta_{L+2G} - \delta_{L+3G} , \\
&\ldots\ldots\ldots\ldots\ldots\ldots\ldots\ldots
\end{aligned} \tag{9.41}$$

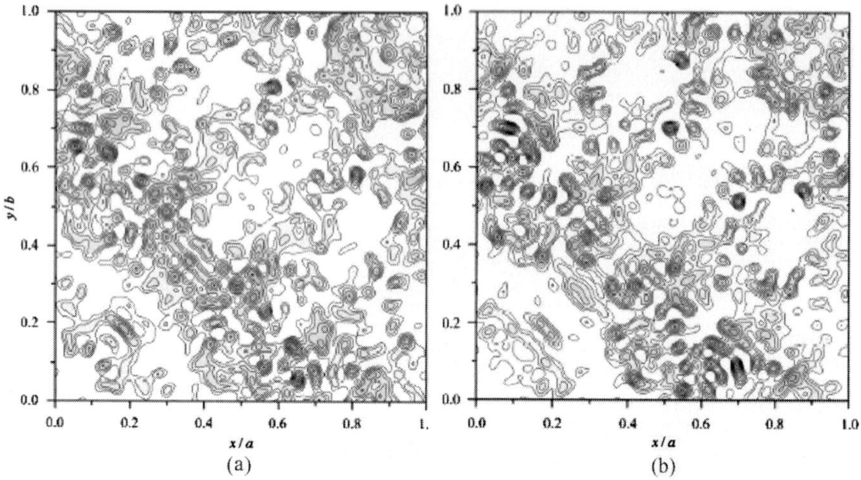

Fig. 9.26. The electron density map of the $z = 0$ basal plane of tetragonal lysozyme calculated by using (**a**) 7360 measured triplet phases and (**b**) 7360 calculated phases from the PDB [682]

If m triplet phases are measured in a single column, there will be only $m + 2$ individual phases, including the G-reflection phase $\delta_{G\text{–}O}$ common to all columns. Thus, for each additional triplet along a column, only one unknown individual phase is added to that column. This fact makes the reference-beam/stereoscopic imaging advantageous over the Renninger scan technique, where an additional three-wave diffraction usually generates two unknown individual phases. By following this recursion procedure, 1085 individual phases, which are used for the electron density calculation, are deduced from the starting phase set [682] consisting of 191 (hk0) centric reflections and the primary (111) reflection.

Third, in order to make the multiple-wave diffraction techniques a really operational and effective method for single-crystal structure determination, some questions need to be answered, such as what is the optimal number of primary (reference) reflections and triplets needed for solving a structure, which primary reflection is most appropriate to generate closely linked triplets, and how the error propagates in the phase extension. Aside from these fundamental questions, the automation for three-wave diffraction data collection, phase retrieval and extension, electron density calculation and structure refinement is also important for high throughput structure analysis.

9.3 Determination of Crystal Lattice Constants and Perfection

9.3.1 Determination of Lattice Constants for Single-Crystals

Multiple-wave diffraction patterns, either the Kossel diagram or Renninger scan, are governed by the lattice constants of the crystal investigated and the X-ray wavelength used. Thus, the crystal lattice constants and the wavelength of the radiation source can be determined or characterized from the angular position of the diffraction pattern or the intensity distribution of diffraction profiles.

A Kossel diagram is composed of many reflection conics for both characteristic and continuous radiation (see Fig. 4.5). The spatial separation in Kossel diagrams represents the range of the wavelength of the radiation. Usually, the Kossel conic of a reflection G can be described by

$$x^2 + y^2 = z^2 \tan^2 \alpha \, , \tag{9.42}$$

where α, the half-apex angle of the Kossel cone is equal to $90° - \theta_G$, θ_G being the Bragg angle of the G reflection. The axis of the cone, the z axis, is along the reciprocal lattice vector $\boldsymbol{g}$ of the G reflection and may or may not be coincident with the incident wave. If $\boldsymbol{g}$ is not along the direction of the incident wave, the coordinates defining the conic with respect to the incident wave can be easily found by using an appropriate transformation of coordinates. As described in Chap. 4, the intersection of the Kossel conics indicates the occurrence of a multiple-wave diffraction. For a given X-ray wavelength λ, according to the geometry, two or three adjacent intersection points on a Kossel diagram can be used to determine the lattice constants of the crystal investigated. The error in the lattice constant measured is about $\pm(2-10) \times 10^{-5}$ Å [81,82,341,343–346,699]. For a non-coplanar coincidental multiple-wave diffraction of a specific wavelength, the lattice constants can be determined directly from the coordinates of the conics. A well-known classic example is the use of the four-wave (000)(220)(133)(313) non-coplanar coincidental diffraction of diamond at $\lambda = 1.540562$ Å for lattice-constant measurements [340]. The accuracy is about $\pm 5 \times 10^{-5}$ Å. However, such a multiple-wave diffraction very seldom occurs for crystals of small molecules of wavelength around 1.5 Å.

The first lattice constant measurement by using the collimated-beam technique was reported by Renninger for a diamond crystal and CuKα_1 radiation [4]. The angular positions of multiple-wave diffraction peaks relative to the symmetry mirrors of the (222) diffraction pattern were measured, giving the lattice constant 3.5668 Å with an accuracy of 10^{-4} Å. Similar results obtained for diamond, silicon and germanium with an improved experimental setup have also been reported [347,348]. From (4.5) for the symmetric Bragg geometry of a primary reflection G, a general expression for the change in the azimuth angular position ψ of multiple-wave diffraction, denoted as $\Delta\beta$, due to the variations of lattice constant perpendicular and parallel to the surface normal is given by

$$\Delta\beta = -(\Delta\ell_n)\frac{1}{\sin\beta}\left(\frac{\lambda}{\cos\theta_G} - \frac{\cos\beta}{\ell_n}\right) - (\Delta\ell_p)\frac{1}{\sin\beta} \times \left[\frac{(\ell_p - g)\,\lambda}{\ell_n\cos\theta_G} - \frac{\lambda^2 g^2\cos\beta}{4\ell_p\cos^2\theta_G}\right], \tag{9.43}$$

where the variation $\Delta\ell_p$ of the component of the reciprocal lattice vector $\boldsymbol{\ell}$ of the secondary reflection L parallel to the reciprocal vector $\boldsymbol{g}$ of the primary reflection is related to the lattice-constant variation $\Delta d_\perp$ perpendicular to the crystal surface as

$$\frac{\Delta\ell_p}{\ell_p} = \Delta\theta_G\cot\theta_G = -\frac{\Delta d_\perp}{d_\perp}. \tag{9.44}$$

$\Delta\ell_n$ is the variation of ℓ normal to $\boldsymbol{g}$, which is proportional to the lattice-constant change parallel to the crystal surface. Substituting (9.44) into (9.43), we can write $\Delta\beta$ as

$$\Delta\beta = -(\Delta\ell_n)\frac{1}{\sin\beta}\left(\frac{\lambda}{\cos\theta_G} - \frac{\cos\beta}{\ell_n}\right) - (\Delta\theta_G)\,\ell_p\frac{\cot\theta_G}{\sin\beta} \times \left[\frac{(\ell_p - g)\,\lambda}{\ell_n\cos\theta_G} - \frac{\lambda^2 g^2\cos\beta}{4\ell_p\cos^2\theta_G}\right]. \tag{9.45}$$

Once $\Delta\beta$ is measured from a Renninger scan and $\Delta d_\perp$ is determined from the measured $\Delta\theta_G$, the variation of lattice constant parallel to the crystal surface can then be calculated according to (9.45).

The lattice constant a of a cubic crystal, based on (4.5), takes the form

$$a = \frac{\lambda\left(\ell^2 - \ell_p g\right)}{2\ell_n\cos\theta_G\cos\beta}. \tag{9.46}$$

Taking the variations of a with respect to θ_G and β leads to

$$\frac{\Delta a}{a} = (\Delta\theta_G)\tan\theta_G + (\Delta\beta)\tan\beta\,. \tag{9.47}$$

For a given primary reflection G, $\Delta a/a$ depends on the β angle due to the presence of the secondary reflection. The smaller the $\tan\beta$ value, the more precise the lattice constant a. Referring to Chap. 4, we can see that the smaller the β angle, the larger the peak width. Therefore, multiple-wave diffractions suitable for lattice constant determination usually have a wider peak width [229, 347, 348]. With this consideration, the use of (513), (551), (117), and (331) as the secondary reflections for silicon, germanium, and diamond with CuKα_1 and CuKα_2 radiation gives the lattice constants of 5.430941 Å, 5.657819 Å, and 3.566986 Å, respectively [347, 348]. The differences between these values and that determined by the two-wave diffraction methods [700–706] are within 10^{-5} Å to 10^{-6} Å.

Dynamical effects may cause the peak position of a multiple-wave diffraction to shift as large as 10^{-6} radians, which is of the order of magnitude of the electric susceptibility χ (see, Fig. 8.26), especially for strong reflections and perfect crystals. For imperfect crystals, the dynamical effects are so small that can be neglected in the determination of lattice constants.

9.3.2 Strain Measurements of Epitaxy Layer on Crystal Substrate

Layered materials are important in fabricating devices for micro-electronics, optoelectronics, and spintronics. For these applications, thin layers can be prepared by ion implantation, ion sputtering, laser-beam ablation, chemical vapor deposition (CVD), liquid phase epitaxy (LPE), vapor phase epitaxy (VPE), molecular-beam epitaxy (MBE), vacuum chemical epitaxy (VCE), metal organic chemical vapor deposition (MOCVD), and many other methods. Strains are often generated due to lattice distortion and lattice mismatch between layers and substrates. The lattice mismatch, usually encountered in epitaxial materials, is due to either the difference in thermal expansion coefficients or the difference in composition between the epitaxial layer and the substrate. The presence of strain may degrade the performance and life-time of the devices made of the layered materials. It is known that a tetragonal lattice distortion is usually encountered in epitaxial layers with respect to a cubic crystal substrate. Many X-ray diffraction techniques have been developed to characterize this lattice distortion. By means of two-wave diffraction, the lattice mismatches parallel and perpendicular to the interfacial boundary, denoted as $\Delta a_{\parallel}$ and $\Delta a_{\perp}$, respectively, can be determined by measuring the angular shifts of the layer diffraction profiles relative to the substrate peaks with a symmetric and an asymmetric Bragg reflection.

Simultaneous determination of $\Delta a_{\parallel}$ and $\Delta a_{\perp}$ is made possible by the use of multiple-wave diffraction. According to (9.43), (9.44), and (9.45), the angular deviations $\Delta\theta_G$ and $\Delta\beta$ measured from a multiple-wave diffraction can determine both of the lattice mismatches: the former gives $\Delta a_{\perp}$ and the latter $\Delta a_{\parallel}$. Experimentally, the divergent-beam technique has been applied to estimate the tetragonal distortion for an LPE (001) InGaAsP/InP by using the six-wave $(000)(006)(2\bar{2}4)(2\bar{2}2)(\bar{2}24)(\bar{2}22)$ diffraction for $\mathrm{CuK}\alpha_1$ radiation [349] and for a GaAlAs/GaAs LPE system by using two three-wave (000)(006)(511) and $(000)(006)(\bar{5}\bar{1}5)$ diffractions [352]. In the six-wave diffraction measurements, the six-wave case is decomposed into two four-wave, $(000)(006)(2\bar{2}4)(2\bar{2}2)$ and $(000)(006)(\bar{2}24)(\bar{2}22)$, cases due to the tetragonal lattice distortion. The two four-wave cases occur at different β angular positions. The separation between the two cases, $\Delta\beta$, which is a function of the compositional concentration of arsenic, gives $\Delta a_{\parallel}$, provided that $\Delta a_{\perp}$ can be estimated from $\Delta\theta_G$. The determined $\Delta a_{\parallel}$ and $\Delta a_{\perp}$ against the arsenic concentration in the liquid composition, X^{ℓ}_{As}, of InGaAsP/InP are plotted in Fig. 9.27 [349, 351].

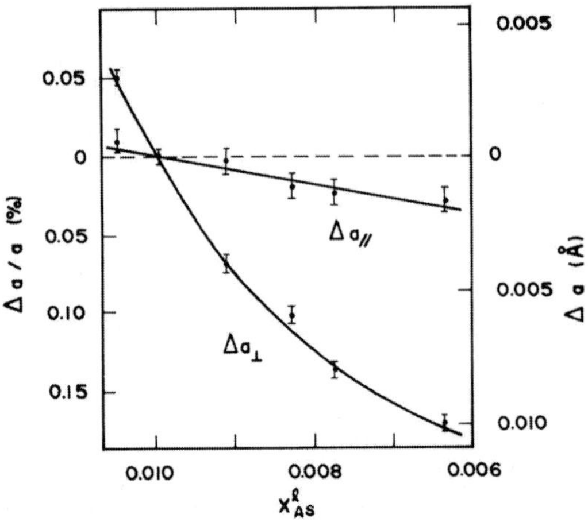

Fig. 9.27. Measured $\Delta a_\perp$ and $\Delta a_\parallel$ for $X^\ell_{Ga} = 0.0007$ and various X^ℓ_{As} [349]

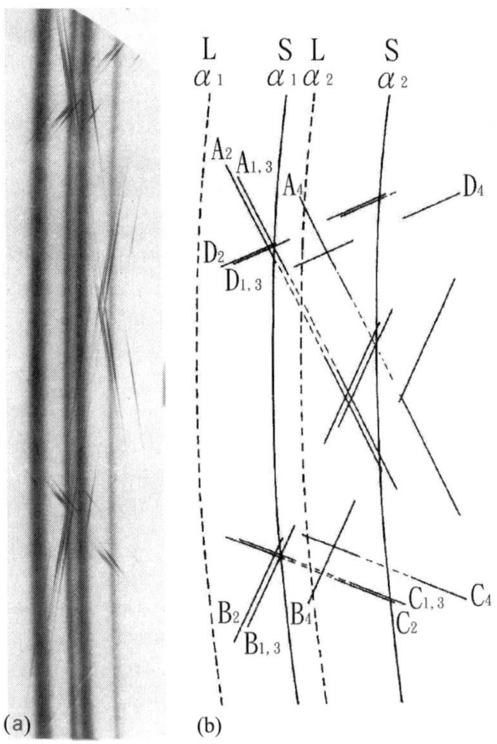

Fig. 9.28. (006) reflection image of $Ga_{0.34}Al_{0.66}As/GaAs$ for CuKα: (**a**) experimental and (**b**) simulation [352, 708]

The determined $\Delta a_{\|}$ is much smaller than $\Delta a_{\perp}$ for various arsenic concentrations. For the GaAlAs/GaAs experiments, the technique of Spooner and Wilson [346] was adopted. Interesting hybrid multiple-wave diffraction images of the combination of diffractions from the epitaxial layer and the substrate were observed. Figure 9.28(a) shows the (006) reflection image of an LPE $Ga_{0.34}Al_{0.66}As/GaAs$, where the inclined crossing streaks are the hybrid multiple-wave diffractions [352].

Hybrid multiple-wave diffraction is sometimes called the modulation of Renninger scan intensity (MORSI) [707] due to the presence of a layer reflection in the multiple-wave diffraction process for the substrate or the presence of a substrate reflection in the diffraction process for the layer. Figure 9.29 shows all possible diffraction situations involving the layer and substrate reflections. Figures 9.29(a) and (b) show that the three-wave diffractions composed of the secondary L and the coupling $H = G - L$ reflection take place solely in the substrate and the layer, respectively. The resultant diffracted wave is reflected back in the direction of the G reflection. These two cases are denoted by SS and LL. The hybrid diffractions SL and LS shown in Figs. 9.29(c) and (d), consist of the secondary L reflection of the substrate and the coupling H reflection from the layer, and vice versa. This kind of hybrid diffraction can be observed in both the divergent-beam Kossel and the Renninger scans.

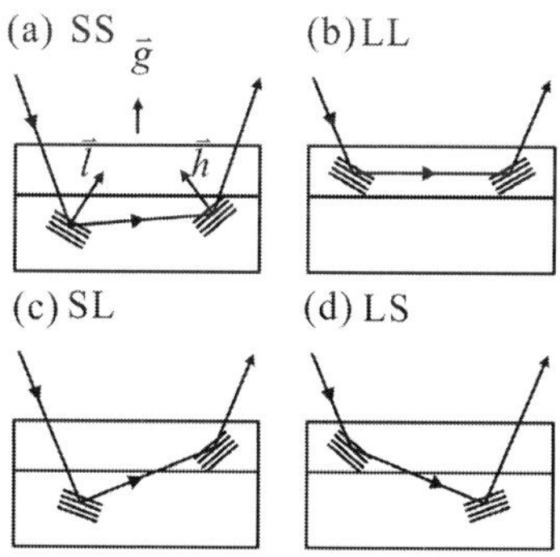

Fig. 9.29. Possible hybrid multiple diffraction: (**a**) Substrate-substrate (SS)(**b**) Layer-layer (LL) (**c**) Substrate-layer (SL) and (**d**) Layer-substrate (LS) [709]

The indexing of a hybrid diffraction in a Kossel diagram can be easily carried out by using the double-wavevector representation [438]. Referring to Fig. 9.30(a) for a two-wave (O, G) diffraction, we can define the double-wavevectors and the sphere of reflection as

$$\boldsymbol{r}_O = -2\boldsymbol{k}_O \,, \qquad \boldsymbol{r}_G = 2\boldsymbol{k}_G \,,$$
$$|\boldsymbol{r}_O| = |\boldsymbol{r}_G| = R = \frac{2}{\lambda} \,, \tag{9.48}$$

where $\boldsymbol{r}_O = \boldsymbol{OM}$, $\boldsymbol{r}_G = \boldsymbol{ON}$, and the wavevectors $\boldsymbol{k}_O = \boldsymbol{CO}$, $\boldsymbol{k}_G = \boldsymbol{CG}$. The quantity R is the radius of the sphere of reflection. The Kossel plane

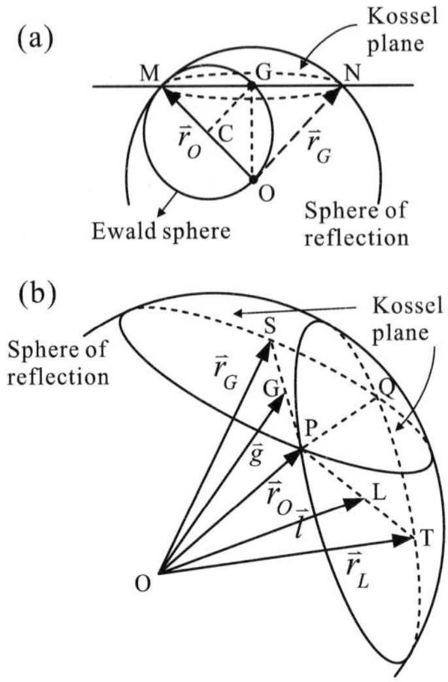

Fig. 9.30. Double-wavevector representation of X-ray diffraction: (**a**) two-wave and (**b**) three-wave cases [438, 708]

cutting the Ewald sphere at point G is perpendicular to the reciprocal lattice vector $\boldsymbol{g}$ and is governed by

$$\boldsymbol{r}_G \cdot \boldsymbol{g} = g^2 \,, \qquad \boldsymbol{r}_O \cdot \boldsymbol{g} = g^2 \,. \tag{9.49}$$

The corresponding Bragg's law takes the alternative form

$$\boldsymbol{r}_G = 2\boldsymbol{g} - \boldsymbol{r}_O \,. \tag{9.50}$$

For a three-wave (O, G, L) case, two Kossel planes, representing the primary reflection G and the secondary reflection L, intersects at points P and Q (Fig. 9.30(b)). For simplicity, the incident double-wavevector $\boldsymbol{r}_O$ is assumed to be $\boldsymbol{OP}$. The Bragg diffraction condition for the reflection L is

$$\boldsymbol{r}_L = 2\boldsymbol{\ell} - \boldsymbol{r}_O \,, \tag{9.51}$$

where $\boldsymbol{\ell} = \boldsymbol{OL}$ is the reciprocal lattice vector of the L reflection. The difference between (9.50) and (9.51) gives

$$\boldsymbol{r}_G = 2(\boldsymbol{g} - \boldsymbol{\ell}) + \boldsymbol{r}_L \,, \tag{9.52}$$

which indicates that, in Fig. 9.30(b), $\boldsymbol{TS} = 2\boldsymbol{LG} = 2(\boldsymbol{g} - \boldsymbol{\ell})$. Namely,

$$\boldsymbol{r}_L \cdot (\boldsymbol{g} - \boldsymbol{\ell}) = -|\boldsymbol{g} - \boldsymbol{\ell}|^2 \,. \tag{9.53}$$

This implies that the three-wave diffraction involves at least two simultaneous diffraction paths: (1) the primary reflection G; (2) the detoured diffraction consisting of the L reflection and the coupling $G - L$ reflection. Both paths yield the resultant diffracted wave along the G reflected direction $\boldsymbol{r}_G$. Figure 9.29 merely shows the second path for three-wave diffractions in layered structures, where $\boldsymbol{h} = \boldsymbol{g} - \boldsymbol{\ell}$. For a four-wave (O, G, L, M) case, the diffraction conditions, (9.48)–(9.53), hold. Therefore, the $\boldsymbol{r}_M$ of the M diffracted wave follows the relations for $\boldsymbol{r}_L$.

The incident and diffracted double-wavevectors, $\boldsymbol{r}_O$ and $\boldsymbol{r}_G$, can be determined by solving the equations (9.48)–(9.53). In comparison with the experimental wavevectors, $\boldsymbol{r}_O$ can be expressed in terms of the incident angle θ with respect to the atomic planes of the primary G reflection and the azimuth angle ψ related to a reference vector $\boldsymbol{v}$ normal to $\boldsymbol{g}$. Similarly, the $\boldsymbol{r}_M$ is represented by the exit angle θ' and the azimuth angle ψ' with respect to $-\boldsymbol{v}$ [708]. These four angles are given by

$$\theta = \sin^{-1}\left(\frac{\boldsymbol{r}_O \cdot \hat{\boldsymbol{n}}}{R}\right), \quad \psi = \tan^{-1}\left[\frac{\boldsymbol{r}_O \cdot (\hat{\boldsymbol{n}} \times \hat{\boldsymbol{v}})}{\boldsymbol{r}_O \cdot \hat{\boldsymbol{v}}}\right];$$
$$\theta' = \sin^{-1}\left(\frac{\boldsymbol{r}_G \cdot \hat{\boldsymbol{n}}}{R}\right), \quad \psi' = \tan^{-1}\left[\frac{\boldsymbol{r}_G \cdot (\hat{\boldsymbol{n}} \times \hat{\boldsymbol{v}})}{\boldsymbol{r}_G \cdot \hat{\boldsymbol{v}}}\right], \tag{9.54}$$

where the unit vector $\hat{\boldsymbol{n}}$ normal to the G planes is along $\boldsymbol{g}$.

By considering these four angles along with the geometry for Kossel diagrams, all the diffraction streaks due to the continuum spectra of the X-ray source, including the hybrid multiple-wave diffractions from thin layers and substrates, can be indexed. Figure 9.28(b) shows the simulated image for the divergent-beam Kossel diagram (Fig. 9.28(a)) of a (006) LPE $\mathrm{Ga_{0.34}Al_{0.66}As/GaAs}$ sample and CuKα radiation [708]. The labels S, L, α_1, and α_2 stand for the substrate, epitaxial layer, CuKα_1, and CuKα_2 of the (006) reflection, respectively. The indices of the streaks refer to the possible hybrid multiple-wave diffractions listed in Table 9.1. The labelled streak $A_{1,3}$ takes two different diffraction paths, A_1 and A_3 (see Table 9.1). Path A_1 involves the $515_S + \bar{5}\bar{1}1_L$ and path A_3 the $511_L + \bar{5}\bar{1}5_S$ hybrid diffraction. Some streaks are purely due to either the substrate or the layer. Such streaks can be easily indexed by following the usual reference-vector indexing procedure (see Sect. 4.1.1); therefore, they are not shown in Fig. 9.28(b). The tetragonal lattice distortions, $\Delta a_{\parallel}$ and $\Delta a_{\perp}$, are determined according to (9.43)–(9.45) during the indexing procedure. If an additional GaAs cap layer, denoted as L1, is grown on top of the LPE sample discussed, then more diffraction streaks will appear in the Kossel diagram. These streaks can

Table 9.1. Possible hybrid multiple diffraction paths for 006 $\mathrm{Ga_{1-x}Al_xAs/GaAs}$

	A	B	C	D
1	$515_S + \bar{5}\bar{1}1_L$	$5\bar{1}5_S + \bar{5}11_L$	$335_S + \bar{3}\bar{3}1_L$	$3\bar{3}5_S + \bar{3}31_L$
2	$515_S + 00\bar{4}_L + \bar{5}\bar{1}5_S$	$5\bar{1}5_S + 00\bar{4}_L + \bar{5}15_S$	$335_S + 00\bar{4}_L + \bar{3}\bar{3}5_S$	$3\bar{3}5_S + 00\bar{4}_L + \bar{3}35_S$
3	$511_L + \bar{5}\bar{1}5_S$	$5\bar{1}1_L + \bar{5}15_S$	$331_L + \bar{3}\bar{3}5_S$	$3\bar{3}1_L + \bar{3}35_S$
4	$511_L + 004_S + \bar{5}\bar{1}1_L$	$5\bar{1}1_L + 004_S + \bar{5}11_L$	$331_L + 004_S + \bar{3}\bar{3}1_L$	$3\bar{3}1_L + 004_S + \bar{3}31_L$

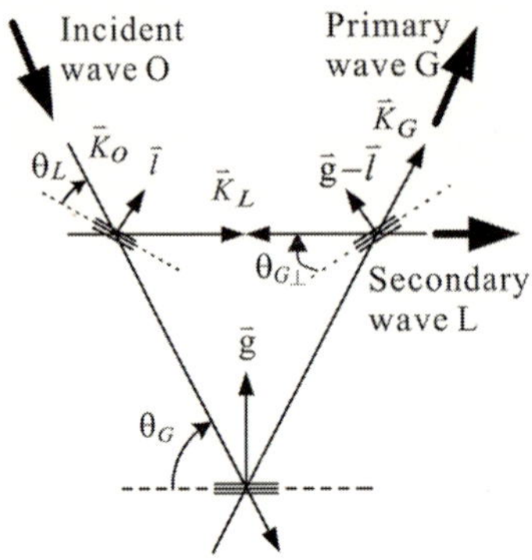

Fig. 9.31. Schematic of the geometry of a three-wave Bragg-surface diffraction [352, 353]

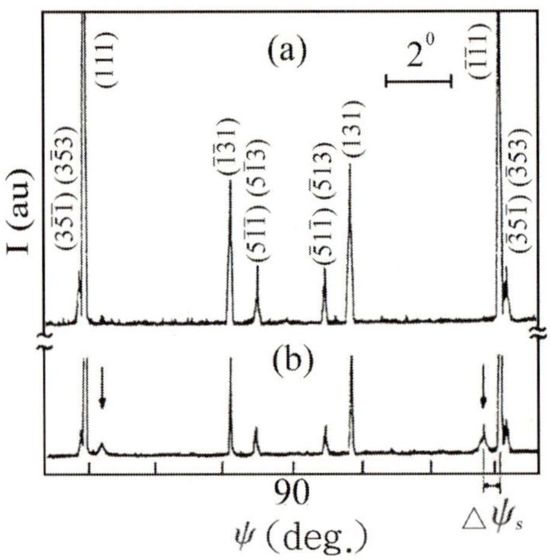

Fig. 9.32. Renninger scan of (002) Si with CuKα_1 for Si substrate (**a**) without and (**b**) with a 0.8 μm GaAs epitaxial layer [709]

be indexed with the same token as for the single-layer sample, except that the subscripts L and S of the reflections in Table 9.1 need to be replaced by L1 and L, respectively.

Hybrid multiple-wave diffraction has also been observed in Renninger scans because of the finite incident beam divergence and crystal mosaic spread [707, 708]. In addition, the special diffraction geometry of a Bragg-surface case makes Bragg-surface hybrid diffraction very sensitive to the perfection of the layered materials, including crystal mosaicity, lattice strain, and defects. Figure 9.31 is the schematic representation of the geometry of a three-wave Bragg-surface diffraction. The incident wave O with wavevector $\boldsymbol{K}_O$ is simultaneously diffracted by the primary reflecting planes G and the

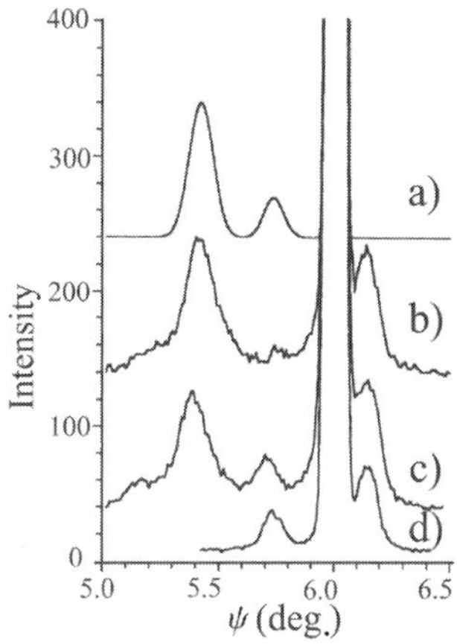

Fig. 9.33. Renninger scan of (002) Si and Cu$K\alpha_1$: (**a**) simulated LS peaks with $a_{\|}(1) = 5.583$ Å and $a_{\|}(2) = 5.663$ Å for samples 1 and 2, (**b**) measured scans for sample 2 with and (**c**) without external bending, (**d**) measured scan for sample 1 [354]

secondary planes L. The secondary wavevector $\boldsymbol{K}_L$ is then diffracted by the coupling planes in the primary wave direction $\boldsymbol{K}_G$.

Figure 9.32 shows the Renninger scans of silicon (002) with Cu$K\alpha_1$ for Si substrates without and with a 0.8 µm GaAs layer. The GaAs layer was grown at 650°C by the VCE technique. A microfocus X-ray source of size 50 µm × 50 µm and a 115 cm long collimator are used, providing a 1 arc minute incident beam divergence. The diffractions with their peaks indexed with a secondary $(hk\ell)$ and $\ell = 1$ are Bragg-surface diffractions, for example $(\bar{1}\bar{1}1)$ and (111). The diffractions with their peaks indicated by arrows near the $(\bar{1}\bar{1}1)$ and (111) are hybrid Bragg-surface diffractions. For the $(\bar{1}\bar{1}1)$, the detoured diffraction path involving the secondary $(111)_L$ and the coupling $(\bar{1}\bar{1}1)_S$ is denoted as $(111)_L + (\bar{1}\bar{1}1)_S$. Since the $(111)_L$ diffracted wave propagates in the epitaxial layer and is nearly parallel to the interface, the hybrid LS Bragg-surface diffraction is most sensitive to the variation in $\Delta a_{\|}$ and the related structural change in the layer and interface.

As an example, Fig. 9.33 shows the Renninger scans of the hybrid diffraction $(111)_L + (\bar{1}\bar{1}1)_S$ for two samples. Sample 1 is a 500 Å VCE GaAs buffer layer on a (001) Si substrate and sample 2 a 1.2 µm VCE GaAs epilayer grown on sample 1. The hybrid $(111)_L + (\bar{1}\bar{1}1)_S$ diffractions from the buffer and the thicker GaAs layers, denoted by LS(1) and LS(2), respectively, and the strong $(111)_S$ three wave diffraction from the substrate are indicated in the simulated diffraction pattern (Fig. 9.33(a)) based on the kinematical theory. The measured Renninger scans of sample 2 with and without external stress are shown in Figs. 9.33(b) and (c). As compared with the diffraction pattern of sample 1 (Fig. 9.33(d)), the angular position of the hybrid LS diffraction peak of sample 2 is different from that of sample 1. This is due to the difference in the lattice constant parallel to the interface, i.e., $a_{\|} = 5.583$ Å for sample

1 and 5.663 Å for sample 2. The interfacial dislocation densities determined from these lattice distortions are in turn different. If all the dislocations are of a 60° type, the dislocation densities estimated from the Burgers vectors are $\rho_1 = 1.40 \times 10^6$/cm for sample 1 and $\rho_2 = 2.14 \times 10^6$/cm for sample 2 [710]. In terms of the critical density $\rho_c = 2.05 \times 10^6$/cm for an inverting lattice curvature, sample 1 is convex while sample 2 is concave to the surface layer. The former is subject to a compressive stress and the latter a tensile stress. It is also known that in a GaAs/Si system the stress generated to accommodate lattice mismatch has an opposite direction to that generated by the difference in the thermal expansion coefficient of the two materials [711]. Moreover, lattice misfit is more relaxed in thicker crystal layers than in thinner ones. Therefore, thick and thin layers show an inversion of curvature when an external stress is applied in the direction opposite to that given by the difference in the thermal expansion coefficient. The original LS peak of the thicker GaAs layer is split into a thicker and a thinner LS peak, as shown in Fig. 9.33(c). Further increase in the stress only broadens the two LS peaks. This peak splitting indicates that the degree of cohesion is larger in the substrate-buffer layer interface than in the buffer-epilayer interface. This is mainly due to that the stress resulting from the difference in the thermal expansion coefficient is compensated by the applied external stress.

Other applications along the same direction include the characterization of InGaP/GaAs systems with hybrid four-wave diffractions [361] and of semiconductor superlattices with three-wave 'effective satellite' [362].

9.3.3 Characterization of Crystal Perfection of Semiconductor Surfaces

The Bragg-surface three-wave diffraction can be equally applied to surfaces of semiconductor crystals, such as silicon, germanium, gallium arsenide, etc.. The 'surface' mentioned here means the surface area with a depth of a few hundreds nanometers. According to the kinematical theory described in Chap. 5, the difference in reflecting power between a three-wave (O, G, L) diffraction and a two-wave (O, G) case is given in (5.36), where G is the primary reflection. This expression is valid for Bragg-surface three-wave diffraction of X-rays. If the primary reflection is forbidden or very weak, then the intensity profile of the primary G reflection for a mosaic crystal with an isotropic Gaussian distribution of mosaic spread is given by [232]

$$I_{G,\mathrm{Mos}}(\theta,\psi) = (Q_{MD} - Q_{GO}) \exp\left(\frac{-\Delta\theta_{LO}^2}{2\eta^2}\right) \exp\left(\frac{-\Delta\theta_{GL}^2}{2\eta^2}\right) + Q_{GO} \exp\left(\frac{-\Delta\theta_{GO}^2}{2\eta^2}\right) , \tag{9.55}$$

where Q_{MD} and Q_{GO} are the peak reflectivities of the three-wave diffraction and the primary reflection G, respectively. $Q_{MD} - Q_{GO}$ is proportional to the $Q_{OL}Q_{LG}$ in (5.36) with $\Delta\theta_{LO} = \Delta\theta_{GL} = 0$, $\Delta\theta_{GL}$ being the angular deviation from the Bragg angle of the G–L reflection. The product of the two exponential functions in the first term on the right of (9.55) is proportional to the number of crystal blocks involved in the detoured diffraction path, first the secondary reflection L in one block and then the coupling reflection G–L in another block. η is the isotropic mosaic spread and

$$\Delta\theta_{MN}(\theta,\psi) = \arccos\{-\lambda \boldsymbol{K}_M(\theta,\psi)(\hat{\boldsymbol{m}} - \hat{\boldsymbol{n}})\} - \theta_{MN} , \tag{9.56}$$

where M and N can be any of the O, G, and L. $\boldsymbol{K}_M$ is the wavevector of the M reflection, which is a function of θ and ψ. The quantity $\hat{\boldsymbol{m}} - \hat{\boldsymbol{n}}$ is the reciprocal lattice unit vector of the $M - N$ reflection.

A quasi-perfect crystal can be considered as being composed of many Guassian distributed large crystal blocks which diffract dynamically. The intensity profile for a quasi-perfect crystal then takes the form

$$I_{G,\mathrm{Perf}}(\theta,\psi) = (Q_{MD} - Q_{GO}) \exp\left(\frac{-\Delta^2}{2\eta^2}\right) + Q_{GO} \exp\left(\frac{-\Delta\theta_{GO}^2}{2\eta^2}\right) , \tag{9.57}$$

where Δ is the approximate intrinsic width of the three-wave dynamical diffraction. The single exponent indicates that the detoured diffraction, the secondary reflection and the coupling reflection take place in a single large crystal block. Hence the reflectivities Q_{MD} and Q_{GO} can be calculated based on the dynamical theory.

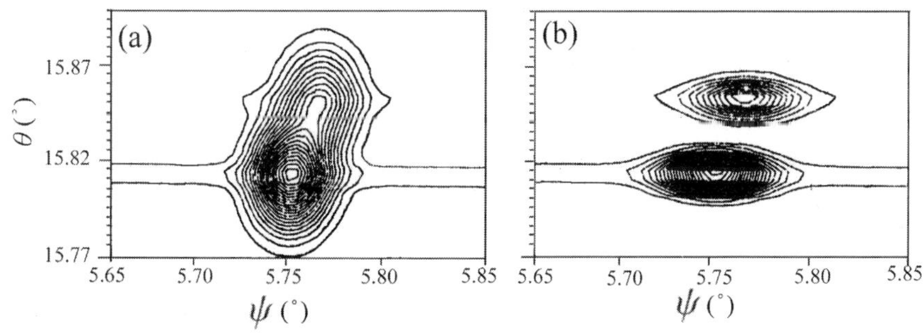

Fig. 9.34. Calculated intensity distributions of GaAs (002) in θ–ψ scans under the three-wave (000)(002)($1\bar{1}1$) condition for CuKα_1 and CuKα_2 : (**a**) mosaic crystal, (**b**) quasi-perfect crystal [355]

The mosaic spread η can be replaced [355] by

$$\eta(\varsigma) = \sqrt{\frac{\cos^2\varsigma}{\eta_\perp^2} + \frac{\cos^2\varsigma}{\eta_\parallel^2}} \tag{9.58}$$

to take care of the anisotropic distribution of the misorientation of the crystal blocks, where $\eta_\perp$ and $\eta_\parallel$ are the mosaic spreads perpendicular and parallel to the crystal surface. ς is the azimuthal position of the wave N around the reciprocal lattice vector $\boldsymbol{M}-\boldsymbol{N}$.

The contribution of the primary reflection G to the intensity profile can be negligibly small when a forbidden or a very weak reflection is chosen as the primary reflection. Under this circumstance, the detoured diffraction path, as shown in Fig. 9.31 lies in the surface region of the crystal investigated. The crystal imperfection of this region should, in principle, affects the diffraction profile, which, in turn, provides the structural information of the crystal surface. Indeed, calculations based on (9.55) and (9.57) for the three-wave $(000)(002)(1\bar{1}1)$ Bragg-surface diffraction in GaAs (001) crystal with CuKα_1 and CuKα_2 show appreciable difference in the intensity distribution (θ, ψ) of the primary (002) reflection (Fig. 9.34). Note that (002) is a weak reflection for GaAs. The input parameters for the calculations are $Q_{MD} = 1520$ counts/sec and $Q_{GO} = 110$ counts/sec and the anisotropic mosaic spreads $\eta_\perp = 17''$ and $\eta_\| = 59''$, which are estimated from the experimental results. As shown in Fig. 9.34(a), for a mosaic crystal, the FWHM's in θ, W_θ $(= 138'')$ is larger than that in ψ, W_ψ $(= 111'')$, which is much larger than the two-wave width $W_\theta^{002} = 43''$. While for a quasi-perfect crystal, $W_\theta \simeq W_\theta^{002} \sim 58''$ is much smaller than $W_\psi = 144''$. This fundamental difference in diffraction width may be useful for the characterization of the imperfection of crystal surface. As examples, two experiments are illustrated in the following:

(a) The effects of surface treatment on the Ge (001) surface. Figure 9.35 shows the (θ, ψ) mappings of the intensity of the primary reflection (002) for the as-cut, mechanical polished, and chemical etched (001) surfaces of a commercial Czochralski germanium under the three-wave $(000)(002)(1\bar{1}1)$ Bragg-surface diffraction condition, where (002) is a space-group symme-

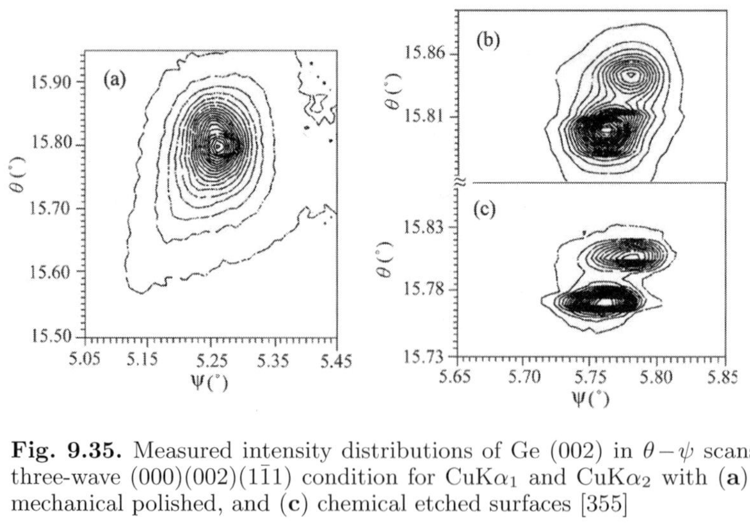

Fig. 9.35. Measured intensity distributions of Ge (002) in $\theta-\psi$ scans under the three-wave $(000)(002)(1\bar{1}1)$ condition for CuKα_1 and CuKα_2 with (**a**) as-cut, (**b**) mechanical polished, and (**c**) chemical etched surfaces [355]

try forbidden reflection. The microfocus X-ray source (CuKα radiation) and the related experimental conditions mentioned in the previous section are employed for the intensity mapping. For the mechanical polishing, a diamond paste of 1 µm is used, and for the chemical etching, the NaOH solution is utilized to remove 10 µm layer from the polished surface. A single peak comprising both the CuKα_1 and CuKα_2 (002) reflections is observed for the as-cut crystal (Fig. 9.35(a)). A preferred misorientation of the crystal blocks at the surface is also shown as clearly as the lower intensity contours indicate. These are the evidence of a highly damaged surface due to crystal cutting. Since $W_\theta > W_\psi$ for the more intense contours, the diffraction is of the kinematical nature. The diffracting crystal contributed to this intensity distribution is a mosaic crystal. After the crystal being mechanically polished, the dynamical diffraction becomes slightly dominant over the kinematical contribution, because $W_\theta < W_\psi$. The crystal surface region can be considered as a mixture of mosaic and quasi-perfect crystals. The upper part of this region is mosaic-like and the lower part is perfect. When the crystal is chemically etched, the crystal surface is more perfect than mosaic, as can be seen from the FWHM values, i.e., $W_\theta(= 43'') \ll W_\psi(= 140'')$.

(b) Lattice damage in the ion implanted GaAs (001) surface. The three-wave (000)(002)(111) Bragg-surface diffraction is used to investigate the lattice damage in a Se^+ ion implanted GaAs (001) surface. (002) is a symmetric Bragg reflection and (111) a surface reflection. The same experimental setup as that of using a microfocus X-ray source is employed to carry out the (θ, ψ) intensity mapping. The effective beam divergence is 63 arcseconds in both vertical and horizontal directions. Figures 9.36(a), (b), and (c) are the intensity distributions of the (002) reflection with 6×10^{14} 80 keV and 15×10^{14} 80 keV, and 3×10^{14} 120 keV Se^+ ion implantation, respectively, for CuKα_1 and CuKα_2 radiation. These three samples are referred to as samples 1, 2, and 3.

In Fig. 9.36(a), only the diffraction peaks of the GaAs lattice matrix appear. While in Figs. 9.36(b) and (c), extra peaks due to the lattice distortion of the implanted layers show up as doublets together with those from the GaAs matrix. This means that no definite implanted layer is formed in sample 1. The measured full widths at half maxima (FWHM) in the θ and ψ scans, denoted as W_θ and W_ψ, are 39″, 40″, and 39″ in W_θ and 196″, 88″, and 266″ in W_ψ of the GaAs matrix for samples 1, 2, and 3, respectively. The corresponding FWHM's of the implanted layers of sample 2 and sample 3 are 60″ and 58″ in W_θ and 83″ and 125″ in W_ψ. In contrast to these widths, the measured widths, W_{GIXD}, from the in-plane (220) GIXD are 70″, 61″, and 74″ for samples 1, 2 and 3, respectively, which are much narrower than the W_ψ values. Therefore, the W_ψ broadening is not only related to the in-plane mosaicity but related to the dynamical diffraction effects (see Fig. 9.35). The latter is governed by the coherence length of the crystal lattice, within which the crystal diffracts coherently. This length scale D can be

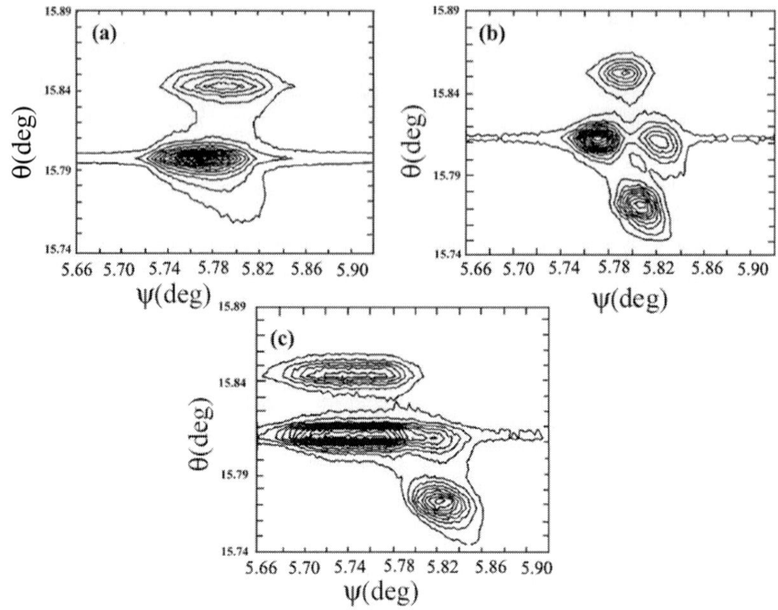

Fig. 9.36. Measured intensity distributions of GaAs (002) in $\theta - \psi$ scans under the Bragg-surface three-wave (000)(002)(111) condition for $\mathrm{CuK}\alpha_1$ and $\mathrm{CuK}\alpha_2$ with (**a**) 6×10^{14} 80 keV, (**b**) 15×10^{14} 80 keV, and (**c**) 3×10^{14} 120 keV Se^+ ion implantation [356]

estimated as usual from the momentum transfer Δq as $1/\Delta q$. The coherent length for a two-wave diffraction is determined from the FWHM of the θ–2θ, while for the Bragg-surface diffraction, the in-plane coherent length D_ψ is given [355] by

$$D_\psi = \frac{0.33\lambda}{\sin\alpha \sin\theta_L}\left(\frac{1}{W_\psi}\right) , \tag{9.59}$$

where α is the angle between the reciprocal lattice vectors $\boldsymbol{\ell}$ and $\boldsymbol{g}$, and θ_L is the Bragg angle of the secondary reflection, namely, the surface reflection. The reciprocal relation between D_ψ and W_ψ implies that the W_ψ broadening reflects the reduction of the coherence length D_ψ due to the ion implantation damage. For sample 1, $W_\psi = 196''$ corresponds to $D_\psi = 280$ nm, which is smaller than $D_\psi = 457$ nm for a perfect GaAs crystal. This confirms that the reduction in D_ψ by the lattice distortion results from the implantation. Accordingly, the $W_{\mathrm{GIXD}} = 70''$ is also broadened compared with the $45''$ of a perfect GaAs. The in-plane coherence lengths D_ψ derived from the GaAS matrix and the implanted layers are respectively 623 nm and 660 nm for sample 2 and 206 nm and 438 nm for sample 3. The coherence length for the implanted layers being longer than that for the matrix is attributed to both the release of the constraints from the matrix and the self-organization in the damaged lattice by local heating. Among the three samples, sample 2 has the longest coherence length, which indicates that most of the lattice damage does not propagate to the matrix because of the self-organization. However, the coherence length of the matrix is shortened for the higher-energy implantation in sample 3, because the lattice damage also occurs in the matrix. In view of the differences between the W_{GIXD} and W_{ψ} values, it is evident that the Bragg-suface three-wave diffraction is indeed more surface sensitive than the conventional grazing incidence two-wave diffraction.

Other investigations on the crystal perfection of porous silicon have been reported. The size of columnar region can also be determined with the same diffraction technique [355].

9.4 Multiple-Wave Diffraction Excited Fluorescence

9.4.1 Multiple X-Ray Standing-Wave Excited Fluorescence (MXSWF)

X-ray fluorescence, described in Chap. 2, is generated from a substance when the incident X-ray photon energy is at or near an absorption edge of a constituent atom of the substance. The emitted fluorescence from that atom is distributed isotropically, except for the direction along the dipole moment excited by the X-rays [712]. If the substance is of the form of a single-crystal and its angular position θ satisfies Bragg's law for a given reflection, say, G, then resonance diffraction, accompanied by X-ray fluorescence, occurs for the given atom. If the crystal happens to be a perfect crystal, then the emitted fluorescence is θ-angle-dependent, i.e., direction-dependent. This directionality of X-ray fluorescence results from the excitation of the atom by the standing-wave formed by the incident and the reflected waves during the diffraction. As described in (6.209), the fluorescence yield, proportional to the wavefield intensity I_F, is directly related to the photoelectric absorption of the atoms, the positions of the atoms, and the reflection phases. Since the absorption depends on the angle of incidence, X-ray fluorescence is also angle-dependent. This kind of photon emission is called the 'X-ray standing-wave excited fluorescence (XSW)', first demonstrated experimentally by Batterman in 1962 for a two-wave reflection geometry [35,713]. Since then, the technique of two-wave X-ray standing-wave fluorescence has been developed to determine the positions of impurity atoms, either interstitial or substitutional, in the surface/interface of semiconductor crystals [714–721], high-Tc superconductors thin films [722, 723], quasi-crystals [724], layered synthetic microstructures (LSM) or organic Langmur-Blodgett (LB) multilayers [725–731]. Detailed references to the theory and applications of X-ray standing-waves can be founded in the recent reviews [732, 733]. Since the fluorescence yield is proportional to the wavefield intensity, the stronger the wavefield intensity, the larger the fluorescence yield. As an illustration the calculated angular distribution of the wavefield intensity at various atomic positions in the (220) atomic planes is shown in Fig. 9.37 for Si (220) and MoKα_1 radiation. The reflectivity of the (220) reflection is also indicated. The crystal thickness considered in the calculation is smaller than the extinction length, so the modification on the wavefield intensities due to absorption is negligibly small. The expected distribution of the fluorescence yield is similar to what shown in Fig. 9.37, provided that the incident photon energy is sufficient to excite the impurity atoms investigated. By comparing the measured fluorescence distributions with the calculated curves, the positions of the impurity atoms relative to the (220) planes can be determined.

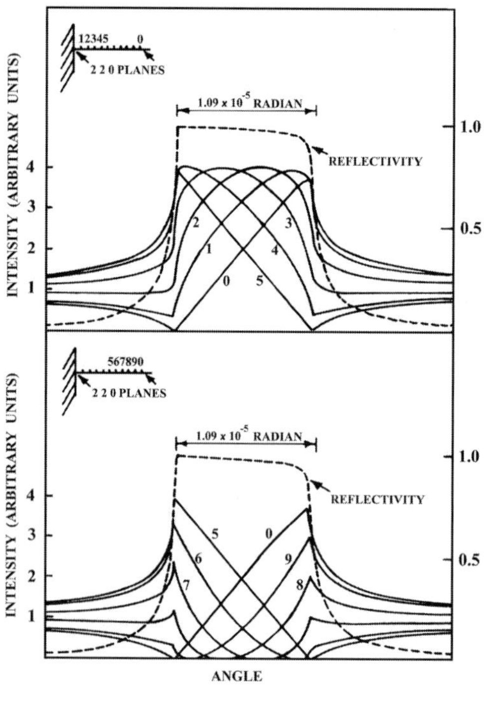

Fig. 9.37. Intensities of X-ray wavefield (*solid line*) at extended (220) lattice positions (0–9) and the reflectivity (*dashed line*) of silicon for MoKα_1 [715]

The experimental setup of measuring this type of X-ray fluorescence usually consists of a well-parallel monochromatic incident beam, a goniometer with an angular resolution of sub-arcsecond, and two detectors, one for reflectivity and the other for fluorescence. The detector for fluorescence is usually placed at a position normal to the crystal surface. This two-wave X-ray standing-wave fluorescence measurement can only give the position of impurity atoms relative to one set of atomic planes. For two- or three-dimensional information of the positions of the impurity atoms, at least two reflections are needed. The use of (111) and (220) of silicon to determine the position of the bromine atom for surface registration is an example [716].

Multiple-wave diffraction, resulting from the two- or three-dimensional periodicity of crystals, is in principle capable of extracting the two- or three-dimensional structural information for impurity atoms on a crystal surface. The three-wave (000)(111)(220) X-ray standing-wave fluorescence experiment for silicon has been proposed with predicted fluorescence distributions calculated based on (6.209) [53]. The first multiple-wave fluorescence measurement was reported by Greiser and Materlik in 1987 by using a three-wave (000)(333)(115) diffraction geometry for germanium, where (333) is parallel to the crystal surface and (115) is an asymmetric Bragg reflection [379]. The experiments were carried out with the ROEMO instrument [734–736] at the Hamburg Synchrotron Radiation Laboratory HASYLAB of DESY.

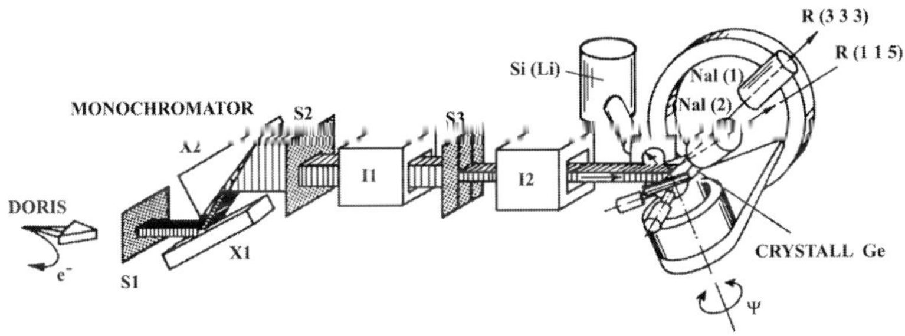

Fig. 9.38. Schematic of the experimental arrangement for three-wave XSW. S1–S3: beam collimators; X1 and X2: monochromator crystals; I1 and I2: ion chambers [379]

Figure 9.38 is the schematic of the experimental setup. The synchrotron radiation from the DORIS storage ring is monochromatized with a slightly dispersive double-crystal monochromator composed of a symmetrical Ge (440) crystal and an asymmetrical Si (333) crystal cut with an asymmetry angle of 21.5° to reduce the vertical divergence. High harmonic contamination is removed with this monochromator. The energy resolution is $\Delta E \simeq 1.5\,\mathrm{eV}$ at 12.8 keV with an emittance angle $\Delta\theta = 1.6\,\mu\mathrm{rad}$. Slits S1, S2, and S3 are used to trim the incident beam. The horizontal divergence reduced by S3 is about 30 µrad. A Ge [111] cut crystal is mounted on a goniometer, providing θ and ψ rotations for the sample. Two Na(Tl) scintillation counters are used to measure the (333) and (115) diffracted waves and an energy dispersive Si(Li) detector is used to collect the GeKα fluorescence radiation.

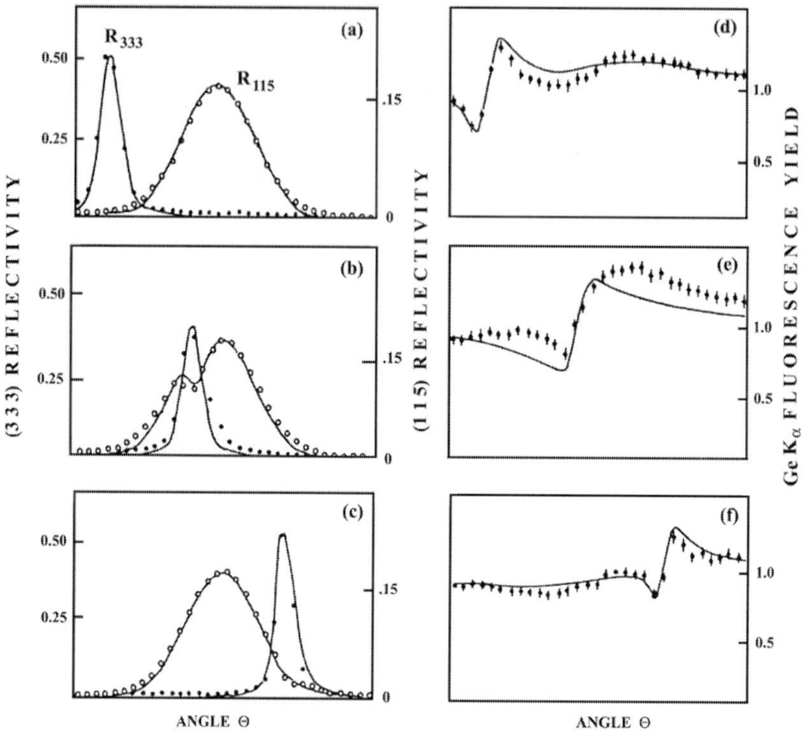

Fig. 9.39. Three-wave (000)(333)(115) rocking curves and XSW fluorescence measurements of Ge: Rocking curves relative to (115) at (**a**) $\Delta\psi = -40\,\mu$rad, (**b**) the exact three-wave position ($\Delta\psi = 0$), and (**c**) $\Delta\psi = 40\,\mu$rad; XSW fluorescence distributions of GeKα at (**d**) $\Delta\psi = -40\,\mu$rad, (**e**) the exact three-wave position ($\Delta\psi = 0$), and (**f**) $\Delta\psi = 40\,\mu$rad. (*solid line*): calculation; otherwise: measurement [379]

The measured rocking curves in the θ rotation for (115) and (333) at three ψ positions with a 40 μrad increment before, at, and after the exact three-wave diffraction point are shown in Figs. 9.39(a), (b), and (c), respectively. The corresponding fluorescence yields measured at the emittance angle $\alpha = 10°$ off the crystal surface are displayed in Figs. 9.39(d), (e), and (f). The calculated curves based on the dynamical calculation are also shown in solid lines. In Fig. 9.39(b), the two rocking curves overlap, indicating the three-wave diffraction situation. Modification in intensity for both (333) and (115) are clearly visible, which is attributed to the three-wave dynamical interaction. As to the fluorescence yield, the asymmetric intensity distributions, first decreasing then increasing, resembling the curve labelled by '0' in Fig. 9.37, are seen for (115) and (333) in Figs. 9.39(d) and (f). The distribution is slightly complicated in the exact three-wave diffraction situation (Fig. 9.39(e)) due to the three-wave interaction. Nevertheless, the intensity asymmetry is still observable. The slightly off fit between the calculated and the measured curves is due to the assumption that the structure factor of the coupling reflection (222) is zero. In any case, the asymmetric distributions of the fluorescence yield indicate that the germanium atoms are in the substitutional sites, which agrees with the known structure of germanium. Further analysis taking care of the extinction and reabsorption of the fluorescence for different α angles reconfirms the position of Ge atoms in the crystal (Fig. 9.40). Namely, in contrast to the calculated curve for interstitial Ge atoms at $\alpha = 0°$, the measured curve of the fluorescence yield is in good agreement with the calculated one for substitutional Ge atoms at $\alpha = 10°$. It seems that the measurements near the three-wave diffraction position are sufficient to give simultaneously the two-dimensional structural information of the atoms interested.

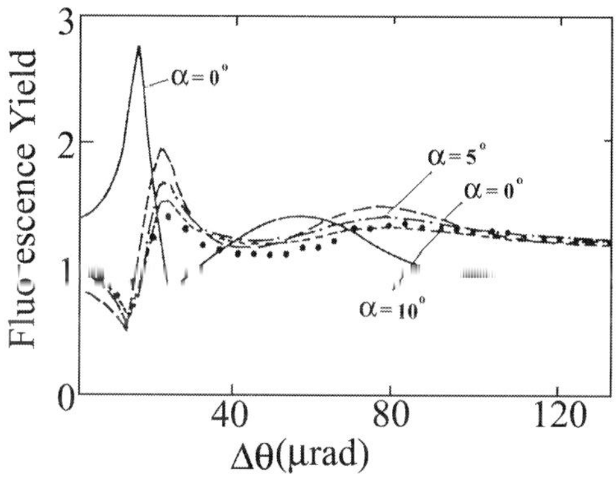

Fig. 9.40. GeKα fluorescence yield near the three-wave (000)(333)(115) diffraction position measured at the take-off angle $\alpha = 10°$ (*dotted curve*), calculated for substitutional Ge atoms at $\alpha = 0°$ (*long-dashed*), 5° (*dashed-dotted*), and 10° (*short-dashed curves*), and calculated for interstitial Ge atoms at $\alpha = 0°$ (*solid curve*) [379]

The interaction of X-rays with matter produces, through secondary processes, fluorescence, photoelectrons, Compton and thermal diffuse scattering. All these by-products can provide the structural information of the matter. During an X-ray standing-wave excitation, the photoelectrons exhibit the same angle-dependent characteristics as fluorescence due to impurity atoms in a crystal. The experiments for the three-wave standing-wave excited photoelectron spectroscopy were carried out in 1992 with the vertical wiggler beamline 14B at Photon Factory, Japan and two high-resolution monochromator systems [737, 738]. The first monochromator system consists of both vertical Si (111) and horizontal Si (220) channel-cut monochromators arranged in a non-dispersive scheme to confine the incident beam to a narrow area so as to improve the energy- and angle-resolution. The layout of the setup is similar to that shown in Fig. 9.38. The second monochromator system is composed of a vertical Si (111) channel-cut crystal followed by a Si (220) crystal plate and a Si (111) cut crystal for a six-wave transmission (Fig. 9.41). The anomalous transmission of the six-wave $(000)(220)(242)(044)(\bar{2}24)(\bar{2}02)$ diffraction in a 5 mm thick silicon perfect crystal ($\mu T \simeq 40$ for $\lambda = 1.15\,\text{Å}$) is employed to produce a high-resolution incident beam, with $\Delta E/E \simeq 1.5 \times 10^{-5}$, for the Si three-wave (000)(111)(220) standing-wave excited photoelectron measurements. Two scintillation counters are used for the (111) and (220) reflections. Photoelectrons are measured by a gas proportional counter with energy resolutions of 15% to 20%. The K-photoelectron yield, integrated over energy, with an average escape depth of about 200 nm, is selected by the discriminator.

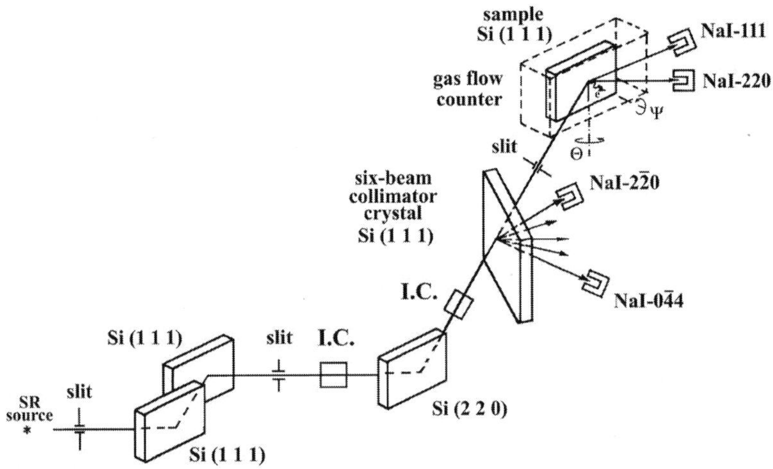

Fig. 9.41. Multi-crystal X-ray optics setup with a six-wave collimator for three-wave XSW photoelectron measurements [737, 738]

Figure 9.42 shows the angular distributions of X-ray reflectivities and photoelectron yields measured with the six-wave collimator/monochromator for (111) and (220) at $\Delta\psi = 30''$ and $\Delta\psi = -1''$ away from the exact three-wave diffraction position. Similar to that in Figs. 9.39(a) and (c), the reflectivities and photoelectron yields show two-wave characteristics at the position off the three-wave interaction point. At or near the exact three-wave diffraction position, the reflectivities are reduced considerably by the three-wave dynamical interaction due to the reallocation of intensity to the transmitted wave. In addition, the two separate asymmetric photoelectron distributions are merged into a single curve with pronounced asymmetry, which is attributed to the substitutional nature of the Si atoms and the triplet phase involving the (111), (220), and $(11\bar{1})$ reflections. The calculated

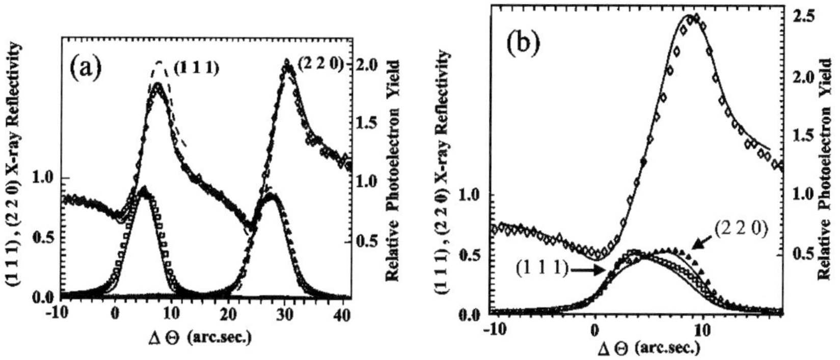

Fig. 9.42. Measured X-ray reflectivities (*lower curves*) and photoelectron yields (*upper curves*) at (**a**) $\Delta\psi = 30''$ and (**b**) $\Delta\psi = -1''$ from the exact three-wave (000)(111)(220) position [737, 738]

reflectivities and photoelectron yields are also shown in the same figure, which agree well with the measured results.

Based on the above examples, the multiple-wave excited X-ray fluorescence and photoelectron are useful tools to identify the positions of the atoms in crystalline materials.

9.4.2 Multiple-Wave Diffraction Anomalous Fine Structure (MDAFS)

X-ray scattering and X-ray absorption are closely related. The scattered intensity and the absorption coefficient are linked through the real part f' and the imaginary part f'' of the anomalous scattering amplitude by the Kramers-Krönig relations [95]. The well-developed extended X-ray absorption fine structure (EXAFS) and near-edge X-ray absorption fine structure (NEXAFS) (or X-ray absorption fine structure (XANES)) can provide short-range order structural information, such as the near-neighbor bond lengths, types, coordination number, and disorders around the specifically excited absorbing atoms for EXAFS and the valence, empty orbital and bonding information for XANES, respectively. On the other hand, X-ray diffraction can provide long-range atomic structural information. In 1992, Stragier et al. [739] and Pickering et al. [740] combined the two analysis techniques to form the so-called 'diffraction anomalous fine structure (DAFS)' method for both the short- and long-range order structure investigation of Cu single-crystals and powders, respectively. Several structural studies of using this technique have since then been reported for thin films, multi-layers, powders, and single-crystals [741–749].

In the conventional DAFS, the intensities of two-wave diffractions versus the photon energies in the vicinity of an absorption edge can be measured. The fine spectral distribution of the diffracted intensity, together with the absorption spectra, gives simultaneously atom-site and wavevector selectivities, which cannot be obtained from the usual EXAFS and NEXAFS. That is to say, DAFS can provide EXAFS- and XANES-like information for a specific subset of atoms selected by the diffraction condition. In addition, DAFS gives atomic scattering factor-like spectroscopic and structural information for the inequivalent sites of a single atomic species within a crystal unit cell. The final goal of DAFS is to obtain the site-dependent spectral distributions of f' and f'' of the excited atom. As is known, the scattering amplitudes from atoms in solids consist of a smooth and an oscillating fine structure. The former is due to a bare free atom, while the latter is from the modulation caused by the scattering from the surrounding atoms. That is, the atomic scattering factor for a given momentum transfer $\boldsymbol{q}$ and photon energy E is

$$f(\boldsymbol{q},E) = f_{OA}(\boldsymbol{q}) + \Delta f'_{OA}(E) + \mathrm{i}\Delta f''_{OA}(E) + \Delta f''_{OA}(E)\chi_A \,, \qquad (9.60)$$

where

$$\chi_A(E) = \chi'_A(E) + \mathrm{i}\chi''_A(E)\,. \qquad (9.61)$$

The first three terms on the right of (9.60) are the contributions from the bare atom, the normal atomic scattering factor, and the real and the imaginary dispersion correction. The last term, analogous to the fine structure function of EXAFS, is related to the oscillating fine structure χ_A function, whose real part χ'_A and imaginary part χ''_A can be defined from (9.60) as

$$\chi'_A(E) = \frac{\Delta f'_{\mathrm{exp}}(E) - \Delta f'_{OA}(E)}{\Delta f''_{OA}(E)}\,,$$
$$\chi''_A(E) = \frac{\Delta f''_{\mathrm{exp}}(E) - \Delta f''_{OA}(E)}{\Delta f''_{OA}(E)}\,. \qquad (9.62)$$

$\Delta f'_{\mathrm{exp}}$ and $\Delta f''_{\mathrm{exp}}$ are the anomalous dispersion corrections obtained directly from DAFS and EXAFS.

The diffracted intensity of the Bragg reflection G, as in the kinematical approach, is given by

$$I(\boldsymbol{q}_G,E) = S \cdot D(E) \cdot A(\boldsymbol{q}_G,E) \cdot Lp(\boldsymbol{q}_G,E)\left|F(\boldsymbol{q}_G,E)\right|^2 \,, \qquad (9.63)$$

where S is a scale factor, D the experimental correction, A the absorption correction, and Lp the Lorentz-polarization factor. For a first-order extended DAFS (EDAFS), i.e., in a wider energy range, the structure factor takes the form

$$F(\boldsymbol{q}_G,E) = F_0\mathrm{e}^{\mathrm{i}\delta_0} + \sum_{j=1}^{N_A} \Delta f''_{OA}\left|\alpha_{Aj}\right|\mathrm{e}^{\mathrm{i}\delta_{Aj}}\chi_{Aj}\,, \qquad (9.64)$$

where the first term on the right covers the scattering of all non-resonant atoms and the Thomson scattering from the resonant atom A. The second term is related to the fine structure resulting from all the resonant A atoms at different site j, and

$$\alpha_{Aj} = C_{Aj} \exp\left(-M_{Aj} q^2\right) \exp\left(\mathrm{i}\delta_{Aj}\right) , \tag{9.65}$$

where C_{Aj} is the occupation factor, the probability for the resonant A atom occupying the j-site. M_{Aj} is the Dedye-Waller factor and $\delta_{Aj} = \boldsymbol{q}_G \cdot \boldsymbol{r}_{Aj}$, $\boldsymbol{r}_{Aj}$ being the position vector of the j-site in the unit cell.

There are two ways to extract the diffraction fine structure: the direct spline method and the iterative Kramer-Krönig method [741]. In the direct spline method, the DAFS data are splined and normalized, then the spline is removed and used as a reference. This normalized fine structure yields

$$\begin{aligned} \chi_q(k) &= \frac{|F_0|}{2\,|\alpha_A|\,\Delta f''_{OA}} \frac{I_{\mathrm{exp}} - I_{0\mathrm{exp}}}{I_{0\mathrm{exp}}} \\ &= \sum_{j=1}^{N_A} w_{Aj} \left[\cos\left(\delta_0 - \delta_{Aj}\right) \chi'_{Aj} + \sin\left(\delta_0 - \delta_{Aj}\right) \chi''_{Aj}\right] , \end{aligned} \tag{9.66}$$

where $k = \hbar[2m(E - E_0)]^{1/2}$ is the modulus of the photoelectron wavevector and $w_{Aj} = |\alpha_{Aj} / \sum_j \alpha_{Aj}|$, E_0 being the binding energy. The first line of the equation is directly related to the corrected measured two-wave diffraction intensity I_{exp} and the smooth diffraction background $I_{0\mathrm{exp}}$ at and off the resonance energies, respectively. The second line of the equation has been derived from (9.63) and (9.60) under a first-order approximation.

A pair of $\Delta f'$ and $\Delta f''$ can be found by the iterative Kramer-Krönig method, which give the best fit to the measured intensity data. Iteration by using the $\Delta f'$ and $\Delta f''$ values of Cromer-Liberman [750] as starting values is necessary to find a correct pair of $\Delta f'$ and $\Delta f''$.

Local structural information can be extracted from the isolated fine structure obtained from one of the two methods mentioned by using the algorithm FEFF or the like [741], which is similar to the EXAFS analysis.

The data analysis procedures of the DAFS mentioned are, however, rather complicated, so only a few research groups who developed the algorithms could perform the analysis. On the other hand, the accuracy of the results obtained from the Hilbert transformation with the Kramers-Krönig relations for truncated measured data may not be guaranteed. For this practical reason, a multiple-wave DAFS (MDAFS) has recently been proposed, where a simple analysis procedure is given and the useful phase information of the structure-factor multiplets is considered [751]. The MDAFS experiment in a three-wave diffraction geometry is briefly described below.

Consider the three-wave $(000)(\bar{2}\bar{2}\bar{2})(\bar{1}3\bar{3})$ diffraction of a GaAs $(\bar{1}\bar{1}\bar{1})$ cut crystal, with the As rich surface facing the X-rays. The $(\bar{2}\bar{2}\bar{2})$ is a symmetric Bragg reflection and $(\bar{1}3\bar{3})$ an asymmetric Bragg reflection. Two scintillation counters are used; one is to measure the $(\bar{2}\bar{2}\bar{2})$ diffracted intensity and the other, placed normal to the crystal surface, to detect the fluorescence yield of AsK. The experimental setup and conditions are the same as that for the resonance phase measurements described in Sect. 9.1.5 [323]. The energy range used covers the AsK edge and the energy resolution is about 0.2 eV. The fluorescence yield and the three-wave diffraction profile for each photon energy are measured. Figure 9.43(a) shows the Renninger scan for $E = 11.866\,\mathrm{keV}$, where the average background is the two-wave intensity of $(\bar{2}\bar{2}\bar{2})$. The asymmetric profile results from the crystallographic phase effect. The intensities at maximum, minimum, and the two-wave diffraction intensity, which determine the visibility R_v of the asymmetry, are plotted against the photon energy in Fig. 9.43(b). Here the visibility, closely related to the triplet phase δ_3, is defined experimentally as

$$R_v = \frac{I_2 - I_{\mathrm{min}}}{I_{\mathrm{max}} - I_2}\,, \tag{9.67}$$

where I_2 is the two-wave intensity background and I_{max} and I_{min} are the maximum and minimum intensities of the three-wave profile.

That the phase-dependence of R_v is closely related to the f' and f'' makes R_v a sensitive parameter for the determination of the spectral distributions of f' and f''. Since f'' can be directly obtained from the absorption spectrum derived from the fluorescence yield after proper absorption corrections, the corresponding f' can be estimated from R_v. The R_v and f'' versus photon energy are shown in Figs. 9.44(a) and 9.45(a), respectively. With the aid of

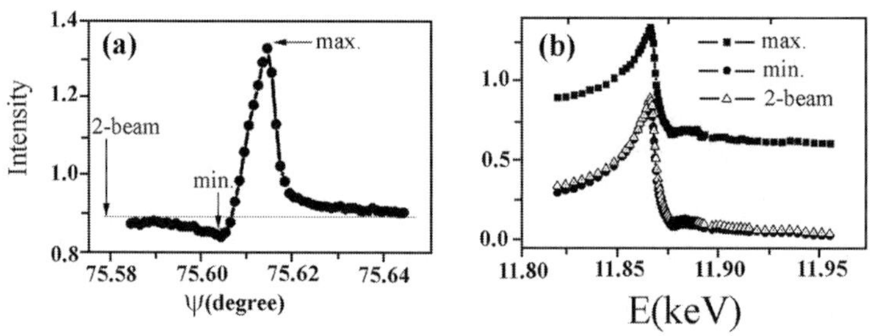

Fig. 9.43. (**a**) Measured GaAs $(\bar{2}\bar{2}\bar{2})$ intensity profile at the three-wave (000) $(\bar{2}\bar{2}\bar{2})(\bar{1}3\bar{3})$ diffraction position. (**b**) Diffraction intensities at the maximum, minimum, and two-wave background of (**a**) versus X-ray photon energy E [751]

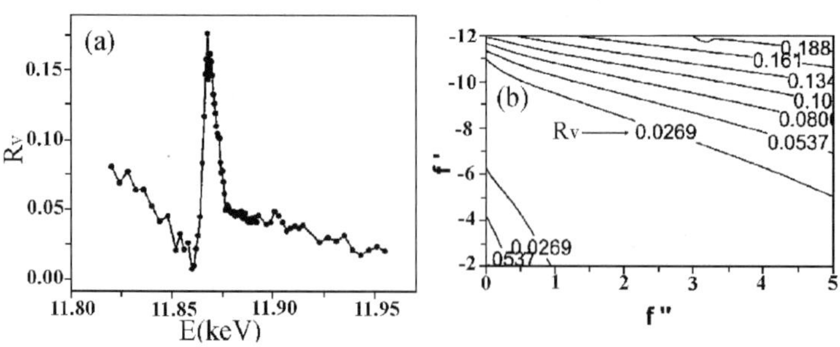

Fig. 9.44. (**a**) Measured intensity ratio R_v versus photon energy. (**b**) The calculated R_v versus f' and f'' for Fig. 9.43 [751]

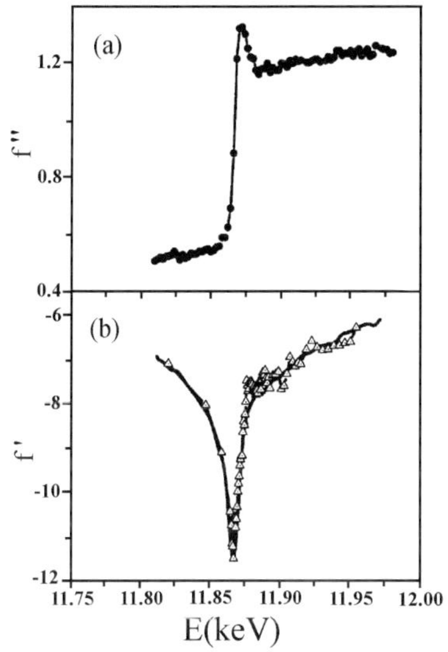

Fig. 9.45. Measured f'' and f' from the fluorescence yield and intensity ratio R_v from Fig. 9.43, respectively [751]

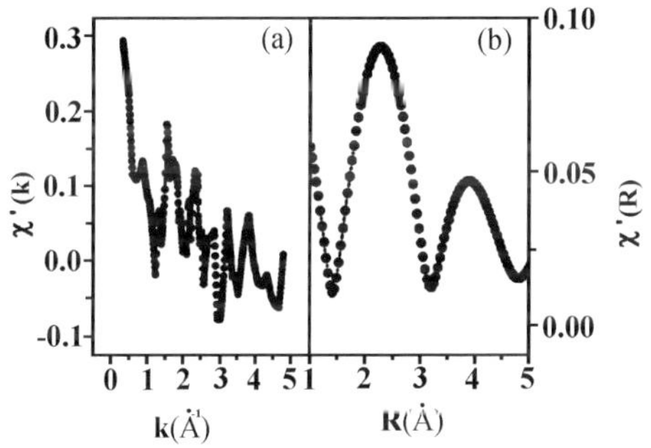

Fig. 9.46. (**a**) The real part of the fine structure function in k-space and (**b**) the radial distribution function $\chi'(R)$ [751]

dynamical calculation, the equi-intensity ratio curves for various combinations of f' and f'' are plotted in Fig. 9.44(b). For a given f'' at energy E, the f' can be easily found in the equi-intensity plot. The determined f' is shown in Fig. 9.45(b). The corresponding fine structure function $\chi(k)$ can be deduced from Fig. 9.45 according to (9.62). Figure 9.46(a) shows the real part of this function. The Fourier transform of $\chi'(k)$ gives the radial distribution function $\chi'(R)$, which yields the distance between the arsenic and the nearest gallium to be about 1.50 Å, very close to the known value, 1.48 Å.

The site selectivity can be easily implemented in the input structure factors for the dynamical calculation. For thin films and layered epitaxial materials, the Born or Bethe approximations can be employed to give the kinematical intensity ratios R_v, which could yield the spectral distributions for f' and f''.

9.5 Studies of Phase Transition/Transformation

Single-crystals under external influences, such as the applied electric and magnetic fields, temperature gradient, pressure gradients, etc., may undergo phase transformation and transition and result in structural modification. Accompanied with the modification, some physical properties of the crystals may also be changed. Since multiple-wave diffraction is capable of providing three-dimensional structural information and is extremely sensitive to small structural variations, many applications of this diffraction technique to study phase transition and transformation have been pursued and reported. The determination of piezoelectric constants for nonlinear optical materials [381, 382] will be described as an example.

9.5.1 Determination of Piezoelectric Constants for Nonlinear Optical Materials

The nonlinear optical properties of chiral organic crystals have been intensively studied in recent years for the use of second harmonic generation and signal modulation by linear electro-optic effects. For this purpose, large crystals with high perfection have been grown. Due to the chirality of their structures, these crystals may also exhibit piezoelectric or pyroelectric effects, leading to the optoacoustic and photorefractive phenomena.

An organic crystal, meta-nitroaniline (mNA), under an applied electric field is subject to X-ray multiple-wave diffraction for measuring piezoelectric constants. The crystal belongs to the orthorhombic system with point group mm2 and lattice constants $a = 6.501$ Å, $b = 19.330$ Å, and $c = 5.082$ Å. The polar axis is along [001].

Strains are developed when the crystal is under a DC voltage or a low-frequency electric field, which indicates an inverse piezoelectric effect. The relation between the strains ε, the piezoelectric constants d's, and the applied electric field E for the mm2 point group is given by [752]

$$\begin{bmatrix} \varepsilon_{xx} \\ \varepsilon_{yy} \\ \varepsilon_{zz} \\ 2\varepsilon_{yz} \\ 2\varepsilon_{zx} \\ 2\varepsilon_{xy} \end{bmatrix} = \begin{bmatrix} 0 & 0 & d_{31} \\ 0 & 0 & d_{32} \\ 0 & 0 & d_{33} \\ 0 & d_{24} & 0 \\ d_{15} & 0 & 0 \\ 0 & 0 & 0 \end{bmatrix} \begin{bmatrix} E_x \\ E_y \\ E_z \end{bmatrix}, \qquad (9.68)$$

where the $\boldsymbol{x}$, $\boldsymbol{y}$, and $\boldsymbol{z}$ are along the basis vectors $\boldsymbol{a}$, $\boldsymbol{b}$, and $\boldsymbol{c}$ and the Voigt notation is used for d's [381,382]. When the electric field is applied along the polar axis $\mathbf{c}$. i.e., $E_x = E_y = 0$ and $E_x = E$, (9.68) leads to

$$\begin{aligned}\varepsilon_{xx} &= d_{31}E = \frac{\Delta a}{a}\,,\\ \varepsilon_{yy} &= d_{32}E = \frac{\Delta b}{b}\,,\\ \varepsilon_{zz} &= d_{33}E = \frac{\Delta c}{c}\,.\end{aligned} \tag{9.69}$$

The variation in the lattice constants can be determined from the angular positions of the multiple-wave diffraction, as described in Sect. 9.3.1. Referring to (4.5), we can write the geometric condition for a three-wave (O, G, L) diffraction occurring in a orthorhombic crystal as

$$\cos\beta = \frac{1}{2}\frac{\ell^2 - \ell_G}{\sqrt{\frac{1}{\lambda^2} - \frac{g^2}{4}}\sqrt{\ell^2 - \frac{\ell_G^2}{g^2}}}\,, \tag{9.70}$$

where

$$\begin{aligned}\ell^2 &= \left(\frac{h_L}{a}\right)^2 + \left(\frac{k_L}{b}\right)^2 + \left(\frac{\ell_L}{c}\right)^2,\\ g^2 &= \left(\frac{h_G}{a}\right)^2 + \left(\frac{k_G}{b}\right)^2 + \left(\frac{\ell_G}{c}\right)^2,\\ \ell_G &= \frac{h_L h_G}{a^2} + \frac{k_L k_G}{b^2} + \frac{\ell_L \ell_G}{c^2}\,.\end{aligned} \tag{9.71}$$

The subscripts G and L refer to the primary reflection G and the secondary reflection L. By properly choosing the primary and secondary reflections, the derivatives of (9.70) with respect to the lattice constants a, b, and c yield the following simple expressions:

$$\begin{aligned}\frac{\Delta a}{a} &= \tan\beta_{h_L00}\Delta\beta_{h_L00} - \frac{\lambda^2 k_G^2}{4b^2 - \lambda^2 k_G^2}\frac{\Delta b}{b}\,,\\ \frac{\Delta c}{c} &= \tan\beta_{00\ell_L}\Delta\beta_{00\ell_L} - \frac{\lambda^2 k_G^2}{4b^2 - \lambda^2 k_G^2}\frac{\Delta b}{b}\,,\\ \frac{\Delta b}{b} &= -\cot\theta_G\Delta\theta_G\,,\end{aligned} \tag{9.72}$$

where the first two equations are obtained, respectively, from the two secondary reflections (h_L00) and $(00\ell_L)$ with a common primary reflection $(0k_G0)$. The third one is obtained from the variation of the Bragg law for the primary reflection $(0k_G0)$ in relation to b.

The piezoelectric constants can be estimated straightforwardly from (9.69) and (9.72), if the changes in $\Delta\beta$, namely, the angular position ψ, can be determined from multiple-wave diffraction experiments.

The multiple-wave diffraction experiments were carried out at the synchrotron radiation source (SRS) wiggler station 16.3, Daresbury Laboratory, U.K.. The angular precision is 0.0001° in θ and 0.0005° in ψ. The samples are rectangular crystal plates with their sides parallel to the $\boldsymbol{a}$, $\boldsymbol{b}$, and $\boldsymbol{c}$.

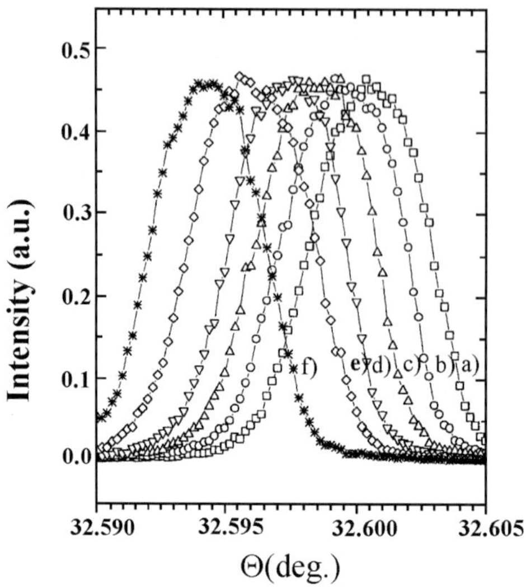

Fig. 9.47. Rocking curves of mNA (0 14 0) for various applied electric fields: (**a**) 0, (**b**) 2×10^5, (**c**) 4×10^5, (**d**) 6×10^5, (**e**) 8×10^5, and (**f**) 10×10^5 V/m [381]

The larger face with an area of about 20 mm × 10 mm is parallel to the (001) planes and the thickness along the [001] direction ranges from 2 to 5 mm. The electric contacts are established with conductive sponges placed between a metal plates and the sample's larger face. A uniform electric field within the crystal is established. The symmetric (0 14 0) Bragg reflection was chosen as the primary reflection for the wavelength $\lambda = 1.4878$ Å. Renninger scans for the three-wave cases (000)(0 14 0)(200) and (000)(0 14 0)(002) and the rocking curve measurements of the symmetric Bragg reflections (080) and (0 14 0) were performed for various electric fields, from 0 to 1.2×10^6 V/m, applied in the [001] direction. Figures 9.47 and 9.48 show the measured rocking curves of (0 14 0) and the three-wave (000)(0 14 0)(002) diffraction profiles, respectively. As clearly seen in Fig. 9.47, the peak positions of the rocking curves shift while the profile shape of the curves remains undistorted. This evidences the lattice strain produced by the piezoelectric effect. The reversible change in lattice constants with respect to the applied electric field has also been detected in these samples. In Fig. 9.48, both the peak position and shape are changed for different E-fields.

The $\Delta b/b$ estimated from the rocking curves and the $\Delta a/a$ and $\Delta c/c$ measured from the three-wave diffraction profiles, based on (9.72), are plotted as a function of the applied electric field in Fig. 9.49. As expected for the piezoelectric effect, the linearity in all the strain versus E plots is observed. The corresponding piezoelectric constants are then determined via (9.69) as $d_{31} = 7.3(1) \times 10^{-11}$ m/V, $d_{32} = 16.5(7) \times 10^{-11}$ m/V, and $d_{33} = 10.3(8) \times 10^{-11}$ m/V. These measured values are larger than those

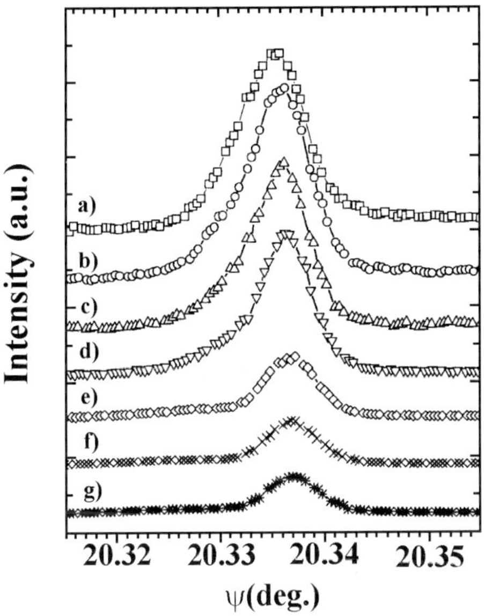

Fig. 9.48. Three-wave (000)(0 14 0)(002) diffraction profiles of mNA for various electric fields: (**a**) 0, (**b**) 1.2×10^5, (**c**) 2.4×10^5, (**d**) 3.6×10^5, (**e**) 4.8×10^5, (**f**) 6×10^5, and (**g**) 12×10^5 V/m [381]

obtained in the measurement of the direct piezoelectric effect [753]. This is because electrode-area effects in the sponges made of polymeric materials can introduce systematic difference between the inverse and direct piezoelectric effects [754].

The same technique has been successfully applied to a monoclinic MBANP [(-)-2-(α-methylbenzylamino)-5-nitropyridine] nonlinear organic crystal and L-arginine hydrochloride monohydrate for the determination of tensor elements of piezoelectric coefficients [382, 385]. Investigation of the habit modification of nearly perfect crystals of KDP by trivalent manganese ions with the same technique has also been pursued [384]. Along the same line of research, the real-time measurements of transverse and longitudinal lattice deformation induced by shock-waves in lithium fluoride single-crystals by using this multiple diffraction technique have recently been demonstrated [755]. It was found that in the measurements with the three-wave (000)(111)(220) and (000)(200)(220) diffractions for $\lambda = 1.54$ Å, the shock compression below 4 GPa along the [111] direction produced no transverse lattice deformation, while the shock compression along the [100] orientation produced equal interplanar spacing changes along the longitudinal and transverse directions with respect to the applied shock-wave.

Other applications include the study of structural change in GaAs crystal under high electric fields [756] and the investigation of the transport property

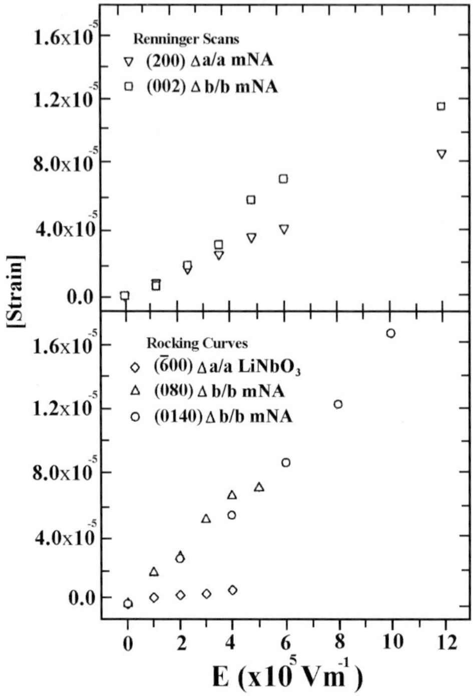

Fig. 9.49. Strain versus the electric field measured from (**a**) three-wave diffractions and (**b**) two-wave rocking curves. The data from $LiNbO_3$ are for comparison [381]

and the internal structure of charge density waves in $K_{0.3}MoO_3$ with an external electric field [757].

9.6 Multiple-Wave X-Ray Optics

Multiple-wave diffraction, as described in Chaps. 3 and 4, involves two angular parameters, i.e., the Bragg angle θ of the primary reflection and the azimuthal angle ψ of rotation around the primary reciprocal lattice vector. In order to generate a multiple-wave diffraction, two diffraction conditions need to be fulfilled: the Bragg law for the primary reflection and the requirement given by (4.5). The latter imposes a more strict constraint on the multiple-wave X-ray optics than on two-wave diffraction cases. Hence, multiple diffraction can, in principle, provide better beam conditioning for the incident and scattered waves in terms of monochromaticity, beam divergence, variable polarization, etc.. The applications derived from these characteristics, such as multiple-wave X-ray monochromator, phase plate, and interferometry, are described below.

9.6.1 Multiple-Wave X-Ray Monochromators

Monochromatic X-radiation is usually desired in most X-ray diffraction and spectroscopy experiments, with either classical Bremsstrahlung or synchrotron radiation. Single- , double-, multi-, and curved-crystal monochromators with flat, channel-cut, or asymmetrically cut crystals are frequently used in two-wave diffraction geometries for this purpose [55,389,547,758–764]. In contrast to a conventional two-wave diffraction monochromator, the dynamical interaction in a multiple-wave monochromator is usually confined to a very small angular range, thus providing an ideal monochromatic incident beam. Multiple-wave monochromators were first proposed by Kottwitz [367–369] and Deslattes [765]. Kottwitz suggested use Umweg-type multiple-wave diffractions for monochromatizing high-flux thermal neutrons. The beam conditioning of a Bragg-type Umweg three-wave (O, G, L) diffraction monochromator relies not only on the Bragg's law of the primary reflection G, but also on the condition given in (4.5), which can be rewritten as

$$\cos\beta = \frac{\lambda\left(\ell^2 - \ell_p g\right)}{2\ell_n \cos\theta_G} \,, \tag{9.73}$$

where g and ℓ are the moduli of the reciprocal lattice vectors of the G and L reflections, respectively (see Fig. 4.3). ℓ_p and ℓ_n are the components of ℓ parallel and perpendicular to $\boldsymbol{g}$; θ_G is the Bragg angle of the G reflection. The β angle, defined in Fig. 4.3 and (4.4), is related to the azimuthal angle ψ around $\boldsymbol{g}$.

The beam parameters of a monochromator include the angular width of acceptance for the incident wave, $W_{\theta,O}$, the angular width of reflection for the diffracted wave, $W_{\theta,G}$, and the spatial widths S_O and S_G of the incident and diffracted waves. For a three-wave monochromator, the angular positions and widths of the incident and the diffracted waves can be directly obtained from the first-order Bethe approximation described in Chap. 7. Similar to that of a two-wave monochromator, the angular width of acceptance $W_{\theta,O}$ and the width of reflection $W_{\theta,G}$ for a three-wave monochromator, according to (7.125), are given by

$$W_{\theta,O} = \frac{2\,|\chi_{GO,LO}|}{\sqrt{b}\sin 2\theta_G} \,, \qquad W_{\theta,G} = \frac{2\sqrt{b}\,|\chi_{GO,LO}|}{\sin 2\theta_G} \,, \tag{9.74}$$

where b is the asymmetry parameter. $\chi_{GO,LO}$ is the modified polarizability defined in (7.66) as

$$\chi^{\sigma}_{GO,LO} = P_\sigma \chi_{G\text{–}O} - \frac{P_1 \chi_{L\text{–}O}\chi_{G\text{–}L}}{2\xi_L} \,. \tag{9.75}$$

For simplicity, we only consider the σ polarization here, thus $P_\sigma = 1$ and $P_1 = d_1^2 + d_2^2$ as given in (7.68). The spatial widths S_O and S_G of the incident and primary diffracted waves satisfy the relation

$$S_O = \frac{S_G}{b} , \tag{9.76}$$

so that Liouville's theorem holds.

In addition, the corresponding angular widths in ψ, $W_{\psi,O}$ and $W_{\psi,G}$, need to be known as well. According to the Bethe approximation, the diffracted intensity distribution I_G of the primary reflection G near the three-wave diffraction position is proportional to the square of the absolute value of $\chi_{GO,LO}$, which is a function of $\Delta\theta$ and $\Delta\psi$ deviated from the exact three-wave diffraction point. In turn, the width $W_{\psi,O}$ can be found from I_G. For simplicity, a further assumption is made that ξ_L depends solely on $\Delta\psi$, according to (7.61),

$$\xi_L = \frac{\chi_O}{2} + \zeta\gamma_L - (\Delta\psi)\cos\theta_G \sin\beta , \tag{9.77}$$

where $\Delta\psi = -\tan\Psi_T$ and $\sin\beta = \cos\eta_L \sin\phi_L$ as in (7.61). This assumption implies that the azimuth rotation is set at $\Delta\theta = 0$. Because it is not possible to obtain an analytical expression for the accommodation ζ for a general three-wave diffraction, the approximation

$$\zeta = C_1 |\chi_O| + \mathrm{i}C_2 |\chi_O| \tag{9.78}$$

is adopted, where the coefficients C's, depending on the strengths of the reflections involved, are usually less than 1. By substituting (9.77) and (9.78) into (9.75), the intensity I_G can be expressed as

$$\begin{aligned} I_G &\propto \left|\chi^{\sigma}_{GO,LO}\right|^2 \\ &= \left|\chi_{G\text{-}O}\right|^2 - \frac{G_o \left|\chi_{O\text{-}L}\chi_{G\text{-}O}\chi_{L\text{-}G}\right| \left(y\cos\delta_3 + C_2\gamma_L |\chi_O| \sin\delta_3\right)}{2\left(y^2 + \gamma_L^2 C_2^2 |\chi_O|^2\right)} \\ &\quad + \frac{G_o^2 \left|\chi_{L\text{-}O}\right|^2 \left|\chi_{L\text{-}G}\right|^2}{4\left(y^2 + \gamma_L^2 C_2^2 |\chi_O|^2\right)^2} , \end{aligned} \tag{9.79}$$

where

$$\begin{aligned} G_o &= \cos^2\beta , \\ y &= \left(\frac{1}{2} + C_1\gamma_L\right) |\chi_O| - (\Delta\psi)\cos\theta_G \sin\beta , \end{aligned}$$

and δ_3 is the triplet phase. Consequently, the width $W_{\psi,O}$ is given by

$$W_{\psi,O} = \frac{\sqrt{3}\,|\gamma_L|\, C_2 \,|\chi_O|}{\cos\theta_G \sin\beta} . \tag{9.80}$$

To determine the width $W_{\psi,G}$ of divergence, the reversed three-wave (O', G', L') situation needs to be considered. According to the optical theorem of reciprocity, the incident wave O' is the diffracted wave G, and the diffracted wave G' is the incident O. If the (O, G, L) is a Bragg-Bragg case, then the reversed (O', G', L') is a Bragg-Laue case. The secondary L'–O' Laue diffraction becomes the coupling $G-L$ reflection. Hence $\gamma_{L'} = -\gamma_L$. Since the two cases are equivalent to each other according to the reciprocity theorem, the azimuthal width of divergence $W_{\psi,G}$ takes the same form as $W_{\psi,O}$:

$$W_{\psi,G} = \frac{\sqrt{3}\,|\gamma_L|\,C_2\,|\chi_O|}{\cos\theta_G \sin\beta} \,. \tag{9.81}$$

The relationship between the angular width, resolution, and dispersion is governed by the Bragg's law and the multiple-wave diffraction condition, (4.5). The derivatives of the two equations give

$$\begin{aligned}\frac{\Delta\lambda}{\lambda} &= (\Delta\theta)\cot\theta_G \qquad \text{and}\\ \frac{\Delta\lambda}{\lambda} &= -(\Delta\theta)\tan\theta_G - (\Delta\beta)\tan\beta\,,\end{aligned} \tag{9.82}$$

which can be represented in the λ-β-θ DuMond diagram, shown in Fig. 9.50. The hatched area is the allowed region in the phase space for a three-wave diffraction monochromator. There are two equations for three variables, i.e., $\Delta\lambda$, $\Delta\theta$, and $\Delta\beta$, in (9.82). To determine the variables, one more relation is needed. This relation comes from the beam divergence of the incident radiation. In the case of synchrotron radiation, the beam divergence is defined by the photon emittance and the beam slits. In general, the variation in the θ angle is conventionally assumed to be along the vertical y-direction perpendicular to the orbital plane of the storage ring, while the β axis is in the x-direction in the orbital plane normal to the orbital trajectory. Since the trajectories of synchrotron photons are ellipses in the vertical $y-y'$ phase space and hyperbolae in the horizontal $x-x'$ phase space (see Fig. 2.2), the three-wave monochromator can be considered as an instrument to transform any point (x_O, x'_O) and (y_O, y'_O) to point (x_G, x'_G) and (y_G, y'_G) by following the

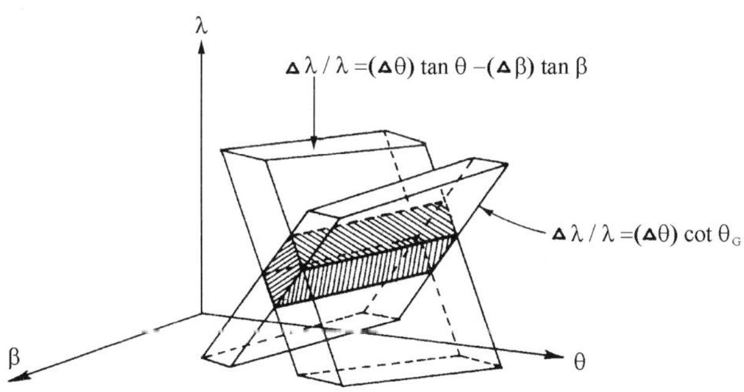

Fig. 9.50. The λ-β-θ DuMond diagram for a three-wave monochromator [766]

relations given by (9.82). The characteristics of the three-wave monochromator with appropriate transformation matrices should be represented in the λ-$x(S_{\psi,O})$-$x'(\Delta\beta)$ and λ-$y(S_{\theta,O})$-$y'(\Delta\theta)$ diagrams together with the λ-$x'(\Delta\beta)$-$y'(\Delta\theta)$ diagram. The quantity $y(S_{\theta,O})$ means that the position y depends on the beam size $S_{\theta,O}$ in the θ plane, namely, y plane, of the incident wave O. Other quantities have similar meanings to $y(S_{\theta,O})$. For illustration, let's consider the transformation from (x_O, x'_O) to (x_G, x'_G). According to (7.122), the peak position is deviated from the Bragg angle θ_G for the incident wave by $\Delta\theta_O$,

$$\Delta\theta_O = \theta_o - \theta_G = \frac{(1-b)\chi_{O,r}}{b\sin 2\theta_G} + \frac{C_{\sigma G,r}}{2\sin 2\theta_G} - \frac{C_{\sigma O,r}}{2b\sin 2\theta_G}\,, \tag{9.83}$$

where θ_o is the incident angle with respect to the atomic planes G. The peak position is deviated from θ_G for the diffracted G wave by $\Delta\theta_G$, which can be obtained in a similar way and is given by replacing b by $1/b$ in (9.83) for an asymmetrically cut crystal as

$$\Delta\theta_G = \theta_g - \theta_G = \frac{(b-1)\chi_{O,r}}{\sin 2\theta_G} + \frac{C_{\sigma G,r}}{2\sin 2\theta_G} - \frac{bC_{\sigma O,r}}{2\sin 2\theta_G}\,, \tag{9.84}$$

where θ_g is the diffracted angle with respect to the atomic planes G. The relation between the two peak positions $\Delta\theta_O$ and $\Delta\theta_G$ is

$$\Delta\theta_G = A\Delta\theta_O\,, \tag{9.85}$$

where

$$A = \frac{\dfrac{(b-1)\chi_{O,r}}{\sin 2\theta_G} + \dfrac{C_{\sigma G,r}}{2\sin 2\theta_G} - \dfrac{bC_{\sigma O,r}}{2\sin 2\theta_G}}{\dfrac{(1-b)\chi_{O,r}}{b\sin 2\theta_G} + \dfrac{C_{\sigma G,r}}{2\sin 2\theta_G} - \dfrac{C_{\sigma O,r}}{2b\sin 2\theta_G}}\,. \tag{9.86}$$

Let $x'_O = \Delta\theta_O$, $x'_G = \Delta\theta_G$, $x_O = S_O$, $x_G = S_G$ and consider the angular increments $\Delta x'_O = \Delta\theta$ and $\Delta x'_G = \Delta\theta$. The total angular variations of the incident and diffracted waves must satisfy (9.85), namely,

$$x'_G - \Delta x'_G = A\,(x'_O - \Delta x'_O)\,, \tag{9.87}$$

with

$$\Delta x'_G = \Delta x'_O = \Delta\theta = -\left[(\Delta\beta)\tan\beta + \left(\frac{\Delta\lambda}{\lambda}\right)\right]\cot\theta_G\,, \tag{9.88}$$

derived from (9.82). From (9.76) and (9.88), the transformation matrix takes the form

$$\begin{bmatrix} x_G \\ x'_G \\ \frac{\Delta\lambda}{\lambda} \end{bmatrix} = \begin{bmatrix} \frac{1}{|b|} & 0 & 0 & 0 \\ 0 & A & (A-1)\tan\beta\cot\theta_G & (A-1)\cot\theta_G \\ 0 & 0 & 0 & 1 \end{bmatrix} \begin{bmatrix} x_O \\ x'_O \\ \Delta\beta \\ \frac{\Delta\lambda}{\lambda} \end{bmatrix}. \tag{9.89}$$

The spread angle Ω_y of the photon emittance in the vertical y direction is much larger than the intrinsic width of the two-wave dynamical diffraction. Therefore, $\Delta\theta$ in (9.82) can be substituted by $\Omega_y + W_{\theta,O}$. This leads to

$$\frac{\Delta\lambda}{\lambda} = (\Omega_y + W_{\theta,O}) \cot\theta_G \,,$$
$$\Delta\beta = -(\Omega_y + W_{\theta,O}) \frac{\tan\theta_G + \cot\theta_G}{\tan\beta} \,. \qquad (9.90)$$

Consequently, the angular widths W_β in β for both the O and G waves are

$$W_{\beta,O} = (\Omega_y + W_{\theta,O}) T_G + W_{\psi,O} \,,$$
$$W_{\beta,G} = (\Omega_y + W_{\theta,G}) T_G + W_{\psi,G} \,, \qquad (9.91)$$

where

$$T_G = \frac{\tan\theta_G + \cot\theta_G}{\tan\beta} \,. \qquad (9.92)$$

The position-angle-wavelength scheme, namely, the λ-$y(S_{\theta,O})$-$y'(\Delta\theta)$ diagram, is shown in Fig. 9.51 as an illustration. The useful region is the shaded volume confined by the Bragg's law of the primary G reflection, the second equation in (9.82), and the elliptical trajectory of the photon emittance. Similar volume can be constructed for the λ-$x(S_{\psi,O})$-$x'(\Delta\beta)$ diagram, where the photon emittance shows a hyperbola trajectory. The constraints imposed on θ and β give the X-ray beam a better degree of parallelism than the beam from a two-wave monochromator.

The above qualitative description of multiple-wave monochromators is based on the Born or Bethe approximation, which is valid for situations off the exact multiple-wave diffraction position. In reality, the monochromator is desired to be operated under its optimal conditions, for example, with maximum diffraction intensity and highest resolution. Therefore, in order to provide quantitative analysis of the monochromator, the situations at the exact multiple diffraction point have to be considered and the exact dynamical calculation needs to be used.

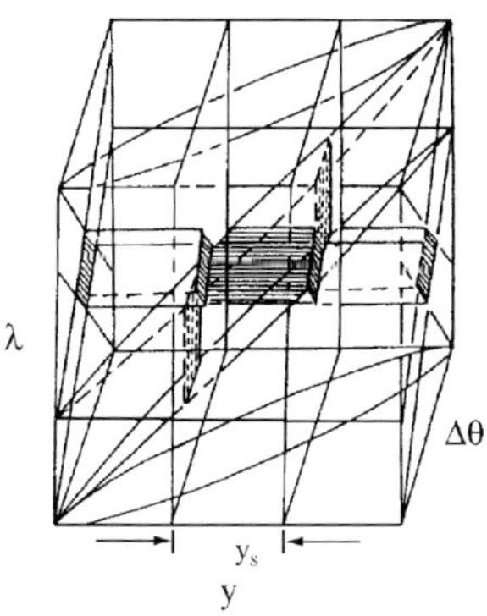

Fig. 9.51. The λ-$y(S_{\theta,O})$-$y'(\Delta\theta)$ diagram for three-wave monochromator [767]

Multiple-wave double-crystal or even multi-crystal monochromators can be designed and implemented for practical use, especially for synchrotron experiments. For double-crystal monochromators, the conventional (+, −) and (+, +) arrangements can be adopted. However, the conventional dispersive and non-dispersive optics are further modified by the two possible settings 'IN' and 'OUT' for a multiple-wave diffraction and the combination of the 'IN' and 'OUT' settings for two-crystal arrangements, i.e., (IN, IN), (OUT, OUT), (IN, OUT), and (OUT, IN). If there are no conventional two-wave double-crystal monochromators present, the characteristics of the (IN, IN) and (OUT, OUT) arrangements are the same for the same reflections involved. So are the (IN, OUT) and (OUT, IN) arrangements. In terms of the β-λ and λ-θ diagrams, the (IN, OUT) arrangement is equivalent to the (+, +) two-wave double-crystal setting, while the (IN, IN) and (OUT, OUT) are equivalent to the (+, −) case.

From the Dumond diagrams for two-wave double-crystal monochromators, where the incident and reflected waves, O_1 and G_1 for the first crystal and O_2 and G_2 for the second crystal, are considered, the angle of acceptance $\Delta\theta_{\mathrm{in}}$, angle of reflection $\Delta\theta_{\mathrm{out}}$, and dispersion are given by [389]

$$\begin{aligned}\Delta\theta_{\mathrm{in}} &= T_1 W_{O2} + |W_{O1} - T_1 W_{G1}| \ ,\\ \Delta\theta_{\mathrm{out}} &= T_2 W_{G1} + |W_{G2} - T_2 W_{O1}| \ ,\\ \frac{\Delta\lambda}{\lambda} &= \frac{1}{T}(W_{G1} + W_{O2}) \ ,\end{aligned} \tag{9.93}$$

where

$$T = \tan\theta_{G1} + \tan\theta_{G2} \ , \qquad T_1 = \frac{\tan\theta_{G1}}{T} \ , \qquad T_2 = \frac{\tan\theta_{G2}}{T} \ . \tag{9.94}$$

The above equations are also valid for the $(+n, -m)$ double-crystal setting, except that $T = |\tan\theta_{G1} - \tan\theta_{G2}|$. m and n stand for different reflections. For the $(+n, -n)$ setting, where the same Bragg reflection is used for the two crystals, the following simple relations hold:

$$\begin{aligned}\Delta\theta_{\mathrm{in}} &= \Omega_y \ ,\\ \frac{\Delta\lambda}{\lambda} &= (\Omega_y + W_{\theta,O1})\cot\theta_{G1} \ .\end{aligned} \tag{9.95}$$

The relation between $\Delta\lambda$ and $\Delta\beta$ for a three-wave monochromator can be obtained from (9.82) as

$$\frac{\Delta\lambda}{\lambda} = \frac{\cot\theta_G \tan\beta}{\tan\theta_G + \cot\theta_G} = \frac{\cot\theta_G}{T_G} \ . \tag{9.96}$$

Using this relation and the DuMond diagrams for three-wave double-crystal settings and following the derivation for two-wave double-crystal monochromators, we obtain the wavelength dispersion and angular widths as [766]

$$\frac{\Delta\lambda}{\lambda} = \frac{W_{\psi,G1} + W_{\psi,O2}}{\left|\dfrac{T_{G1}}{\cot\theta_{G1}} \pm \dfrac{T_{G2}}{\cot\theta_{G2}}\right|} \ ,$$

$$\Delta\theta_{\text{in}} = \frac{W_{\beta,O1}}{T_{G1}} + W_{\theta,O1} \,,$$
$$\Delta\theta_{\text{out}} = \frac{W_{\beta,G2}}{T_{G2}} + W_{\theta,G2} \,, \tag{9.97}$$

where the minus sign of $\pm$ in the first equation is for the (IN, IN) and (OUT, OUT) settings, while the plus sign is for the (OUT, IN) and (IN, OUT) ones. If the same three-wave diffraction is involved in the two crystals, (9.97) is valid for the (IN, OUT) and (OUT, IN) settings, but it's not for the (IN, IN) and (OUT, OUT) cases. This is because the denominator of the first equation of (9.97) can be zero when the minus sign is adopted. For this situation, the relations

$$\frac{\Delta\lambda}{\lambda} = \left(\frac{\Delta\lambda}{\lambda}\right)_1 \,,$$
$$\Delta\theta_{\text{in}} = W_{\theta,O1} \,,$$
$$\Delta\theta_{\text{out}} = W_{\theta,G2} \,. \tag{9.98}$$

should be used. The subscripts 1 and 2 mean the first and the second crystals in a double-crystal setting. The dispersion becomes more complicated with the presence of a conventional two-wave double-crystal monochromator; however, all the optical characteristics are calculable from the dynamical theory. Hence, on the other hand, this complication provides flexibility in designing new monochromators. Another advantage is the easy tuning of dispersion via a slight rotation in β for the multiple-wave (IN, OUT) monochromator, where the width in $\Delta\lambda$ depends on the azimuthal angle. The disadvantages of this type of monochromators are: (1) additional rotation mechanism being needed for the β or ψ, (2) the trade-off of the intensity for the angular width of acceptance, and (3) the relatively complicated geometry. The proposed three-wave monochromators include the use of the $(000)(222)(11\bar{1})$, $(000)(222)(113)$, and $(000)(222)(\bar{1}\bar{1}1)$ three-wave diffractions for Ge [766,767], the multiple diffractions involving the (0003) and (0001) forbidden reflections for hcp Be and the multiple-wave diffraction with the forbidden reflection (002) for Ge, and many others [367–369]. Note that the intensity of the three-wave (222) reflection is weaker than that of the two-wave (111) reflection as is expected.

Multiple-wave multi-crystal monochromators were also proposed [768]. One of the multiple-wave coplanar monochromators is shown for three-crystal and four-crystal arrangements in Fig. 9.52. A high degree of monochromaticity with $\Delta\lambda/\lambda$ ranging from 10^{-7} to 10^{-8} and the collimation of 0.1 to 0.01 arcsec could be achieved.

The use of a bent perfect crystal with a Umweg multiple-wave diffraction as a monochromator for thermal neutrons was demonstrated in 1979 [370–373]. The crystal was elastically deformed so that the integrated intensity of the primary reflection was increased according to Takagi's dynamical theory for imperfect crystals [128, 129]. An increase of more than twenty-fold in the primary (222) reflected intensity of the four-wave (000)(222)(153)

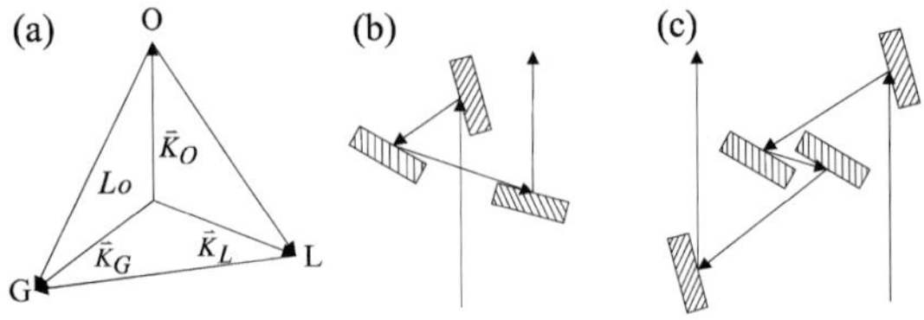

Fig. 9.52. Schematic of three-wave (O, G, L) coplanar multi-crystal monochromators: (**a**) in-plane diffraction geometry, (**b**) three-crystal setting, and (**c**) four-crystal setting. *Lo* is the three-wave Lorentz point and $\boldsymbol{K}$'s are the wavevectors inside the crystals [768]

$(\bar{3}11)$ diffraction was observed for a bent silicon crystal with the radius of curvature equal to 33 m [373].

The anomalous transmission of X-rays occurred in a transmission-type multiple-wave diffraction is another means of producing well-collimated monochromatic radiation. The crystal absorption off the exact multiple-wave diffraction position confines the transmitted beam to a very small angular and spatial region. A symmetric six-wave case, $(000)(220)(\bar{2}02)(044)(242)$ $(\bar{2}24)$, in which the six reciprocal lattice points form a regular hexagon has been extensively investigated as far as the anomalous transmission is concerned [52, 53, 184, 213–215, 217, 225, 769–772]. According to the dynamical calculations [53, 184, 213, 770, 771], in the total of twelve modes of propagation, two of them possess nearly zero absorption at the exact six-wave position. For a thick perfect crystal, like Ge and Si, only these two modes survive after the X-ray waves propagating through the crystal. The rest of the modes are suppressed by the crystal absorption. Because of these two residual modes, an intense, extremely parallel, and monochramatic transmitted beam is expected. Experimentally, this collimated beam with intensity enhancement has been observed for the six-wave case in a perfect germanium crystal and CuKα_1 [217] and in a 3 mm thick silicon crystal with synchrotron radiation at 1.54 Å [225].

The anomalous transmission in this six-wave diffraction has recently been reexamined experimentally [772]. According to the spherical-wave dynamical theory for multiple diffraction [197, 773], the anomalous transmission effect is often smeared out by X-ray focusing and defocusing effects. Direct measurement using a high-resolution diffractometric arrangement in both vertical and horizontal directions, or, θ and ψ, is preferred and is advantageous over topographic observation.

The six-wave diffraction experiment was carried out with the same experimental setup as that shown in Fig. 9.41, except that the sample is now a (111) cut perfect silicon crystal plate 3 mm thick. As shown in Fig. 9.41, the collimator is a 5 mm thick silicon crystal plate of the same crystal orientation as the sample. The anomalous transmission of the six-wave case through this collimating crystal was utilized to produce a highly collimated and parallel monochromatic incident wave for the sample crystal. The θ rotation for the Bragg angle of the (044) reflection and the azimuth ψ rotation around the reciprocal lattice vector of (044) are kept at 0.1 arcsec/step. The σ polarized incident radiation with respect to the (044) was chosen. The wavelength used was $\lambda = 0.93\,\text{Å}$ and the corresponding μT was about 12. The (000), (220), and (202) transmitted waves were measured with point detectors. Figure 9.53 shows the measured and calculated intensity distributions of the (000) wave against $\Delta\theta$ deviated from the exact six-wave diffraction point for $\Delta\psi = -2.8''$, 0 and $4.4''$. The convolution with the incident beam divergences was employed in the calculation. The experimental results are in good agreement with the theoretical ones. The intensity at the exact six-wave point, $\Delta\psi = 0$ and $\Delta\theta = 0$, is about 3.3 times stronger than that of the (220) reflection at $\Delta\psi \neq 0$. This intensity enhancement at the exact six-wave diffraction position can be further increased by increasing the crystal thickness. Of course, the absolute intensity of the exact transmission wave will drop. In any case, the transmitted wave with about 2 arcsec divergence nearly in the vertical and horizontal directions can be achieved. This arrangement could be used as a monochromator for other high-resolution experiments, especially when high intense sources, such as synchrotron radiation from undulators are used. The application of this kind of monochromator to the multiple-wave experiments for standing-wave excited fluorescence and photoelectrons, shown in the previous section, is an example.

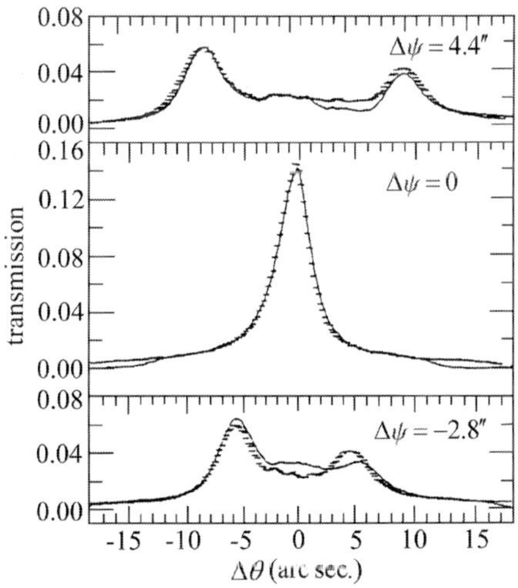

Fig. 9.53. Measured (*cross line*) and calculated (*solid line*) curves of the (000) wave in a six-wave case [772]

9.6.2 Multi-Wave Phase Plate and X-Ray Polarimeter

The vector characteristics of X-rays have been utilized in many fields of research as useful tools of probing the detailed structures of matter, including magnetic structures and properties, atomic-site dependent electronic structures, and many others [174, 485, 487, 488, 774]. The increasing demand on the polarization tunability of X-ray sources has led to the development of cross undulators [775], elliptical wigglers [378], and crystal diffraction optics [776–778]. As to the diffraction optics, X-ray phase plate using the phase difference between the σ and π diffracted waves of a two-wave Bragg reflection in a perfect crystal has been proposed and used to produce elliptically or circularly polarized X-rays [378, 779, 780]. The phase difference of 18° of Ge (220) reflection and five times consecutive reflections were employed to provide circularly polarized light.

Multiple-wave diffraction is a phase-sensitive scattering process. Its intensity distribution and scattering matrix are phase dependent. The σ and π components of the wavefields inside the diffraction crystal are entangled in most cases, except for the coplanar multiple-wave diffraction. The phase difference between the two polarized wave components of a general multiple-wave diffraction can, in principle, be very large compared to that of two-wave cases. Thus, it is advantageous to use a non-coplanar multiple-wave diffraction in designing a phase plate for changing an unpolarized or linearly polarized X-ray wave into an elliptically or circularly polarized wave. Theoretical studies have indicated the feasibility of making multiple-wave X-ray phase plate for this purpose [377].

Figure 9.54 shows the calculated phase difference between the σ- and π-polarized components and the diffracted intensities of the primary (222) reflection of the three-wave Ge $(000)(222)(\bar{1}\bar{1}3)$ diffraction for $\lambda = 1.54056$ Å. The intensities of the σ- and π-polarized waves are comparable and the phase

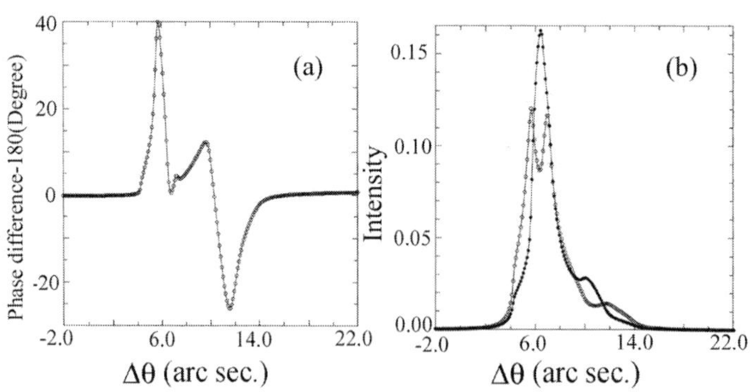

Fig. 9.54. Dynamical calculations for the three-wave Ge $(000)(222)(\bar{1}\bar{1}3)$ diffraction and $\lambda = 1.54056$ Å: (**a**) phase difference between the σ and π polarized components and (**b**) the diffracted intensities of the primary (222) reflection [377]

difference is about 30° at $\Delta\theta = 6''$ and $\Delta\psi = 0$. While the two component waves have almost identical amplitudes and nearly zero phase difference at $\Delta\theta = 7''$ and $\Delta\psi = 0$. Detailed calculation indicates that an elliptically polarized wave is generated at $\Delta\theta = 5.85''$ and $\Delta\psi = 0$, and it becomes linearly polarized at $\Delta\theta = 6.8''$ and $\Delta\psi = 0$. The phase difference can be as large as 90°, for example, at $\Delta\theta = 5.1''$ and $\Delta\psi = -4.7''$ for the three-wave diffraction (000)(222)($5\bar{3}\bar{1}$) [377]. Since the phase difference and the intensities of the two component waves are so sensitive to the small change of the angular position, of the order of a fractional arcsecond, the realization of this kind of X-ray phase plate requires a proper monchromatization scheme. The proposed six-wave transmission monochromator just mentioned in the previous section serves this purpose.

The linear polarization state of the produced polarized X-ray wave by a phase plate is usually characterized by a single- or multi-bounce Bragg reflection with the Bragg angle $\theta_G = 45°$, the so-called linear polarization analyzer [94, 380, 781]. The reason is that the π components of X-rays are completely suppressed due to $\cos 2\theta_G = 0$. To characterize a circularly polarized radiation, a circular phase plate, together with a linear polarization analyzer, is frequently adopted. As just described, multiple-wave diffraction, a phase sensitive diffraction process, generally produces mixed polarized waves. In principle, this diffraction can, in turn, be utilized to measure the polarization of a circularly or elliptically polarized radiation source [782, 783]. By following the principles of optics [502], the intensity and polarization of an incident X-ray plane wave $\boldsymbol{D} = [D_\sigma, D_\pi \exp(i\varepsilon)]$ can be described in terms of the four Stokes parameters as [376, 784–786]

$$\begin{aligned} s_0 &= D_\sigma^2 + D_\pi^2 , & s_1 &= D_\sigma^2 - D_\pi^2 , \\ s_2 &= 2D_\sigma D_\pi \cos\varepsilon , & s_3 &= 2D_\sigma D_\pi \sin\varepsilon . \end{aligned} \tag{9.99}$$

s_0 represents the total intensity. s_1, s_2, and s_3 are the intensity differences between the 0° and 90° linear polarizations, between the 45° and −45° linear polarizations, and between the right- and the left-handed circular polarizations, respectively. Or, the polarization of the incident X-ray wave can also be represented by the Poincarè vector of polarization defined as

$$\boldsymbol{P} = (P_1, P_2, P_3) = \left(\frac{s_1}{s_0}, \frac{s_2}{s_0}, \frac{s_3}{s_0}\right) . \tag{9.100}$$

Then the corresponding polarization density matrix is

$$\tilde{\varrho} = \frac{1}{2}\begin{bmatrix} 1 + P_1 & P_2 - \mathrm{i}P_3 \\ P_2 + \mathrm{i}P_3 & 1 - P_1 \end{bmatrix} . \tag{9.101}$$

When a two-wave (O, G) diffraction takes place, the polarization density matrix of the diffracted wave is transformed to

$$\tilde{\varrho}' = \tilde{T}\tilde{\varrho}\tilde{T}^\dagger , \tag{9.102}$$

where $\tilde{T}$ is the transformation matrix defined as

$$\begin{bmatrix} D_{\sigma G} \\ D_{\pi G} \end{bmatrix} = \tilde{T} \begin{bmatrix} D_{\sigma O} \\ D_{\pi O} \end{bmatrix} , \tag{9.103}$$

and $\tilde{T}^{\dagger}$ is the Hermitian matrix of $\tilde{T}$. The diffracted intensity I_G is then equal to the trace of $\tilde{\varrho}'$, i.e.,

$$I_G = \mathrm{Tr}\tilde{\varrho}' , \tag{9.104}$$

which is related to the Poincarè vector.

The diffracted wavefields and intensities of a three-wave (O, G, L) diffraction can also be expressed in terms of the the Poincarè vector by the same token. Although exact dynamical calculations can provide the transformation matrix for a given polarization of the incident wave and given diffraction geometry, an analytical expression directly linking the diffracted intensity and the Poincarè vector would be more useful for the estimation of X-ray polarization. For this purpose, the Bethe or the Born approximation facilitates the finding of appropriate transformation matrix $\tilde{T}$. For example, with the Born approximation given in Chap. 7, the matrix $\tilde{T}$ derived from (7.202) is given by

$$\begin{bmatrix} D_{\sigma G} \\ D_{\pi G} \end{bmatrix} = c_0 \tilde{T} \begin{bmatrix} D_{\sigma O} \\ D_{\pi O} \end{bmatrix} , \tag{9.105}$$

where c_0 is the scalar coefficient of the cross-product given in (7.202) and

$$\tilde{T} = \begin{bmatrix} 1 & 0 \\ 0 & \cos 2\theta_G \end{bmatrix} - A_L(\psi) \exp(\mathrm{i}\delta_3)\, \tilde{B} . \tag{9.106}$$

The resonance function $A_L(\psi)$ and the matrix $\tilde{B}$ are

$$A_L(\psi) = \frac{k^2}{k_L^2 - k^2} \approx \frac{k}{2\,(\boldsymbol{\ell} \cdot \hat{\boldsymbol{\sigma}}_O) \cos\theta_G\,(\psi - \psi_0)} ,$$

$$\tilde{B} = \begin{bmatrix} B_{\sigma\sigma} & B_{\pi\sigma} \\ B_{\sigma\pi} & B_{\pi\pi} \end{bmatrix} = \frac{1}{k^2} \left| \frac{\chi_{G-L}\chi_{L-O}}{\chi_{G-O}} \right|$$

$$\times \begin{bmatrix} k_L^2 - (\boldsymbol{\ell} \cdot \hat{\boldsymbol{\sigma}}_O)^2 & -(\boldsymbol{k}_L \cdot \hat{\boldsymbol{\pi}}_O)(\boldsymbol{\ell} \cdot \hat{\boldsymbol{\sigma}}_O) \\ -(\boldsymbol{k}_L \cdot \hat{\boldsymbol{\pi}}_G)(\boldsymbol{\ell} \cdot \hat{\boldsymbol{\sigma}}_O) & k_L^2 (\hat{\boldsymbol{\pi}}_G \cdot \hat{\boldsymbol{\pi}}_O) - (\boldsymbol{k}_L \cdot \hat{\boldsymbol{\pi}}_G)(\boldsymbol{\ell} \cdot \hat{\boldsymbol{\pi}}_O) \end{bmatrix} , \tag{9.107}$$

where $\boldsymbol{k}$'s are the wavevectors of the diffracted waves and ψ_0 is the azimuthal angle of the exact three-wave diffraction point. By following (9.104), the diffracted intensities of both σ and π components are given by

$$\begin{aligned} I_{\sigma G} &= -A_L(\psi)\,[(1 + P_1)\, B_{\sigma\sigma} \cos\delta_3 + P_2 B_{\pi\sigma} \cos\delta_3 - P_3 B_{\pi\sigma} \sin\delta_3] , \\ I_{\pi G} &= -A_L(\psi) \cos 2\theta \big[(1 - P_1)\, B_{\pi\pi} \cos\delta_3 + P_2 B_{\sigma\pi} \cos\delta_3 \\ &\quad + P_3 B_{\sigma\pi} \sin\delta_3 \big] . \end{aligned} \tag{9.108}$$

By choosing appropriate three-wave diffractions with the known triplet phases δ_3, the Poincarè vector can be determined without difficulty.

As an illustration, Fig. 9.55 shows the Renninger scans of the three-wave (000)(442)(511), (000)(442)(151), and (000)(442)(153) cases of GaAs with left- and right-handed elliptically polarized incident radiation at $\lambda = 1.3$ Å [376]. (442) is the primary reflection. The left- and right-handed elliptical polarized radiations were extracted at 0.11 mrad below and above the orbital plane of the CHESS synchrotron facility. The profile asymmetry reflects the effect of the triplet phases, which are 83°, 83°, and −82° for the (511), (151), and (153) three-wave cases, respectively. The intensity difference $\Delta I = I(\Delta\psi) - I(-\Delta\psi)$ versus $1/|\Delta\psi|$ for the three-wave cases are shown in Fig. 9.56, where $\Delta\psi = \psi - \psi_0$. The slopes of the best fit lines, according to (9.108), give $\boldsymbol{P} = (0.77 \pm 0.18, 0.05 \pm 0.40, -0.57 \pm 0.07)$ and $|\boldsymbol{P}| = 0.96 \pm 0.14$, which is very close to complete polarization, i.e., $|\boldsymbol{P}| = 1$ [376].

It should be noted that the transformation matrix is slightly different for different approximations adopted. In addition, in the characterization of circular polarized waves, choosing the three-wave cases with δ_3 close to ±90° can lead to more accurate polarization values.

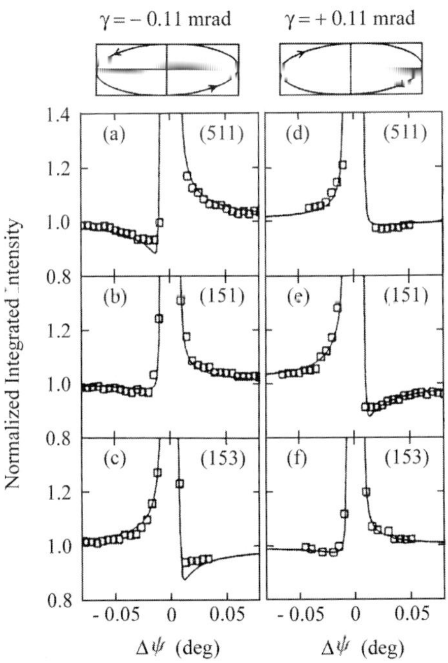

Fig. 9.55. Renninger scans of the three-wave (000)(442)(511), (000)(442)(151), and (000)(442)(153) cases of GaAs and $\lambda = 1.3$ Å:(**a**)–(**c**) right- and (**d**)–(**f**) left-handed elliptically polarized incident radiation [376]

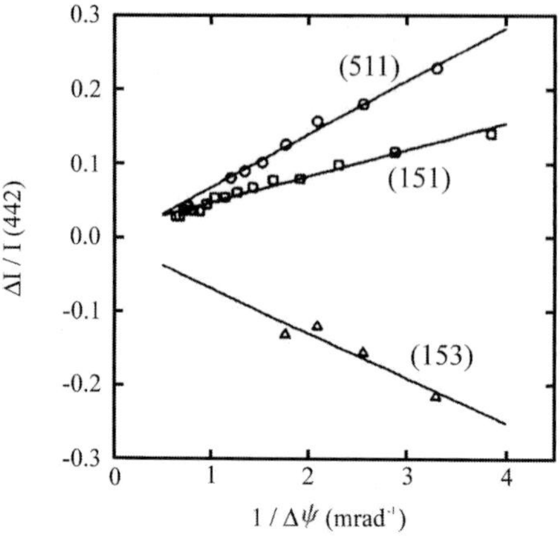

Fig. 9.56. Measured intensity difference $\Delta I = I(\Delta\psi) - I(-\Delta\psi)$ versus $1/|\Delta\psi|$ for the three-wave (000)(442)(511), (000)(442)(151), and (000)(442)(153) cases of GaAs measured from Fig. 9.55 [376]

9.7 Other Applications

Other applications including the multiple-wave X-ray interferometer, topography, and plasma diagnosis are briefly described below.

9.7.1 Multiple-Wave X-Ray Interferometers

The Bonse-Hart X-ray interferometer has been widely used for the precise measurement of atomic scattering factors, real and imaginary parts, lattice constants, X-ray wavelength, and many others [55, 161, 162, 787–795]. The Laue (LLL) interferometer is most frequently used, where three Laue reflections involved serve as an X-ray beam splitter, a mirror, and a beam analyzer, respectively. It, however, suffers crystal absorption and beam broadening due to the Borrmann fan effects. Even though the Bragg (BBB) interferometer involving three Bragg reflections can improve the absorption situation, the angular shift of Bragg peak and beam broadening are still a problem.

The advantage of providing a well-collimated beam from the multiple-wave diffraction has led to the development of a three-wave coplanar X-ray interferometer [53, 375]. The geometry of this special interferometer is shown in Fig. 3.8, where the three-wave (000)(440)(404) coincident coplanar diffraction of Si at $\lambda = 1.6628$ Å is adopted. Because of the involved coplanar reflections, the visibility of the interference is improved by suppressing the correlation between the σ and π diffracted waves inside the crystal. The spatial resolutions of the interference fringes are also improved to give 0.5 μm horizontal and 5 μm vertical by the use of total reflection in this diffraction geometry. These improved optical properties, especially the phase sensitivity of three-wave diffraction, make this kind of interferometer very useful for phase-contrast imaging.

9.7.2 Multiple-Wave X-Ray Topography

X-ray topography is one of many powerful tools for mapping crystal defects onto photographs or two-dimensional detectors. Various transmission and reflection geometries and multi-crystal arrangements have been adopted for this imaging purpose. As for the determination of the Burgers vector of dislocations, several topographs are usually needed for unambiguous defect analysis. Since simultaneous reflections occur in a multiple-wave diffraction, multiple-wave topography is an ideal way for simultaneous recording of the intensity distributions of two or more reflections from a crystal. This certainly facilitates the determination of the Burgers vector.

Multiple-wave topographs for liquid phase epitaxy films have been obtained experimentally by using a five-wave diffraction [374]. However, due to the complicated intensity distribution at the exact multiple-wave diffraction point, the defect structural information may be masked by the dynamical diffraction effects. Therefore, the diffraction slightly off the exact multiple-wave point might be more appropriate than at the exact point for X-ray topographic investigation. Similar situation has been encountered in multiple-wave X-ray standing-wave fluorescence. In other words, the multiple-wave dynamical diffraction effects are purposely avoided for simultaneous topographic imaging and fluorescence measurements.

9.7.3 Plasma Diagnosis and Spectral Characterization of Radiation Sources

The angular position (θ, ψ) of a multiple-wave diffraction depends on the ratio between the wavelength λ and lattice constant d, where θ is the Bragg angle of the primary reflection and ψ the azimuthal angle around the reciprocal lattice vector of the primary reflection. In the beginning of this chapter, we showed that for a given wavelength the precise determination of lattice constants can be achieved. If a crystal with known lattice constants is used for generating multiple-wave diffraction from an unknown radiation source, the angular distribution of the diffraction intensity, in principle, will exhibit the spectrum of the radiation source.

Two three-wave diffractions, (000)(200)(111) and (000)(200) $(1\bar{1}1)$, of germanium have been applied to low-energy plasma diagnosis for the characterization of X-radiation from hot plasma generated from tokamak [363–365]. Similar application to the determination of X-ray photon energy from an undulator using twenty-four-wave back diffraction has also been reported [590], which has been described in Chap. 8.

9.8 Prospects of Future Development

The future development and possible applications of multiple-wave diffraction can be expected to follow the current trends of the development of the theories and applications described in this and previous chapters. As has already been stated, multiple-wave diffraction requires simultaneous occurrence of reflections from a periodic arrangement with or without modulation structures. And the intense and energy tunable X-ray synchrotron sources are most suited for performing this particular diffraction technique. The development of this multiple-wave method in relation to scattering, diffraction, and spectroscopy may be useful to tackle minuscule structural changes, both atomic and electronic, encountered in condensed matter physics and materials sciences on one hand. The method may, on the other hand, provide solutions to unsolved problems and a means of perfecting the current techniques for structure determination and material characterization. A brief description of possible future development is given below:

(a) Crystal-structure determination using multiple-wave diffraction. The intensity correction due to extinction and the determination of reflection phase are two major problems for crystal-structure determination. The kinematical approach for multiple-wave diffraction takes care of the secondary extinction automatically in the power-transfer equations. While the primary extinction is considered in the dynamical theory. In addition, the phase of X-ray reflection can be determined on a quantitative basis. Therefore, information about the intensity, namely the amplitude, and the phase of a structure-factor can be extracted from the intensity distributions of a multiple-wave diffraction. Moreover, it has already been possible to collect a large number of triplet phases. The next step is to optimize the experimental procedures for data collection, phase extension, and intensity correction so that an operational method for crystal structure determination can be realized.

(b) Determination of interface structure. The atomic arrangement of a surface can be determined by various means, such as scanning tunnelling microscope (STM), atomic force microscope (AFM), X-ray grazing incidence diffraction, helium scattering, etc.. The determination of the structure of an interface between a thin film and a substrate crystal is not so easy as that of surfaces. Although the very intense X-ray beams from undulators being confined to a very tiny spot may be useful in probing interfaces, most of the X-ray photons are wasted and the structural information from both the thin film and substrate might be mingled. In principle, there might be a better way of using most of photons to reveal the interface structure. One of the possible candidates serving this purpose is the Bragg-surface three-wave diffraction. The 'surface' diffracted wave propagating along the surface of a substrate may convey the structural information of the interface between a thin film and the substrate. If such a diffraction technique can be developed, then the structure of the interface between the substrate and other objects, like quantum dots and nano particles, may be probed in great detail.

(c) Determination of fine electronic structure, including orbital hybridization, charge and orbital ordering. The minuscule anisotropic electronic structure, which cannot be detected with an ordinary two-wave diffraction, may be revealed with resonance multiple-wave diffraction involving space-group symmetry forbidden reflections. In this case, high brilliance X-ray sources from undulators or the fourth generation synchrotron facility are needed for providing sufficient X-ray photons. Phase shift due to resonance could provide detailed information about the charge and orbital engaged in the material system.

(d) Direct investigation of the dynamics of density-wave under external excitation. The phase of structure-factor triplets involving charge-density-wave (CDW) fractional reflections has already been demonstrated with the multiple-wave diffraction technique. When an electric field is applied, the CDW as well as the host crystal lattice will be affected. If the strength of the applied field is sufficiently large, then the change of the host lattice may result in a phase shift on the triplet phase because of the symmetry breaking. This phase change could be useful to reveal the transition of the CDW from the so-called creeping state into the plastic flow disordered state, and eventually enters the sliding-solid phase [796]. If this kind of investigation can be realized, the other transport properties in modulated structures, like the stripes in high T_C superconductors, could be measured.

(e) Development of multiple-wave X-ray resonators. X-ray cavity is the first optical component required for constructing an X-ray or a γ-ray laser system. The anomalous transmission in multiple-wave diffraction can be utilized for making X-ray resonators with thick crystals. In addition, the beam-size confinement in multiple-wave diffraction could be advantageous in producing quasi-coherent X-ray beams with a very small divergence and dispersion. There are of course some difficulties to be overcome, such as the lattice strain, crystal surface finishing, thermal stability, crystal handling, and the requirement of ultra-high energy resolution for the incident radiation.

References

1. E. Wagner: Physik. Z. **21**, 632 (1920)
2. O. Berg: Wiss. Veröff. Siemens-Konz. **5**, 89 (1926)
3. G. Mayer: Z. Krist. **66**, 585 (1928)
4. M. Renninger: Z. Krist. **106**, 141 (1937) Naturwissenschaften **25**, 43 (1937)
5. Y. Cauchois, H. Hulubei, J. Weigle: Helv. Physica Acta **10**, 218 (1937)
6. A. Pabst: Am. Mineral. **24**, 566 (1939)
7. C.J. Davisson, F.E. Haworth: Phys. Rev. **66**, 351 (1944)
8. W. Kossel: Nachr. Akad. Wiss. Göttingen, Math. Phys. **1**, 229 (1935)
9. W. Kossel: Ann. Phys. (Leipzig) **25**, 512; **26**, 533 (1936)
10. W. Kossel: Ergebn. Exakten Naturwiss. **16**, 295 (1937)
11. W. Kossel, H. Voges: Ann. Phys. (Leipzig) **23**, 677 (1935)
12. H. Bethe: Ann. Phys. (Leipzig) **87**, 55 (1928)
13. H. Raether: Z. Phys. **78**, 527 (1932)
14. J.A. Darbyshire, E.R. Cooper: Proc. Roy. Soc. **A152**, 104 (1935)
15. W.N. Lipscomb: Acta Cryst. **2**, 193 (1949)
16. W. Kossel, G. Möllenstedt: Ann. Phys. (Leipzig) **36(5)**, 113 (1939)
17. S. Miyake, K. Kambe: Acta Cryst. **7**, 216 (1954)
18. K. Kambe, S. Miyake: Acta Cryst. **7**, 218 (1954)
19. K. Kambe: J. Phys. Soc. Jpn. **12**, 13, 25 (1957)
20. H.J. Hay: Atomic Energy Research Establishment. Report No. R2982, Harwell, Berkshire, England (1959)
21. R.R. Spencer, J.R. Smith: Bull. Amer. Phys. Soc. **4**, 245 (1959)
22. D.A. O'Connor, J. Sosnowski: Acta Cryst. **14**, 292 (1961)
23. V.P. Duggal, K.R. Rao, C.L. Thapar, V. Singh: Proc. Indian Acad. Sci. **53**, 59 (1961)
24. R.M. Moon, C.G. Shull: Bull. Amer. Phys. Soc. **6**, 261 (1961)
25. G. Borgonovi, G. Caglioti: Nuovo Cim. X **24**, 1174 (1962)
26. R.M. Moon, C.G. Shull: Acta Cryst. **17**, 805 (1964)
27. G. Borrmann, W. Hartwig: Z. Krist. **121**, 401 (1965)
28. G. Borrmann: Phys. Z. **42**, 157 (1941)
29. G. Borrmann: Z. Phys. **127**, 297 (1950)
30. M. von Laue: Ergeb. Exakt. Naturwiss. **10**, 133 (1931)
31. J.C. Slater: Rev. Mod. Phys. **30**, 197 (1958)
32. R.W. James: Solid State Phys. **15**, 53 (1962)
33. P.P. Ewald: J. Phys. Soc. Jpn. Suppl. B-II **17**, 48 (1962)
34. P.P. Ewald (ed.): *Fifty Years of X-Ray Diffraction* (A. Oosthoek's Uitgeversmaatschappij N.V., Utrecht 1962)
35. B.W. Battermann, H. Cole: Rev. Mod. Phys. **36**, 681 (1964)
36. G. Borrmann: Z. Krist. **120**, 143 (1964)
37. P.P. Ewald: Rev. Mod. Phys. **37**, 48 (1965)
38. P.P. Ewald: Acta Cryst. **A25**, 103 (1969)

39. A. Authier: In: *Advances in Structure Research by Diffraction Methods*, ed. by R. Brill, R. Mason (Pergamon, Oxford 1970)
40. N. Kato: In: *X-Ray Diffraction*, ed. by L.V. Azaroff (McGraw-Hill, New York 1974)
41. A. Authier: In: *Diffraction and Imaging Techniques in Material Science*, ed. by S. Amelinckx, R. Gevers, G. Remaut, J. Van Landuy (North-Holland, Amsterdam 1978) pp. 715–756
42. P.P. Ewald: Acta Cryst. **A35**, 1 (1979)
43. P.P. Ewald: *On the Foundations of Crystal Optics*. Monograph **10**, ed. by H.J. Juretschke (American Crystallographic Association, Buffalo, New York 1991)
44. A. Authier: Cryst. Res. Technol. **33** 517 (1998)
45. A. Authier, C. Malgrange: Acta Cryst. **A54**, 806 (1998)
46. A. Authier, S. Lagomarsino, B.K. Tanner (ed.): *X-Ray and Neutron Dynamical Diffraction: Theory and Applications*. NATO Series, Series B, Physics, Vol. 357 (Plenum Press, New York, London 1996)
47. M. von Laue: *Röntgenstrahl-Interferenzen* (Akadem. Verlagsges., Leipzig 1944)
48. W.H. Zachariasen: *Theory of X-Ray Diffraction in Crystals* (John Wiley & Sons, New York 1945)
49. R.W. James: *The Optical Principle of the Diffraction of X-Rays* (Bell, London 1950)
50. M. von Laue: *Röntgenstrahl-Interferenzen* (Akademische, Frankfurt 1960)
51. L.H. Schwartz, J.B. Cohen:*Diffraction from Materials* (Academic, New York 1977)
52. Z.G. Pinsker: *Dynamical Scattering of X-Rays in Crystals*, Springer Ser. Solid-State Sci., Vol. 3 (Springer, Berlin, Heidelberg, New York 1978)
53. S.-L. Chang: *Multiple Diffraction of X-Rays in Crystals* (Springer, Berlin, Heidelberg 1984)
54. D.W.J. Cruickshank, H.J. Juretschke, N. Kato (ed.): *P.P. Ewald and his Dynamical Theory of X-Ray Diffraction* (IUCr/Oxford University Press, New York 1992)
55. A. Authier: *Dynamical Theory of X-Ray Diffraction* (IUCr/Oxford University Press, Oxford, New York 2001)
56. W. Friedrich, P. Knipping, M. von Laue: Ann. Phys. (Leipzig) **41** 971 (1913)
57. P.P. Ewald: Ann. Phys. (Leipzig) **49**, 1 (1916)
58. P.P. Ewald: Ann. Phys. (Leipzig) **49**, 117 (1916)
59. P.P. Ewald: Ann. Phys. (Leipzig) **54**, 519 (1917)
60. P.P. Ewald: Z. Krist. **A97**, 1 (1937)
61. M. von Laue: Ann. Phys. **41**, 989 (1913)
62. W.L. Bragg: Nature (London) **90**, 410 (1912)
63. W.L. Bragg: Proc. Cambridge Phil. Soc. **17**, 43 (1913)
64. W.L. Bragg: Proc. Roy. Soc. (London) **A89**, 248 (1913)
65. W.H. Bragg, W.L. Bragg: Proc. Roy. Soc. (London) **A88**, 428 (1913)
66. C.G. Darwin: Phil. Mag. **27**, 315 (1914)
67. C.G. Darwin: Phil. Mag. **27**, 675 (1914)
68. J.A. Prins: Z. Phys. **63**, 477 (1930)
69. M. Kohler: Ann. Phys. (Leipzig) **18**, 265 (1933)
70. C.G. Darwin: Phil. Mag. **43**, 800 (1922)
71. M. Kohler: Sitzungsber. Preuss. Akad. Math. Physikal. Kl., 334 (1935)
72. G. Molière: Ann. Phys. (Leipzig) **35**, 272 (1939)
73. G. Molière: Ann. Phys. (Leipzig) **35**, 297 (1939)
74. G. Molière: Ann. Phys. (Leipzig) **36**, 265 (1939)

75. Y.H. Ohtsuki: J. Phys. Soc. Jpn. **19**, 2285 (1964)
76. Y.H. Ohtsuki: J. Phys. Soc. Jpn. **20**, 374 (1965)
77. H. Wagenfeld: Phys. Rev. **144**, 216 (1966)
78. M. Ashkin, M. Kuriyama: J. Phys. Soc. Jpn. **21**, 1549 (1966)
79. P.H. Dederichs: Phys. Kondens. Mater. **5**, 347 (1966)
80. P.H. Dederichs: Phys. Status Solidi **23**, 377 (1967)
81. J.P. Hannon, G.T. Trammell: Phys. Rev. **169**, 315 (1968)
82. J.P. Hannon, G.T. Trammell: Phys. Rev. **186**, 306 (1969)
83. A.M. Afanas'ev, Y. Kagan: Acta Cryst. **A24**, 163 (1968)
84. Y.H. Ohtsuki, S. Yanagawa: J. Phys. Soc. Jpn. **21**, 326, 502 (1966)
85. M. Kuriyama: J. Phys. Soc. Jpn. **23**, 1369 (1967)
86. M. Kuriyama: Phys. Status Solidi **24**, 743 (1967)
87. M. Kuriyama: J. Phys. Soc. Jpn. **25**, 846 (1968)
88. M. Kuriyama: Acta Cryst. **A25**, 682 (1969)
89. M. Kuriyama: Acta Cryst. **A26**, 56 (1970)
90. M. Kuriyama: Acta Cryst. **A27**, 273 (1971)
91. M. Kuriyama: Acta Cryst. **A28**, 588 (1972)
92. H. Sano, K. Ohtaka, Y.H. Ohtsuki: J. Phys. Soc. Jpn. **27**, 1254 (1969)
93. M. Sauvage, C. Malgrange, J.-F. Pétroff: J. Appl. Cryst. **16**, 17 (1983)
94. D.P. Siddons, M. Hart, Y. Amemiya, J.B. Hastings: Phys. Rev. Letters **64**, 1967 (1990)
95. G. Materlik, C.J. Sparks, K. Fischer (eds.): *Resonant Anomalous X-Ray Scattering* (North-Holland, Amsterdam, London, New York, Tokyo 1994)
96. D.H. Templeton, L.K. Templeton: Acta Cryst. **A38**, 62 (1982)
97. V.E. Dmitrienko: Acta Cryst. **A39**, 29 (1983)
98. G. Borrmann: In: *Beitr. Phys. Chem. 20 Jahrhunderts* (Vieweg & Sohn, Braunschweig 1959) pp. 262-282
99. G. Borrmann: Naturwiss. **42**, 67 (1955)
100. A. Authier: C. R. Acad. Sci. Paris **251** 2003 (1960)
101. A. Authier: Bull. Soc. Fr. Minér. Crist. **84**, 51 (1961)
102. A.R. Lang, G. Kowalski, A.P.W. Makepeace, M. Moore: Phil. Mag. **B53** L53 (1986)
103. N. Kato, A.R. Lang: Acta Cryst. **12**, 787 (1959)
104. C. Malgrange, A. Authier: C. R. Acad. Sci. Paris **261**, 3774 (1965)
105. S. Kikuta, K. Kohra: J. Phys. Soc. Jpn. **25**, 923 (1968)
106. S. Kikuta, K. Kohra: J. Phys. Soc. Jpn. **25**, 924 (1968)
107. K. Kohra, S. Kikuta: Acta Cryst. **A24**, 200 (1968)
108. M. Renninger: Acta Cryst. **A24**, 143 (1968)
109. K. Sugii, T. Matsushita, K. Kohra: J. Phys. Soc. Jpn. **31**, 1295 (1971)
110. B.W. Batterman, G. Hildebrandt: Phys. Stat. Sol. **23**, K147 (1967)
111. B.W. Batterman, G. Hildebrandt: Acta Cryst. **A24**, 150 (1968)
112. H. Hashizume, K. Nakayama, T. Matsushita, K. Kohra: J. Phys. Soc. Jpn. **29**, 806 (1970)
113. K. Nakayama, H. Hashizume, K. Kohra: J. Phys. Soc. Jpn. **30**, 893 (1970)
114. M. Hart, A.D. Milne: Phys. Stat. Sol. **26**, 185 (1968)
115. V. Mocella, J.P. Guigay, Y. Epelboin, J. Härtwig, J. Baruchel: J. Phys. D: Appl. Phys. **32**, A88 (1999)
116. V. Mocella, Y. Epelboin, J.P. Guigay,: Acta Cryst. **A56**, 308 (2000)
117. N. Kato: Acta Cryst. **13**, 349 (1960)
118. N. Kato: Z. Naturforsch. **15a**, 369 (1960)
119. N. Kato: Acta Cryst. **14**, 526 (1961)
120. N. Kato: Acta Cryst. **14**, 627 (1961)
121. N. Kato: J. Appl. Phys. **39**, 2225 (1968)

122. N. Kato: J. Appl. Phys. **39**, 2231 (1968)
123. P. Penning, D. Polder: Philips Res. Repts. **16**, 419 (1961)
124. P. Penning, D. Polder: Acta Cryst. **17**, 950 (1961)
125. N. Kato: J. Phys. Soc. Jpn. **18**, 1785 (1963)
126. N. Kato: J. Phys. Soc. Jpn. **19**, 67 (1964)
127. N. Kato: J. Phys. Soc. Jpn. **19**, 971 (1964)
128. S. Takagi: Acta Cryst. **15**, 1311 (1962)
129. S. Takagi: J. Phys. Soc. Jpn. **26**, 1239 (1969)
130. D. Taupin: Bull. Soc. Fr. Minér. Crist. **87**, 469 (1964)
131. D. Taupin: Acta Cryst. **23**, 25 (1967)
132. N. Kato: Acta Cryst. **16**, 276 (1963)
133. N. Kato: Acta Cryst. **16**, 282 (1963)
134. K. Kambe: Z. Naturforsch. **20a**, 770 (1965)
135. M. Wilkins: Phys. Status Solidi **13**, 529 (1966)
136. F. Balibar, A. Authier: Phys. Status Solidi **21**, 413 (1967)
137. F.N. Chukhovskii, A.A. Shtolberg: Phys. Status Solidi **41**, 815 (1970)
138. A.M. Afanas'ev, V.G. Kohn: Acta Cryst. **A27**, 421 (1971)
139. A. Howie, M.J. Whelan: Proc. Roy. Soc. **A263**, 217 (1961)
140. U. Bonse: Z. Phys. **177**, 385 (1964)
141. N. Kato: Acta Cryst. **A32**, 453 (1976)
142. N. Kato: Acta Cryst. **A32**, 458 (1976)
143. N. Kato: Acta Cryst. **A35**, 9 (1979)
144. N. Kato: Acta Cryst. **A36**, 171 (1980)
145. N. Kato: Acta Cryst. **A36**, 763 (1980)
146. N. Kato: Acta Cryst. **A36**, 770 (1980)
147. W.F. Berg: Naturwissenshaften **19**, 391 (1931)
148. G.N. Ramachandran: Proc. Ind. Acad. Sci. **A19**, 280 (1944)
149. C.S. Barrett: Trans. Metal. Soc. AIME **161**, 15 (1945)
150. N. Wooster, W.A. Wooster: Nature (London) **155**, 786 (1945)
151. A.R. Lang: Acta Met. **5**, 358 (1957)
152. A.R. Lang: J. Appl. Phys. **29**, 597 (1958)
153. H. Barth, R. Hosemann: Z. Naturforsch. **13a**, 792 (1958)
154. G. Borrmann, W. Hartwig, H. Irmler: Z. Naturforsch. **13a**, 423 (1958)
155. U. Bonse: Z. Phys. **153**, 278 (1958)
156. U. Bonse, E. Kappler: Z. Naturforsch. **13a**, 348 (1958)
157. J.B. Newkirk: J. Appl. Phys. **29**, 996 (1958)
158. J.B. Newkirk: Phys. Rev. **110**, 1465 (1958)
159. D.K. Bowen, B.K. Tanner: *High Resolution X-ray Diffraction and Topography* (Taylor & Francis Ltd, London 1998)
160. A.R. Lang: In: *International Tables of Crystallography*, Vol. C, ed. by A.J.C. Wilson, E. Prince (Kluwer Academic Publishers, Dordrecht 1999) pp. 113–123
161. U. Bonse, M. Hart: Appl. Phys. Lett. **6**, 155 (1965)
162. U. Bonse, M. Hart: Appl. Phys. Lett. **7**, 99 (1965)
163. B.W. Batterman: Phys. Rev. **A133**, 759 (1964)
164. B.W. Batterman: Phys. Rev. Lett. **22**, 703 (1969)
165. W.C. Marra, P. Eisenberger, A.Y. Cho: J. Appl. Phys. **50**, 6927 (1979)
166. G.H. Vineyard: Phys. Rev. B **26**, 4146 (1982)
167. A.M. Afanas'ev, M.K. Meldonyan: Acta Cryst. **A39**, 207 (1983)
168. S.R. Andrews, R.A. Cowley: J. Phys. C. **18**, 6427 (1986)
169. I.K. Robinson: Phys. Rev. B **33**, 3830 (1986)
170. P.M. Platzmann, N. Tzoar: Phys. Rev. B **2**, 3556 (1970)
171. F. de Bergevin, M. Brunnel: Phys. Lett. A **39**, 141 (1972)

172. M. Brunnel, F. de Bergevin: In: *Proc. Int. Conf. Magn. ICM-73, Vol. II.* (Nauka, Moscow 1974) pp. 127–130
173. D.H. Templeton, L.K. Templeton: Phys. Rev. B **3**, 1167 (1994)
174. V.A. Belyakov, V.A. Dmitrienko: Sov. Phys. Usp. **32**, 697 (1989)
175. E. Lamla: Ann. Phys. (Leipzig) **36**, 194 (1939)
176. N. Kato: Acta Cryst. **11**, 885 (1958)
177. E.J. Saccocio, A. Zajac: Acta Cryst. **18**, 478 (1965)
178. E.J. Saccocio, A. Zajac: Phys. Rev. **A139**, 255 (1965)
179. G. Hildebrandt: Phys. Status Solidi **15**, K131 (1966)
180. G. Hildebrandt: Phys. Status Solidi **24**, 245 (1967)
181. G. Hildebrandt: Acta Cryst. **A25**, 209 (1969)
182. G. Hildebrandt: Phys. Status Solidi **(a)15**, K83 (1973)
183. G. Hildebrandt: Krist. und Techn. **13**, 1095 (1978)
184. T. Joko, A. Fukuhara: J. Phsy. Soc. Jpn. **22**, 597 (1967)
185. P.P. Ewald, Y. Heno: Acta Cryst. **A24**, 5 (1968)
186. Y. Heno, P.P. Ewald: Acta Cryst. **A24**, 16 (1968)
187. P. Penning: Advan. X-Ray Anal. **10**, 67 (1967)
188. P. Penning: Philips Res. Rep. **23**, 12 (1968)
189. P. Penning, D. Polder: Philips Res. Rep. **23**, 1 (1968)
190. A.M. Afanas'ev, V.G. Kohn: Phys. Stat. Sol. **(a)28**, 61 (1975)
191. V.G. Kohn: Sov. Solid State Phys. **20**, 1152 (1975)
192. A.M. Afanas'ev, V.G. Kohn: Acta Cryst. **A32**, 308 (1976)
193. A.M. Afanas'ev, V.G. Kohn: Acta Cryst. **A33**, 178 (1977)
194. A. Dalisa, A. Zajac, C.H. Ng: Phys. Rev. **168**, 859 (1968)
195. W. Uebach, G. Hildebrandt: Z. Krist. **129**, 1 (1969)
196. V.G. Kohn: Sov. Solid State Phys. **18**, 2538 (1976)
197. V.G. Kohn, A.H. Toneyan: Acta Cryst. **A42**, 441 (1986)
198. Yu S. Terminasov, L.V. Tuzov: Usp. Fiz. Nauk. **83**, 223 (1964) [English transl.: Sov. Phys.-Usp. **7**, 734 (1964)
199. A. Zajac, E.J. Saccocio, J. Bindell: Bull. Amer. Phys. Soc. **9**, 50 (1964)
200. A. Zajac, A. Dalisa: Acta Cryst. **A21**, 27 (1966)
201. W. Uebach, G. Hildebrandt: Z. Krist. **129**, 1 (1968)
202. B. Post: Acta Cryst. **A25**, 94 (1969)
203. M. Umeno: Phys. Status Solidi **(a)2**, K203 (1970)
204. T.I. Borodina, A.A. Katsnelson, V.I. Kissin: Phys. Status Solidi **(a)3**, 105 (1970)
205. S. Balter, R. Feldman, B. Post: Phys. Rev. Lett. **27**, 307 (1971)
206. R. Feldman, B. Post: Phys. Status Solidi **(a)12**, 273 (1972)
207. M. Umeno: Phys. Status Solidi **(a)11**, 510 (1972)
208. V.I. Iveronova, A.A. Katsnelson, T.I. Borodina, I.G. Sapcova: Phys. Status Solidi **(a)11**, 39 (1972)
209. T.C. Huang, B. Post: Acta Cryst. **A29**, 35 (1973)
210. T.C. Huang, M.H. Tillinger, B. Post: Z. Naturforsch. **28a**, 600 (1973)
211. W. Uebach: Z. Naturforsch. **28a**, 1214 (1973)
212. T.I. Borodina, V.I. Iveronova, A.A. Katsnelson, T.K. Runova: Phys. Status Solidi **(a)28**, 365 (1975)
213. M. Umeno, G. Hildebrandt: Phys. Status Solidi **(a)31**, 583 (1975)
214. M. Umeno: Phys. Status Solidi **(a)37**, 561 (1976)
215. M. Umeno: Phys. Status Solidi **(a)38**, 701 (1976)
216. T.I. Borodina, V.I. Iveronova, A.A. Katsnelson, T.K. Runova: Sov. Krist. **20**, 490 (1975) [English transl.: Sov. Phys.-Cryst. **20**, 300 (1975)
217. S.A. Kshevetsky, I.P. Mikhailyuk: Sov. Krist. **21**, 381 (1976) [English transl.: Sov. Phys.-Cryst. **21**, 209 (1976)

218. V.D. Kos'mik, S.A. Kshevetsky, M.L. Kshevetskaya, I.P. Mikhailyuk, M.V. Ostapovich: Sov. Krist. **21**, 899 (1976) [English transl.: Sov. Phys.-Cryst. **21**, 514 (1976)
219. B. Post, S.L. Chang, T.C. Huang: Acta Cryst. **A33**, 90 (1977)
220. I.P. Mikhailyuk, V.D. Kos'mik, S.A. Kshevetsky: Sov. Krist. **22**, 1270 (1977)
221. I.P. Mikhailyuk, S.A. Kshevetsky, V.D. Kos'mik, M.V. Ostapovich: Sov. Krist. **23**, 399 (1978)
222. I.P. Mikhailyuk, S.A. Kshevetsky, V.D. Kos'mik, M.V. Ostapovich: Sov. Krist. **23**, 403 (1978)
223. I.P. Mikhailyuk, M.V. Ostapovich, S.A. Kshevetsky: Sov. Krist. **25**, 853 (1980)
224. R. Høier, A. Aanestad: Acta Cryst. **A37**, 787 (1981)
225. S.-L. Chang: Z. Naturforsch. **37a**, 501 (1982)
226. B. Post, P.W. Wang, T. Hom: Z. Naturforsch. **37a**, 528 (1982)
227. O. Pacherová, R. Bubáková: Z. Naturforsch. **37a**, 617 (1982)
228. S.A. Kshevetskii, Yu.P. Stetsko, S.A. Shelud'ko: Sov. Phys. Crystallogr. **30**, 270 (1985)
229. S.A. Kshevetskii, Yu.P. Stetsko: Sov. Tech. Phys. Lett. **11**, 576 (1985)
230. W.H. Zachariasen: Acta Cryst. **18**, 705 (1965)
231. W.H. Zachariasen: Acta Cryst. **18**, 1139 (1965)
232. S. Caticha-Ellis: Acta Cryst. **A25**, 666 (1969)
233. P. Coppens: Acta Cryst. **A24**, 253 (1968)
234. P.R. Prager: Acta Cryst. **A27**, 563 (1971)
235. C.B.R. Parente, S. Caticha-Ellis: Jpn. J. Appl. Phys. **13**, 1501 (1974)
236. K. Tanaka, Y. Saito: Acta Cryst. **A31**, 841 (1975)
237. B. Post: Acta Cryst. **A32**, 292 (1976)
238. S.-L Chang: Acta Cryst. **A38**, 41 (1982)
239. Y. Soejima, S. Okazaki, T. Matsumoto: Acta Cryst. **A41**, 128 (1985)
240. N.B. Bolotina, T.S. Chernaya, A.M. Golubev: Sov. Phys. Cryst. **35(2)**, 167 (1990)
241. C.B.R. Parente, V.L. Mazzocchi, F.J.F. Pementel: J. Appl. Cryst. **27**, 463 (1994)
242. J.M. Sasaki, L.P. Cardoso, C. Campos, K.J. Roberts, G.F. Clark, E. Pantos, M.A. Sacilotti: J. Appl. Cryst. **29**, 325 (1996)
243. E. Rossmanith: Z. Krist. **213**, 563 (1998)
244. E. Rossmanith: J. Appl. Cryst. **32**, 355 (1999)
245. E. Rossmanith: J. Appl. Cryst. **33**, 921 (2000)
246. R. Colella: Acta Cryst. **A30**, 413 (1974)
247. V.G. Kohn: Phys. Status Solidi **(a)54**, 375 (1979)
248. S.-L. Chang: Appl. Phys. Lett. **34**, 239 (1979)
249. S.-L. Chang: Acta Cryst. **A35**, 543 (1979)
250. S.-L. Chang: Acta Cryst. **A38**, 516 (1982)
251. K. Hümmer, H.W. Billy: Acta Cryst. **A38**, 841 (1982)
252. J. Hartwig: Krist. Tech. **13**, 1117 (1978)
253. S.A. Stepanov, A.P. Ulyanenkov: Acta Cryst. **A50**, 579 (1994)
254. X.R. Huang, M. Dudley, J.Y. Zhao: Acta Cryst. **A57**, 68 (2001)
255. Yu.P. Stetsko, S.-L. Chang: Acta Cryst. **A53**, 28 (1997)
256. M. Hart, A.R. Lang: Phys. Rev. Lett. **7**, 120 (1961)
257. B. Post: Phys. Rev. Lett. **39**, 760 (1977)
258. H. Jagodzinski: Acta Cryst. **A36**, 104 (1980)
259. L.D. Chapman, D.R. Yoder, R. Colella: Phys. Rev. Lett. **46**, 1578 (1981)
260. S.-L. Chang: Appl. Phys. **A26**, 221 (1981)
261. S.-L. Chang: Phys. Rev. Lett. **48**, 163 (1982)
262. H.J. Juretschke: Phys. Rev. Lett. **48**, 1487 (1982)

263. H.J. Juretschke: Phys. Lett. **92A**, 183 (1982)
264. S.L. Chang, F.S. Han: Acta Cryst. **A38**, 414 (1982)
265. G. Thorkildsen, F. Mo: 7th European Crystallography Meeting, Abstr. **1**, 3 (1982)
266. F.S. Han, S.L. Chang: Acta Cryst. **A39**, 98 (1983)
267. R. Hoier, K. Marthinsen: Acta Cryst. **A39**, 854 (1983)
268. B. Post: Acta Cryst. **A39**, 711 (1983)
269. P.P. Gong, B. Post: Acta Cryst. **A39**, 719 (1983)
270. H.J. Juretschke: Acta Cryst. **A40**, 379 (1984)
271. B. Post, J. Nicolosi, J. Ladell: Acta Cryst. **A40**, 684 (1984)
272. J.Z. Tischler, B.W. Batterman: Phys. Rev. **B30**, 7060 (1984)
273. M.C. Schmidt, R. Colella: Phys. Rev. Lett. **55**, 715 (1985)
274. J.Z. Tischler, Q. Shen, R. Colella: Acta Cryst. **A41**, 451 (1985)
275. S.-W. Luh, S.-L. Chang: Acta Cryst. **A44**, 662 (1988)
276. S.-L. Chang, Y.S. Huang, C.H. Chao, M.T. Tang, Y.P. Stetsko: Phys. Rev. Lett. **80**, 301 (1998)
277. S.L. Chang, J.A.P. Valladares: Appl. Phys. **A37**, 57 (1985)
278. S.L. Chang: Phys. Rev. **B33**, 5848 (1986)
279. K. Hümmer, H.W. Billy: Acta Cryst. **A42**, 127 (1986)
280. B. Post, P.P. Gong, L. Kern, J. Ladell: Acta Cryst. **A42**, 178 (1986)
281. H.J. Juretschke: Acta Cryst. **A42**, 405 (1986)
282. H.J. Juretschke: Acta Cryst. **A42**, 449 (1986)
283. J.Z. Tischler, B.W. Batterman: Acta Cryst. **A42**, 510 (1986)
284. K. Marthinsen, R. Hoier: Acta Cryst. **A42**, 484 (1986)
285. Q. Shen: Acta Cryst. **A42**, 525 (1986)
286. Q. Shen, R. Colella: Acta Cryst. **A42**, 533 (1986)
287. F. Mo, B.C. Hauback, G. Thorkildsen: Acta Chem. Scand. **A42**, 130 (1988)
288. K. Hümmer, H. Bondza, E. Weckert: Acta Cryst. **A43**, C222 (1987)
289. E. Weckert, H. Bondza, K. Hümmer: Acta Cryst. **A43**, C264 (1987)
290. S.A. Kshevetskii, Yu.P. Stetsko, S.A. Shelud'ko: Sov. Phys. Krist. **32**, 177 (1987)
291. S.-L. Chang, M.T. Tang: Acta Cryst. **A44**, 1068 (1988)
292. M.T. Tang, S.-L. Chang: Acta Cryst. **A44**, 1073 (1988)
293. E. Weckert, K. Hümmer: Acta Cryst. **A46**, 387 (1990)
294. Q. Shen, R. Colella: Nature **329**, 232 (1987)
295. K. Hümmer, E. Weckert, H. Bondza: Acta Cryst. **A45**, 182 (1989)
296. S.-L. Chang, M.T. Huang, M.T. Tang, C.H. Lee: Acta Cryst. **A45**, 870 (1989)
297. S.C. Chien, M.T. Tang, S.-L. Chang: Acta Cryst. **A45**, 703 (1989)
298. K. Hümmer, E. Weckert: Acta Cryst. **A46**, 534 (1990)
299. K. Hümmer, H. Bondza, E. Weckert: Z. Krist. **195**, 169 (1991)
300. K. Hümmer, W. Schwegle, E. Weckert: Acta Cryst. **A47**, 60 (1991)
301. S.-L. Chang, H.E. King, Jr., M.T. Huang, Y. Gao: Phys. Rev. Lett. **67**, 3113 (1991)
302. M.T. Huang, S.-L. Chang: Acta Cryst. **A48**, 541 (1992)
303. M.T. Huang, C.M. Wang, S.-L. Chang: Acta Cryst. **A50**, 342 (1994)
304. K. Hümmer, E. Weckert: Acta Cryst. **A51**, 431 (1995)
305. E. Weckert, K. Hümmer: Acta Cryst. **A53**, 108 (1997)
306. S.-L. Chang: Acta Cryst. **A54**, 886 (1998)
307. F. Mo, R.H. Mathiesen, P.M. Alzari, J. Lescar, B. Rasmussen: In*Materials Structure in Chemistry, Biology, Physics and Technology*, Vol. 5, ed. by R. Kuzel, J. Lhotha, L. Dobiasova (Czech and Slovak Crystallographic Associaton, Praha 1998) p. 484

308. R.H. Mathiesen, F. Mo, A. Eikenes, T. Nyborg, H.B. Larsen: Acta Cryst. **A54**, 338 (1998)
309. Q. Shen: Phys. Rev. Lett. **80**, 3268 (1998)
310. S.-L. Chang, C.-H. Chao, Y.-S. Huang, Y.-C. Jean, H.-S. Sheu, F.-J. Liang, H.-C. Chien, C.-K. Chen, H.-S. Yuan: Acta Cryst. **A55**, 933 (1999)
311. E. Weckert, K. Holzer, K. Schroer, J. Zellner, K. Hümmer: Acta Cryst. **D55**, 1320 (1999)
312. Yu.P. Stetsko, S.-L. Chang: Acta Cryst. **A55**, 457 (1999)
313. Yu.P. Stetsko, S.-L. Chang: Acta Cryst. **A55**, 683 (1999)
314. Q. Shen: Phys. Rev. Lett. **83**, 4784 (1999)
315. Q. Shen, X.-R. Huang: Phys. Rev. B **63**, 174102 (2001)
316. S.-L. Chang, Yu.P. Stetsko, Y.-S. Huang, C.-H. Chao, F.-J. Liang, C.-K. Chen: Phys. Lett. **A264**, 328 (1999)
317. K. Holzer, E. Weckert, K. Schroer: Acta Cryst. **D56**, 322 (2000)
318. Q. Shen, S.Kycia, I. Dobrianov: Acta Cryst. **A56**, 264 (2000)
319. Q. Shen, S. Kycia, I. Dobrianov: Acta Cryst. **A56**, 268 (2000)
320. Q. Shen: Phys. Rev. **B61**, 8593 (2000)
321. C.M. Weeks, H. Xu, H.A. Hauptman, Q. Shen: Acta Cryst. **A56**, 280 (2000)
322. Yu.P. Stetsko, H.J. Juretschke, Y.-S. Huang, C.-H. Chao, C.-K. Chen, S.-L. Chang: Acta Cryst. **A56**, 394 (2000)
323. Yu.P. Stetsko, G.-Y. Lin, Y.-S. Huang, C.-H. Chao, S.-L. Chang: Phys. Rev. Lett. **86**, 2026 (2001)
324. Yu.P. Stetsko, H.J. Juretschke, Y.-S. Huang, Y.R. Lee, T.C. Lin, S.-L. Chang: Acta Cryst. **A57**, 359 (2001)
325. C.-M. Wang, C.-H. Chao, S.-L. Chang: Acta Cryst. **A57**, 420 (2001)
326. C.-H. Chao, C.-Y. Hung, Y.-S. Huang, C.-H. Ching, Y.-R. Lee, Y.-C. Jean, S.-C. Lai, Yu.P. Stetsko, H. Yuan, S.-L. Chang: Acta Cryst. **A58**, 33 (2002)
327. G. Thorkildsen: Acta Cryst. **A43**, 361 (1987)
328. K. Hümmer, E. Weckert, H. Bondza: Acta Cryst. **A46**, 393 (1990)
329. M.T. Tang, S.-L. Chang: Phys. Lett. A. **143/08**, 405 (1990)
330. Q. Shen, K.D. Finkelstein: Phys. Rev. Lett. **65**, 3337 (1990)
331. K. Hümmer, W. Schwegle, E. Weckert: Acta Cryst. **A45**, 60 (1991)
332. E. Weckert, R. Müller, J. Zellner, I. Zezegers, R. Loris: Z. Krist. **217**, 651 (2002)
333. A.S. Soares, D.L.D. Caspar, E. Weckert, A. Héroux, K. Hölzer, K. Schroer, J. Zellner, D. Schneider, W. Nolan, R.M. Sweet: Acta Cryst. **D59**, 1716 (2003)
334. S.L. Morelhao, S. Kycia: Phys. Rev. Lett. **89**, 015501-1 (2002)
335. S.L. Morelhao: Acta Cryst. **A59**, 470 (2003)
336. G. Thorkildsen, H.B. Larsen, E. Weckert, D. Semmingsen: J. Appl. Cryst. **36**, 1324 (2003)
337. V.M. Kaganer, M. Albrecht, A. Hirnet, M. Gierer, W. Moritz, B. Jenichen, K.H. Ploog: Phys. Rev. B **61**, R16 355 (2000)
338. Y.-S. Huang, C.-H. Chao, Yu.P. Stetsko, Y.R. Lee, C.H. Ching, C.Y. Huang, G.Y. Lin, T.C. Lin, S.-L. Chang: Phys. Rev. B **64**, 085406-1 (2001)
339. T.L. Lee, R. Felici, K. Hirano, B. Cowie, J. Zegenhagen, R. Colella: Phys. Rev. B **64**, 201316 (2001)
340. K. Lonsdale: Philos. Trans. **A240**, 219 (1947)
341. D.R. Schwartzenberger: Philos. Mag. **47**, 1242 (1959)
342. B.H. Heise: J. Appl. Phys. **33**, 938 (1962)
343. E.T. Peters, R.E. Ogilvie: Trans. AIME **233**, 89 (1965)
344. B.J. Isherwood, C.A. Wallace: Nature **212**, 173 (1966)
345. W.G. Morris: J. Appl. Phys. **39**, 1813 (1968)
346. F.J. Spooner, C.G. Wilson: J. Appl. Cryst. **6**, 132 (1973)

347. B. Post: J. Appl. Cryst. **8**, 452 (1975)
348. T. Hom, W. Kiszenik, B. Post: J. Appl. Cryst. **8**, 457 (1975)
349. S.-L. Chang: Appl. Phys. Lett. **37**, 819 (1980)
350. B.R. Brown, M.A.G. Halliwell, B.J. Isherwood: J. Microscopy **118**, 375 (1980)
351. S.-L. Chang: Acta Cryst. **A37**, 876 (1981)
352. B.J. Isherwood, B.R. Brown, M.A.G. Halliwell: J. Cryst. Growth **54**, 449 (1981)
353. S.L. Morelhão, L.P. Cardoso: J. Appl. Phys. **73**, 4218 (1993)
354. S.L. Morelhão, L.P. Cardoso: Solid State Commun. **88**, 465 (1993)
355. S.L. Morelhão, L.P. Cardoso: J. Appl. Cryst. **29**, 446 (1996)
356. M.A. Hayashi, S.L. Morelhão, L.H. Avanci, L.P. Cardoso, J.M. Sasaki, L.C. Kretly, S.-L. Chang: Appl. Phys. Lett. **71**, 2614 (1997)
357. S.L. Morelhão, L.H. Avanci, M.A. Hayashi, L.P. Cardoso, S.P. Collins: Appl. Phys. Lett. **73**, 2194 (1998)
358. S.L. Morelhão, E. Abramof: J. Appl. Cryst. **32**, 871 (1999)
359. S.L. Morelhão, L.H. Avanci, M.A. Hayashi, L.P. Cardoso: Mater. Res. Soc. Proc. **561**, 87 (1999)
360. L.H. Avanci, S.L. Morelhão: Acta Cryst. **A56**, 507 (2000)
361. M.A. Hayashi, L.H. Avanci, L.P. Cardoso, J. Bettini, M.M.G. de Carvalho, S.L. Morelhão, S.P. Collins: J. Synch. Rad. **6**, 29 (1999)
362. S.L. Morelhão, L.H. Avanci, A.A. Quivy, E. Abramof: J. Appl. Cryst. **35**, 69 (2002)
363. B.S. Fraenkel: Appl. Phys. Lett. **36**, 341 (1980)
364. B.S. Fraenkel: X-Ray Spectrom. **9**, 189 (1980)
365. B.S. Fraenkel: Appl. Phys. Lett. **41(3)**, 234 (1982)
366. C.S.G. Cousins, L. Gerward, J. Staun Olsen: Phys. Status Solidi **(a)48**, 113 (1978)
367. D.A. Kottwitz: Acta Cryst. **A24**, 117 (1968)
368. D.A. Kottwitz: Phys. Rev. **175**, 1056 (1968)
369. D.A. Kottwitz: Acta Cryst. **A25**, 459 (1969)
370. P. Mikula, R.T. Michalec, M. Vrána, J. Vávra: Acta Cryst. **A35**, 962 (1979)
371. P. Mikula, M. Vrána, R.T. Michalec, J. Kulda, J. Vávra: Phys. Status Solidi **(a)60**, 549 (1980)
372. M. Vrána, P. Mikula, R.T. Michalec, J. Kulda, J. Vávra: Acta Cryst. **A37**, 459 (1981)
373. P. Mikula, J. Kulda, M. Vrána, R.T. Michalec, J. Vávra: Nucl. Instrum. Meth. **197**, 563 (1982)
374. S.-L. Chang: Appl. Phys. Lett. **40(9)**, 793 (1982)
375. W. Graeff, U. Bonse: Z. Phys. **B27**, 19 (1977)
376. Q. Shen, K.D. Finkelstein: Phys. Rev. B **45**, 5075 (1992)
377. L.-Y. Jang, M.-T. Huang, S.-L. Chang: Nucl. Instr. and Meth. in Phys. Res. A **372**, 543 (1996)
378. T. Ishikawa, K. Hirano, K. Kanzaki, S. Kikuta: Rev. Sci. Instrum. **63**, 1098 (1992)
379. N. Greiser, G. Materlik: Z. Phys. B – Condensed Matter **66**, 83 (1987)
380. G. Materlik, P. Suortti: J. Appl. Cryst. **17**, 7 (1984)
381. L.H. Avanci, L.P. Cardoso, S.E. Girdwood, D. Pugh, J.N. Sherwood, K.J. Roberts: Phys. Rev. Lett. **81**, 5426 (1998)
382. L.H. Avanci, L.P. Cardoso, J.M. Sasaki, S.E. Girdwood, K.J. Roberts, D. Pugh, J.N. Sherwood: Phys. Rev. B **61**, 6507 (2000)
383. J. Stahn, A. Pucher, U. Pietsch, J. Zellner, E. Weckert: Acta Cryst. **A55**, 1034 (1999)

384. X. Lai, K.J. Roberts, L.H. Avanci, L.P. Cardoso, J.M. Sasaki: J. Appl. Cryst. **36**, 1230 (2003)
385. J.M.A. Almeida, M.A.R. Miranda, C.M.R. Remédios, F.E.A. Melo, P.T.C. Freire, J.M. Sasaki, L.P. Cardoso, A.O. dos Santos, S. Kycia: J. Appl. Cryst. **36**, 1348 (2003)
386. K. Huang: *Statistical Mechanics* (John Wiley & Sons, New York, London, Sydney 1963) p. 76, p. 141
387. J.W.M. DuMond: Phys. Rev. **52**, 872 (1937) J.W.M. DuMond, J.P. Youtz: J. Appl. Phys. **11**, 357 (1940)
388. S. Krinsky, M.L. Perlman, R.E. Watson: 'Characteristics of Synchrotron Radiation and of its Sources'. In: *Handbook on Synchrotron Radiation*, Vol. 1A, ed. by E.-E. Koch (North-Holland, Amsterdam, New York, Oxford 1983) pp. 65–171
389. T. Matsushita, H. Hashizume: 'X-Ray Monochromators'. In: *Handbook on Synchrotron Radiation*, Vol. 1A, ed. by E.-E. Koch (North-Holland, Amsterdam, New York, Oxford 1983) pp. 261–314
390. E.-E. Koch (ed.): *Handbook on Synchrotron Radiation*, Vol. 1A (North-Holland, Amsterdam, New York, Oxford 1983)
391. E. Merzbacher: *Quantum Mechanics* (John Wiley & Sons, New York 1961)
392. A. Messiah: *Quantum Mechanics* (North-Holland, Amsterdam 1966)
393. J.J. Sakurai: *Advanced Quantum Mechanics* (Addison-Wesley, New York 1967)
394. J.D. Jackson: *Classical Electrodynamics* (John Wiley & Sons, New York 1967)
395. B. Crasemann (ed.): *Atomic Inner-Shell Physics* (Plenum, New York 1985)
396. H.A. Kramers, W. Heisenberg: Z. Phys. **31**, 681 (1925)
397. A.H. Compton, S.K. Allison: *X-Rays in Theory and Experiment*, 2nd edn. (D. Van Nostrand, Princeton 1935)
398. L.V. Azaroff: *Elements of X-Ray Crystallography* (McGraw-Hill, New York 1968)
399. R.L. Mössbauer: Z. Phys. **151**, 124 (1958)
400. D.P. Siddons, J.B. Hastings, G. Faigel, J.R. Grover, P.E. Haustein, L.E. Berman: Rev. Sci. Instrum. **60**, 1649 (1989)
401. R. Kronig: J. Opt. Soc. Am. **12**, 547 (1926)
402. H.A. Kramers: *Collected Scientific Papers* (North-Holland, Amsterdam 1956)
403. P.M. Morse, H. Feshbach: *Methods of Theoretical Physics* (McGraw-Hill, New York 1953) p. 944
404. N.F.M. Henry, K. Lonsdale (eds.): *International Tables for X-Ray Crystallography*, Vol. 1 (Kynoch, Birmingham 1968)
405. P.P. Ewald (eds.): *Fifty Years of X-Ray Diffraction* (A. Oosthoek's Uitgeversmaatschappij N.V., Utrecht 1962)
406. P. Debye: Verh. Dtsch. Phys. Gesell. **15**, 678 (1913) Ann. D. Physik **43**, 49 (1914)
407. P. Debye: Phys. Z. **31**, 419 (1930)
408. I. Waller: Z. Phys. **17**, 398 (1923)
409. R.W. James: *The Optical Principle of the Diffraction of X-Rays* (Bell, London 1950)
410. R.D. Burbank: Acta Cryst. **19**, 957 (1965)
411. D. Unangst, W. Melle: Acta Cryst. **A31**, 234 (1975)
412. R.S. Williamson, I. Fankuchen: Rev. Sci. Instrum. **30**, 908 (1959)
413. H. Cole, F.W. Chambers, H.M. Dunn: Acta Cryst. **15**, 138 (1962)
414. S.-L. Chang, S. Caticha-Ellis: Acta Cryst. **A34**, 825 (1978)
415. E. Rutherford, E.N. da C. Andrade: Philos. Mag. **28**, 263 (1914)
416. W.H. Bragg : Nature **43**, 31 (1914)
417. H. Seemann: Ann. Phys. (Leipzig) **51**, 391 (1916)

418. H. Seemann: Ann. Phys. (Leipzig) **53**, 461 (1917)
419. H. Seemann: Phys. Z. **20**, 169 (1919)
420. W. Gerlach: Phys. Z. **22**, 557 (1921)
421. W. Gerlach: Verh. Phys. Med. Ges. Würzburg **56**, 55 (1921)
422. H. Seemann: Ann. Phys. (Leipzig) **6**, 793 (1930) **7**, 633 (1930)
423. H. Seemann, O. Kantorowicz: Naturwissenschaften **18**, 526 (1930)
424. B. Hess: Z. Krist. **97**, 197 (1937)
425. B. Hess: Z. Krist. **104**, 294 (1942)
426. W. Linnki: Nature **124**, 946 (1929)
427. W. Linnki: Z. Phys. **61**, 220 (1930)
428. G. Borrmann: Naturwissenschaften **23**, 591 (1935)
429. G. Borrmann: Ann. Phys. (Leipzig) **27**, 669 (1936)
430. R. Castaing, A. Guinier: C. R. Acad. Sci. Paris **232**, 1948 (1951)
431. J.L. McCall, K.L. Strabel, J.S. Duerr: J. Appl. Phys. **38**, 2695 (1967)
432. B.A. Newman, S. Weissmann: J. Appl. Cryst. **1**, 139 (1968)
433. T. Imura, S. Weismann, J.J. Slade Jr.: Acta Cryst. **15**, 786 (1962)
434. H. Yakowitz: J. Appl. Phys. **37**, 4455 (1966)
435. T. Ellis, L.G. Naani, A. Shrier, S. Weissmann, G.E. Padawer, N. Hosokawa: J. Appl. Phys. **35**, 3364 (1964)
436. T. Fujiwara: J. Sci. Hiroshima Univ. **A7**, 179 (1937)
437. T. Fujiwara: J. Sci. Hiroshima Univ. **A9**, 233 (1939)
438. B.J. Isherwood, C.A. Wallace: Acta Cryst. **A27**, 119 (1971)
439. E.W. Nuffield: *X-Ray Diffraction Methods* (John Wiley & Sons, New York 1966)
440. R.E. Hanneman, R.E. Ogilvie, A. Modrezjewski: J. Appl. Phys. **33**, 1429 (1962)
441. R. Tixier, C. Waché: J. Appl. Cryst. **3**, 466 (1968)
442. J. Frazer, G. Arrhenius: In: *Proceedings of the IVth Congress on X-Ray Optics and Microanalysis* (Hermann, Paris 1966) p. 516
443. K.J.H. Mackay: In: *Proceedings of the IVth Congress on X-Ray Optics and Microanalysis* (Hermann, Paris 1966) p. 544
444. P.L. Ryder, H. Halbig, W. Pitsch: In: *Proceedings of the VIIth Congress on X-Ray Optics and Microanalysis*, ed. by G. Moellenstedt, K. H. Gaukle (Springer, Berlin, Heidelberg, New York 1969) p. 388
445. B.D. Cullity: *Elements of X-Ray Diffraction* (Addison-Wesley, Reading 1978)
446. S. Kikuchi: Proc. Imp. Acad. Jpn. **4**, 354 (1928)
447. G.H. Stout, L.H. Jensen: *X-Ray Structure Determination*, 2nd edn. (John Wiley & Sons, New York 1989)
448. C. Giacovazzo, H.L. Monaco, D. Viterbo, F. Scordari, G. Gilli, G. Zanotti, M. Catti: *Fundamentals of Crystallography*, ed. by C. Giacovazzo (Oxford University Press, New York 1992)
449. H.L. Yakel, I. Fankuchen: Acta Cryst. **15**, 1188 (1962)
450. H.L. Yakel, I. Fankuchen, J.W. Jeffery, A. Whitaker: Acta Cryst. **19**, 963 (1965)
451. A. Santoro, M. Zocchi: Acta Cryst. **21**, 293 (1966)
452. R. Hoier, A. Aanestad: Acta Cryst. **A37**, 787 (1981)
453. S.-L. Chang: Cryst. Rev. **1**, 87 (1987)
454. S.-L. Chang, M. T. Tang: Acta Cryst. **A44**, 1065 (1988)
455. W.A. Hendrickson: Science **254**, 51 (1991)
456. Q. Shen: Phys. Rev. B **59**, 11109 (1999)
457. S.-L. Chang, C.H. Chao, Y.S. Huang, Y.C. Jean, H.S. Sheu, F.J. Liang, H.C. Chien, C.K. Chen, H.S. Yuan: Submitted to Nature on Feb. 28, 1998 (unpublished)

458. S.-L. Chang: 'Direct Determination of X-ray Reflection Phases Using Multiple Diffraction: Theory and Experiments', 3rd Conference of the Asian Crystallographic Association at Bangi, Malaysia (1998), paper PL-3
459. C.H. Chao, Y.S. Huang, F.J. Liang, H.C. Chien, C.K. Chen, S.-L. Chang, Y.C. Jean, H.S. Sheu, H.S. Yuan: 'Direct Phase Determination for Macromolecular Crystals by Stereoscopic Multi-Beam Imaging', 3rd Conference of the Asian Crystallographic Association at Bangi, Malaysia (1998), paper 15P7
460. E.K. Kov'ev, A.V. Andreev, M.I. Deigen: Sov. Phys. Solid State **26**, 1927 (1984)
461. Z. Otwinowski, W. Minor: Methods Enzymol. **276**, 307 (1997)
462. M. Tegze, G. Faigel, S. Marchesini, M. Belakhovsky, A.I. Chumakov: Phys. Rev. Lett. **82**, 4847 (1999)
463. G. Thorkildsen, F.H. Mathiesen, H.B. Larsen: J. Appl. Cryst. **32**, 943 (1999)
464. G. Thorkildsen, H.B. Larsen, R.H. Mathiesen, F. Mo: J. Appl. Cryst. **33**, 49 (2000)
465. T.-S. Gau, Y.-C. Jean, K.-Y. Liu, C.-H. Chung, C.-K. Chen, S.-C. Lai, C.-H. Shu, Y.-S. Huang, C.-H. Chao, Y.-R. Lee, C.T. Chen, S.-L. Chang: Nucl. Instrum. & Meth. in Phys. Res. **A466**, 569 (2001)
466. M. von Laue: Ann. Phys. (Leipzig) **41**, 971 (1913)
467. W.C. Hamilton: Acta Cryst. **10**, 629 (1957)
468. W.H. Zachariasen: Acta Cryst. **23**, 558 (1967)
469. W.H. Zachariasen: Acta Cryst. **A24**, 421 (1968)
470. E. Rossmanith: J. Appl. Cryst. **36**, 1467 (2003)
471. P.J. Becker, P. Coppens: Acta Cryst. **A30**, 129, 148 (1974)
472. P.J. Becker, P. Coppens: Acta Cryst. **A31**, 417 (1975)
473. S.W. Wilkins: Phil. Trans. Roy. Soc. (London) **299**, 275 (1981)
474. T.M. Sabine: Acta Cryst. **A44**, 368 (1988)
475. T.M. Sabine: In: *International Tables of Crystallography*, Vol. C, ed. by A.J.C. Wilson, E. Prince (Kluwer Academic Publishers, Dordrecht 1999) pp. 603–607
476. H.-C. Hu: Acta Cryst. **A53**, 484 (1997)
477. H.-C. Hu: Acta Cryst. **A53**, 493 (1997)
478. M.M. Woolfson, H.-F. Fan: *Physical and Non-Physical Methods of Solving Crystal Structures* (Cambridge University Press, Cambridge 1995)
479. J. Nicolosi: Ph.D thesis, Polytechnic Institute of New York (1982)
480. J.P. Hannon: In: *Resonant Anomalous X-Ray Scattering*, ed. by G. Materlink, G.J. Sparks, K. Fischer (Elsevier Science B.V., Amsterdam 1994) p.565
481. D.H. Templeton, L.K. Templeton: Acta Cryst. **A36**, 237 (1980)
482. V.E. Dmitrienko: Acta Cryst. **A40**, 89 (1984)
483. V.E. Dmitrienko, E.N. Ovchinnikova, K. Ishida: JETP Lett. **69**, 938 (1999)
484. V.E. Dmitrienko, E.N. Ovchinnikova: Acta Cryst. **A56**, 340 (2000)
485. D.H. Templeton, L.K. Templeton: Acta Cryst. **A41**, 133 (1985)
486. D.H. Templeton, L.K. Templeton: Acta Cryst. **A42**, 478 (1986)
487. A. Kirfel, A. Petcov, K. Eichhorn: Acta Cryst. **A47**, 180 (1991)
488. K.D. Finkelstein, Q. Shen, S. Shastri: Phys. Rev. Lett. **69**, 1612 (1992)
489. S. Miyake, Y.H. Ohtsuki: Acta Cryst. **A30**, 103 (1974)
490. E. Yablonovitch: Phys. Rev. Lett. **58**, 2059 (1987)
491. E. Yablonovitch, T.J. Gmitter: Phys. Rev. Lett. **63**, 1950 (1989)
492. E. Yablonovitch, T.J. Gmitter, K.M. Leung: Phys. Rev. Lett. **67**, 2295 (1991)
493. S. John: Phys. Rev. Lett. **58**, 2486 (1987)
494. J.S. McCalmont, M.M. Sigalas, G. Tuttle, K.-M. Ho, C.M. Soukolis: Phys. Rev. Lett. **68**, 2759 (1996)
495. K.M. Leung, Y.F. Liu: Phys. Rev. Lett. **65**, 2646 (1990)

496. K.M. Ho, C.T, Chan, C.M. Soukoulis: Phys. Rev. Lett. **65**, 3152 (1990)
497. H.S. Sözüer, J.W. Haus: Phys. Rev. B **45**, 13962 (1992)
498. N. Kato: J. Phys. Soc. Jpn. **7**, 397 (1952)
499. N. Kato, T. Katagawa, T. Saka: Krystallogr. **16**, 1110 (1971)
500. B. Okkerse: Philips Res. Repts. **17**, 464 (1962)
501. H. Jeffreys, B. Jeffreys: *Methods of Mathematical Physics* (Cambridge University Press, Cambridge 1956) p. 503
502. M. Born, E. Wolf: *Principles of Optics*, 6th edn. (Pergamon, New York 1983)
503. H. Schlangenotto: Z. Phys. **203**, 17 (1967)
504. R.S. Martin, J.H. Wilkinson: Numer. Math. **12**, 349 (1968)
505. F.R. Gantmacher: *The Theory of Matrices* (Chelsea, New York 1960)
506. I. Waller: Ann. Phys. **79**, 261 (1929)
507. F. Miller: Phys. Rev. **47**, 209 (1935)
508. J.D. Jackson: *Classical Electrodynamics*, 2nd edn. (John Wiley & Sons, New York 1974)
509. H. Yoshioka: J. Phys. Soc. Jpn. **12**, 618 (1957)
510. L.I. Schiff: *Quantum Mechanics*. (McGraw-Hill, New York 1968)
511. A.J. Fetter, J.D. Walecka: *Quantum Theory of Many-Particle Systems*. (McGraw-Hill, New York 1971)
512. J.C. Inkson: *Many-Body Theory of Solids*. (Plenum Press, New York 1984)
513. G. Thorkildsen, H.B. Larsen: Acta Cryst. **A54**, 120 (1998)
514. H.B. Larsen, G. Thorkildsen: Acta Cryst. **A54**, 129 (1998)
515. H.B. Larsen, G. Thorkildsen: Acta Cryst. **A54**, 137 (1998)
516. G. Thorkildsen, H.B. Larsen, E. Weckert: Acta Cryst. **A57**, 389 (2001)
517. A. Authier, C. Malgrange, M. Tournarie: Acta Cryst. **A24**, 126 (1968)
518. P. Becker: Acta Cryst. **A33**, 667 (1977)
519. P. Becker, F. Dunstetter: Acta Cryst. **A40**, 241 (1984)
520. J. Bremer, G. Thorkildsen: Acta Cryst. **A42**, 191 (1986)
521. M. Gell-Mann, M.L. Goldberger: Phys. Rev. **91**, 398 (1955)
522. W. Tobocman: *Theory of Direct Nuclear Reactions*. (Oxford University Press, New York 1961)
523. G. Thorkildsen, H.B. Larsen: Acta Cryst. **A58**, 252 (2002)
524. C. Kittel: In: *Introduction to Solid State Physics* (Wiley, New York 1971) p.697
525. P. Farwig, H.W. Schürmann: Z. Physik **204**, 489 (1967)
526. S. Kishino: J. Phys. Soc. Jpn. **31**, 1168 (1971)
527. S. Kishino, K. Kohra: Jpn. J. Appl. Phys. **10**, 551 (1971)
528. S. Kishino, A. Noda, K. Kohra: J. Phys. Soc. Jpn. **33**, 158 (1972)
529. T. Bedynska: Phys. Stat. Sol. (a) **19**, 365 (1973)
530. T. Bedynska: Phys. Stat. Sol. (a) **25**, 405 (1974)
531. F. Fustichelli: Phil. Mag. **31**, 1 (1975)
532. J. Härtwig: Phys. Stat. Sol. (a) **37**, 417 (1976)
533. O. Brümmer, H.R. Höche, J. Nieber: Phys. Stat. Sol. (a) **33**, 587 (1976)
534. O. Brümmer, H.R. Höche, J. Nieber: Phys. Stat. Sol. (a) **37**, 529 (1976)
535. J. Härtwig: Phys. Stat. Sol. (a) **42**, 495 (1977)
536. J. Härtwig: Exper. Technik der Physik (Berlin) **26**, 131 (1978)
537. J. Härtwig: Exper. Technik der Physik (Berlin) **26**, 447 (1978)
538. J. Härtwig: Exper. Technik der Physik (Berlin) **26**, 535 (1978)
539. O. Brümmer, H.R. Höche, J. Nieber: Phys. Stat. Sol. (a) **46**, K131 (1978)
540. J. Härtwig: Acta Cryst. **A37**, 802 (1981)
541. O. Brümmer, H.R. Höche, J. Nieber: Z. Naturforsch. (a) **37**, 519 (1982)
542. H.R. Höche, J. Nieber: Phys. Stat. Sol. (a) **108**, 489 (1988)

543. H.R. Höche, J. Nieber, M. Clausnitzer, G. Materlik: Phys. Stat. Sol. (a) **105**, 53 (1988)
544. H.-H. Hung, S.-L. Chang: Acta Cryst. **A45**, 823 (1989)
545. A.M. Afanas'ev, R.M. Imamov, E.Kh. Mukhamedzhanov: Crystal. Rev. **3**, 157 (1992)
546. H.-H. Hung, S.-L. Chang: Europhys. Lett. **23(6)**, 415 (1993)
547. V. Holý: In: *X-ray and Neutron Dynamical Diffraction: Theory and Applications, NATO ASI Series, Series B, Physics, Vol. 357*, ed. by A. Authier, S. Lagomarsino, B.K. Tanner (Plenum Press, New York and London 1996) pp. 33–42
548. A. Authier: Cryst. Res. Technol. **33**, 517 (1998)
549. A. Zeilinger, T.J. Beatty: Phys. Rev. B **27**, 7239 (1983)
550. P.A. Aleksandrov, A.M. Afanas'ev, S.A. Stepanov: Kristallografiya **29**, 197 (1984)
551. A.M. Afanas'ev, S.M. Afanas'ev, P.A. Aleksandrov, R.M. Imamov, E.M. Pashaev: Phys. Stat. Sol. (a) **86**, K1 (1984)
552. A.M. Afanas'ev, R.M. Imamov, A.V. Maslov, E.M. Pashaev: Phys. Stat. Sol. (a) **84**, 73 (1984)
553. A.L. Golovin, R.M. Imamov, S.A. Stepanov: Acta Cryst. **A40**, 225 (1984)
554. A.L. Golovin, R.M. Imamov, E.A. Kondrashikina, V. Novikov: Sov. Phys. Technical Physics **30**, 834 (1985)
555. P.L. Cowan: Phys. Rev. B **32**, 5437 (1985)
556. H.R. Höche, O. Brümmer, J. Nieber: Acta Cryst. **A42**, 585 (1986)
557. O. Sakata, H. Hashizume: Rep. Res. Lab. Eng. Mat. Tokyo Inst. Techn. **12**, 45 (1987)
558. O. Sakata, H. Hashizume: Jpn. J. Appl. Phys. **27**, L1976 (1988)
559. O. Sakata, H. Hashizume: Rep. Res. Lab. Eng. Mat. Tokyo Inst. Techn. **16**, 27 (1991)
560. H. Hashizume, O. Sakata: Colloques de Physique **50**, C7-225 (1989)
561. H. Hashizume, O. Sakata: Rev. Sci. Instrum. **60**, 2373 (1989)
562. S.M. Durbin, T. Gog: Acta Cryst. **A45**, 132 (1989)
563. T. Jach, P.L. Cowan, Q. Shen, M.J. Bedzyk: Phys. Rev. B **39**, 5739 (1989)
564. A.M. Afanas'ev, O.G. Melikyan: Phys. Stat. Sol. (a) **122**, 459 (1990)
565. A.M. Afanas'ev, A.V. Esayan: Phys. Stat. Sol. (a) **126**, 303 (1991)
566. T.-P. Tseng, S.-L. Chang: Acta Cryst. **A46**, 567 (1990)
567. X. Huang, M. Dudley: Acta Cryst. **A59**, 163 (2003)
568. K. Kohra, T. Matsushita: Z. Naturforsch. (a) **27**, 484 (1972)
569. O. Brümmer, H.R. Höche, J. Nieber: Phys. Stat. Sol. (a) **53**, 565 (1979)
570. A. Steyerl, K.A. Steinhauser: Zeit. Phys. B **34**, 221 (1979)
571. A. Caticha, S. Caticha-Ellis: Phys. Rev. B **25**, 971 (1982)
572. W. Graeff, G. Materlik: Nucl. Instrum. Meth. **195**, 97 (1982)
573. R. Colella, A. Luccio: Optics Commun. **50**, 41 (1984)
574. V.I. Kushnir, E.V. Suvorov: JETP **44**, 205 (1986)
575. B. Dorner, E. Burkel, J. Peisl: Nucl. Instrum. & Meth. in Phys. Res. A **246**, 450 (1986)
576. D.P. Woodruff, D.L. Seymour, C.F. Conville, C.E. Riley, M.D. Crapper, N.P. Prince, R.G. Jones: Phys. Rev. Lett. **58**, 1460 (1987)
577. H. Hashizume, T. Nakahata: Jpn. J. Appl. Phys. **27**, L1568 (1988)
578. A. Caticha, N. Caticha, S. Caticha-Ellis: Appl. Phys. Lett. **54**, 887 (1989)
579. A. Caticha, S. Caticha-Ellis: Phys. Stat. Sol. (a) **119**, 47 (1990)
580. A. Caticha, S. Caticha-Ellis: Phys. Stat. Sol. (a) **119**, 643 (1990)
581. C. Giles, C. Cusatis: Appl. Phys. Lett. **59**, 641 (1991)

582. H. Hashizume, M. Sugitama, T. Niwa, O. Sakata, P.L. Cowan: Rev. Sci. Instrum. **63**, 1142 (1992)
583. A. Caticha: Phys. Rev. B **49**, 33 (1994)
584. Yu.P. Stetsko, S.A. Kshevetskii, I.P. Mikhailyuk: Sov. Tech. Phys. Lett. **14**, 13 (1988); Yu.P. Stetsko: Ph.D. dissertation, Chernovtsy State University, Ukraine (1990)
585. L. De Caro, L. Tapfer: Phys. Rev. B **55**, 105 (1997)
586. Yu.V. Shvyd'ko, E. Gerdau, J. Jäschke, O. Leupold, M. Lucht, H.D. Rüter: Phys. Rev. B **57**, 4968 (1998)
587. Yu.V. Shvyd'ko, M. Lerche, J. Jäschke, M. Lucht, E. Gerdau, M. Gerken, H.D. Rüter H.-C. Wille, P. Becker, E.E. Alp, W. Sturhahn, J. Sutter, T.S. Toellner: Phys. Rev. Lett. **85**, 495 (2000)
588. K.D. Liss, R. Hock, M. Gomm, B. Waibel, A. Margerl, M. Krisch, M. Tucoulou: Nature **404**, 371 (2000)
589. K.-D. Liss, R. Hock, M. Gomm, B. Waibel, A. Magerl, M. Krisch, R. Tucoulou: Proc. SPIE **4143**, 78 (2001)
590. J.P. Sutter, E.E. Alp, M.Y. Hu, H. Sinn, W. Stuhrbahn, T.S. Toellner, G. Bortel, R. Colella: Phys. Rev. B **63**, 94111-1 (2001)
591. Yu.V. Shvyd'ko: *X-Ray Optics, High Energy Resolution Application.* (Springer, Berlin 2004)
592. Yu.V. Shvyd'ko, M. Lerche, H.-C. Wille, E. Gerdau, M. Lucht, H.D. Rüter, E.E. Alp, R. Khachatryan: Phys. Rev. Lett. **90**, 013904 (2003)
593. A.Y. Nikulin, A. Souvorov, K. Goodden, K. Tamasaku, T. Ishikawa: J. Phys. D: Appl. Phys. **36**, A87 (2003)
594. P.P. Ewald: Acta Cryst. **A24**, 5 (1968)
595. B. Post: Acta Cryst. **A35**, 17 (1979)
596. C.Y. Jen, S.-L. Chang: Acta Cryst. **A48** 650 (1992)
597. I.K. Robinson, W.K. Waskiewicz, P.H. Fuoss, J.B. Stark, R.L. Johnson: Phys. Rev. B **33**, 7013 (1986)
598. I.K. Robinson, R.T. Tung, R. Feidenhans'l: Phys. Rev. B **38**, 3632 (1988)
599. N. Kashiwagura, Y. Kahihara, M. Sakata, J. Harada, S.W. Wilkins, A.W. Stevenson: Jpn. J. Appl. Phys. **26**, L2026 (1987)
600. I.K. Robinson: Acta Cryst. **A54**, 772 (1998)
601. S.G.J. Mochrie, D.M. Zehner, B.M. Ocko, D. Gibbs: Phys. Rev. Lett. **64**, 2925 (1990)
602. G.A. Held, J.L. Jordan-Sweet, P.M. Horn, A. Mak, L.C. Feldman: Solid State Commun. **72**, 39 (1989)
603. H. Dosch, L. Mailänder, H. Reichert, J. Peisl, R.L. Johnson: Phys. Rev. B **43**, 13172 (1991)
604. K.S. Liang, E.B. Sirota, K.L. D'Amico, G.J. Hughes, S.K. Sinha: Phys. Rev. Lett. **59**, 2447 (1987)
605. R. Colella: Phys. Rev. B **34**, 13827 (1991)
606. A. Caticha: Phys. Rev. B **47**, 76 (1993)
607. S. Nakatani, T. Takahashi: Surf. Sci. **311**, 433 (1994)
608. T.-S. Gau, S.-L. Chang: Phys. Lett. A **196**, 223 (1994)
609. T.-S. Gau, S.-L. Chang: Acta Cryst. **A51**, 920 (1995)
610. H.-C. Chien, T.-S. Gau, S.-L. Chang, Yu.P. Stetsko: Acta Cryst. **A55**, 677 (1999)
611. R. Feidenhans'l: Surf. Sci. Rep. **10**, 105 (1989)
612. I.K. Robinson, D.J. Tweet: Rep. Prog. Phys. **55**, 599 (1992)
613. S.L. Chang: Phys. Status Solidi A **47**, 717 (1977)
614. J. Harada: Ultramicroscopy **52**, 233 (1993)
615. S.-W. Luh, S.-L. Chang: Acta Cryst. **A47**, 502 (1991)
616. S.L. Morelhão, L.H. Avanci: Acta Cryst. **A57**, 192 (2001)

617. Yu.P. Stetsko, Y.-R. Lee, M.-T. Tang, S.-L. Chang: Acta Cryst. **A60**, 64 (2004)
618. H.J. Juretschke: Cryst. Res. Technol. **33**, 569 (1998)
619. A.A. Katsnelson, V.I. Kissin, N.A. Polyakova: Sov. Cryst. **14**, 965 (1969)
620. A.A. Katsnelson, T.I. Borodina, V.I. Kissin: Phys. Status Solidi (a) **3**, 105 (1970)
621. A.A. Katsnelson, V.I. Iveronova, T.I. Borodina, I.G. Sapkova: Phys. Status Solidi (a) **11**, 39 (1972)
622. A.A. Katsnelson, V.I. Iveronova, T.I. Borodina, T.K. Runova: Phys. Status Solidi (a) **28**, 365 (1975)
623. Q. Shen, D. Pringle, M. Szebenyi, J. Wang: Rev. Sci. Instrum. **73**, 1646 (2001)
624. S.-L. Chang, Yu.P. Stetsko, Y.-R. Lee: Zeit Krist. **217**, 662 (2002)
625. A.M. Afanas'ev, A.V. Zozulya, M.V. Koval'chuk, M.A. Chuev: JETP Lett. Vol. 75 **7**, 309 (2002)
626. J. Laval: Rev. Mod. Phys. **30**, 222 (1958)
627. D.A. O'Connor: Proc. Phys. Soc. **91**, 917 (1967)
628. R. Kohler, W. Mohling, H. Peibst: Phys. Stat. Sol. **41**, 75 (1970); **61(b)**, 173 (1974)
629. D.G. Carlson, A. Segmüller, E. Mosekilde, H. Cole, J.A. Armstrong: Appl. Phys. Lett. **18**, 330 (1971)
630. D.G. Carlson, A. Segmüller: Phys. Rev. Lett. **28**, 175 (1972)
631. O. Ishibashi, M. Kitamura, A. Odajima: Phys. Lett. A **44**, 371 (1973)
632. S.D. Leroux, R. Colella, R. Bray: Phys. Rev. Lett. **35**, 230 (1975)
633. S.D. Leroux, R. Colella, R. Bray: Phys. Rev. Lett. **37**, 1056 (1976)
634. A.M. Afanas'ev, S.L. Azizian: Acta Cryst. **A37**, 125 (1980)
635. A.M. Afanas'ev, P.A. Aleksandrov, R.M. Imamov: *X-ray Diffraction Diagnostics of Submicron Layers.* (Nauka, Moscow, 1989)
636. S.-L. Chang, J.H. Horng, S.W. Luh, M.T. Tang, S.S. Pan, M. Sasaki: Acta Cryst. **A44**, 63 (1988)
637. S.-L. Chang, Y.S. Tsai, M.T. Huang: Phys. Lett. A **177**, 61 (1993)
638. Y.-S. Tsai, M.-T. Huang, S.-L. Chang: Chin. J. Phys. **32**, 641 (1994)
639. H. Burzlaff, K. Hümmer: Acta Cryst. **A44**, 506 (1988)
640. J.M. Bijvoet, A.F. Peerdeman, A.F. van Bommel: Nature **168**, 271 (1951)
641. D. Rogers: *Theory and Practice of Direct Methods in Crystallography*, ed. by M.F.C. Ladd, R.A. Palmer (Plenum Press, New York 1980) pp. 23–91
642. J. C. Spence, J.M. Zuo, M. O'Keeffe, K. Marthinsen, R. Hoier: Acta Cryst. **A50**, 647 (1994)
643. H. Burzlaff, J. Lange, H. Zimmermann: Acta Cryst. **A51**, 91 (1995)
644. R. Colella: Acta Cryst. **A51**, 438 (1995)
645. P.M. de Wolff, T. Janssen, A. Nanner: Acta Cryst. **A37**, 625 (1981)
646. R.E. Peierls: *Quantum Theory of Solids.* (Oxford University Press, New York 1955)
647. H. Fröhlich: Proc. Roy. Soc. London Ser. A **223**, 296 (1954)
648. A.W. Overhauser: Phys. Rev. Lett. **4**, 462 (1960)
649. A.W. Overhauser: Phys. Rev. **128**, 1437 (1960)
650. G. Grüner: *Density Waves in Solids.* (Perseus, Cambridge, MA 2000)
651. Gy, Hutiry, J. Sólyom: *Charge Density Waves in Solids.* (Springer-Verlag, Berlin 1985)
652. A.W. Overhauser: Advances in Physics **27**, 343 (1978)
653. G.F. Giuliani, A.W. Overhauser: Phys. Rev. B **26**, 1660 (1982)
654. C.-H. Du, M.-T. Tang, Yu.P. Stetsko, Y.-R. Ru, C.-W. Wang, C.-W. Cheng, S.-L. Chang: Acta Cryst. **A60**, 219 (2004)

655. S.-L. Chang, C.-H. Du, M.-T. Tang, Yu.P. Stetsko, Y.-R. Lee, T.-T. Lin, S.-C. Weng, W.-S. Sun: 'Direct Phase Determination of Fractional Reflection in Charge-Density-Wave Crystals Using X-Ray Multiple Diffraction'. In: *Sagamore XIV Conference at Broome, Australia, Aug. 13–18, 2003.* Paper SThP-03 (2003)
656. D.E. Moncton, J.D. Axe, F.J. DiSalvo: Phys. Rev. B **16**, 801 (1977)
657. D. Shechtman, I. Blech, D. Gratias, J.W. Cahn: Phys. Rev. Lett. **53**, 1951 (1984)
658. H. Lee, R. Colella, L.D. Chapman: Acta Cryst. **A49**, 600 (1993)
659. I.S. Elfimov, N.A. Skorikov, V.I. Anisimov, G.A. Sawatzky: Phys. Rev. Lett. **88**, 015504 (2002)
660. S.A. Darst, M. Ahlers, P.H. Meller, E.W. Kubalek, R. Blankenberg, H.O. Ribi, H. Ringsdorf, R.D. Kornberg: Biophy. J. **59**, 387 (1991)
661. R.D. Kornberg, S.A. Darst: Curr. Opin. Struct. Biol. **1**, 642 (1991)
662. J.H. Fendler: Chem. Mater. **8**, 1616 (1996)
663. S. Dietrich, A. Haase: Phys. Rep. **260**, 1 (1995)
664. L.D. Marks, E. Bengu, C. Collazo-Davila, D. Grozea, E. Landree, C. Leslie, W. Simkler: Surf. Rev. Lett. **5**, 1087 (1998)
665. L.D. Marks, E. Landree: Acta Cryst. **A54**, 296 (1998)
666. L.D. Marks: Phys. Rev. B **60**, 2771 (1999)
667. X. Torrelles, J. Rius, F. Boscherini, S. Heun, B.H. Mueller, S. Ferrer, J. Alvarez, C. Miravitlles: Phys. Rev. B **57**, 4281 (1998)
668. X. Torrelles, J. Rius, C. Miravitlles, S. Ferrer: Surf. Sci. **423**, 338 (1999)
669. Y.-S. Huang, C.-S. Chao, Yu.P. Stetsko, Y.-R. Lee, S.-L. Chang: In: *18th IUCr. Congress and General Assembly, Glasgow, 1999* Paper P06.11.003 (1999)
670. H. Baltes, Y. Yacoby, R. Pindak, R. Clarke, L. Pfeiffer, L. Berman: Phys. Rev. Lett. **79**, 1285 (1997)
671. Y. Yacoby: Solid State Commun. **91**, 529 (1994)
672. E. Weckert, R. Müller, J. Zellner, I. Zegers, R. Loris: Z. Kristallogr. **214**, 1 (1999)
673. H. Vogler, A. Iglesias, W. Moritz, H. Over: Phys. Rev. B **57**, 2315 (1998)
674. E. Weckern, W. Schwegle, K. Hümmer: Proc. R. Soc. London Ser. A **442**, 33 (1993)
675. F. Mo, R.H. Mathiesen, B.C. Hauback, E.T. Adman: Acta Cryst. **D52**, 893 (1996)
676. K. Hölzer, E. Weckert, K. Schroer, K. Hümmer: Z. Krist. Suppl. **12**, 257 (1997)
677. R.H. Mathiesen, F. Mo: Acta Cryst. **D53**, 262 (1997)
678. K. Hölzer, K. Hümmer, K. Schroer, E. Weckert, J. Zellner: Z. Krist. Suppl. **15**, 78 (1998)
679. C.M. Weeks, H. Xu, H.A. Hauptman, Q. Shen: Acta Cryst. **A56**, 280 (2000)
680. F. Mo, R.H. Mathiesen, P.M. Alzari, J. Lescar, B. Rasmussen: Acta Cryst. **D58**, 1780 (2002)
681. D. Pringle, Q. Shen: J. Appl. Cryst. **36**, 29 (2003)
682. Q. Shen, J. Wang: Acta Cryst. **D59**, 809 (2003)
683. Q. Shen: Acta Cryst. **A59**, 335 (2003)
684. M.F.C. Ladd, R.A. Palmer (eds.): *Theory and Practice of Direct Methods in Crystallography* (Plenum, New York 1980)
685. C. Giacovazzo: *Direct Methods in Crystallography.* (Academic, London 1980)
686. H. Schenk (ed.): *Direct Methods for Solving Crystal Structures* (Plenum Press, New York 1991)
687. H. Hauptmann, J. Karle: *Solution of the Phase Problem I. The Centrosymmetric Crystal,* ACA Monograph **3** (Polycrystal Service, Pittsburgh 1953)
688. D. Sayre: Acta Crystallogr. **5**, 60 (1952)

689. B.K. Vainshtein, V.M. Fridkin, V.L. Indenbom: *Modern Crystallography*, Vol. II, Springer Ser. Solid-State Sci. Vol. **21** (Springer, Berlin, Heidelberg, New York 1982)
690. G. Bricogne: Acta Cryst. **A40**, 410 (1984)
691. G. Bricogne: Acta Cryst. **A44**, 517 (1988)
692. G. Bricogne, C.J. Gilmore: Acta Cryst. **A46**, 284 (1990)
693. E. Prince, L. Sjölin, R. Alenljung: Acta Cryst. **A44**, 216 (1988)
694. S. Doublié, S. Xiang, C.J. Gilmore, G. Bricogne, W.W.C. Jr. Carter: Acta Cryst. **A50**, 164 (1994)
695. C.J. Gilmore: Acta Cryst. **A52**, 561 (1996)
696. W.P. Su: Acta Cryst. **A51**, 845 (1995)
697. C.I. Chou, T.K. Lee: Acta Cryst. **A58**, 42 (2002)
698. M.C. Vaney, S. Maignan, M. Riès-Kautt, A. ducruix: Acta Cryst. **D52**, 505 (1996)
699. B.H. Heise: J. Appl. Phys. **33**, 938 (1962)
700. W.L. Bond: Acta Cryst. **13**, 814 (1960)
701. R. Deslattes, A. Henins: Phys. Rev. Lett. **31**, 972 (1973)
702. A. Segmüller: Adv. X-Ray Anal. **13**, 155 (1970)
703. M.E. Straumanis, E.Z. Aka: J. Am. Chem. Soc. **73**, 5643 (1951)
704. A. Smakula, J. Kalnajs: Phys. Rev. **99**, 1937 (1955)
705. A. Cooper: Acta Cryst. **15**, 578 (1962)
706. D.N. Batchelder, R.O. Simmons: J. Appl. Phys. **36**, 2864 (1965)
707. B.J. Greenberg, J. Ladell: appl. Phys. Lett. **50**, 436 (1987)
708. S.L. Morelhão, L.P. Cardoso: J. Cryst. Growth **110**, 543 (1991)
709. S.L. Morelhão, L.P. Cardoso, J.M. Sasaki, M.M.G. de Carvalho: J. Appl. Phys. **70(5)**, 2589 (1991)
710.
711. W. Stolz, F.E.G. Guimaraes, K. Ploog: J. Appl. Phys. **63**, 492 (1988)
712. S.-L. Chang: 'X-Ray Fluorescence Imaging'. In: *Encyclopedia of Imaging Science and Technology*, ed. by J.P. Hornak (John Wiley & Sons, New York 2002)
713. B.W. Batterman: Appl. Phys. Lett. **1**, 68 (1962)
714. P.L. Cowan, J.A. Golovchenko, M.F. Robbins: Phys. Rev. Lett. **44**, 1680 (1980)
715. M.J. Bedzyk, W.M. Gibson, J.A. Golovchenko: J. Vac. Sci. Tech. **20**, 634 (1982)
716. J.A. Golovchenko, J.R. Patel, D.R. Kaplan, P.L. Cowan, M.J. Bedzyk: Phys. Rev. Lett. **49**, 560 (1982)
717. A. Kazimirov, J. Zegenhagen, M. Cardona: Science **282**, 930 (1998)
718. A. Lessmann, M. Schuster, H. Riechert, S. Brennan, A. Munkholm, G. Materlik: J. Phys. D: Appl. Phys. **32**, A65 (1999)
719. J. Falta, D. Bahr, G. Materlik, B.H. Müller, M. Horn-von Hoegen: Appl. Phys. Lett. **68**, 1394 (1996)
720. A.M. Afanasev, R.M. Imamov, A.V. Maslov, V.G. Mokerov, E.M. Pashaev, A.B. Vavilov, A.S. Ignatev, G.Z. Nemzev, A.A. Zaitsev: Kristallographiya **38**, 58 (1993)(Engl. Transl. Cryst. Rep. **38**, 319 (1993)
721. M. Schuster, A. Lessmann, A. Munkholm, S. Brennan, G. Materlik, H. Riechert: J. Phys. D: Appl. Phys. **28**, A206 (1995)
722. M.V. Kovalchuk, A.Ya. Kreines, Yu.A. Osipian, V.V. Kvardakov, V.A. Somenkov: Pisma Zh. Ekso. Teor. Fiz. **65**, 703 (1997) (Engl. Transl. JETP Lett. **65**, 735 (1997)
723. J. Zegenhagen, T. Siegrist, E. Fontes, L.E. Berman, J.R. Patel: Solid State Comm. **93**, 763 (1995)

724. T. Jach, Y. Zhang, R. Colella, M. de Boissieu, M. Boudard, A.I. Goldman, T.A. Lograsso, D.W. Delanney, S. Kycia: Phys. Rev. Lett. **82**, 2904 (1999)
725. T.W. Barbee, W.K. Warburton: Mater. Lett. **3**, 17 (1984)
726. T. Nakagiri, K. Sakai, A. Iida, T. Ishikawa, T. Matsushita: Thin Solid Films **133**, 219 (1985)
727. A. Iida, T. Matsushita, T. Ishikawa: Jpn. J. Appl. Phys. Part 2 **24**, L675 (1985)
728. M.J. Bedzyk, D.H. Bilderback, G.M. Bommarito, M. Caffrey, J.S. Schildkraut: Science **241**, 1788 (1988)
729. S.I. Zheludeva, M.V. Kovalchuk, N.N. Novikova, I.V. Bashelhanov, N.N. Salaschenko, A.D. Akhsakhalyan, Yu.Ya. Platonov: Rev. Sci. Instrum. **63**, 1027 (1992)
730. S.I. Zheludeva, M.V. Kovalchuk, N.N. Novikova, A.N. Sosphenov: J. Phys. D: Appl. Phys. **26**, A206 (1993)
731. S.I. Zheludeva, M.V. Kovalchuk, N.N. Novikova, A.N. Sosphenov, N.E. Malysheva, N.N. Salaschenko, A.D. Akhsakhalyan, Yu.Ya. Platonov: Physica B **198**, 259 (1994)
732. I.A. Vartanyants, M.V. Kovalchuk: Rep. Prog. Phys. **64**, 1009 (2001)
733. J.C. Woicik, E.J. Nelson, D. Heskett, J. Warner, L.E. Berman, B.A. Karlin, I.A. Vartanyants, M.Z. Hasan, T. Kendelewicz, Z.X. Shen, P. Pianetta: Phys. Rev. B **64**, 125115 (2001)
734. G. Materlik, J. Zegenhagen: Phys. Lett. A **104**, 47 (1987)
735. A. Krolzig, G. Materlik, J. Zegenhagen: Nucl. Instrum. Meth. **208**, 613 (1982)
736. A. Krolzig, G. Materlik, M. Swars, J. Zegenhagen: Nucl. Instrum Meth. **219**, 430 (1987)
737. A.Yu. Kazimirov, M.V. Kovalchuk, I.Yu. Kharitonov, L.V. Samoilova, T. Ishikawa, S. Kikuta: Rev. Sci. Instrum. **63(1)**, 1019 (1992)
738. A.Yu. Kazimirov, M.V. Kovalchuk, V.G. Kohn, I.Yu. Kharitonov, L.V. Samoilova, T. Ishikawa, S. Kikuta, K. Hirano: Phys. Stat. Sol. (a) **135**, 507 (1993)
739. H. Stragier, J.O. Cross, J.J. Rehr, L.B. Sorensen, C.E. Bouldin, J.C. Woicik: Phys. Rev. Lett. **69**, 3064 (1992)
740. I.J. Pickering, M. Sansone, J. Marsch, G.N. George: J. Am. Chem. Soc. **115**, 6302 (1993)
741. L.B. Sorensen, J.O. Cross, M. Newville, B. Ravel, J.J. Rehr, H. Stragier, C.E. Boulin, J.C. Woicik: In: *Resonant Anomalous X-Ray Scattering*, ed. by G. Materlink, G.J. Sparks, K. Fischer (Elsevier Science B.V., Amsterdam 1994) p. 389
742. H. Renevier, J.L. Hodeau, P. Wolfers, S. Andrieu, J. Weigelt, R. Frahm: Phys. Rev. Lett. **78**, 2775 (1997)
743. J. Vacínová, J.L. Hodeau, P. Wolfers, J.P. Lauriat, E. Elkaïm: J. Synchrotron Radiat. **2**, 236 (1995)
744. J.L. Hodeau, J. Vacínová: J. Synchrotron Radiat. News **9**, 15 (1996)
745. D.C. Meyer, K. Richter, A. Seidel, J. Weigelt, R. Frahm, P. Paufler: J. Synchrotron Radiat. **5**, 1275 (1998)
746. M.G. Proietti, H. Renevier, J.L. Hodeau, J. Garcia, J.F. Berar, P. Wolfers: Phys. Rev. B **59**, 5479 (1999)
747. J.C. Woicik, J.O. Cross, C.E. Bouldin, B. Ravel, J.G. Pellegrino, B. Steiner, S.G. Bompadre, L.B. Sorensen, K.E. Miyano, J.P. Kirkland: Phys. Rev. B **58**, R4215 (1998)
748. B. Ravel, C.E. Bouldin, H. Renevier, J.L. Hodeau, J.F. Berar: Phys. Rev. B **60**, 778 (1999)

749. J.L. Hodeau, V. Favre-Nicolin, S. Bos, H. Renevier, E. Lorenzo, J.F. Berar: Chem. Rev. **101**, 1843 (2001)
750. D.T. Cromer, D. Liberman: J. Chem. Phys. **53**, 1891 (1970)
751. S.-L. Chang, Y.-R. Lee, Yu.P. Stetsko, S.-Y. Cheng, G.-G. Lin, W.-S. Sun: 'Three-Beam Diffraction Anomalous Fine Structure of Gallium Arsenide'. In: *AsCA'03/Crystal-23 Conference at Broome, Australia, Aug. 10–13, 2003.* Paper ATu1B-3 (2003)
752. J.F. Nye: *Physical Properties of Crystals* (Clarendon Press, Oxford 1985)
753. A.L. Bain, N. El-Korashy, S. Gilmour, R.A. Pethric, J.N. Sherwood: Philos. Mag. B **66**, 293 (1992)
754. S.C. Abrahams: Acta Cryst. **A50**, 658 (1994)
755. P.A. Rigg, Y.M. Gupta: Phys. Rev. B **63**, 094112 (2001)
756. J. Stahn, A. Pucher, U. Pietsch, J. Zellner, E. Weckert: Acta Cryst. **A55**, 1034 (1999)
757. C.-H. Du, M.-T. Tang, Yu.P. Stetsko, Y.-R. Lee, J.-J. Lee, S.-L. Chang: 'The Study of the Phases of CDWs Under the Application of Electric Fields Using Multiple X-Ray Diffraction'. In: *Sagamore XIV Conference at Broome, Australia, Aug. 13–18, 2003.* Paper SThP-12 (2003)
758. A. Freund: In: *Synchrotron Radiation in Structural Biology*, ed. by R.M. Sweet, A.D. Woodhead (Plenum Press, New York 1989) pp. 255–292
759. A. Freund: In: *Neutron and Synchrotron Radiation for Condensed Matter Studies, Vol. 1*, ed. by J. Baruchel, J.-L. Hodeau, M.S. Lehmann, J.-R. Regnard, C. Schlenker (Springer-Verlag, Berlin 1993) pp. 79–93
760. B.W. Batterman, D.H. Bilderback: In: *Handbook on Synchrotron Radiation, Vol. 3*, ed. by G.Brown, D.E. Moncton (elsevier Science B.V., Amsterdam 1991) pp. 105-153
761. C. Malgrange, D. Ferret: Nucl. Instrum. Meth. Phys. Res. A **314**, 285 (1992)
762. M. Hart: In: *X-Ray and Neutron Dynamical Diffraction: Theory and Applications, NATO ASI Series, Series B, Physics, Vol. 357*, ed. by A. Authier, S. Lagomarsino, B.K. Tanner (Plenum Press, New York, London 1996) pp. 73–90
763. M. Hart, L.E. Berman: Acta Cryst. **A54**, 850 (1998)
764. W. Treimer: Cryst. Res. Technol. **33**, 643 (1998)
765. R.D. Deslattes: Appl. Phys. Lett. **12**, 133 (1968)
766. C.H. Chen, S.-L. Chang: Nucl. Instrum. Meth. A **306**, 584 (1991)
767. C.-M. Lo, S.-L. Chang: J. Physique **C9**, 75 (1987)
768. Yu.P. Stetsko, I.M. Rarenko, G.V. Leskova: Surf. Invest. **12**, 519 (1997)
769. T.C. Huang, M.H. Tillinger, B. Post: Z. Naturforsch. **28a**, 600 (1973)
770. A.M. Afanas'ev, V.G. Kohn: Acta Cryst. **A33**, 178 (1974)
771. V.G. Kohn: Sov. Cryst. **32**, 844 (1987)
772. A.Yu. Kazimirov, M.V. Kovalchuk, V.G. Kohn, T. Ishikawa, S. Kikuta, K. Hirano: Europhys. Lett. **24(3)**, 211 (1993)
773. V.G. Kohn: Sov. Solid State Phys. **19**, 3567 (1977)
774. M. Blume, D. Gibbs: Phys. Rev. B **37**, 1779 (1988)
775. J. Bahrdt, A. Gaupp, W. Gudat, M. Mast, K. Molter, W.B. Peatman, M. Scheer, Th. Schroeter, Ch. Wang: Rev. Sci. Instr. **63**, 339 (1992)
776. S. Annaka, K. Kohra, M. Ando: Jpn. J. Appl. Phys. **20**, L463 (1981)
777. K. Hirano, K. Izumi, T. Ishikawa, S. Annaka, S. Kikuta: Jpn. J. Appl. Phys. **30**, 407 (1981)
778. O. Brümmer, Ch. Eisenschmidt, H.R. Hoche: Acta Cryst. **A40**, 394 (1984)
779. D.M. Mills: Nucl. Instrum. Meth. A **266**, 531 (1988)
780. B. Batterman: Phys. Rev. B **45**, 12677 (1992)
781. M. Hart: Philos. Mag. B **38**, 41 (1978)

782. S.D. Shastri, K.D. Finkelstein, Q. Shen, B.W. Batterman, D.A. Walko: Rev. Sci. Instrum. **66(2)**, 1581 (1995)
783. K. Hirano, T. Mori, A. Iida, R. Colella, S. Sasaki, Q. Shen: Jpn. J. Appl. Phys. **35**, 5550 (1996)
784. Q. Shen: Acta Cryst. **A49**, 605 (1993)
785. Q. Shen, K.D. Finkelstein: Rev. Sci. Instrum. **64(12)**, 3451 (1993)
786. Q. Shen, S. Shastri, K.D. Finkelstein: Rev. Sci. Instrum. **66(2)**, 1610 (1995)
787. U. Bonse, M. Hart: Appl. Phys. Lett. **7**, 238 (1965)
788. U. Bonse, M. Hart: Z. Phys. **188**, 154 (1965)
789. M. Hart: Proc. Roy. Soc. A **346**, 1 (1975)
790. U. Bonse, W. Graeff, R. Teworte, H. Rauch: Phys. Stat. Sol. (a) **43**, 487 (1977)
791. M. Hart: Nucl. Instrum. Meth. **172**, 209 (1980)
792. W. Treimer: Kristall und Technik **13**, 1105 (1978)
793. S.A. Werner: Phys. Today **33**, 24 (1980)
794. D.K. Bowen: In: *X-Ray and Neutron Dynamical Diffraction: Theory and Applications, NATO ASI Series, Series B, Physics, Vol. 357*, ed. by A. Authier, S. Lagomarsino, B.K. Tanner (Plenum Press, New York, London 1996) pp. 381–410
795. R. Colella: In: *X-Ray and Neutron Dynamical Diffraction: Theory and Applications, NATO ASI Series, Series B, Physics, Vol. 357*, ed. by A. Authier, S. Lagomarsino, B.K. Tanner (Plenum Press, New York, London 1996) pp. 369–860
796. L. Balents, M.P.A. Fisher: Phys. Rev. Lett. **75**, 4270 (1995) G. Materlink, G.J. Sparks, K. Fischer

Figure Acknowledgements

The author is indebted to the publishers indicated below for their kind permission to reproduce in the text the following figures (with the original author's consent):

1. Fig. 2.5: Reprinted from *Handbook on Synchrotron Radiation*, Vol. 1A, ed. by E.-E. Koch (North-Holland, Amsterdam, New York, Oxford 1983), T. Matsushita, H. Hashizume: 'X-Ray Monochromators', pp. 261-314, copyright (1983) by Elsevier
2. Figs. 4.2, 4.7, 9.3: Reprinted from S.-L. Chang: Acta Cryst. **A54**, 886 (1998), copyright (1998) by International Union of Crystallography
3. Figs. 4.3–4.6, 4.8: Reprinted from S.-L. Chang: *Multiple Diffraction of X-Rays in Crystals* (Springer, Berlin, Heidelberg 1984), copyright (1984) by Springer-Verlag
4. Fig. 4.9: Reprinted from Q. Shen: Phys. Rev. B **59**, 11109 (1999), copyright (1999) by American Physical Society
5. Fig. 4.10: Reprinted from Q. Shen: Phys. Rev. Lett. **80**, 3268 (1998), copyright (1998) by American Physical Society
6. Figs. 4.11, 9.4, 9.24: Reprinted from C.-H. Chao, C.-Y. Hung, Y.-S. Huang, C.-H. Ching, Y.-R. Lee, Y.-C. Jean, S.-C. Lai, Yu.P. Stetsko, H. Yuan, S.-L. Chang: Acta Cryst. **A58**, 33 (2002), copyright (2002) by International Union of Crystallography
7. Fig. 4.12: Reprinted from Q. Shen, S.Kycia, I. Dobrianov: Acta Cryst. **A56**, 264 (2000), copyright (2000) by International Union of Crystallography
8. Figs. 4.13, 4.14, 4.15: Reprinted from S.-L. Chang, C.-H. Chao, Y.-S. Huang, Y.-C. Jean, H.-S. Sheu, F.-J. Liang, H.-C. Chien, C.-K. Chen, H.-S. Yuan: Acta Cryst. **A55**, 933 (1999), copyright (1999) by International Union of Crystallography
9. Fig. 4.17: Reprinted from M. Tegze, G. Faigel, S. Marchesini, M. Belakhovsky, A.I. Chumakov: Phys. Rev. Lett. **82**, 4847 (1999), copyright (1999) by American Physical Society
10. Figs. 4.18, 9.11: Reprinted from E. Weckert, K. Hümmer: Acta Cryst. **A53**, 108 (1997), copyright (1997) by International Union of Crystallography

11. Fig. 4.19: Reprinted from G. Thorkildsen, F.H. Mathiesen, H.B. Larsen: J. Appl. Cryst. **32**, 943 (1999), copyright (1999) by International Union of Crystallography
12. Fig. 4.20: Reprinted from T.-S. Gau, Y.-C. Jean, K.-Y. Liu, C.-H. Chung, C.-K. Chen, S.-C. Lai, C.-H. Shu, Y.-S. Huang, C.-H. Chao, Y.-R. Lee, C.T. Chen, S.-L. Chang: Nucl. Instrum. & Meth. in Phys. Res. **A466**, 569 (2001),copyright (2001) by Elsevier
13. Fig. 4.21: Reprinted from D. Pringle, Q. Shen: J. Appl. Cryst. **36**, 29 (2003), copyright (2003) by International Union of Crystallography
14. Figs. 4.22, 4.23: Reprinted from Q. Shen, S. Kycia, I. Dobrianov: Acta Cryst. **A56**, 268 (2000), copyright (2000) by International Union of Crystallography
15. Fig. 5.2: Reprinted from E. Rossmanith: J. Appl. Cryst. **36**, 1467 (2003), copyright (2003) by International Union of Crystallography
16. Figs. 6.2–6.6, 6.16: Reprinted from S.-L. Chang: *Multiple Diffraction of X-Rays in Crystals* (Springer, Berlin, Heidelberg 1984), copyright (1984) by Springer-Verlag
17. Fig. 7.1: Reprinted from Yu.P. Stetsko, S.-L. Chang: Acta Cryst. **A53**, 28 (1997), copyright (1997) by International Union of Crystallography
18. Figs. 7.6, 7.7, 7.8: Reprinted from S.-W. Luh, S.-L. Chang: Acta Cryst. **A44**, 662 (1988), copyright (1988) by International Union of Crystallography
19. Fig. 7.9: Reprinted from G. Thorkildsen, H.B. Larsen: Acta Cryst. **A54**, 120 (1998), copyright (1998) by International Union of Crystallography
20. Figs. 8.6, 8.8: Reprinted from H.-H. Hung, S.-L. Chang: Europhys. Lett. **23(6)**, 415 (1993), copyright (1993) by EDP Sciences
21. Figs. 8.11, 8.12, 8.13: Reprinted from H.-H. Hung, S.-L. Chang: Acta Cryst. **A45**, 823 (1989), copyright (1989) by International Union of Crystallography
22. Figs. 8.16–8.29: Reprinted from S.-L. Chang: *Multiple Diffraction of X-Rays in Crystals* (Springer, Berlin, Heidelberg 1984), copyright (1984) by Springer-Verlag
23. Figs. 8.33–8.36: Reprinted from T.-P. Tseng, S.-L. Chang: Acta Cryst. **A46**, 567 (1990), copyright (1990) by International Union of Crystallography
24. Figs. 8.37, 8.39–8.42: Reprinted from T.-S. Gau, S.-L. Chang: Acta Cryst. **A51**, 920 (1995), copyright (1995) by International Union of Crystallography
25. Fig. 8.38: Reprinted from H.-C. Chien, T.-S. Gau, S.-L. Chang, Yu.P. Stetsko: Acta Cryst. **A55**, 677 (1999), copyright (1999) by International Union of Crystallography
26. Figs. 8.44, 8.45, 9.6: Reprinted from Yu.P. Stetsko, H.J. Juretschke, Y.-S. Huang, C.-H. Chao, C.-K. Chen, S.-L. Chang: Acta Cryst. **A56**, 394 (2000), copyright (2000) by International Union of Crystallography

27. Figs. 8.46–8.49, 9.7, 9.9: Reprinted from Yu.P. Stetsko, H.J. Juretschke, Y.-S. Huang, Y.R. Lee, T.C. Lin, S.-L. Chang: Acta Cryst. **A57**, 359 (2001), copyright (2001) by International Union of Crystallography
28. Fig. 8.50: Reprinted from B. Post, S.L. Chang, T.C. Huang: Acta Cryst. **A33**, 90 (1977), copyright (1977) by International Union of Crystallography
29. Figs. 8.51: Reprinted from S.-L. Chang, J.H. Horng, S.W. Luh, M.T. Tang, S.S. Pan, M. Sasaki: Acta Cryst. **A44**, 63 (1988), copyright (1988) by International Union of Crystallography
30. Fig. 8.52: Reprinted from Y.-S. Tsai, M.-T. Huang, S.-L. Chang: Chin. J. Phys. **32**, 641 (1994), copyright (1994) by Physical Society of the Republic of China
31. Figs. 8.58–8.61: Reprinted from J.P. Sutter, E.E. Alp, M.Y. Hu, H. Sinn, W. Stuhrbahn, T.S. Toellner, G. Bortel, R. Colella: Phys. Rev. B **63**, 94111-1 (2001), copyright (2001) by American Physical Society
32. Fig. 9.1: Reprinted from K. Hümmer, E. Weckert, H. Bondza: Acta Cryst. **A46**, 393 (1990), copyright (1990) by International Union of Crystallography
33. Figs. 9.2, 9.12: Reprinted from E. Weckert, K. Hümmer: Acta Cryst. **A53**, 108 (1997), copyright (1997) by International Union of Crystallography
34. Fig. 9.5: Reprinted from Q. Shen, K.D. Finkelstein: Phys. Rev. Lett. **65**, 3337 (1990), copyright (2001) by American Physical Society
35. Fig. 9.10: Reprinted from A.M. Afanas'ev, A.V. Zozulya, M.V. Koval'chuk, M.A. Chuev: JETP Lett. Vol. 75 **7**, 309 (2002), copyright (2002) by MAIK "Nauka/Interperiodica"
36. Fig. 9.11: Reprinted from C.-H. Du, M.-T. Tang, Yu.P. Stetsko, Y.-R. Ru, C.-W. Wang, C.-W. Cheng, S.-L. Chang: Acta Cryst. **A60**, 219 (2004), copyright (2004) by International Union of Crystallography
37. Figs. 9.13–9.16: Reprinted from Yu.P. Stetsko, G.-Y. Lin, Y.-S. Huang, C.-H. Chao, S.-L. Chang: Phys. Rev. Lett. **86**, 2026 (2001), copyright (2001) by American Physical Society
38. Figs. 9.17–9.19: Reprinted from S.-L. Chang, Y.S. Huang, C.H. Chao, M.T. Tang, Y.P. Stetsko: Phys. Rev. Lett. **80**, 301 (1998), copyright (1998) by American Physical Society
39. Figs. 9.20, 9.21: Reprinted from Y.-S. Huang, C.-S. Chao, Yu.P. Stetsko, Y.-R. Lee, C.-H. Ching, C.-Y. Hung, G.-Y. Lin, T.-C. Lin, S.-L. Chang: Phys. Rev. B **64**, 085406 (2001), copyright (2001) by American Physical Society
40. Figs. 9.22, 9.23: Reprinted from V.M. Kaganer, M. Albrecht, A. Hirnet, M. Gierer, W. Moritz, B. Jenichen, K.H. Ploog: Phys. Rev. B **61**, R16 355 (2000), copyright (2000) by American Physical Society
41. Fig. 9.25: Reprinted from K. Holzer, E. Weckert, K. Schroer: Acta Cryst. **D56**, 322 (2000), copyright (2000) by International Union of Crystallography

42. Fig. 9.26: Reprinted from Q. Shen, J. Wang: Acta Cryst. **D59**, 809 (2003), copyright (2003) by International Union of Crystallography
43. Fig. 9.27: S.-L. Chang: Appl. Phys. Lett. **37**, 819 (1980), copyright (1980) by American Institute of Physics
44. Figs. 9.28, 9.30: Reprinted from S.L. Morelhão, L.P. Cardoso: J. Cryst. Growth **110**, 543 (1991), copyright (1991) by Elsevier
45. Figs. 9.29, 9.32: Reprinted from S.L. Morelhão, L.P. Cardoso, J.M. Sasaki, M.M.G. de Carvalho: J. Appl. Phys. **70(5)**, 2589 (1991), copyright (1991) by American Institute of Physics
46. Fig. 9.31: Reprinted from S.L. Morelhão, L.P. Cardoso: J. Appl. Phys. **73**, 4218 (1993), copyright (1993) by American Institute of Physics
47. Fig. 9.33: S.L. Morelhão, L.P. Cardoso: Solid State Commun. **88**, 465 (1993), copyright (1993) by Elsevier
48. Figs. 9.34, 9.35: Reprinted from S.L. Morelhão, L.P. Cardoso: J. Appl. Cryst. **29**, 446 (1996), copyright (1996) by International Union of Crystallography
49. Fig. 9.36: Reprinted from M.A. Hayashi, S.L. Morelhão, L.H. Avanci, L.P. Cardoso, J.M. Sasaki, L.C. Kretly, S.-L. Chang: Appl. Phys. Lett. **71**, 2614 (1997), copyright (1997) by American Institute of Physics
50. Figs. 9.38–9.40: Reprinted from N. Greiser, G. Materlik: Z. Phys. B – Condensed Matter **66**, 83 (1987), copyright (1987) by Springer-Verlag
51. Fig. 9.41: Reprinted from A.Yu. Kazimirov, M.V. Kovalchuk, V.G. Kohn, I.Yu. Kharitonov, L.V. Samoilova, T. Ishikawa, S. Kikuta, K. Hirano: Phys. Stat. Sol. (a) **135**, 507 (1993), copyright (1993) by Wiley-VCH
52. Fig. 9.42: Reprinted from A.Yu. Kazimirov, M.V. Kovalchuk, I.Yu. Kharitonov, L.V. Samoilova, T. Ishikawa, S. Kikuta: Rev. Sci. Instrum. **63(1)**, 1019 (1992), copyright (1992) by American Institute of Physics
53. Fig. 9.47–9.49: L.H. Avanci, L.P. Cardoso, S.E. Girdwood, D. Pugh, J.N. Sherwood, K.J. Roberts: Phys. Rev. Lett. **81**, 5426 (1998), copyright (1998) by American Physical Society
54. Fig. 9.50: Reprinted from C.H. Chen, S.-L. Chang: Nucl. Instrum. Meth. A , **306**, 584 (1991), copyright (1991) by Elsevier
55. Fig 9.53: Reprinted from A.Yu. Kazimirov, M.V. Kovalchuk, V.G. Kohn, T. Ishikawa, S. Kikuta, K. Hirano: Europhys. Lett. **24(3)**, 211 (1993), copyright (1993) by EDP Sciences
56. Fig. 9.54: Reprinted from L.-Y. Jang, M.-T. Huang, S.-L. Chang: Nucl. Instru. and Meth. in Phys. Res. A **372**, 543 (1996), copyright (1996) by Elsevier
57. Figs. 9.55, 9.56: Reprinted from Q. Shen, K.D. Finkelstein: Phys. Rev. B **45**, 5075 (1992), copyright (1992) by American Physical Society

Index

Springer Series in Solid-State Sciences

Editors: M. Cardona P. Fulde K. von Klitzing H.-J. Queisser

1 **Principles of Magnetic Resonance** 3rd Edition By C. P. Slichter
2 **Introduction to Solid-State Theory** By O. Madelung
3 **Dynamical Scattering of X-Rays in Crystals** By Z. G. Pinsker
4 **Inelastic Electron Tunneling Spectroscopy** Editor: T. Wolfram
5 **Fundamentals of Crystal Growth I** Macroscopic Equilibrium and Transport Concepts By F. E. Rosenberger
6 **Magnetic Flux Structures in Superconductors** 2nd Edition By R. P. Huebener
7 **Green's Functions in Quantum Physics** 2nd Edition By E. N. Economou
8 **Solitons and Condensed Matter Physics** Editors: A. R. Bishop and T. Schneider
9 **Photoferroelectrics** By V. M. Fridkin
10 **Phonon Dispersion Relations in Insulators** By H. Bilz and W. Kress
11 **Electron Transport in Compound Semiconductors** By B. R. Nag
12 **The Physics of Elementary Excitations** By S. Nakajima, Y. Toyozawa, and R. Abe
13 **The Physics of Selenium and Tellurium** Editors: E. Gerlach and P. Grosse
14 **Magnetic Bubble Technology** 2nd Edition By A. H. Eschenfelder
15 **Modern Crystallography I** Fundamentals of Crystals Symmetry, and Methods of Structural Crystallography 2nd Edition By B. K. Vainshtein
16 **Organic Molecular Crystals** Their Electronic States By E. A. Silinsh
17 **The Theory of Magnetism I** Statics and Dynamics By D. C. Mattis
18 **Relaxation of Elementary Excitations** Editors: R. Kubo and E. Hanamura
19 **Solitons** Mathematical Methods for Physicists By. G. Eilenberger
20 **Theory of Nonlinear Lattices** 2nd Edition By M. Toda
21 **Modern Crystallography II** Structure of Crystals 2nd Edition By B. K. Vainshtein, V. L. Indenbom, and V. M. Fridkin
22 **Point Defects in Semiconductors I** Theoretical Aspects By M. Lannoo and J. Bourgoin
23 **Physics in One Dimension** Editors: J. Bernasconi and T. Schneider
24 **Physics in High Magnetics Fields** Editors: S. Chikazumi and N. Miura
25 **Fundamental Physics of Amorphous Semiconductors** Editor: F. Yonezawa
26 **Elastic Media with Microstructure I** One-Dimensional Models By I. A. Kunin
27 **Superconductivity of Transition Metals** Their Alloys and Compounds By S. V. Vonsovsky, Yu. A. Izyumov, and E. Z. Kurmaev
28 **The Structure and Properties of Matter** Editor: T. Matsubara
29 **Electron Correlation and Magnetism in Narrow-Band Systems** Editor: T. Moriya
30 **Statistical Physics I** Equilibrium Statistical Mechanics 2nd Edition By M. Toda, R. Kubo, N. Saito
31 **Statistical Physics II** Nonequilibrium Statistical Mechanics 2nd Edition By R. Kubo, M. Toda, N. Hashitsume
32 **Quantum Theory of Magnetism** 2nd Edition By R. M. White
33 **Mixed Crystals** By A. I. Kitaigorodsky
34 **Phonons: Theory and Experiments I** Lattice Dynamics and Models of Interatomic Forces By P. Brüesch
35 **Point Defects in Semiconductors II** Experimental Aspects By J. Bourgoin and M. Lannoo
36 **Modern Crystallography III** Crystal Growth By A. A. Chernov
37 **Modern Chrystallography IV** Physical Properties of Crystals Editor: L. A. Shuvalov
38 **Physics of Intercalation Compounds** Editors: L. Pietronero and E. Tosatti
39 **Anderson Localization** Editors: Y. Nagaoka and H. Fukuyama
40 **Semiconductor Physics** An Introduction 6th Edition By K. Seeger
41 **The LMTO Method** Muffin-Tin Orbitals and Electronic Structure By H. L. Skriver
42 **Crystal Optics with Spatial Dispersion, and Excitons** 2nd Edition By V. M. Agranovich and V. L. Ginzburg
43 **Structure Analysis of Point Defects in Solids** An Introduction to Multiple Magnetic Resonance Spectroscopy By J.-M. Spaeth, J. R. Niklas, and R. H. Bartram
44 **Elastic Media with Microstructure II** Three-Dimensional Models By I. A. Kunin
45 **Electronic Properties of Doped Semiconductors** By B. I. Shklovskii and A. L. Efros
46 **Topological Disorder in Condensed Matter** Editors: F. Yonezawa and T. Ninomiya

Springer Series in Solid-State Sciences

Editors: M. Cardona P. Fulde K. von Klitzing H.-J. Queisser

Springer Series in Solid-State Sciences

Editors: M. Cardona P. Fulde K. von Klitzing H.-J. Queisser

90 **Earlier and Recent Aspects of Superconductivity**
Editors: J. G. Bednorz and K. A. Müller

91 **Electronic Properties of Conjugated Polymers III** Basic Models and Applications
Editors: H. Kuzmany, M. Mehring, and S. Roth

92 **Physics and Engineering Applications of Magnetism** Editors: Y. Ishikawa and N. Miura

93 **Quasicrystals** Editors: T. Fujiwara and T. Ogawa

94 **Electronic Conduction in Oxides** 2nd Edition
By N. Tsuda, K. Nasu, F. Atsushi, and K. Siratori

95 **Electronic Materials**
A New Era in Materials Science
Editors: J. R. Chelikowsky and A. Franciosi

96 **Electron Liquids** 2nd Edition By A. Isihara

97 **Localization and Confinement of Electrons in Semiconductors**
Editors: F. Kuchar, H. Heinrich, and G. Bauer

98 **Magnetism and the Electronic Structure of Crystals** By V.A. Gubanov, A.I. Liechtenstein, and A.V. Postnikov

99 **Electronic Properties of High-T_c Superconductors and Related Compounds**
Editors: H. Kuzmany, M. Mehring, and J. Fink

100 **Electron Correlations in Molecules and Solids** 3rd Edition By P. Fulde

101 **High Magnetic Fields in Semiconductor Physics III** Quantum Hall Effect, Transport and Optics By G. Landwehr

102 **Conjugated Conducting Polymers**
Editor: H. Kiess

103 **Molecular Dynamics Simulations**
Editor: F. Yonezawa

104 **Products of Random Matrices** in Statistical Physics By A. Crisanti, G. Paladin, and A. Vulpiani

105 **Self-Trapped Excitons**
2nd Edition By K. S. Song and R. T. Williams

106 **Physics of High-Temperature Superconductors**
Editors: S. Maekawa and M. Sato

107 **Electronic Properties of Polymers**
Orientation and Dimensionality of Conjugated Systems Editors: H. Kuzmany, M. Mehring, and S. Roth

108 **Site Symmetry in Crystals**
Theory and Applications 2nd Edition
By R. A. Evarestov and V. P. Smirnov

109 **Transport Phenomena in Mesoscopic Systems** Editors: H. Fukuyama and T. Ando

110 **Superlattices and Other Heterostructures**
Symmetry and Optical Phenomena 2nd Edition
By E. L. Ivchenko and G. E. Pikus

111 **Low-Dimensional Electronic Systems**
New Concepts
Editors: G. Bauer, F. Kuchar, and H. Heinrich

112 **Phonon Scattering in Condensed Matter VII**
Editors: M. Meissner and R. O. Pohl

113 **Electronic Properties of High-T_c Superconductors**
Editors: H. Kuzmany, M. Mehring, and J. Fink

114 **Interatomic Potential and Structural Stability**
Editors: K. Terakura and H. Akai

115 **Ultrafast Spectroscopy of Semiconductors and Semiconductor Nanostructures**
2nd Edition By J. Shah

116 **Electron Spectrum of Gapless Semiconductors**
By J. M. Tsidilkovski

117 **Electronic Properties of Fullerenes**
Editors: H. Kuzmany, J. Fink, M. Mehring, and S. Roth

118 **Correlation Effects in Low-Dimensional Electron Systems**
Editors: A. Okiji and N. Kawakami

119 **Spectroscopy of Mott Insulators and Correlated Metals**
Editors: A. Fujimori and Y. Tokura

120 **Optical Properties of III–V Semiconductors**
The Influence of Multi-Valley Band Structures
By H. Kalt

121 **Elementary Processes in Excitations and Reactions on Solid Surfaces**
Editors: A. Okiji, H. Kasai, and K. Makoshi

122 **Theory of Magnetism**
By K. Yosida

123 **Quantum Kinetics in Transport and Optics of Semiconductors**
By H. Haug and A.-P. Jauho

124 **Relaxations of Excited States and Photo-Induced Structural Phase Transitions**
Editor: K. Nasu

125 **Physics and Chemistry of Transition-Metal Oxides**
Editors: H. Fukuyama and N. Nagaosa

Printing and Binding: Strauss GmbH, Mörlenbach